KINDLING 6

ADVANCES IN BEHAVIORAL BIOLOGY

Recent Volumes in This Series

Volume 39 THE BASAL GANGLIA III
Edited by Giorgio Bernardi, Malcolm B. Carpenter, Gaetano Di Chiara, Micaela Morelli, and Paolo Stanzione

Volume 40 TREATMENT OF DEMENTIAS: A New Generation of Progress
Edited by Edwin M. Meyer, James W. Simpkins, Jyunji Yamamoto, and Fulton T. Crews

Volume 41 THE BASAL GANGLIA IV: New Ideas and Data on Structure and Function
Edited by Gérard Percheron, John S. McKenzie, and Jean Féger

Volume 42 CALLOSAL AGENESIS: A Natural Split Brain?
Edited by Maryse Lassonde and Malcolm A. Jeeves

Volume 43 NEUROTRANSMITTERS IN THE HUMAN BRAIN
Edited by David J. Tracey, George Paxinos, and Jonathan Stone

Volume 44 ALZHEIMER'S AND PARKINSON'S DISEASES: Recent Developments
Edited by Israel Hanin, Mitsuo Yoshia, and Abraham Fisher

Volume 45 EPILEPSY AND THE CORPUS CALLOSUM 2
Edited by Alexander G. Reeves and David W. Roberts

Volume 46 BIOLOGY AND PHYSIOLOGY OF THE BLOOD–BRAIN BARRIER: Transport, Cellular Interactions, and Brain Pathologies
Edited by Pierre-Olivier Couraud and Daniel Scherman

Volume 47 THE BASAL GANGLIA V
Edited by Chihoto Ohye, Minoru Kimura, and John S. McKenzie

Volume 48 KINDLING 5
Edited by Michael E. Corcoran and Solomon Moshé

Volume 49 PROGRESS IN ALZHEIMER'S AND PARKINSON'S DISEASES
Edited by Abraham Fisher, Israel Hanin, and Mitsuo Yoshida

Volume 50 NEUROPSYCHOLOGY OF CHILDHOOD EPILEPSY
Edited by Isabelle Jambaqué, Maryse Lassonde, and Olivier Dulac

Volume 51 MAPPING THE PROGRESS OF ALZHEIMER'S AND PARKINSON'S DISEASE
Edited by Yoshikuni Mizuno, Abraham Fisher, and Israel Hanin

Volume 52 THE BASAL GANGLIA VII
Edited by Louise F. B. Nicholson and Richard L. M. Faull

Volume 53 CATECHOLAMINE RESEARCH: From Molecular Insights to Clinical Medicine
Edited by Toshiharu Nagatsu, Toshitaka Nabeshima, Richard McCarty, and David S. Goldstein

Volume 54 THE BASAL GANGLIA VI
Edited by Ann M. Graybiel, Mahlon R. Delong, and Stephen T. Kitai

Volume 55 KINDLING 6
Edited by Michael E. Corcoran and Solomon L. Moshé

A Continuation Order Plan is available for this series. A continuation order will bring delivery of each new volume immediately upon publication. Volumes are billed only upon actual shipment. For further information please contact the publisher.

KINDLING 6

Edited by

Michael E. Corcoran
University of Saskatchewan
Saskatoon, Saskatchewan, Canada

and

Solomon L. Moshé
Albert Einstein College of Medicine
Yeshiva University
Bronx, New York, USA

Springer

Michael E. Corcoran
University of Saskatchewan
Saskatoon, Saskatchewan
Canada
corcoran@admin.usask.ca

Solomon L. Moshé
Albert Einstein College of Medicine
Yeshiva University
Bronx, New York
USA
moshé@aecom.yu.edu

Library of Congress Cataloging-in-Publication Data

International Conference on Kindling (6th: 2004: Victoria, B.C.)
Kindling 6
p. cm. — (Advances in behavioral biology; v. 55)
"Proceedings of the Sixth International Conference on Kindling, held June 4–6, 2004, in Victoria, British Columbia, Canada"—T.p. verso.
Includes bibliographical references and index.
ISBN 0-387-24380-1
1. Kindling (Neurology)—Congresses. I. Corcoran, Michael E. II. Moshé, Solomon L. III. Title. IV. Series.

RC372.5.I498 2004
616.8'5307—dc22
2005040250

ISBN-10: 0-387-24380-1 e-ISBN: 0-387-26144-3 Printed on acid-free paper.
ISBN-13: 978-0387-24380-1

Proceedings of the Sixth International Conference on Kindling, held June 4–6, 2004, in Victoria, British Columbia, Canada

Printed in the United States of America (EB)

9 8 7 6 5 4 3 2 1

springeronline.com

DEDICATION

For Dan McIntyre and Ron Racine

PREFACE

This book contains the contents of the Sixth International Conference on Kindling, which was held in Victoria, British Columbia, in early June 2004. The Conference brought together a group of leading epilepsy researchers from around the world whose research is concerned with the molecular, anatomical, and physiological substrates of kindling, as well as the age-dependency, behavioral consequences, sensitivity to drugs and other treatments, and clinical relevance of kindling. There was an exciting mix of established researchers – those we have previously referred to as the "stalwarts of kindling" – as well as a number of investigators who have more recently entered the field. That so many of the latter are relatively junior colleagues is a testament to the dynamism and the growth in research on this important preparation of experimental epilepsy.

The conference began with a tribute to Dr. Frank Morrell, who was one of the early kindlers and an eminent neurologist, delivered by Dr. Jerome Engel Jr. Dr. Rafael Gutierrez then delivered a fond reminiscence of Dr. Augusto Fernandez-Guardiola, his mentor and also a prominent kindling researcher, who died shortly before the Conference. Both of these men will be missed, and their contributions to our understanding of epilepsy and epileptogenesis cannot be over-stated.

We organized Kindling 6 into loose themes. Each presentation was followed by a discussion and question period, and there were several General Discussions scheduled during the course of the Conference. As in past Kindling Conferences, we have included these discussions in the book. Much of the magic at Kindling 6 occurred in these informal discussions, and we hope that you will enjoy reading them as much as we enjoyed participating in them.

Kindling 6 is dedicated to Drs. Dan McIntyre and Ron Racine, who were among the original kindlers and who continue to produce provocative, informative, and cutting edge research. Dan McIntyre was a graduate student in Graham Goddard's laboratory at the University of Waterloo when Goddard was performing the initial work on kindling. He performed some of the seminal research on the neuropharmacology and behavioral consequences of kindling, and more recently he has been exploring the molecular, physiological, and behavioral differences between the Fast Kindling and Slow Kindling strains of rat. Ron Racine was a graduate student at McGill University when he first learned of Goddard's then unpublished discovery, and he immediately applied an elegant electrophysiological analysis to kindling that is reminiscent of Morrell's earlier work on the mirror focus. He subsequently generated paradigmatic research on the anatomical, electrophysiological, neurochemical, and behavioral changes associated with kindling, always in search of the fundamental mechanisms of epileptogenesis. We cannot think of anyone more representative of the outstanding researchers in this field than Dan McIntyre and Ron Racine, and it is a pleasure to thank them for their contributions.

We thank the following sponsors for very generously providing the financial or logistical support that made Kindling 6 possible:

UCB Pharma
Ortho-McNeil
Glaxo-Welcome
Novartis
Pfizer
University of Saskatchewan
Albert Einstein College of Medicine, Montefiore Medical Center

Several colleagues served as discussants at the Conference, and we gratefully acknowledge their participation. They are: Györgi Buzsaki, Rutgers University; Karen Gale, Georgetown University; Tony Phillips, University of British Columbia; Phil Schwartzkroin, University of California, Davis; and Claude Wasterlain, UCLA School of Medicine. Their important contributions will be evident to all reading the book. We are also grateful that the following people were able to attend as observers: Drs. Alan Blau and William Seidel, Ortho-McNeil; and Dr. Randall Stewart, National Institute of Health.

The outstanding efforts of a number of people assisted critically in the organization of the Conference and in its smooth operation: Janet Corcoran played an indispensable role in the initial planning and in making the hotel arrangements, and George Schandl and Lindsay Timmermans of the Coast Harbourside Hotel and Marina ensured that the food was hot, the air conditioning was cold, and the weather was cooperative. Tonya McGowan prepared graphics, including the delightful logo of the meeting proudly displayed on the souvenir t-shirts participants received. John Howland, Aaron Sheerin, and Ken Wolfe kept the computers and data projectors running, and magically were able to avoid any major technical meltdowns. Deanna Wolfe and Ken Wolfe transcribed the tapes of the discussion and question periods, and Michele Vircavs did the monumental job of organizing the chapters and related material for publication, including insuring that the formatting was correct. All of their efforts resulted in what a number of participants called one of the best meetings they had ever attended. Finally, of course, the participants themselves did a first rate job, and *they* made Kindling 6 the success that it was. The only way the meeting could have been better would have been if Calgary had beaten Tampa Bay for the Stanley Cup.

Michael E. Corcoran and Solomon L. Moshé

CONTENTS

PREFACE vii

THEME 1: PATHOGENESIS OF KINDLING

DEVELOPMENTAL PATTERNS

1. DEVELOPMENTAL TEMPORAL LOBE EPILEPSY IN AMYGDALA KINDLED KITTENS: AN UPDATE 1
Margaret N. Shouse, John C. Scordato, and Paul R. Farber

2. PRENATAL BETAMETHASONE EXPOSURE SUPPRESSES KINDLING EPILEPTOGENESIS IN IMMATURE RATS 11
Libor Velišek

3. DEVELOPMENT OF KINDLING IN IMMATURE FAST AND SLOW KINDLING RATS 19
Jana Velišková, Jeremy Asnis, Fred A. Lado, and Dan C. McIntyre

4. NEONATAL HEAT-INDUCED CONVULSIONS AFFECT BEHAVIOURS IN NEONATAL, JUVENILE AND ADULT RATS 27
Deborah Saucier, Avril Keller, Aaron Sheerin, R. Daniel LaPorte, and Jerome Y. Yager

5. EXPERIMENTAL ABSENCE VERSUS AMYGDALOID KINDLING 37
Filiz Yılmaz Onat, Esak Eşkazan, and Rezzan Aker

6. PENTYLENETETRAZOL-INDUCED KINDLING AS A MODEL OF ABSENCE AND CONVULSIVE FORMS OF EPILEPSY 49
Alexey A. Shandra and Leonid Godlevsky

ELECTROPHYSIOLOGY

7. SODIUM CURRENT PROPERTIES IN DIFFERENT MODELS OF EPILEPSY 61
Wytse J. Wadman, Rogier Min, and Guangchun Sun

8. KINDLING THE GABAERGIC PHENOTYPE OF THE GLUTAMATERGIC GRANULE CELLS 71
Rafael Gutiérrez, Gisela Gómez-Lira, Carmen Vivar, Jasmín Maqueda, Mónica Ramirez, Mónica Lamas, Mario Treviño, and Héctor Romo-Parra

9. HIPPOCAMPAL KINDLING AND $GABA_B$ RECEPTOR FUNCTIONS 81
L. Stan Leung, Xinhuai Liu, Kevin J. Canning, and Bixia Shen

10. ALTERED INTERACTION BETWEEN THE ENTORHINAL CORTEX AND HIPPOCAMPUS IN AMYGDALA KINDLED RATS 91
Uwe Heinemann, Doris Albrecht, Achim Behr, and Tengis Gloveli

ANATOMY, MORPHOLOGY, AND NEURAL CIRCUITRY

11. NATURE AND CONSEQUENCES OF SEIZURES ORIGINATING IN THE BRAINSTEM 99
Shigeru Chiba, Takahiro Ishimoto, Yuji Ishimaru, Yoshiyuki Tamura, and Akihiko Nunomura

12. INVOLVEMENT IN THE CLAUSTRUM AND VENTROMEDIAL THALAMUS IN EPILEPTOGENESIS 115
Aaron Sheerin, Xia Zhang, Deb Saucier, and Michael Corcoran

13. FROM ULTRASTRUCTURE TO NETWORKS: KINDLING-INDUCED CHANGES IN NEOCORTEX 125
G. Campbell Teskey, Corey Flynn, Marie-H. Monfils, Crystal D. Goertzen, Luke C. Henry, and Pamela A. Valentine

14. KINDLING AS A TOOL FOR STUDYING THE ROLE OF SUBCORTICAL STRUCTURES IN LIMBIC SEIZURES 137
Edward H. Bertram and John Williamson

15. KINDLING, EPILEPSY, AND THE PLASTICITY OF NETWORK SYNCHRONIZATION 147
Thomas P. Sutula

GENERAL DISCUSSION 1 161

GENES, SPECIES, AND STRAINS

16. ELECTROPHYSIOLOGICAL AND ANATOMICAL DIFFERENCES, BEHAVIORAL COMORBIDITIES AND GENE EXPRESSION IN FAST AND SLOW KINDLING RAT STRAINS 171
Dan C. McIntrye and Krista L. Gilby

17. GENE EXPRESSION CHANGES IN KINDLING 181
Don Liang and Thomas N. Seyfried

18. A PROTEOMIC APPROACH TO THE MOLECULAR ANALYSIS OF KINDLING 189
John T. Slevin, Sidney W. Whiteheart, and Thomas C. Vanaman

SYNAPTIC PHARMACOLOGY AND NEUROCHEMISTRY

19. AMYGDALA METABOTROPIC GLUTAMATE RECEPTORS AND KINDLING 199
Patricia Shinnick-Gallagher

20. INDUCTION OF B_1 BRADYKININ RECEPTORS IN THE KINDLED BRAIN 209
Michele Simonato, Manuela Mazzuferi, Donata Rodi, Anna Binaschi, Simona Mantovani, Brice Ongali, and Rejean Couture

21. GALANIN AND KINDLING 219
Merab Kokaia

22. KINDLING, NEUROTROPHINS AND AXON-GUIDANCE FACTORS 229
Ronald J. Racine, Margaret Fahnestock, and Bin Xu

23. CONDITIONAL DELETION OF TRKB PREVENTS EPILEPTOGENESIS IN THE KINDLING MODEL 241
James O. McNamara, Xiao-Ping He, and Robert Kotloski

24. NEUROPEPTIDE Y AND ITS RECEPTORS IN KINDLING EPILEPTOGENESIS 249
Cristina Richichi, Ramla Benmaamar, Marco Gobbi, En-Ju D. Lin, Matthew J. During, Gunther Sperk, and Annamaria Vezzani

THEME 2: BEHAVIORAL CONSEQUENCES OF KINDLING

25. CONTRIBUTION OF PRE KINDLING AFFECTIVE STATE TO HEMISPHERIC DIFFERENCES IN THE EFFECTS ON ANXIETY OF BASOLATERAL AMYGDALA KINDLING 263
Robert Adamec, Tanya Shallow, Jacqueline Blundell, and Paul Burton

26. EFFECTS OF KINDLING ON SPATIAL MEMORY: CHARACTERISTICS AND MECHANISMS 273
Darren K. Hannesson, Ken Wolfe, and Michael E. Corcoran

27. A POTENTIAL ROLE FOR THE HIPPOCAMPUS IN THE EXPRESSION OF KINDLING-INDUCED FEAR 285
Lisa E. Kalynchuk and Amanda J. Wintink

28. DOES INADVERTENT CONDITIONING CONTRIBUTE TO THE MAJOR FEATURES OF KINDLING? 295
Steven J. Barnes and John P. J. Pinel

THEME 3: DRUGS AND INTERVENTIONS AGAINST KINDLING

29. CONTINGENT TOLERANCE AND CROSS TOLERANCE TO ANTICONVULSANT EFFECTS IN AMYGDALA-KINDLED SEIZURES: MECHANISTIC AND CLINICAL IMPLICATIONS 305
R.M. Post, Z-J Zhang, S.R.B. Weiss, G. Xing, and K. Obeng

30. MECHANISMS OF PHARMACORESISTANCE IN THE PHENYTOIN-RESISTANT KINDLED WISTAR RAT 315
Heidrun Potschka and Wolfgang Löscher

31. DEVELOPMENT OF NEW ANTICONVULSANTS USING THE KINDLING MODEL 325
W. McIntyre Burnham, Deborah Lonsdale, Afshin Shahzamani, Claudia Perez-Cruz, and Heather E. Edwards

32. CANNABINOIDS AND KINDLING 333
Michael E. Corcoran, Deborah Saucier, Xia Zhang, Aaron Sheerin, and Kenenth Wolfe

33. LOW FREQUENCY SINE WAVE STIMULATION DECREASES THE INCIDENCE OF KINDLED SEIZURES 343
Jeffrey H. Goodman

GENERAL DISCUSSION 2 355

THEME 4: CLINICAL RELEVANCE OF KINDLING FOR OUR UNDERSTANDING OF EPILEPSY IN PATIENTS

34. PREDISPOSED SUSCEPTIBILITY IN PRIMATE EPILEPTOGENESIS AND ANTI-EPILEPTOGENEIS 361
Juhn A. Wada

35. NATURAL HISTORY OF MESIAL TEMPORAL LOBE EPILEPSY WITH HIPPOCAMPAL SCLEROSIS: HOW DOES KINDLING COMPARE WITH OTHER COMMONLY USED ANIMAL MODELS 371
Jerome Engel, Jr.

36. CLINICAL EVIDENCE OF EPILEPSY-RELATED PLASTICITY 385
Giuliano Avanzini and Silvana Franceschetti

37. PATHOLOGICAL SENSITIZATION OF THE DOPAMINE SYSTEM IN EXPERIMENTAL EPILEPTOGENESIS: IMPLICATION FOR THE MECHANISM OF EPILEPTIC PSYCHOSIS 399
Kiyoshi Morimoto and Mitsumoto Sato

INDEX 413

DEVELOPMENTAL TEMPORAL LOBE EPILEPSY IN AMYGDALA KINDLED KITTENS: An update

Margaret N. Shouse, John C. Scordato and Paul R. Farber*

1. INTRODUCTION

In 1990, we published the first report of amygdala kindling in kittens and the first description of a spontaneous seizure model in immature animals.[3] We found that spontaneous epilepsy was more likely to occur after amygdala kindling in preadolescents than in adults and that the post-kindling onset was most rapid in the youngest animals. This update is intended to fill some gaps in the initial report on the ontogeny of temporal lobe epilepsy.[4]

2. METHODS

The study population was increased from 24 to a total of 58. The largest expansion was in the youngest age group, and, unlike the previous report, gender distribution was equal in each of the 3 age groups. Females reach puberty at ~7 months and males at ~1 year. Age at initial after discharge (AD) and gender distribution was as follows: young preadolescents from 2.5-5-month-olds (n=30; n=15 male and 15 female); older preadolescents from 5.5 to 6.5-month-olds (n=10, 5 male and 5 female) and adults $\geq$ 1 year (n= 18; 9 male and 9 female).

A modified kindling procedure was used in all preadolescents (n=40) and in 12 of 18 adults. Multiple stimulations per day were used to establish AD thresholds at the beginning of kindling. The initial stimulus was 100 uA and the increment was 100 uA. The interstimulus interval was one minute instead of the usual 24h. Threshold was defined as the stimulus intensity (mA) required to elicit the first AD. Afterward, once daily stimulation at initial AD threshold was used to obtain the first stage 6 seizure, which is a generalized tonic-clonic convulsion (GTC) in cats. After initial kindling, threshold was re-established in one day, as at the beginning of kindling.

* M. N. Shouse, J. C. Scordato and P. R. Farber, VAGLAHS. M. N. Shouse, Department of Neurobiology, UCLA School of Medicine, Los Angeles CA. Correspondence:, 2242 S. Bentley Ave. # 4, Los Angeles CA 90064-1940

Post-kindling follow-up in the initial series lasted from 3 days to 9 months, with the shortest follow-up in the youngest animals (2 months maximum). The follow-up was characterized by sporadic polygraphic and behavioral observation methods of spontaneous seizure detection, variable numbers of evoked seizure trials and, in the youngest kittens, by undetected bouts of convulsive status which were frequently terminal. Three changes were made to address these limitations during the 4-month post-kindling follow-up in 33 of the new cats: (1) 5-day per week routine EEGs and continuous 24h video recordings in the vivarium were used to improve seizure detection. (2) One-day post-kindling seizure threshold tests were conducted 5 days per week with a one-month lapse in threshold trials scheduled at specific intervals. The objective was to evaluate the effects of frequent or recent elicited GTCs on onset or persistence of spontaneous seizures. The distribution of suspended post-kindling threshold tests was (a) a 28-day lapse which began after the first post-kindling threshold test [n= 11 preadolescents ≤5.0months, 3 older preadolescents and 2 adults] and (b) a 28-day lapse begun one month post-kindling [n=11 preadolescents ≤5.0months, 3 older preadolescents and 3 adults]. (3) Use of anticonvulsants, usually nembutal, was extended to include frequent GTCs (>1 per hour) as opposed to convulsive status (> 1 GTC within 5 min or a GTC lasting >5min).

3. RESULTS

3.1 Kindling Development

Tables 1 and 2 present findings on a number of standard kindling parameters as a function of age and gender. The youngest animals continue to exhibit high individual variability but differed significantly from older animals on several variables. Gender was not a factor in kindling development in any age group.

3.1.1 Kindling Thresholds

Table 1 compares AD thresholds (mA) at the beginning of kindling to GTC thresholds obtained at the end of kindling as well as one month and 2-to-4 months post-kindling. When compared to older animals, kittens in the youngest age group had significantly higher thresholds at the beginning and end of kindling as well as one month later. Thresholds subsided to levels of older animals two-to-four months later when animals reached the age of older adolescents or adults.

3.1.2 Kindling Rates

Table 2 shows that youth also influenced kindling rate. When compared to older animals, the youngest kittens displayed significantly fewer ADs with stage 1-2 clinical seizures, a similar number of generalized ADs with stage 3-5 clinical seizures and significantly faster overall kindling rates, defined as the total number of ADs to the first GTC.

Table 1. Kindling development in kittens and adult cats: AD or GTC thresholds (mA0 at the beginning, end and 1-4 months after kindling as a function of age and gender. From [4].
AD, after discharge; GTC, generalized tonic-clonic convulsion. Values ± S.E.M. Range is given in parentheses.

Age at initial AD (months)	Initial AD thresholds	GTC thresholds at the end of kindling	GTC thresholds one month post-kindling	GTC thresholds 2-4 months post-kindling
2.5-5.0 (n=30)	3.98±0.8* (0.5-20)	2.0±0.6* (0.3-18)	1.2±0.2*+ (0.3-3.0)	0.7± 0.1+ (0.1-1.8)
Male (n=15)	3.8±1.0	1.9±0.7	1.2±0.2	0.8±0.1
Female (n=15)	4.0±1.3	2.1±0.8	1.1±0.3	0.7±0.2
5.5-6.5 (n=10)	1.2±0.3 (0.3-4.0)	1.0±0.1 (0.3-1.5)	0.8±0.1 (0.2-1.3)	0.8±0.2 (0.1-1.4)
Male (n=5)	1.2±0.5	1.2±0.3	0.8±0.2	0.9±0.3
Female (n=15)	1.1±0.6	0.9±0.2	0.8±0.3	0.8±0.2
≥12 (n=18)	1.1±0.3 (0.1-1.5)	0.7±0.3 (0.1-1.4)	0.6±0.2 (0.1-1.0)	0.7±0.3 (0.2-1.2)
Male (n=9)	1.3±0.4	0.7±0.3	0.7±0.3	0.6±0.3
Female (n=9)	1.0±0.3	0.7±0.2	0.6±0.3	0.7±0.4

*p <.05 from older kittens and adults
+n=26 surviving kittens

Table 2. Kindling development in kittens and adult cats: Kindling rates as a function of age and gender. Values are means ± S.E.M. Range is in parentheses. From [4].

Age at initial AD (months)	No. ADs with Stage 1-2 seizures	No ADs with Stage 3-5 seizures	Kindling rate: total no. ADs to first stage 6
2.5-5.0 (n=30)	3.0±0.5* (0-10)	11.2±1.1 (1-31)	14.2±1.2* (4-37)
Male (n=15)	2.8±0.6	11.0±1.2	13.8±1.3
Female (n=15)	3.2±0.7	11.4 ±1.3	14.6±1.4
5.5-6.5 (n=10)	11.3±1.3 (5-21)	13±1.2 (5-18)	25.3±2.4 (10-36)
Male (n=5)	10.6±0.4	11.6±1.8	25.0±1.6
Female (n=15)	12.0±2.7	14.4±1.5	25.6±4.9
≥12 (n=18)	10.0±0.9 (0-17)	13.9±1.0 (2-22)	23.9±0.6 (16-22)
Male (n=9)	10.4±0.9	13.6±0.9	24.0±0.7
Female (n=9)	9.7±1.0	14.1±1.0	23.8±0.5

*p <.001 from older adolescents and adults

3.1.3 Kindling Rrogression

ADs in the youngest kittens could be focal but were usually generalized from the outset. Initial ADs in older animals can also be generalized from the outset, but it is far less common. The progression of the other electroclinical signs during evoked seizures was similar regardless of age, as previously reported.[3]

3.2 Spontaneous Seizures

Table 3 lists the incidence and types of spontaneous seizures as well as some characteristics of the clinical course. The data confirm and extend previous results, as follows. Youngest animals are most likely to develop spontaneous epilepsy, defined as seizures occurring at least 1h after an evoked GTC. Seventeen of 30 (57%) of the youngest cats developed spontaneous seizures as compared to 2 of 10 (20%) of the older adolescents and 1 of 18 (6%) of the adults. Gender did not seem to be a factor since, overall, half the epileptic animals were males and half were females.

The subtle nature of some EEG and clinical seizure manifestations makes it likely that numerous seizure types still went unnoticed, and onset of spontaneous epilepsy still cannot be conclusively documented in spite of the increased surveillance. Early detection and treatment of frequent GTCs (>1 per hour) did, however, improve survival rates in the

Table 3. Spontaneous seizure types and associated behaviours. From [4].

Age at initial AD (months)	No. of cats with spontaneous seizures (> 1h after evoked seizure)	No. of cats per spontaneous seizure type	No. and/or range (min/max) of seizures per spontaneous seizure type	Mean ± S.E.M. density of stage 3 and 6 seizures (no. per detection day)	Transient behaviours associated with dense stage 6 seizure clusters
2.5-5.0 (n=30)	17 (8 male, 9 female)	Stage 3 (n=4) Stage 6 (n=15) Focal subclinical (n=2) 'Catnip' (n=2)	Stage 3: n=12 (2-5) Stage 6: n=171 (1-36) Focal subclinical: n=14 (2-12) 'Catnip': 1-2 days	Stage 3: 1.9±0.3 Stage 6: 3.2±0.6*	Atonia; Reduced mobility; social isolation; 'psychic' blindness
5.5-6.5 (n=10)	2 (1 male, 1 female)	Stage 6	Stage 6 (n=15)	Stage 6: 2.3±1.3	None
>12 (n=18)	1 male	Stage 6	Stage 6 (n=3)	Stage 6: 1.5±0	None

*p <.05 from preadolescents and adults

youngest kittens. The longer follow-up in newly added, youngest animals (4-month follow-up vs. 3 days-to-2 months in the youngest cats from the initial series) revealed a multifaceted post-kindling seizure disorder with a catastrophic clinical course.

3.2.1. Spontaneous Seizure Types

We detected a larger variety and number of EEG and clinical seizure manifestations and a much higher coincidence of multiple seizures types in young animals than previously reported.[3] Two types of spontaneous seizures resembled either stage 3 or stage 6 evoked kindled seizures, but there could be differences from the evoked seizure pattern. Kittens with a history of "all-or-none" generalized AD could have focal onset spontaneous stage 3 or 6 seizures, and, conversely, kittens with focal ADs could display spontaneous GTCs with generalized onset.

The other two seizure types were less frequently detected and very different from evoked seizures with respect to EEG and clinical manifestations. Repetitive, brief (3-10sec) or long-lasting (1.5 min) focal seizures were detected in kindled amygdala or in the lateral geniculate nucleus of thalamus and did not have clinical accompaniment. Nearly continuous multifocal and generalized EEG seizures were also detected, sometimes with overt clinical signs (e.g., fast pulse, staring, purring and jackknife seizures like infantile spasms) and sometimes with little noticeable clinical accompaniment (continuous rotary tail wagging). EEG discharge with these 'catnip' seizures was so pervasive that waking and SWS were indistinguishable. The seizures stopped only when the animal entered REM sleep or was alerted by physical manipulation.

3.2.2 Clinical Course In Cats With Spontaneous Epilepsy

When compared to older preadolescents and adults, young kittens showed 1) an increased number of spontaneous seizures, 2) a progressive increase in the number and density of seizure clusters, usually with increasingly less time between seizure clusters and 3) frequent bouts of convulsive status epilepticus which could occur at any time but tended to increase with dense seizure clusters towards the end of follow-up.

Behavioral anomalies were observed in the video tapes from the vivarium and also during routine EEGs or seizure threshold tests. These strange behaviors could accompany dense seizure clusters but were not associated with ictal EEG discharge. The sequelae ranged from sensory or motor deficits ('psychic' blindness, atonic episodes, restricted mobility) to social isolation and placidity. These behavioral deficits could be life-threatening because the animals appeared reluctant to move, even to search for food or water, and did not seem to discriminate food from other objects such as litter.

Odd behaviors were also more likely to accompany frequent evoked seizures in younger animals without detected spontaneous seizures than in older animals with or without detected spontaneous seizures. Examples are fluctuations in temperament (docility to irritability), sexual dysfunction (repeated homosexual assaults, reduced sexual interest or dysmenhoria) and post-ictal bulimia.

The behaviors of young kindled animals resemble an admixture of symptoms of the youth related syndrome called acquired epileptic aphasia[5] and the non-age related syndrome called Kluver Bucy syndrome[6]. Both are affiliated with human temporal lobe epilepsy[5, 6] and/or insult to the temporal lobe in humans, monkeys and cats.[1, 5, 6]

Table 4. Factors in the onset and/or maintenance of spontaneous seizures. Failed inhibition was indexed by peri-stimulus seizure activity during seizure threshold trials. This refers to focal, multifocal or generalized EEG discharge other than AD that occurred either during the 1-min interstimulus intervals before convulsions or, later, after the end of the evoked convulsions (post-ictal); clinical accompaniment ranged from stage 1-6 seizures. From [4].

Age at initial AD (months)	Failed inhibition during threshold trials	Stimulus intensity during kindling	Convulsive status epilepticus	Post-kindling latency to onset of 1st spontaneous seizure	Latency to spontaneous seizures after last evoked convulsion	No. of evoked convulsions to onset of spontaneous seizure
2.5-5.0 (n=17)	Dense inter-stimulus or post-ictal discharge <96 h before spontaneous seizures	1.5-20mA*	12 of 17 cats; total=36	0-2.5 months*	1-72h (n=15) 0.75 months (n=1)	12.9±2*
5.5-13.0 (n=3)	Sporadic	0.7-1.8mA	1 of 3 cats; total=1	2.5-4 months	24-72 h	39±1

*p <.05 from older preadolescents and adults; + see text on suspended threshold trials.

3.3 Factors In The Onset and Maintenance of Spontaneous Epilepsy

Table 4 summarizes a number of potential factors contributing to the ontogeny of spontaneous epilepsy after amygdala kindling. Cats are divided into two age groups as nearly all factors differentiated onset and/or maintenance in these groups.

3.3.1. Failed Inhibition During Threshold Tests

The absence of refractory periods is indexed by brief repetitive seizures (focal or generalized EEG discharges with eye twitches, etc.) to subconvulsive stimuli during the 1-min interstimulus intervals prior to the stimulus evoking the GTC and also by recurrent seizures instead of post-ictal depression after evoked convulsions. These events were typical of threshold tests in young kittens but were less frequently seen in older adolescents and were rarely if ever noticed in adults. A two-to three-fold increase in one or both of these peri-stimulus indices of reduced inhibition was seen 24-96h prior to the first detected spontaneous seizure and again before subsequent seizure clusters in the youngest kittens. Reduced inhibition also characterizes increased seizure susceptibility in rat pups, as indexed by reduced kindling 'antagonism' and by the absence of inter-site differences in limbic system kindling [4].

3.3.2. Stimulus Intensity During Kindling

The significantly higher stimulus intensity needed to evoke AD or GTCs in young kittens may have contributed to onset in the youngest age group especially since thresholds remained high during the one month post-kindling period when onset of spontaneous seizures most often occurred. The decline in thresholds 2-4 months after kindling makes this factor less likely to contribute to maintenance of the seizure disorder, but the possibility of an electrolytic lesion can not be excluded as a factor in onset or maintenance of spontaneous epilepsy in the young.

3.3.3. Convulsive Status Epilepticus

The frequent bouts of status epilepticus in young kittens might have resulted in global cells loss, which in turn could have sustained seizure activity after onset. In rare cases, histology was obtained from epileptic kittens early in follow-up but failed to show more gliosis or cell loss than that seen in non-epileptic, older animals. This is consistent with reports that rat pups are less susceptible than mature rats to seizure-related brain damage or to synaptic reorganization (e.g.,[4]). We can not confirm this observation in most of our kittens because euthanasia was delayed until the animals matured and had a considerable seizure history.

3.3.4. Post-kindling Latency to Onset of Spontaneous Epilepsy

Youth was associated with faster post-kindling onset. Age at initial AD was significantly correlated with post-kindling latency to onset of spontaneous epilepsy ($r=.0.59$, $p<.05$) regardless of whether or when the 28-day lapses in evoked seizure trials occurred after kindling.

3.3.5. Latency to Spontaneous Seizures After Evoked GTC

All but one of the spontaneous seizures were detected from 1-72h after evoked GTCs. Youngest kittens could have shorter and in one case longer post-ictal latencies than older animals. Recently evoked seizures seemed important to maintenance of spontaneous epilepsy at all ages.

3.3.6. Number of Elicited GTCs Prior To Spontaneous Seizure Onset

Youngest kittens sustained significantly fewer evoked seizure trials after kindling than did older animals prior to the onset of spontaneous seizures. This difference suggested that the number of elicited seizures is more important to onset in older than in younger animals.

3.3.7. Suspension of Post-kindling Threshold Tests

A 28-day lapse in elicited GTCs occurred immediately vs. one month after kindling in 33 cats. Three of 11 youngest preadolescents (27 %) and 1 of 5 older preadolescents and adults (25%) had detected onset of spontaneous epilepsy with prolonged post-kindling latencies (2.5 months±0) when the seizure trials were suspended immediately after kindling. In contrast, 8 of 11 (73%) of the youngest preadolescents had spontaneous seizure onset when seizure trials were suspended one month after kindling. Seven of 8 kittens had onset prior to the delayed suspension (mean post-kindling onset = 17±2 days). Only one kitten had his first spontaneous seizure during the delayed suspension of evoked seizure trials. Onset occurred 1.75 months after kindling and 3 weeks after the lapse in seizure threshold trials began. Overall, the timing of frequent evoked seizure trials *early* in the post-kindling course was most closely associated with the incidence and rapid post-kindling onset of spontaneous seizures in the young.

With one exception, no spontaneous seizures were detected >72h after the suspensions began. Behavioral anomalies associated with spontaneous seizure clusters also ended soon after seizure clusters, usually within 3-to-30 days. The behaviors associated with evoked seizures (e.g., sexual dysfunction) did *not* necessarily remit during lapses in evoked seizure trials. Continued evoked seizures thus seemed important to maintenance of the spontaneous seizure disorder and related behavioral anomalies.

4. CONCLUSIONS

We confirmed and extended our initial report as follows: (1) Young animals are far more likely than adults to exhibit spontaneous epilepsy, indexed by seizures that occur ≥1h after stimulus-evoked seizures and that can persist to adulthood; (2) The youngest animals are most likely to exhibit accelerated kindling rates and rapid post-kindling onset of spontaneous epilepsy with a catastrophic clinical course; and (3) Early detection and treatment (nembutal) of frequent GTCs or convulsive status epilepticus improved survival rates and at the same time revealed a complex clinical picture. The profile included a variety of EEG and/or clinical seizure manifestations and a progressive increase in the number and density of seizure clusters. Odd behavioral sequelae could accompany dense seizure clusters and ranged from sensory or motor deficits ('psychic' blindness, atonic episodes, restricted mobility) to social isolation and placidity. Developmental deterioration associated with spontaneous epilepsy need not be gender-related and is substantially enhanced by frequent or recent evoked seizures. The onset of spontaneous epilepsy and the post-kindling progression can be stopped or minimized by the suspension of evoked seizure trials early in the post-kindling course

5. ACKNOWLEDGEMENTS

Supported by the Department of Veterans Affairs.

6. REFERENCES

1. Baron-Cohen, S., Ring, H.A., Bullmore, E.T., Wheelright, S. Ashwin, C. and Williams, S.C.R. The amygdala theory of autism. *Neuroscience and Biobehavioral Reviews* 24 (2000) 355-364.

2. Haas, Kurt Z., Sperber, E.F., Benenati, B, Stanton, P.K. and Moshe, S.L. Idiosyncrasies of limbic kindling in developing rats. In Corcoran, M.E. and Moshe, S.L., eds. Kindling 5. Plenum Press; New York, 1998, pp. 15-24.
3. Shouse, M.N., King, A., Langer, J, Vreeken, T., King, K., and Richkind, M. The ontogeny of feline temporal lobe epilepsy: kindling a spontaneous seizure disorder in kittens. *Brain Research* 525 (1990)215-224.
4. Shouse, M.N., Scordato, J.C. and Farber, P.R. Ontogeny of feline temporal lobe epilepsy in amygdala kindled kittens: An update. Submitted, 2004.
5. Stefanatos, G.A., Kinsbourne, M. and Wasserstein, J. Acquired epileptiform aphasia: A dimensional view of Landau-Kleffner Syndrome and the relation to autistic spectrum disorders. *Child Neuropsychology* 8: 195-228, 2002.
6. Varon, D. Pritchard, P.B., Wagner, M.T. and Topping, K. Transient Kluver-Bucy syndrome following complex partial status epilepticus. *Epilepsy & Behavior* 4 (2003) 348-351.

Discussion

Wasterlain: For those of us who are not kittenologists would you tell us how this relates to the critical periods of brain development in a kitten and how this relates to the time when you can generate function enryoplia and you know what stage of development that you can start seeing kindling. And also your top priority isn't 2 ½ months why but why not. Have you not tried earlier than 2 ½ months and if not why not?
Shouse: There is a lot of complications if you want to work with kittens. One is that it is really difficult to work with kittens that are not weaned, and about the earliest that you can obtain weaned kittens is at 2 months. The critical period that I am talking about is defined by is soon after weaning. Operationally that's the time at which you can get the uniquely high threshold and they disappear 2 to 4 months later when all of the sudden they are either adolescents or maybe even adults. I cannot tell you that I can relate this to some critical period of development.
Wada: Most intriguing study. Am I correct that you start stimulating these young kittens they end up with multiple pattern of spontaneous seizures? Does that imply the development of independent dischage outside of the kindled amygdala, multiple patterns of epiloptogensis?
Shouse: It may develop, but they can for example originate in the amygdala on the same side at the onset, all the seizures can come from the contralateral amygdala, or they can develop seizures that have nothing to do with it. They can develop all of those.
Wada: So they are entirely independent from the kindled amygdala activity and the dissemination is dependent on the timing of the critical period?
Shouse: Yes, it is dependent on it in the sense that we never see anything like that in the older animals. With this protocol it is only observed in the younger animals.
Wada: How many kindling stimulations do you give?
Shouse: We had to stop after 3 days and in the early ones we had to stop immediately because we got so much status and the animals did not live 24hrs after the seizure.
Engel: To clarify , are you saying that in some of the cats that have spontaneous seizures, the seizures develop from where you stimulated? My understanding from Pinel's early work is that this is a transynaptic phenomenon, that seizure always come from an area that is one synapse removed from where you stimulate.
Shouse: Yes, in the younger kittens almost all of them started where we stimulated.
Engel: Perhaps John Pinel can help: Is over kindling transynaptic in the adult or does it come from the stimulated area?

Pinel: With cortical kindling, we always got the seizures starting at other sites and usually with no involvement of the cortical site. Full generalized seizures would run their course with no involvement of the kindling site. But with amygdaloid kindling we did not have that many electrodes; so we can't say where the seizure started. However, we can say definitely some of the seizures did not start in the site of stimulation. They were starting somewhere else and we weren't picking up the activity for several seconds and sometimes never picking up the activity.
Shouse: This implies that it is the same. We get a lot of these seizures starting from where we stimulated. It is a weird thing that the seizure can, for no apparent reason, start outside.

PRENATAL BETAMETHASONE EXPOSURE SUPPRESSES KINDLING EPILEPTOGENESIS IN IMMATURE RATS

Libor Velíšek*

1. INTRODUCTION

Administration of corticosteroids is common in obstetrics. Corticosteroids are frequently used in pregnant women between 24-34 gestation weeks, if there is a danger of imminent delivery with subsequent respiratory distress syndrome of the newborn.[1] In this situation, one course of corticosteroid therapy is applied, usually consisting of 2-4 injections of betamethasone or dexamethasone spread over 24-48 hours. Follow-up studies in humans have demonstrated that this therapy regimen significantly improves the chance of the premature newborn for survival with no long-term side effects.

However, if there is no delivery within 7-10 days after cessation of the first course of corticosteroid therapy and the risk still prevails, courses of corticosteroid therapy are very often repeated over and over.[2, 3] Thus, some women may have received up to 8-11 corticosteroid therapy courses during the third trimester of pregnancy. The effect of these repeated corticosteroid exposures on the long-term development of offspring is unknown as no long-term prospective follow-up studies are available.[4, 5]

Studies in monkeys have demonstrated that repeated administration of corticosteroids (dexamethasone) during the last third of pregnancy may have deteriorating effect on the development of hippocampus.[6] Smaller hippocampi with fewer pyramidal cell were reported[6, 7]. In rats, we observed increased susceptibility to kainic acid-induced seizures as well as behavioral deficits in simple motor tasks after prenatal exposure to two injections of hydrocortisone (umpublished data).

In this study, we determined the effects of prenatal exposure to betamethasone on development of kindling in the immature rats.

* Libor Velíšek, Departments of Neurology and Neuroscience, Albert Einstein College of Medicine, Einstein/Montefiore Epilepsy Management Center, 1410 Pelham Parkway South, Bronx, NY 10461, USA.

Table 1. Seizure stages for hippocampal kindling in postnatal day 15-16 rats[a]

Stage	Description
0	Behavioral arrest
1	Mouth clonus
2	Head bobbing
3	Unilateral forelimb clonus
3.5	Alternating forelimb clonus
4	Bilateral forelimb clonus
5	Bilateral forelimb clonus + rearing & falling
6	Wild running & jumping & vocalization
7	Tonus

[a] Modified according to Haas et al.[8]

2. METHODS

Animals: Timed pregnant Sprague-Dawley rats were delivered on day 8 of pregnancy (E8) from Taconic Farms (Germantown, NY). Rats were randomly assigned to either betamethasone- or saline-treated groups.

Model: Pregnant rats received two injections of betamethasone (0.4 mg/kg each) in 1 ml of saline 10 hours apart on E15. We chose E15 as the embryonic day with the most vulnerable development of hippocampal formation as documented in previous studies.[9, 10] Further, two separate injections of corticosteroids here represent the minimal possible (i.e., two) repetitions of corticosteroid courses.

Delivery occurred on E22-23. Day of delivery was considered as postnatal day (PN) 0. On PN1, rat pups were checked for sex and weighted. Litter size was kept at no more than 10 pups and if possible of equal proportion of males and females. On PN13, rats were anesthetized with ketamine/xylazine mixture (50/7 mg/kg ip) and twisted stainless steel stimulating electrode (Plastics One, Roanoke, VA) was stereotaxically inserted in the dorsal hippocampus. We chose precise localization of the electrode in the molecular layer of the dentate gyrus to activate major input of the hippocampus. Stereotaxic coordinates were anteroposterior 3.2 mm, lateral 1.5 mm, and depth 3.2 mm from bregma.[11] On PN15, stimulator and EEG recording device were connected and afterdischarge (AD) threshold was determined using ascending technique with 50 μA steps. Once the AD threshold has been determined, kindling stimulations were initiated. Rats were stimulated every 20 minutes with 2 s trains of 50 Hz biphasic constant current pulses with duration of 2 ms and intensity of 400 μA. Rats received 15 kindling stimulations on PN15 and additional 10 stimulations on PN16. We evaluated AD duration and seizure stage modified according to Haas et al.[8] (see Table 1).

After completing kindling stimulations, the rats were anesthetized with pentobarbital (50 mg/kg ip), decapitated, brains removed and frozen in chilled methylbutane at –35°C. Cryostat sections were cut at 30 μm, mounted on slides and stained with cresyl violet to locate electrode tips using light microscopy.

Statistics: Two-way ANOVA was used to evaluate AD thresholds (factors: sex, prenatal treatment. Two-way ANOVA for repeated measures was run on AD duration and seizure stage data (main factors: sex, prenatal treatment; repeated factor: stimulation number). Level of significance was always preset to $P<0.05$.

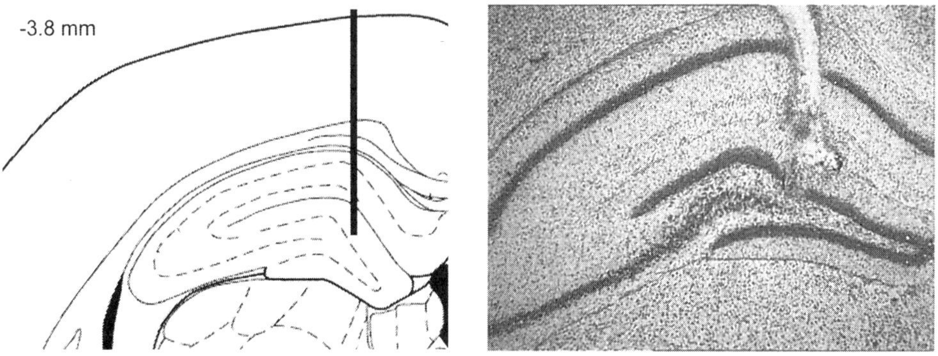

Figure 1. Localization of stimulating electrodes in the dentate gyrus. *Left panel* shows part of a coronal plate from Paxinos and Watson, 1998[11] at –3.8 mm from bregma, at the level, where the electrode was found in the dentate gyrus. Black bar indicates stimulating electrode tract as shown on the histological image below. *Right panel* is an example of a histological section illustrating hippocampus and dentate gyrus with the electrode tract ending in the molecular layer of the dentate gyrus, i.e., our desired position for the stimulating electrode.

3. RESULTS

3.1. Histology

Only those rats with electrode tips in the molecular layer of the dentate gyrus were used for further data evaluation. An example of the position of the electrode tip is illustrated in Figure 1.

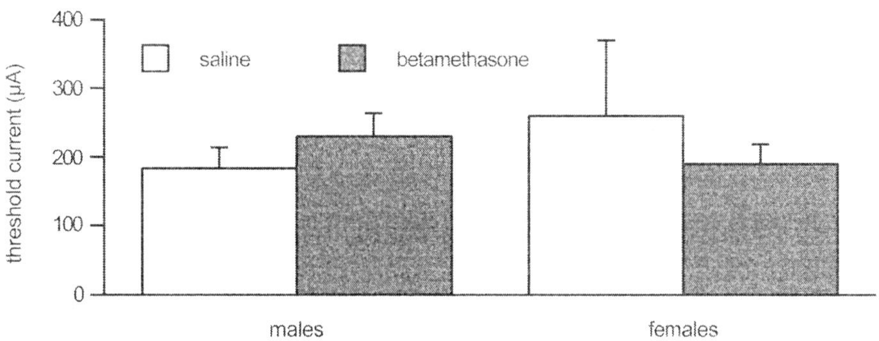

Figure 2. Afterdischarge (AD) threshold (mean ± SEM). *Open columns* – repeated prenatal saline exposure (2x 1 ml/kg ip on E15), *closed columns* – repeated prenatal betamethasone exposure (2x 0.4 mg/kg ip on E15). AD threshold in μA. There was no significant difference based on prenatal exposure or sex of the offspring.

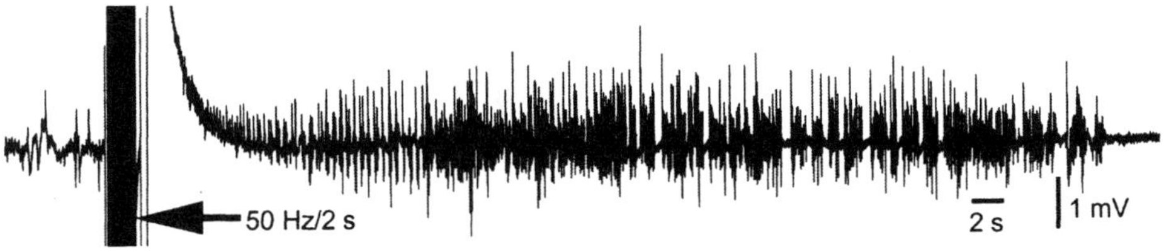

Figure 3. An example of an afterdischarge recorded from the dentate gyrus on PN15. Stimulation artifacts marked by an arrow (50 Hz/ 2 s) were truncated for clarity. Time mark 2 s, calibration 1 mV.

3.2. Afterdischarge Threshold

Stimulation current required for eliciting AD (AD threshold) ranged between 180-240 μA. There was no significant difference in AD thresholds based on sex or prenatal exposure (Figure 2; two-way ANOVA, P always > 0.05).

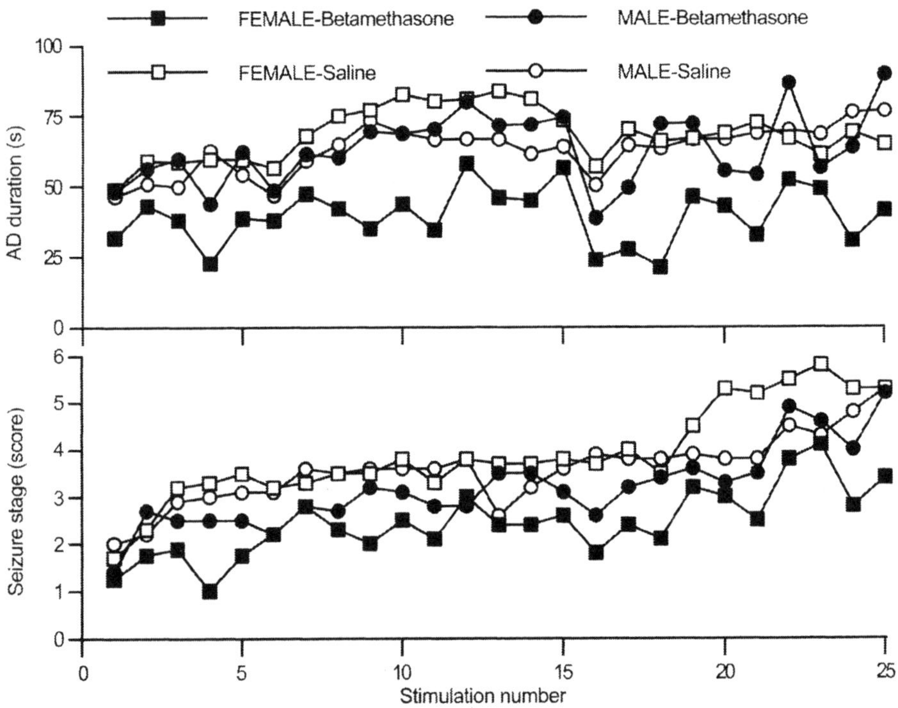

Figure 4. Afterdischarge duration (top) and seizure stage (bottom). *Open symbols* indicate prenatal saline exposure while *closed symbols* prenatal betamethasone exposure. *Circles* represent males and *squares* are females. Means for individual stimulations are presented (SEMs were eliminated for clarity). There was a significant effect of prenatal exposure on AD duration (top; main effect): Betamethasone exposure resulted in shortened ADs especially in female offspring (interaction of main effects). Prenatal exposure to betamethasone significantly decreased severity of seizures (bottom; determined as seizure score; main effect).

3.3. Afterdischarge Duration

Figure 3 illustrates an example of an AD recorded in the dentate gyrus. Two way analysis of variance for repeated measures revealed that there was a significant effect of prenatal exposure on the AD duration ($F_{(1,12)}$=7.723, P<0.05; Figure 4 top). ADs in prenatally betamethasone-exposed rats were shorter than in saline-exposed rats. There was no effect of sex ($F_{(1,12)}$=3.294; P>0.05). However, there was a significant interaction between sex and prenatal exposure ($F_{(1,12)}$=8.297; P<0.05). Additional analysis showed that prenatally betamethasone-exposed females displayed significantly shorter afterdischarges than all other groups. While there was a small difference between the duration of the AD after the first and last stimulation, the ADs indeed increased with time as indicated by a significant difference in the repeated factor (stimulation number; $F_{(24,288)}$=3.052; P<0.05). The decrease in AD duration after stimulation #15 is due to our kindling paradigm as 15 stimulations were delivered on PN15 and additional 10 on PN16. There was no interaction between the repeated factor and main factors.

3.4. Seizure Stage

Prenatal exposure had a significant effect on the seizure score ($F_{(1,12)}$=6.110; P<0.05; Figure 4 bottom). Betamethasone-exposed rats experienced less severe seizures after kindling stimulations than saline-exposed rats. There were no effects of sex on seizure score ($F_{(1,12)}$=0.295; P>0.05). Similarly, no significant interaction between prenatal exposure and sex was discovered ($F_{(1,12)}$=2.144; P>0.05). There was an increase in seizure stage over time as suggested by a significant difference in the repeated factor (stimulation number; $F_{(24,288)}$=12.583, P<0.05). There was no interaction between the repeated factor and main factors.

4. DISCUSSION

Our results indicate that repeated prenatal exposure to betamethasone on embryonic day 15 in the rat decreased kindling epileptogenesis as assessed on postnatal days 15 and 16. Prenatal betamethasone reduced severity of seizures as indicated by decreased seizure scores and also decreased AD duration particularly in female offspring.

Thus, prenatal bethamethasone does not have adverse postnatal effects in terms of enhanced epileptogenesis and seizure susceptibility but rather seems to decrease epileptogenesis in the immature brain. This is in contrast to our previous finding of enhanced developmental seizure susceptibility after prenatal exposure to hydrocortisone (unpublished data). However, reports in humans also indicate that prenatal betamethasone may have beneficial effects compared to other corticosteroids such as dexamethasone[12]. Therefore it is possible, that postnatal adverse effects after prenatal corticosteroid exposure may be a function of the specific corticosteroid used for prenatal exposure. The effects of prenatal betamethasone exposure on AD duration were significantly affected by sex of the offspring. This finding is corroborated by sexual differences found in receptor systems for glucocorticoids and mineralocorticoids in guinea pigs. These receptor systems are significantly affected by prenatal betamethasone exposure.[13]

One of the explanations why we were unable to demonstrate any adverse effects of repeated prenatal betamethasone exposure on postnatal epileptogenesis is the limited number and restricted period of prenatal betamethasone exposure in our treatment paradigm (only 2 injections over a single embryonic day). Alternatively, betamethasone may be relatively free of long-term adverse effects unlike other corticosteroids. Although the sex of the subjects did not alter seizure stage development, the modification of the AD duration after prenatal betamethasone exposure by the sex indicates that sex of the subjects should be considered in steroid hormone effects.

5. ACKNOWLEDGMENTS

The author wishes to thank Ms. Zunju Hu for outstanding technical assistance and Dr. Jana Velíšková for helpful comments. Supported by NS-41366, NS-36238, and NS-20253 from NIH, and by Heffer Family Medical Foundation.

6. REFERENCES

1. B. A. Banks, A. Cnaan, M. A. Morgan, J. T. Parer, J. D. Merrill, P. L. Ballard, and R. A. Ballard, Multiple courses of antenatal corticosteroids and outcome of premature neonates. North American Thyrotropin-Releasing Hormone Study Group, *Am. J. Obstet. Gynecol.* **181**(3), 709-717 (1999).
2. N. P. French, R. Hagan, S. F. Evans, M. Godfrey, and J. P. Newnham, Repeated antenatal corticosteroids: size at birth and subsequent development, *Am. J. Obstet. Gynecol.* **180**(1 Pt 1), 114-121 (1999).
3. P. C. Ng, G. W. Wong, C. W. Lam, C. H. Lee, T. F. Fok, M. Y. Wong, and K. C. Ma, Effect of multiple courses of antenatal corticosteroids on pituitary- adrenal function in preterm infants, *Arch. Dis. Child Fetal Neonatal Ed.* **80**(3), F213-216 (1999).
4. D. A. Guinn, Repeat courses of antenatal corticosteroids: the controversy continues, *Am. J. Obstet. Gynecol.* **190**(3), 585-587 (2004).
5. M. J. Lee, J. Davies, D. Guinn, L. Sullivan, M. W. Atkinson, S. McGregor, B. V. Parilla, K. Hanlon-Lundberg, L. Simpson, J. Stone, D. Wing, K. Ogasawara, and J. Muraskas, Single versus weekly courses of antenatal corticosteroids in preterm premature rupture of membranes, *Obstet. Gynecol.* **103**(2), 274-281 (2004).
6. H. Uno, L. Lohmiller, C. Thieme, J. W. Kemnitz, M. J. Engle, E. B. Roecker, and P. M. Farrell, Brain damage induced by prenatal exposure to dexamethasone in fetal rhesus macaques. I. Hippocampus, *Dev. Brain Res.* **53**(2), 157-167 (1990).
7. H. Uno, S. Eisele, A. Sakai, S. Shelton, E. Baker, O. DeJesus, and J. Holden, Neurotoxicity of glucocorticoids in the primate brain, *Horm. Behav.* **28**(4), 336-348 (1994).
8. K. Haas, E. F. Sperber, and S. L. Moshé, Kindling in developing animals: expression of severe seizures and enhanced development of bilateral foci *Dev. Brain Res.*, **56**, 275-280 (1990).
9. I. M. Germano and E. F. Sperber, Transplacentally induced neuronal migration disorders: an animal model for the study of the epilepsies, *J. Neurosci. Res.* **51**(4), 473-488 (1998).
10. N. Chevassus-au-Louis, A. Rafiki, I. Jorquera, Y. Ben-Ari, and A. Represa, Neocortex in the hippocampus: An anatomical and functional study of CA1 heterotopias after prenatal treatment with methylazoxymethanol in rats, *J. Comp. Neurol.* **394**, 520-536 (1998).
11. G. Paxinos and C. Watson, *The Rat Brain in Stereotaxic Coordinates*, Fourth ed. (Academic Press, San Diego, 1998).
12. A. H. Jobe and R. F. Soll, Choice and dose of corticosteroid for antenatal treatments, *Am. J. Obstet. Gynecol.* **190**(4), 878-881 (2004).
13. D. Owen and S. G. Matthews, Glucocorticoids and sex-dependent development of brain glucocorticoid and mineralocorticoid receptors, *Endocrinology* **144**(7), 2775-2784 (2003).

Discussion

Schwartzkroin: Did you give only 1 injection, and what is the rational for giving the injections at that particular time?

Velisek: Two injections two hrs apart morning/evening of E15. We gave the injections on E15 because we know from our other studies that administration at day gives peak effect. A second reason is that we know from other studies that this is the peak development period of the hippocampus, and if you do something during this period one has a great chance to effect the hippocampus and its function. We gave 2 injections because we were trying to model a repeat injections paradigm.
Schwartzkroin: Do you have any data about baseline changes in either behavior or EEG in these animals?
Velisek: I did not report it here, but there are changes.
Wasterlain: We have long known that one of the things steroids do is differentiate cells early, and am I right to think that at E15 the pyramidal cells are still multiplying? If you differentiate pyramidal cells early they might have less potential for multiplying, and they might also have a different development. What happens to the structure of the hippocampus as a result of the metralmethazone at a critical time?
Velisek: I cannot tell you right now. But you are absolutely right that if there is a sudden stop in differentiation of hippocampal cells, it might lead to a decrease in epileptogenesis. We did not do stereologic cell counts; rather we did cell number estimates in a completely different model of prenatal exposure also on E15, and we found a decreased cellularity in the hippocampus and entorhinal cortex and in those rats we found a significantly decrease epileptogenesis.
Burnham: Heather Edwards and I looked at the effect of prenatal stress using bright lights and confining the animal, and we found quite different effects. We found that the animals were more prone to kindling after birth.
Velisek: Yes, and actually we did other experiments not with metralmethazone but with hydrocortizone and found completely different results.
Potschka: One, is there anything known about the placental penetration of corticosteroids in humans and in rats? Two, patients with anxiety and depression are know to have a disturbed HPA axis and increased endogenous glucocorticoid levels. Is there anything known about whether children from women with anxiety disorders or depression have an increased risk of developing epilepsy?
Velisek: Yes, it is known from a British study performed in 1976, of of mothers who suffered an extreme stress during the third trimester. These kids were followed for 6 years, and there was a significant change in behavior in those kids up to age 6 years.

DEVELOPMENT OF KINDLING IN IMMATURE FAST AND SLOW KINDLING RATS

Jana Velíšková, Jeremy Asnis, Fred A. Lado, and Dan C. McIntyre*

1. INTRODUCTION

Several epileptic syndromes have a genetic background. Thus, genetic models of epilepsy in rodents provide a useful tool to study basic mechanisms of seizures.[1] Using selective breeding technique, two new strains of rats with differential rates of amygdala kindling were produced and named "FAST" and "SLOW" kindling rats.[2, 3] Although the FAST and SLOW kindling rats have differences in kindling paradigms in several brain regions such as hippocampus, piriform and perirhinal cortices, the most striking differences were found during amygdala kindling. Besides the faster kindling rate, amygdala stimulation in FAST rats resulted in significantly longer afterdischarges (ADs) both prior to and after the kindling compared to the SLOW rats.[2, 3] The differences in the excitability of these two strains were determined in adult rats.

In the present study we examined whether: (1) FAST and SLOW kindling rats also differ in susceptibility to flurothyl seizures, a model of primary generalized seizures and (2) the disparity in amygdala kindling between the FAST and SLOW rats is present already during development. In this paper, we present preliminary data addressing these issues.

2. METHODS

We used FAST and SLOW male rats from established colonies in the Carleton University Life Science facility. The rats were then bred and raised in our animal facility accredited by the American Association for Accreditation of Laboratory Animal Care. Rats were housed at a constant temperature (23°C) and relative humidity (60%) with a fixed 12-h light-dark cycle with free access to food and water. Rats were weaned and

* Jana Velíšková, Departments of Neurology and Neuroscience, Albert Einstein College of Medicine, Bronx, NY 10461, USA. Jeremy Asnis and Fred A. Lado, Department of Neurology, Albert Einstein College of Medicine, Bronx, NY, 10461. Dan C. McIntyre, Department of Psychology, Institute of Neuroscience, Carleton University, Ottawa, Ontario K1S 5B6, Canada.

housed 5 per cage at postnatal day (PN) 21. All animal procedures were in accord with guidelines for the care and treatment of laboratory animals of the National Institutes of Health.

2.1. Flurothyl Seizure Testing

Flurothyl vapors are used to induce convulsions.[4] Flurothyl is continuously (rate of 20 μl/min) delivered into an air-tight chamber (9.34 l) *via* a precise microinfusion pump (WPI, Inc.). The advantage of continuous flurothyl infusion is that seizures always occur. Two seizure types develop after flurothyl exposure: clonic and tonic-clonic seizures. In adult rats, clonic seizures consist of head and forelimb clonus and are characterized by preservation of the righting reflex. Tonic-clonic seizures begin with wild running followed by loss of righting reflex and then, after a short tonic contraction of all limbs, long-lasting clonic seizures occur. In PN 15 rats, flurothyl-induced clonic seizures are very brief and within a couple of seconds progress into long-lasting tonic-clonic seizures. We recorded the latency to the onset (measured from the beginning of flurothyl delivery) of the first clonic and first tonic-clonic seizure. Because flurothyl was infused at a constant rate, the latency to the onset of seizures allowed us to calculate the amount of flurothyl necessary to induce the seizures; i.e., the flurothyl seizure threshold for clonic or tonic-clonic seizures. The seizure threshold inversely reflects seizure susceptibility. Thus, the higher the seizure threshold, the lower the seizure susceptibility.

2.2. Surgery

FAST and SLOW kindling pups (PN13) were operated under anesthesia with the i.p. mixture of ketamine (70 mg/kg) and xylazine (10 mg/kg)[5]. A bipolar twisted wire electrode (Plastics One) was stereotactically lowered into the left amygdala (coordinates in mm with respect to bregma: anterior-posterior = 5.3; lateral = 4.0; depth = 7.0 (from the skull); incisor bar at -3.5 mm)[6]. The electrode was then anchored to the skull with two screws and dental acrylic cement. The pups were allowed 2 days for recovery.

2.3. Kindling

The kindling paradigm for pups described previously[7, 8] was modified for the kindling of SLOW and FAST rat pups. Each kindling stimulus consisted of a 2 s train of 2 ms bipolar square pulses delivered at 50 Hz. In a preliminary experiment, we determined that the shortest interstimulus interval required to induce an afterdischarge (AD) is 45 minutes for both SLOW and FAST rat pups, which then was used in this experiment. Due to a high and variable afterdischarge threshold prior to kindling, especially in SLOW kindling pups, the current for each kindling stimulation was set at 1 mA. The pups display a unique behavior during kindling[7, 8] compared to adult rats. We modified our original kindling scale for this experiment and the stages are described in Table 1.

We compared following parameters between the SLOW and FAST kindling rats:

1. AD threshold (minimum stimulus intensity necessary to induce an AD);
2. Kindling rate (number of stimulations to achieve first stage 5;
3. AD duration.

Table 1. Kindling-induced behavioral stages in PN 15 rats

Stage	Behavioral expression
0	Behavioral arrest
1	Chewing
2	Head bobbing
3	Unilateral forelimb clonus
4	Alternating forelimb clonus
5	Climbing, jumping and swimming-like bilateral forelimb clonus
6	Hindlimb clonus, wild running

Rats received up to 20 stimulations in two consecutive days (maximum 10 each day). At the end of the experiment, rats were sacrificed under deep urethane anesthesia; brains were quickly removed, frozen in methylbutane (-20^{0}C) and stored in a freezer (-70^{0}C). Frozen coronal sections were cut throughout the electrode tract, stained with thionin, and examined under a light microscope for electrode placement. Only rats with the electrode tip in the basal amygdala complex were used. Rats with any blood lesion around the track were excluded.

2.4. Statistical Analysis

Flurothyl seizure thresholds in FAST and SLOW kindlers were compared using Student's t-test. For kindling, Student's t-test was used to determine AD threshold and kindling rate (number of stimulations to achieve first stage 5). The AD durations following individual stimulations were calculated using Analysis of Variance (ANOVA) with repeated measures. The level of significance was set to $P \leq 0.05$.

3. RESULTS

3.1. Flurothyl Seizure Threshold

In adult rats, there was a significant difference in the seizure threshold between the FAST and SLOW kindlers. The seizure thresholds for both clonic and tonic-clonic seizures were lower in the FAST compared to the SLOW rats (Table 2). In contrast in PN 15 rats, there was no difference in flurothyl seizure onset between the FAST and SLOW kindlers (Table 2).

Table 2. Features of flurothyl and kindling seizures in FAST and SLOW rats

	Adult rats		Immature (PN 15) rats	
	FAST	SLOW	FAST	SLOW
Flurothyl threshold	Low	High	No difference	
Kindling rate	Fast[3]	Slow[3]	Fast	Slow

3.2. Kindling in Immature FAST and SLOW Kindling Rats

3.2.1. Afterdischarge Threshold

The mean AD threshold in the FAST kindlers, which ranged between 200-400 μA, was significantly lower compared to the SLOW kindlers, with a range between 350-500 μA (Student's t-test, $P \leq 0.05$).

3.2.2. Kindling Rate

Analysis of the rate of kindling showed a significant difference between FAST and SLOW kindling rat pups. In the FAST kindlers, the first stage 5 was achieved following about 10 stimulations. In the SLOW kindlers, the kindling rate was slower, and to achieve stage 5, the pups required 17 or more stimulations (Figure 1 and Table 2; Student's t-test, $P \leq 0.05$).

3.2.3. Afterdischarge Duration

The AD duration did not differ between FAST and SLOW kindling pups regardless of the kindling stage (Figure 2; repeated measures ANOVA, $P \leq 0.05$).

4. DISCUSSION

In this study, we tested whether the two strains of rats (FAST and SLOW) originally selected based on differential kindling rates can be distinguished also in a model of primary generalized seizures. Interestingly, only in adult rats was the flurothyl seizure threshold lower in FAST compared to SLOW rats. In PN 15 rats, no difference in seizure threshold was found. This prompted us to evaluate in PN 15 rats whether distinct kindling rates are already present during development.

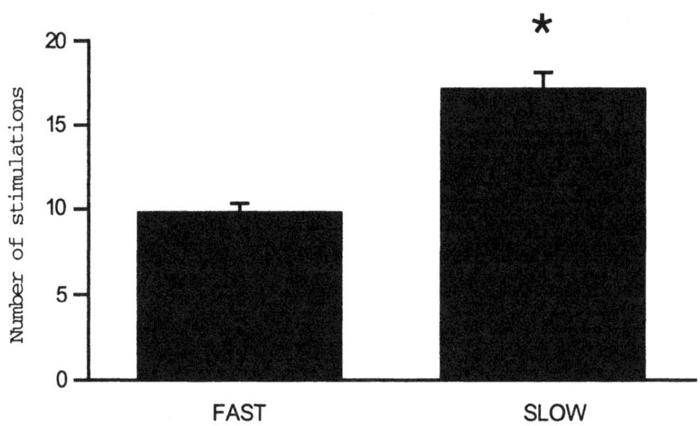

Figure 1. The kindling rate in PN 15 FAST and SLOW rats. The FAST rats required less kindling stimulations (mean ± SEM) to achieve first stage 5 than the SLOW rats.

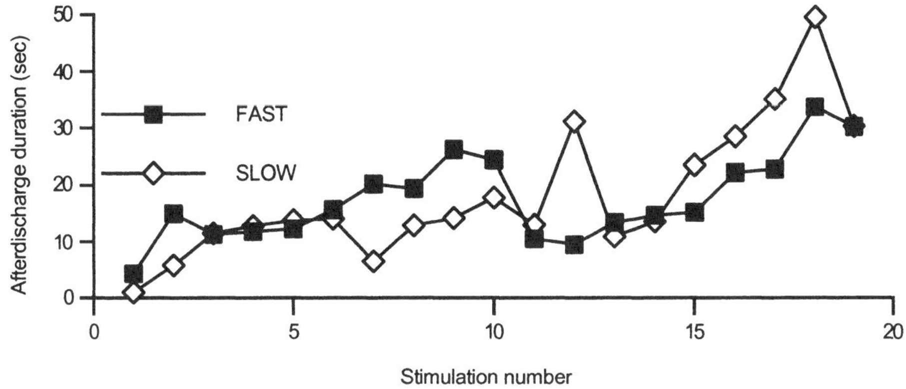

Figure 2. Afterdischarge (AD) duration in PN 15 FAST and SLOW rats. *Closed symbols* represent the FAST rats, *open symbols* indicate the SLOW rats. There was no difference in AD duration between the FAST and SLOW rats during the kindling paradigm.

In the previous study, McIntyre et al. showed that adult FAST kindlers have significantly higher local excitability within several structures (amygdala, hippocampus, piriform and perirhinal cortices) known to be involved in seizure generation compared to the SLOW kindlers. This suggests that the excitation:inhibition ratio in the FAST kindlers is increased, while it is decreased in the SLOW kindlers, and this may predispose to changes in general seizure susceptibility.[3] Accordingly our data show that flurothyl-induced seizures occur faster in the FAST rats than in the SLOW rats. The mechanism, by which flurothyl induces seizures is unclear. Flurothyl has been shown to block both acetylcholine[9] and $GABA_A$[10] receptors in similar concentrations, which produce seizures. In this case, the acetylcholine system may not be involved as there are no differences between the FAST and SLOW kindling rats in the number of cholinergic neurons in the forebrain[3]. On the other hand, there are differences between the FAST and SLOW rats in the GABA system. Accordingly, lower seizure thresholds have been reported for pentylenetetrazol, bicuculline and picrotoxin ($GABA_A$ receptor antagonists).[11] The involvement of the GABA system seems to be specific because strychnine (a glycine antagonist) was ineffective in differentiating between FAST and SLOW rats.[11] The higher susceptibility of adult FAST rats may be related to the differential expression of $GABA_A$ receptor subunits. The adult FAST rats have high expression of $\alpha 2$, $\alpha 3$ and $\alpha 5$ subunits for the $GABA_A$ receptor compared to the SLOW rats with rather high levels of the $\alpha 1$ subunit.[12] It has been suggested that flurothyl has high affinity for $GABA_A$ receptors containing the $\alpha 2$ subunit.[13] This may explain the higher susceptibility of adult FAST rats to flurothyl-induced seizures.

However at PN 15, the susceptibility to flurothyl seizures was not different between FAST and SLOW rats. This may be due to the fact that during development both strains have high levels of the $\alpha 2$ subunit, as $GABA_A$ receptors in the brain are present in "immature" form and contain mainly the $\alpha 2$ subunit.[14] Later with maturation the $\alpha 2$ subunit for the $GABA_A$ receptor is replaced with the $\alpha 1$ subunit;[14] however, this happens only in the SLOW rats, as the FAST adults keep the "immature" form of $GABA_A$ receptors.[12]

Preliminary data also suggest that in terms of susceptibility to kindling the two strains can be distinguished already by PN 15. We propose that during kindling, systems other than the GABA neurotransmission may be involved in producing the differential effects. Overall, the most interesting finding is that in these strains of rats "bred for differential susceptibility to kindling," the kindling differences are present already by PN 15. Although there is differential susceptibility to other seizures, i.e. flurothyl, this susceptibility is developmentally regulated.

5. ACKNOWLEDGEMENT

The authors thank Hong Wang and Zunju Hu for excellent technical work. This work was supported by NIH grants NS-20253, NS-41340, a CURE Foundation Research Grant and Natural Sciences and Research Council Grant.

6. REFERENCES

1. J. L. Noebels, The biology of epilepsy genes, *Annu. Rev. Neurosci.* **26**, 599-625 (2003).
2. R. J. Racine, M. Steingart, and D. C. McIntyre, Development of kindling-prone and kindling-resistant rats: selective breeding and electrophysiological studies, *Epilepsy Res.* **35**(3), 183-195 (1999).
3. D. C. McIntyre, M. E. Kelly, and C. Dufresne, FAST and SLOW amygdala kindling rat strains: comparison of amygdala, hippocampal, piriform and perirhinal cortex kindling, *Epilepsy Res.* **35**(3), 197-209 (1999).
4. J. Velíšková and S. L. Moshé, Sexual dimorphism and developmental regulation of substantia nigra function, *Ann. Neurol.* **50**, 596-601 (2001).
5. J. Velíšková, L. Velíšek, M. L. Nunes, and S. L. Moshé, Developmental regulation of regional functionality of substantia nigra $GABA_A$ receptors involved in seizures, *Eur. J. Pharmacol.* **309**, 167-173 (1996).
6. G. Paxinos and C. Watson, *The rat brain in stereotaxic coordinates*, 4th ed. (Academic Press, New York, 1998), pp. 238.
7. K. Haas, E. F. Sperber, and S. L. Moshé, Kindling in developing animals: expression of severe seizures and enhanced development of bilateral foci, *Dev. Brain Res.* **56**(2), 275-280 (1990).
8. E. F. Sperber, J. Velíšková, B. Benenati, and S. L. Moshé, Enhanced epileptogenicity of area tempestas in the immature rat, *Dev. Neurosci.* **20**(6), 540-545 (1998).
9. E. I. Eger, 2nd, D. Gong, Y. Xing, D. E. Raines, and P. Flood, Acetylcholine receptors and thresholds for convulsions from flurothyl and 1,2-dichlorohexafluorocyclobutane, *Anesth. Analg.* **95**(6), 1611-1615, table of contents (2002).
10. M. Wakamori, Y. Ikemoto, and N. Akaike, Effects of two volatile anesthetics and a volatile convulsant on the excitatory and inhibitory amino acid responses in dissociated CNS neurons of the rat, *J. Neurophysiol.* **66**(6), 2014-2021 (1991).
11. R. J. Racine, M. Steingart, Y. Bureau, and D. C. McIntyre, Differential sensitivity of genetically Fast vs. Slow kindling rats strains to GABAergic convulsive agents, *Neuropharmacology* **45**(7), 918-924 (2003).
12. M. O. Poulter, L. A. Brown, S. Tynan, G. Willick, R. William, and D. C. McIntyre, Differential expression of alpha1, alpha2, alpha3, and alpha5 GABAA receptor subunits in seizure-prone and seizure-resistant rat models of temporal lobe epilepsy, *J. Neurosci.* **19**(11), 4654-4661 (1999).
13. M. D. Krasowski, Differential modulatory actions of the volatile convulsant flurothyl and its anesthetic isomer at inhibitory ligand-gated ion channels, *Neuropharmacology* **39**(7), 1168-1183 (2000).
14. J.-M. Fritschy, J. Paysan, A. Enna, and H. Mohler, Switch in the expression of rat GABAA-receptor subtypes during postnatal development: An immunohistochemical study, *J. Neurosci.* **14**, 5302-5324 (1994).

Discussion

Schwartzkroin: What's the background from which the fast and slow kindlers were derived?
Veliskova: They were Wistar and Long-Evans rats.
Schwartzkroin: So have you looked at a normal non-separated Long-Evans or Wistar?
Veliskova: That's why I said that the experiments are far from complete, because we have to look at these rats and all the stimulation paradigms to compare. So there will be many more control groups than the experimental groups in the end.
Schwartzkroin: We know that there are strain differences among rats in seizure susceptibilities. The other issue I have is in respect to the disassociation between the behavior and the afterdischarge. You said the behavioral abnormalities persisted. What electrode locations were you recording from that you defined your afterdischarge? Might there be afterdischarge in another region?
Veliskova: Absolutely, we did not implant multiple electrodes in these rats. We recorded from the site of the stimulation, the amygdala. So this is a disassociation within the amygdala. It can be ongoing of course and I expect that it will be ongoing within the cortex and other brain structures.
McNamara: Comment and a question. I wonder whether there will be a relationship between low seizure threshold for a given stimulus, whether it is electroshock or PTZ or whatever, and the rate of epileptogenesis? It seems to me that there's not clearly a 1:1 relationship between low threshold to electroshock seizure and the likelihood of becoming epileptogenic. What is the mechanism by which flourothyl tiggers a seizure?
Veliskova: The mechanism is unclear; there is some evidence that it is something to do with $GABA_A$ receptors. But its not known for sure how the flourothyl is working. They are primary generalized seizure because the rats are inhaling the flourothyl and they have the seizures. But the exact mechanism is unknown.
Engel: We know the mechanisms for seizures are different than the mechanism for epileptogenesis. In the human, there's lots of data that suggest that the immature brain is more likely to have a seizure given an insult but less likely to develop epilepsy as a result of that insult. Animal models don't always show the same thing.
Veliskova: Even in these rats, although the statistic do not show significance, the fast kindlers had higher thresholds than the Sprague-Dawley rats, but tended to jump to the higher stages after 2 or 3 stimulations. They had stage 3- 3.5 very easily after 2 or 3 stimulations. The threshold does not definitely correlate with the epileptogenesis. The slow kindlers had a very high threshold but developed epileptogenesis very slowly. Thus, there is no correlation in these rats.
Engel: But Dr. Shouse showed that the kittens were more likely to develop spontaneous seizures than the adults, which is opposite to what we see in humans.
Veliskova: The fast kindlers are showing basically the same as the kittens. They have high threshold, but epileptogenesis is faster.
Sutula: Sympathies with experiments that get more complicated than what you anticipated. One of the terms that you used in the conclusion was that perhaps relative difference in maturity might be an explanation for some of the differences. I think back to Dr. Shouse's talk on the effect of developmental age and the confusion of

developmental age on her study. I think the whole area is complicated because there's some much change that can occur between day21 and 23. These experiments are hopelessly confounded by individual variation as well as developmental differences that are very substantial. I think that it is a horribly difficult thing to work on, and it is not surprising to me that you got a confusing series of results.
Veliskova: Yes, we are planning to test these rats at day 17 instead of day 15 to have this 2 day difference for the developmental difference. Maybe we will see a better pattern and correlation with the Sprague-Dawley rats. Because even stage 4 the could not reach so easily – while the P15 Sprague-Dawley rats raise on the hindlimbs and they have beautiful forelimb clonus, the slow and fast rats had trouble raising nicely.straight up.
McIntrye: Let me just make a comment and see if everyone else has had the same experience. These animals Ron Racine selected before there were pathogen free animals. And one of the thing we noticed when all of the breeders went to pathogen free animals, the kindling rates dropped in half for every strain that they did that to. So it appears that every strain changed hugely from the breeder as a consequence of caesarean derivations, suggesting that something was lost in the unnatural production that perhaps stressed nervous systems and changed animals. We have a whole different animal now that we are working with in terms of the commercial availability versus these fast and slow guys.
Veliskova: I absolutely agree. We actually have absolutely different responses in rats from Charles River and from Taconic and both are Sprague-Dawley rats. It is very important to keep one supplier for all ordering the rats, because we normally order rats from Taconic, but when we order from Charles River we get completely different results.
Unidentified: Just to follow up on that, it is really clear that even if you think that you are getting the same strain from different suppliers or if you buy a strain and breed them in your laboratory, they change. So the strain issue becomes a huge one in comparison.

NEONATAL HEAT-INDUCED CONVULSIONS AFFECT BEHAVIOURS IN NEONATAL, JUVENILE AND ADULT RATS

Deborah Saucier, Avril Keller, Aaron Sheerin, R. Daniel LaPorte, and Jerome Y.Yager*

1. INTRODUCTION

Febrile convulsions are a common form of childhood seizure, occurring in approximately 2-5 % of infants and young children.[1] Simple febrile convulsions are the most common, defined as a generalized seizure of short duration (<15 minutes) that occurs during a febrile illness,[2, 3] in a child between the ages of 6 months to 6 years. The significance of simple febrile convulsions is debated, with many considering them to be benign, as they do not result in gross neuropathology[4] nor do they increase the risk of partial-complex epilepsy (e.g., [5 2, 6]) or cognitive impairments.[7-9] Conversely, some researchers have observed that in adulthood, individuals who have experienced simple febrile convulsions have a slightly higher risk of partial-complex epilepsy in later life (e.g., [3, 10]) and may exhibit cognitive deficits, including: decreased ability to sustain attention; deficits in some types of learning and non-verbal memory; delayed recognition; and decreased visuomotor skills (e.g., [1, 11]). As such, the question of whether febrile convulsions enhance susceptibility to seizure disorders, cognitive impairments or other pathologies remains unresolved.

In order to examine the relationship among febrile convulsions, neuropathologies and behavioural change, investigators have developed paradigms in rat pups that simulate febrile convulsions in infants.[12] Typically, rat pups are heated to ~43C, which results in convulsive behaviours and seizure activity.[13] There are age-dependent changes in susceptibility to heat-induced convulsions (HC), with peak susceptibility to HC occurring sometime between the 2nd and 3rd week of life (e.g., [12, 14, 15]), an age roughly equivalent to the brain development of a 6-12 month old human infant.[16] Further, HC are associated

* Deborah Saucier, Avril Keller, Aaron Sheerin, Daniel LaPorte, University of Saskatchewan, Saskatoon, Sk., S7N 5A5. Jerome Y.Yager, Stollery Children's Hospital, 2C3.V50 Walter C. Mackenzie Health Centre, Edmonton, AB, T6G 2R7.

with enhanced susceptibility to kindling and to seizures induced with pentlyenetetrazol[15] or kainic acid.[17] Although a single HC increased Timm staining in portions of the hippocampus,[18] no neurodegeneration or apoptosis was observed.[18, 19] Thus, HC in the rat appears to produce long-lasting changes in neural plasticity without producing obvious pathologies.

Consistent with the research involving children, investigations of the long-term behavioural consequences of febrile convulsions are also inconclusive, with some researchers finding no long-term behavioural effects (e.g., [19, 20]) and others finding mild impairments in cognitive function in later life (e.g., [16, 21]). However, with the exception of Nealis et al.[16] (who found no effect of febrile convulsions on reflexive behaviours in neonates), few studies have examined the effects of febrile convulsions on the behaviours of neonatal or juvenile pups. Further, no studies have employed a longitudinal design to determine whether or not changes in early behaviours are associated with later behavioural change.

2. EFFECTS OF HC ON BEHVIOURS IN NEONATES

It is inferred that behavioural changes in HC pups result from alterations in the nervous system, brought about by the convulsion. However, alterations in maternal care can also affect later behaviour. Simple separation from the dam can result in increased anxiety behaviour in pups,[22] which can still be observed in adulthood.[23] Quality of maternal care can also affect social behaviours, especially when the quality is poor.[24] Thus, we sought to investigate whether HC resulted in alterations of mother-pup interactions.

Pups are not passive and they actively pursue interactions with their dams. Pups signal their dams by making ultrasonic vocalisations (USVs) when they are cold and in certain social situations.[25-27] USVs will draw lactating females from their nests to engage in searching behaviours for the calling pup.[28] Although USVs decrease in prevalence as the pup matures, they also can be characterized into 10 categories that change with maturity.[29] To date, no one has investigated whether or not febrile convulsions affect USVs, and whether alterations in USVs result in altered mother-pup interactions.

2.1 Method

2.1.1 Heat-Induced Convulsions

On postnatal day 7 (P7), 79 rat pups were selected at random (controlling for sex) and assigned into either the convulsed (C+) or the control (c-) group (male C+: n = 16; male C- n = 18; female convulsed n = 18; female control n = 17). To induce convulsions, pups were placed under a heat lamp (GE 300W) and monitored for convulsions and core body temperature. When either the core body temperature reached 43C or behavioural manifestations of convulsion were evident (i.e., loss of upright posture and tonic – clonic movements) pups were removed from the heat and placed in a lined recovery cage. All pups exhibited convulsions of at least 210 sec duration (mortality, n=1 male). C- pups were treated similarly, but out of direct heat.

2.1.2 Mother-Pup Interactions

On postnatal day 10 (P10) and postnatal day 12 (P12), mother-pup interactions were observed by randomly distributing an entire litter into a 40 cm by 40 cm by 30 cm translucent box. The dam was introduced and all behaviours were videotaped for 10 minutes. We scored 4 types of mother-pup interactions: anogenital licking; grooming; nursing; and retrieval. Anogenital licking was considered as only licking in the anogenital region, whereas grooming was considered as licking anywhere but the anogenital region. An instance of nursing was recorded when a pup successfully latched onto the teat, and the dam allowed it to feed. Retrieval was measured when the dam picked up a pup in her mouth and returned it to the nest. The total number of behaviours was summed for analyses.

2.1.3 Ultrasonic Vocalizations

On P10 and P12, ultrasonic vocalisations were monitored (Petterson model D100 Ultrasound Detector) and digitally recorded (PIII 450 computer). Pups were individually placed in the isolation chamber (20 cm by 20 cm by 25 cm clear jar) for 2 minutes. Each two-minute vocalisation sample was analysed (Avisoft) as in Brudzynski[29] for quantity and structure.

2.2. Results

2.2.1 Maternal Behaviours

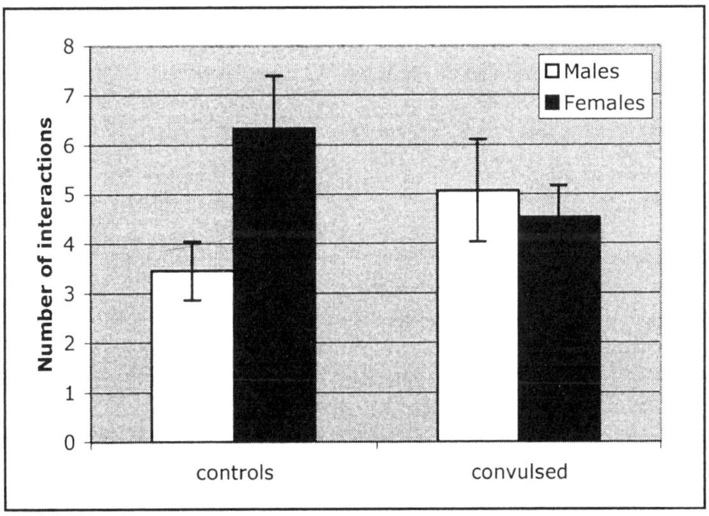

Figure 1. Mean number of interactions with dam (± SEM). There was a trend for a sex by treatment interaction, whereby: C- females interacted more with their dam than both C+ females and C- males; C+ males interacted more with their dam than C+ females, $F(1, 52) = 3.426, p = .070$. Bars represent means (± SEM).

2.2.2 Ultrasonic Vocalisations

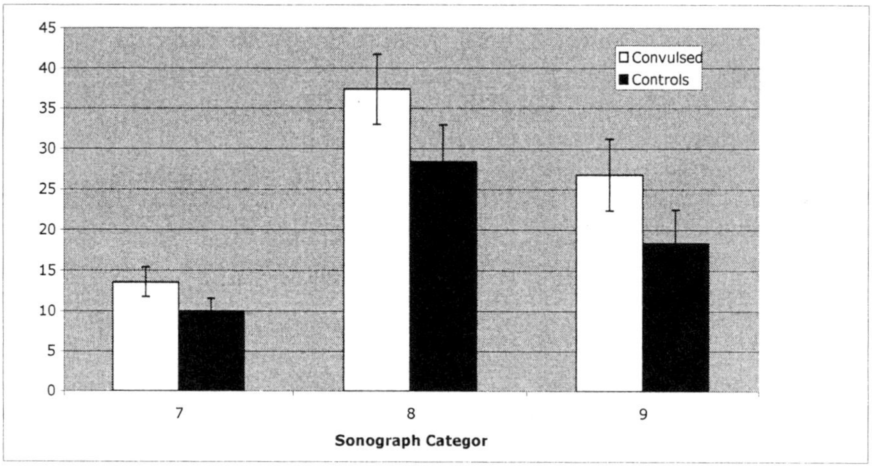

Figure 2. C+ pups exhibited significantly more ultrasonic vocalizations than C- pups, especially in sonograph categories 7 [F (1, 141) = 5.160, p = 0.025], 8 [F (1, 141) = 4.235, p = 0.041], and 9 [F (1, 141) = 4.018, p = 0.047]. Bars represent means (± SEM).

2.2.3 Correlations

There was a significant and positive relation between total number of mother-pup interactions and sum of all categories of USVs, $r_s = .279, p = 0.031$ (Figure 2). However, this correlation was only significant for the HC- pups, $r_s(30)$=.410, p=.024, as HC+ pups $r_s(30)$=.067, p=.727.

2.3 Discussion of the effects of heat-induced convulsions on neonatal rats

There may be subtle sex differences in how mothers respond to convulsed pups, with male C+ pups making more vocalizations and interacting more with the dams than female C+ pups. Further, C+ female pups, despite calling more frequently than their littermate controls, received the low amounts of maternal care. Given that it has been suggested that febrile convulsions have the greatest effect in females (e.g., [10, 11]), slight differences in mother-pup interactions may have important effects on later cognitive ability. In the strictest sense, however, despite making more ultrasonic vocalizations, C+ pups did not receive significantly greater care from their dams. Thus, although there was a relation between USVs and mother-pup interactions, this did not hold for the C+ pups. Further, as sonograph categories 7, 8, and 9 are thought to represent relatively less efficient means of communication, the significant increase in categories 7,8,9 for the C+ pups suggests that C+ pups tended to vocalise in a less efficient manner.[29] This may reflect decreased physical and neural maturation, at least as measured by ability to produce complex, mature USVs.

3. LONGITUDINAL EFFECTS OF HEAT INDUCED CONVULSIONS ON BEHAVIOURS IN JUVENILES AND ADULTS

3.1 Method

3.1.1 Heat-Induced Convulsions

As in 2.1.1, on P7, 78 rat pups were selected and assigned into groups: male C+: n = 25; male C- n = 18; female convulsed n = 13; female control n = 22. However, as ~1/3 of children who exhibit febrile convulsions may go on to have another febrile convulsion,[30] we gave the pups 2 febrile convulsions, the first on P7 and the second on postnatal day P9. All pups in the C+ group had convulsions lasting at least 210 sec (range 210-360 sec) on P7 and again on P9.

3.1.2 Play Behaviours in the Juvenile

Beginning on P30, juvenile play behaviours were monitored. Rats were housed individually for 48 hours. Groups of 4 rats were then reintroduced to each other in a clear box (54 cm by 30 cm by 21 cm) and 4 play periods (10 minutes each; 48 hrs apart) were observed. Although play behaviours were classified individually (wrestling, pinning, nosing, pursuing, and pouncing), the total number of play behaviours was summed for analyses.

3.1.2 Behaviours in the Adult

Beginning on P60 rats were tested in the elevated plus maze (a + shaped maze, with arms 55 cm long and 12 cm wide elevated 60 cm from the floor). The elevated plus maze had two opposing arms that were enclosed and two opposing arms that were open to the room. The number of entries and time (sec) spent in each of the 4 arms was recorded.

Beginning on P70, rats were then given an open-field object recognition task, in which rats were placed in a circular arena (98 cm diameter) containing 2 unique objects made of Lego ™. Over the course of 3 consecutive days, the rats were placed individually in the open field for 5 minutes and their behaviours were recorded. Rats were then given a 48 hr break and reintroduced to the open field, although 1 object had been replaced with a novel object. Again their behaviours were recorded.

3.2 Results

3.2.1 Behaviours in the Juvenile and Adult

There were no significant differences between the C+ and C- groups on any of the measure of juvenile play behaviours, all F values <2.216, all p values > .120. For the open-field object recognition task, although the female C+ rats spent the most time examining the objects ($M = 64.846 \pm 25.062$ *SD*) followed closely by the male C- rats ($M = 62.111 \pm 22.712$ *SD*), and the C- females ($M = 56.227 \pm 18.870$ *SD*) and the C+ males ($M = 55.640 \pm 20.488$ *SD*) spent the least time examining the objects, this did not reach significance, $F(1,74)=2.284$, $p=.067$.

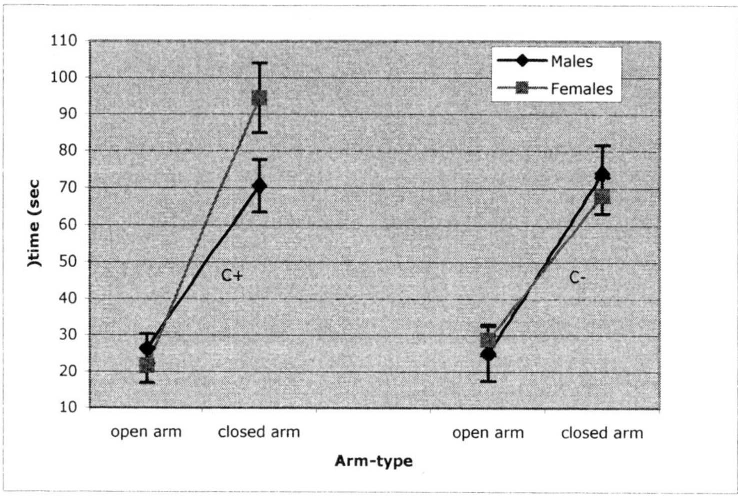

Figure 3. In the elevated plus maze, C+ females spent significantly more time in the closed arm than any other group, $F(1,74)=4.180$, $p=.044$. Bars represent the means (± SEM).

3.2.4 Correlations

Although there were no significant differences between the C+ and the C- groups for the behaviours, the longitudinal design of the study allowed us to investigate possible relations among behaviours. Only correlations with the task that demonstrated significant differences among the groups, the elevated plus maze, are reported.

For the juvenile play behaviours, the entire group exhibited significant relationships between time spent in the closed arm and: the total number of pouncing interactions r_s $(78)=-.240$, $p=.017$; and the time spent pursuing other pups, $r_s(78)=-.199$, $p=.041$. This correlation with the instances of pouncing was primarily due to the C+ group, r_s $(38)=-.318$, $p=.026$; C-: r_s $(40)=-.164$, *n.s.* Conversely, the correlation with pursuing was primarily due to the C- group, r_s $(40)=-.267$, $p=.048$; C+: r_s $(38)=-.157$, *n.s.*. Interestingly, for the C+ group there was a significant correlation between the time spent in the open arm of the elevated plus maze and: time spent investigating the novel objects in the open field, r_s $(38)=.279$, $p=.045$ (C-: r_s $(38)=.008$, $p=.481$); and the number of times that the C+ rats touched the novel objects, r_s $(38)=.451$, $p=.002$ (C-: r_s $(38)=-.130$, $p=.231$).

3.3 Conclusions

Consistent with other studies, febrile convulsions increased time in the closed arm of the elevated plus maze, behaviour associated with enhanced anxiety. Although we observed no significant differences in juvenile play behaviours for the groups, for the C+ group, there were significant correlations among juvenile play behaviours and preferences in the elevated plus maze. In adulthood, C+ rats that interacted more frequently with the novel objects in the open field object recognition task were also more likely to spend time in the open arm of the elevated plus maze. These results were unique to the C+ group, suggesting a more global change in the behaviour of the C+ pups that was only marginally observed using our measures.

4. GENERAL CONCLUSIONS

The C+ pups exhibited significant changes in the behaviour as neonates, in interactions with their mothers, and in adult behaviours associated with enhanced anxiety. Although not measured in the same rats, the decreased USV maturity of the C+ group and the reduced responsiveness of the dams may relate to the enhanced anxiety behaviours in adulthood. Given that there were significant relationships between juvenile play behaviours, open-field object recognition behaviours and arm preference in the elevated plus maze, it is possible that behavioral change following HC may be subtle and pervasive. Further, the delay in maturation in USVs may suggest that there is a longer term developmental delay in other behaviours.

This study is the first to observe changes in USVs in rat pups experiencing febrile seizures. The findings clearly have important implications as to the relationship between febrile convulsions early in life and the more subtle developmental disabilities later in childhood or early adulthood. Though conclusions have been reached in longitudinal studies of children who have experienced febrile convulsions, that their intellectual capacity is unaffected,[7, 31] closer scrutiny of these data suggests that abnormalities of attention and impulsivity later in childhood, may be related to early onset febrile convulsions.[7, 9, 31] Further studies are clearly indicated to identify and correlate alterations in neuroanatomical development that may provide substrate to these alterations in attention, and to determine, most importantly if recurrent febrile seizures enhance the likelihood of attentional difficulties, and whether treatment prevents it.

5. ACKNOWLEDGEMENTS

This research was supported by discovery grants from NSERC and the Heart and Stroke Foundation of Canada to DMS and by scholarships from NSERC to AJK and AHS. We would also like to thank Jennifer Yelland, Amber LaFrenz and Amanda Lisoway for their help with these projects.

6. REFERENCES

1. W. Kolfen, K. Pehle, and S. Konig, Is the long-term outcome of children following febrile convulsions favorable? *Dev. Med. Child Neurol.* **40**, 667-671 (1998).
2. F.U. Knudsen, Febrile seizures: treatment and prognosis, *Epilepsia* **41**, 2-9 (2000).
3. R.J. Baumann, Prevention and management of febrile seizures, *Paediatr. Drugs* **3**, 585-592 (2001).
4. K.E. VanLandingham, E.R. Heinz, J.E. Cavazos, and D.V. Lewis, Magnetic resonance imaging evidence of hippocampal injury after prolonged focal febrile convulsions, *Ann. Neurol.* **43**, 413-426 (1998).
5. J.F. Annegers, W.A. Hauser, S.B. Shirts, and L.T. Kurland, Factors prognostic of unprovoked seizures after febrile convulsions, *N. Engl. J. Med.* **316**, 493-498 (1987).
6. K. Lee, M. Diaz, and J.C. Melchior, Temporal lobe epilepsy -- not a consequence of childhood febrile convulsions in Denmark, *Acta Neurol. Scand.* **63**, 231-236 (1981).

7. Y.C. Chang, N.W. Guo, S.T. Wang, C.C. Huang, and J.J. Tsai, Working memory of school-aged children with a history of febrile convulsions: a population study, *Neurology* **57**, 37-42 (2001).
8. J.H. Ellenberg, D.G. Hirtz, and K.B. Nelson, Do seizures in children cause intellectual deterioration? *N. Engl. J. Med.* **314**, 1085-1088 (1986).
9. C.M. Verity, R. Greenwood, and J. Golding, Long-term intellectual and behavioral outcomes of children with febrile convulsions, *N. Engl. J. Med.* **338**, 1723-1728 (1998).
10. K. Kanemoto, N. Takuji, J. Kawasaki, and I. Kawai, Characteristics and treatment of temporal lobe epilepsy with a history of complicated febrile convulsion, *J. Neurol. Neurosurg. Psychiatry* **64**, 245-248 (1998).
11. S.J. Wallace, and A.M. Cull, Long-term psychological outlook for children whose first fit occurs with fever, *Dev. Med. Child Neurol.* **21**, 28-40 (1979)
12. T.Z. Baram, A. Gerth, and L. Schultz, Febrile seizures: an appropriate-aged model suitable for long-term studies, *Brain Res. Dev. Brain Res.* **98,** 265-270 (1997).
13. D. Holtzman, K. Obana, and J. Olson, Hyperthermia-induced seizures in the rat pup: a model for febrile convulsions in children, *Science* **213,** 1034-1036 (1981).
14. G.L. Holmes, and Y. Ben-Ari, Seizures in the developing brain: perhaps not so benign after all, *Neuron* **21,** 1231-1234 (1998).
15. J.A. McCaughran, Jr., and N. Schechter, Experimental febrile convulsions: long-term effects of hyperthermia-induced convulsions in the developing rat, *Epilepsia* **23**, 173-183 (1982).
16. J.G. Nealis, N.P. Rosman, T.J. De Piero, and E.M. Ouellette, Neurologic sequelae of experimental febrile convulsions, *Neurology* **28**, 246-250 (1978).
17. C. Dube, K. Chen, M. Eghbal-Ahmadi, K. Brunson, I. Soltesz, and T.Z. Baram, Prolonged febrile seizures in the immature rat model enhance hippocampal excitability long term, *Ann. Neurol.* **47**, 336-344 (2000).
18. W. Jiang, T.M. Duong, and N.C. de Lanerolle, The neuropathology of hyperthermic seizures in the rat, *Epilepsia* **40**, 5-19 (1999).
19. Y.C. Chang, A.M. Huang, Y.M. Kuo, S.T. Wang, Y.Y Chang, and C.C. Huang, Febrile seizures impair memory and cAMP response-element binding protein activation, *Ann. Neurol.* **54**, 706-718 (2003).
20. J. Werboff, and J. Havlena, Febrile Convulsions in Infant Rats, and Later Behavior, *Science* **142**, 684-685 (1963).
21. R.A. Kornelsen, F. Boon, L.S. Leung, and D.P. Cain, The effects of a single neonatally induced convulsion on spatial navigation, locomotor activity and convulsion susceptibility in the adult rat, *Brain Res.* **706**, 155-159 (1996).
22. K. Janus, Effects of early separation of young rats from the mother on their open-field behavior, *Physiol. Behav.* **40,** 711-715 (1987).
23. M. Kalinichev, K.W. Easterling, P.M. Plotsky, and S.G. Holtzman, Long-lasting changes in stress-induced corticosterone response and anxiety-like behaviors as a consequence of neonatal maternal separation in Long-Evans rats, *Pharmacol. Biochem. Behav.* **73**,131-140 (2002).
24. B. Zimmerberg, A.J. Rosenthal, and A.C. Stark, Neonatal social isolation alters both maternal and pup behaviors in rats, *Dev. Psychobiol.* **42**, 52-63 (2003)
25. H.N. Shair, S.A. Brunelli, J.R. Masmela, E. Boone, and M.A. Hofer, Social, thermal, and temporal influences on isolation-induced and maternally potentiated ultrasonic vocalizations of rat pups, *Dev. Psychobiol.* **42**, 206-222 (2003).
26. H.N. Shair, J.R. Masmela , S.A. Brunelli, and M.A. Hofer, Potentiation and inhibition of ultrasonic vocalization of rat pups: regulation by social cues, *Dev. Psychobiol.* **30**, 195-200 (1997).
27. G. Sokoloff, and M.S. Blumberg, Thermogenic, respiratory, and ultrasonic responses of week-old rats across the transition from moderate to extreme cold exposure, *Dev. Psychobiol.* **30**, 181-194 (1997).
28. J.T. Allin, and E.M. Banks, Functional aspects of ultrasound production by infant albino rats (Rattus norvegicus), *Anim. Behav.* **20**, 175-185 (1972).
29. S.M. Brudzynski, P. Kehoe, and M. Callahan, Sonographic structure of isolation-induced ultrasonic calls of rat pups, *Dev. Psychobiol.* **34**, 195-204 (1999).
30. L. Velisek, and S.L. Moshe, Temporal lobe epileptogenesis and epilepsy in the developing brain: bridging the gap between the laboratory and the clinic. Progression, but in what direction? *Epilepsia* **44**, 51-59 (2003).
31. Y.C. Chang, N.W. Guo, C.C. Huang, S.T. Wang, and J.J. Tsai, Neurocognitive attention and behavior outcome of school-age children with a history of febrile convulsions: a population study, *Epilepsia* **41**, 412-420 (2000).

Discussion

Engel: This is very important, because febrile convulsions are very common in humans and the potential relation with later epilepsy. There was a study done with guinea pigs. The researcher induced febrile convulsions and demonstrated that the guinea pigs had unilateral hippocampal sclerosis, and it was on the down side when they were lying on one side after a convulsion. I wonder if you could elaborate a little more on what you saw in the pathology and whether it was unilateral or bilateral.

Saucier: Well, I can only tell you about the 13 rats that we have the pathology on, and we saw enhanced Timm staining in the hippocampus, but it was a very subtle difference.

Avanzini: How long did you follow them up before killing, and do you plan to have some longer followup and relate it to spontaneous seizures?

Saucier: In the second study we followed them for 150 days to do the behaviors and the kindling, we didn't monitor spontaneous EEG other than 5 min prior to the kindling, and we saw no seizures that either our animal techs reported or any spontaneous AD. We only kindled the amygdala; so kindling the hippocampus could change what we saw.

Gale: I compliment you on taking the effort to look what's at going on with the animals in terms of their interaction with the mothers. What happened to the body weights of the animals -- did they show any failure to thrive?

Saucier: There was no failure to thrive. Their body weights were actually within 0.2 grams of each other.

Gale: The bigger concern that I have is that this a stressful event that may have nothing to do with their having a seizure, and as you mentioned there is also the heating component. Did you include any controls that they were heated to the point where did they did not have seizures so that you could separate the heating event from having seizures?

Saucier: That is an excellent point. What we did with the control group was separate them for the same duration in litter- matched pairs and we kept them at nest temperature; so it wasn't 40C, it was about 31C. It's not as good of control as you are suggesting . I was thinking of pretreating with a benzodiazapine and heating them up to that temperature. The benzodiazapine adds another complication, but at least you could see whether you are getting similar white matter changes.

Gale: It would be nice to have the heating, to see whether it may be hyperthermia that has some effects as opposed to the seizures.

Saucier: Yes, yes.

Kalynchuk: Deb, I am also interested in the maternal behaviours that the mothers show with the pup that have the convulsions. Michael Meaney has show with stressed rats that the behaviors of the mother change a lot. And it was interesting that you show that the number of interactions that the mother has with those pups does not change. So one thing that I suggest is that you look at the nature of the interactions, because what he has shown is that with stressed rats, the mother will spend as much time with them as they will with the non-stressed rats, but the nature of the interactions is much different. Have you looked at the nature of the interactions, for example whether the mothers were engaged in arch-back nursing as opposed to flat-back nursing? Those kinds of changes will actually effect the emotional behavior of the pups later in life.

Saucier: We have all of the data for that. You are absolutely right – the nature of the interactions was rather different.

Adamec: My comment is somewhat related to Lisa's. You alluded to the fact that differences in the adult plus maze anxieties in the females might be related to their behaviors when they were younger. But you have a choice: It could be due to the changes in the mothers' behavior toward the pups, or it could be the fact the pups are already exhibiting this because ultrasonic calls in fact have been used to measure anxiety in young animals. Have you tried correlating the pups' younger behavior to their adult behaviors in the plus maze?
Saucier: We are running that study right now.
Wasterlain: You failed to distinguish between males and females in 5 to 10 percent of cases, similar to the human failure rate. The Purkinje cells in the cerebellum are very sensitive to heat, and the question is if you looked at the cerebellum.
Saucier: We did not look at the histopathology at the cerebellum, although we did look at neonatal reflex behaviour before and after febrile convulsions and there were no significant differences in the rats' ability to grasp a rod to right themselves.
Phillips: The points I wanted to raise have to do more with mining as much as you can out of this data set. When you are looking at the cognitive and emotional behaviors of the animals and you are doing it as a ongoing study, are you going to be look at tasks that are more related to the hippocampus?
Saucier: Yes, we've put rats in the water maze, and we are going to be doing very different object locations memory tasks, which may be a bit frontal too because they have aspects of working memory.
Phillips: I wouldn't rely solely on the plus maze to look at the anxiety related behaviors. There are very good tests of social interaction that might build on what you are already seeing.
Saucier: I wonder if some of this is similar to what one sees with infants with damage to the left hemisphere, who will reorganize their right hemisphere for speech at the cost of space. What we keep seeing in the rats is that motor function is often preserved. These rats are dumb and are socially kind of strange, and I haven't been able to access this in a really good way.
Moshé: Did you control for seizure, to address the question of whether the seizure produced has nothing to do with hyperthermia?
Saucier: We are going to run a PTZ group.

EXPERIMENTAL ABSENCE VERSUS AMYGDALOID KINDLING

Filiz Yılmaz Onat, Esat Eşkazan, and Rezzan Aker*

1. INTRODUCTION

Neural mechanisms underlying convulsive events are believed to be distinctly different from those of absence seizures.[1] Typical absence epilepsy has been suggested to be related to a predominance of inhibitory activity, in contrast to generalized or focal convulsive seizures where an excess of excitatory activity is present.[2] Likewise, drugs that exacerbate seizure activity, clinically and in animal models, support the hypothesis concerning distinct differences between convulsive and non-convulsive epileptic events. It has long been known that carbamazepine, oxcarbazepine and phenytoin are successfully used in the treatment of partial and secondary generalized seizures, whereas typical absence seizures are clearly exacerbated by carbamazepine and phenytoin.[3, 4] Similarly, despite the fact that vigabatrin is a highly effective anticonvulsive agent against partial seizures,[5] two patients with idiopathic generalized absence epilepsy in whom vigabatrin increased the frequency and severity of absence and absence status were reported by Panayiotopoulos et al.[6] On the other hand, absence seizures in both animal models and humans respond to ethosuximide and trimethadion, which are ineffective against partial seizures.[3, 7] Models of convulsive events as well as non-convulsive seizures offer several unique opportunities to understand the pathophysiology of epileptogenesis in animals and perhaps, by extrapolation, in humans.

Limbic kindling, an animal model of temporal lobe epilepsy, addresses focal seizure development as well as secondary generalization mechanisms due to its progressive manner.[8-10] The fundamental concept in kindling is that the daily application of a subthreshold stimulation to certain subcortical structures, particularly limbic areas, produces growing and progressively spreading afterdischarges on EEG, and eventually triggers convulsive motor responses.

Absence epilepsy is characterized by generalized non-convulsive seizures with complete loss of conciousness. Increased synchronization and GABAergic inhibition within thalamo-cortico-thalamic circuits involving ventrobasal thalamus, reticular thalamic nucleus, and, as shown recently, the perioral region of the primary somatosensorial cortex are implicated in the generation of spike and wave discharges.[2] Among the animal models, Genetic Absence Epilepsy Rats from Strasbourg (GAERS) are a selected inbred strain of Wistar rats developed in

* Filiz Yılmaz Onat, Marmara University, School of Medicine, Istanbul, Turkey, 34668. Esat Eşkazan, Emeritus, Istanbul University, Cerrahpaşa Medical School, Istanbul, Turkey, 34303. Rezzan Aker, Marmara University, School of Medicine, Istanbul, Turkey, 34668.

Strasbourg by Vergnes et al.[11] GAERS is a well defined animal model of absence epilepsy sharing with humans many pharmacological and clinical characteristics of typical absences.[12]

In the present study, we aimed to explore the progression of kindling in GAERS in the first part. The second part of the study was designed to investigate the effect of suppression of absence seizures on kindling development in GAERS.

2. MATERIALS AND METHODS

Experiments were carried out with non-epileptic control Wistar rats and GAERS aged 4-10 months. The animals were housed in a temperature-controlled room (20 ± 3 ^{0}C) with a 12-h light:dark cycle. All animals were allowed free access to commercial rat pellets and tap water. The rats were housed in groups of four per cage. The experimental protocol was approved by the Animal Care and Use Committee of Marmara University (14.2003.Mar).

2.1. Surgery

One week before the kindling experiments, the animals were anesthetized with ketamine (100 mg/kg, intraperitoneally (i.p.)) and chlorpromazine (0.5 mg/kg, i.p.). The head of each animal was placed in a stereotaxic instrument (Stoelting Model 51600). The scalp was longitudinally incised, and the skull was leveled between lambda and bregma. Bipolar twisted stainless-steel electrodes insulated except at the tip for stimulation and recording were implanted bilaterally into the right and left basolateral amygdala (2.6 mm posterior, 4.8 mm lateral and 8.5 mm ventral from the bregma). Coordinates were obtained from the stereotaxic atlas of Paxinos and Watson,[13] and bregma was used as the reference point. Stainless steel screws were used for extradural ground and recording electrodes; they were placed bilaterally in the skull over the frontal and occipital cortices. Electrodes were connected by insulated wires to a microconnector for EEG recordings and were fixed to the skull with dental acrylic. Animals were allowed to recover from surgery for at least 7 days before stimulation.

2.2. Kindling

On the day of the experiment, the animals were placed in the plexiglass cages. Following a one hr stabilization period, basal EEG was recorded for 30 min. Then, to determine the afterdischarge threshold, the rats were stimulated with an initial stimulus of 50 μA (biphasic square wave pulses of 80 Hz, each 1 msec in duration, for a total duration of 2 seconds) and continued with 50 μA increments until an initial afterdischarge was obtained. The nucleus of the basolateral amygdala (BLA) was stimulated twice daily at the current afterdischarge threshold. Seizure stages observed after each stimulus were classified using Racine's[9] standard 5-stage scale: Stage 1, facial movements; stage 2, rhythmic head movements, head nodding; stage 3, unilateral forelimb clonus; stage 4, bilateral forelimb clonus and rearing; stage 5, falling and tonic-clonic convulsion. If none of the animals reached stage 5 seizures in kindling development groups, electrical stimulation was terminated following the 25th stimulus. Thus the maximum number of stimulations was 25 in both groups. Electrical activity in the stimulated region, contralateral amygdale, and cortex were recorded with a PowerLab System before and after each stimulus. Afterdischarge duration was the total duration of spikes in the EEG of the BLA electrode, including the stimulation period.

2.3. Experimental protocol

2.3.1. Experiment 1

Kindling was performed in Wistar control animals (n=11) and GAERS (n=10). Electrical activity of the stimulated region, contralateral amygdale, and cortex was amplified (through BioAmp ML 136) and recorded with a PowerLab 8S System running Chart v.5, (ADI Instruments U.K.) before and after each stimulus

.

2.3.2. Experiment 2

Kindling was performed in GAERS pretreated with ethosuximide at the dose of 100 mg/kg (n=9). After the recording of basal spike-wave activity on the EEG in GAERS, the initial afterdischarge threshold was determined on the first day of the experiment. Then, animals received ethosuximide in order to evaluate the effect of suppression of absence seizures on behavioral seizure severity and afterdischarge duration. EEG from the left and right frontoparietal cortices was recorded in GAERS for 20 min before and 1 hr after ethosuximide injection, and for 20 min after the kindling stimulus.

2.4. Drugs

Ethosuximide was obtained from Sigma Chemical Company and dissolved in saline. Ketamine and chlorpromazine were kindly provided by Eczacıbaşı A.Ş., Turkey. Ethosuximide was administered in a volume of 1 ml/kg and delivered i.p. 1 hr before each electrical stimulation.

2.5. Histological verification

Following experiments, the animals were decapitated to determine electrode placements. The brains were placed in a formaline/sucrose mixture and 50 μm thick frozen sections were taken by a cryostat and stained with thionine. Only the animals with correct electrode placements were included in the study.

2.5. Data analysis

The results were expressed as "mean ± S.E.M." Data were statistically evaluated by repeated measures analysis of variance followed by post-hoc Tukey's multiple comparison test to analyse the change in a treatment group. Two-way ANOVA followed by post-hoc Bonferroni test was used to compare differences in means of different groups. The level of statistical significance was considered to be $p < 0.05$ and $p < 0.01$.

Table 1. Mean afterdischarge threshold intensities in Wistar, GAERS and GAERS pretreated with ethosuximide 100 mg/kg, i.p. (GEK 100) groups[a]

	WISTAR	GAERS	GEK 100
Afterdischarge threshold (μA)	158.3 ± 25.34	122.5 ± 11.46	150 ± 11.79

[a] Data are expressed as mean ± SEM. There is no statiscally significant difference between the groups by one-way ANOVA.

3. RESULTS

3.1. Kindling in Wistar control animals and GAERS

The mean baseline afterdischarge threshold of the Wistar control group was not significantly different than that seen in the GAERS group, as shown in Table 1. In response to afterdischarge threshold stimulation during the early phase of kindling, no differences emerged in reaching stage 2 seizures in the Wistar and GAERS groups (Fig 1). The mean number of stimulations for the development of the first stage 5 seizure was 12.5 ± 0.6 and the mean duration of afterdischarges during the first stage 5 seizure was 37.2 ± 5.4 sec in non-epileptic Wistar animals. In contrast to the motor and generalized seizures of Wistar rats, GAERS never showed stage 3, 4, or 5 seizures after application of 25 stimulations (Fig 1). Therefore, all GAERS were stimulated until the maximum number of stimulations (25) was reached. Two-way ANOVA revealed a significant difference between the Wistar and the GAERS groups ($p < 0.01$). Afterdischarge durations in the GAERS group were significantly shorter than those seen in Wistar control animals ($p < 0.01$; Fig. 2).

3.1. Kindling in GAERS pretreated with ethosuximide

Ethosuximide did not have any significant effect on the mean baseline afterdischarge threshold (Table 1). The mean seizure stages were 2.7 ± 0.2 and 2.6 ± 0.2 at the 15th and 20th stimulations, respectively, in GAERS pretreated with 100 mg/kg of ethosuximide, (Fig 1). As shown in Fig. 3, even though ethosuximide effectively suppressed both spike-wave activity on the EEG and behavioral manifestations of absence epilepsy, the group did not progress to the stage 5 generalized seizure state, even after 25 stimulations, compared to Wistar control animals. On the other hand, some increase in afterdischarge duration was observed in the ethosuximide group (Fig. 2).

4. DISCUSSION

In the first part of our study, only non-epileptic control animals reached a stage 5 generalized convulsive seizure state, while GAERS failed to progress beyond stage 2, even after 25 kindling stimulations. However, the occurrence of focal afterdischarge and the speed of development to a stage 2 seizure state were similar in Wistar control and GAERS groups, indicating that absence seizures do not affect the initiation of

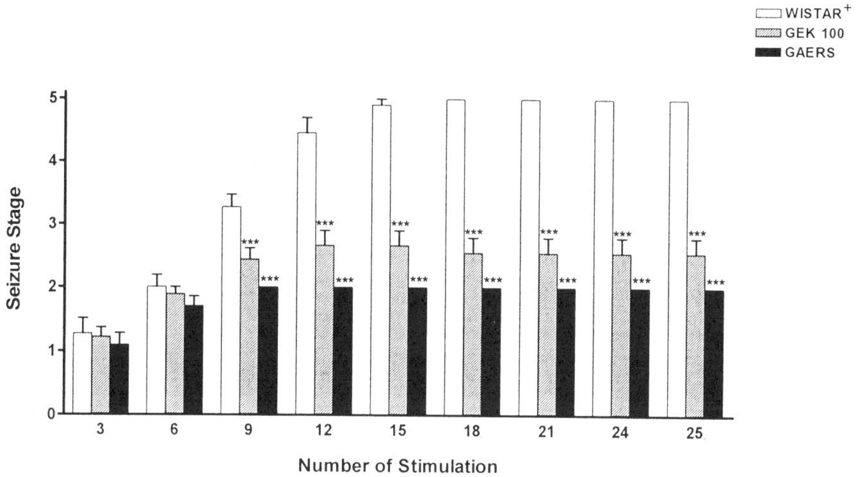

Figure 1. The effect of repeated kindling stimulation on seizure stage in control Wistar, GAERS, and GAERS pretreated with 100 mg/kg ethosuximide i.p. (GEK 100) groups. Data are expressed as mean±SEM. ***P<0.001, significant differences between GAERS and GEK 100 groups in comparison to Wistar group by two-way ANOVA with post-hoc Bonferonni test. + All animals in the Wistar group reached stage 5 seizure state at the 16th stimulation, and they did not receive further stimulations; the bars on 18-21-24 and 25th stimulations are replicates of 16th stimulus data.

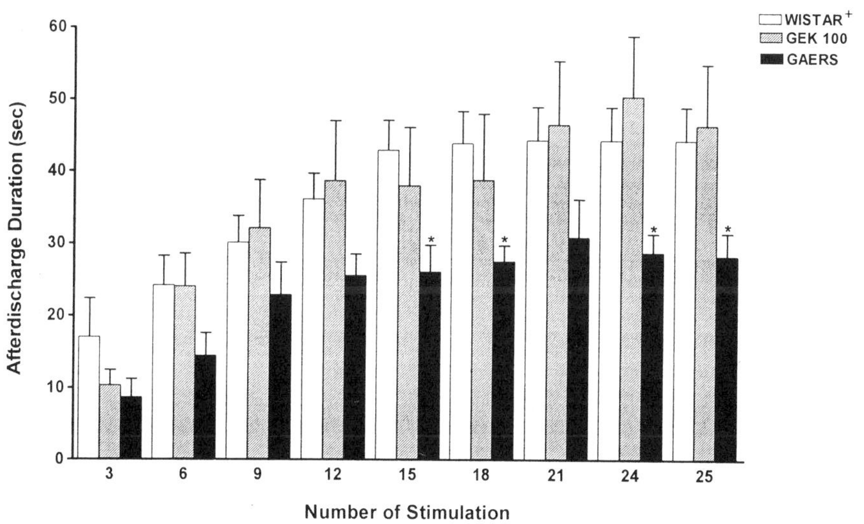

Figure 2. The effect of repeated kindling stimulation on afterdischarge duration in control Wistar, GAERS and GAERS pretreated with 100 mg/kg ethosuximide i.p. (GEK 100) groups. Data are expressed as mean±SEM. *P<0.05, significant differences of GAERS group in comparison to Wistar group by two-way ANOVA with post-hoc Bonferonni test. + All animals in the Wistar group reached stage 5 seizure state at the 16th stimulation and they did not receive further stimulations; the bars on 18-21-24 and 25th stimulations are replicates of 16th stimulus data.

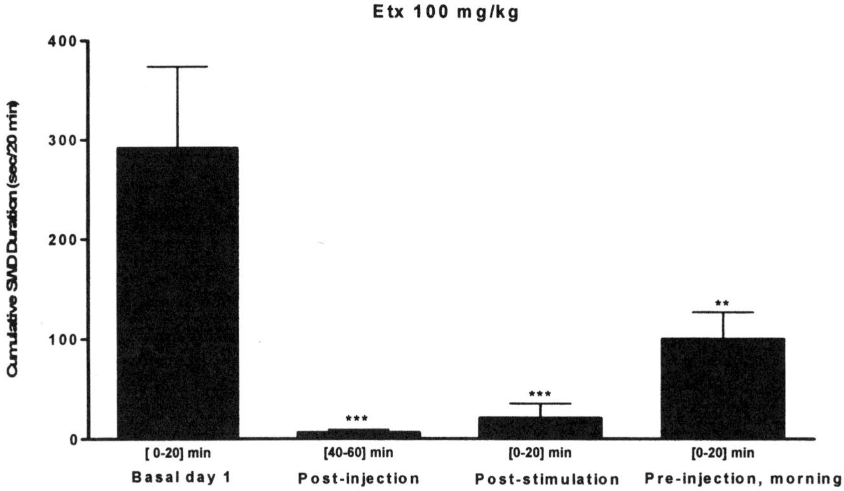

Figure 3. The effect of 100 mg/kg i.p. injection of ethosuximide on spike-wave discharge duration (SWD). The first column shows the basal cumulative SWD at the 1st day of the experiments before the first stimulus and ethosuximide injection. The second column shows SWD between 40-60 min after ethosuximide injection, which is before the stimulation. The third column shows SWD after electrical stimulation. The last column shows SWD before the first daily injection of ethosuximide. Data are expressed as mean±SEM. $^{**}p<0.01$, $^{***}p<0.001$, significant differences compared to basal period by repeated meseaures ANOVA with post-hoc Tukey's multiple comparison test.

limbic partial non-motor seizures. In our laboratory, it was previously reported[14] that afterdischarge durations in the cortex of non-epileptic Wistar rats and GAERS at the 15th stimulation were 42.0 ± 4.3 and 26.1 ± 3.5 sec, respectively, as shown in Fig. 4. Taken together, the failure to develop stage 3-5 seizures and the short afterdischarge durations in the GAERS group in spite of repeated daily electrical stimulation suggests that an inhibitory mechanism might be involved in the transformation of a limbic focal seizure to motor and generalized seizures in rats with absence epilepsy. The reason for failure in response to amygdaloid electrical kindling in GAERS is likely related to neural networks involved in absence epilepsy. Though the importance of the reciprocal thalamocortical network in absence epilepsy has long been accepted, there is little evidence indicating the involvement of limbic structures. Metabolic activity, as measured by local rates of glucose utilization, was found to be increased in limbic structures of GAERS including the hippocampus.[15] Likewise, in a very recent study in which the glutamate neurotransmitter content of the hippocampus of GAERS was determined using an immunocytochemical technique at the electron microscopic level, glutamate density was found to be significantly increased compared to the control group.[16] These results are in agreement with the study reported by Richards et al.,[17] showing an increased basal hippocampal extracellular glutamate levels in GAERS. Besides the involvement of limbic structures in the genetic absence epilepsy model, the evidence for involvement of the thalamus in limbic seizure activity was also obtained from a number of studies,[18, 19] including thalamic hypometabolism ipsilateral to the abnormal temporal lobe on positron

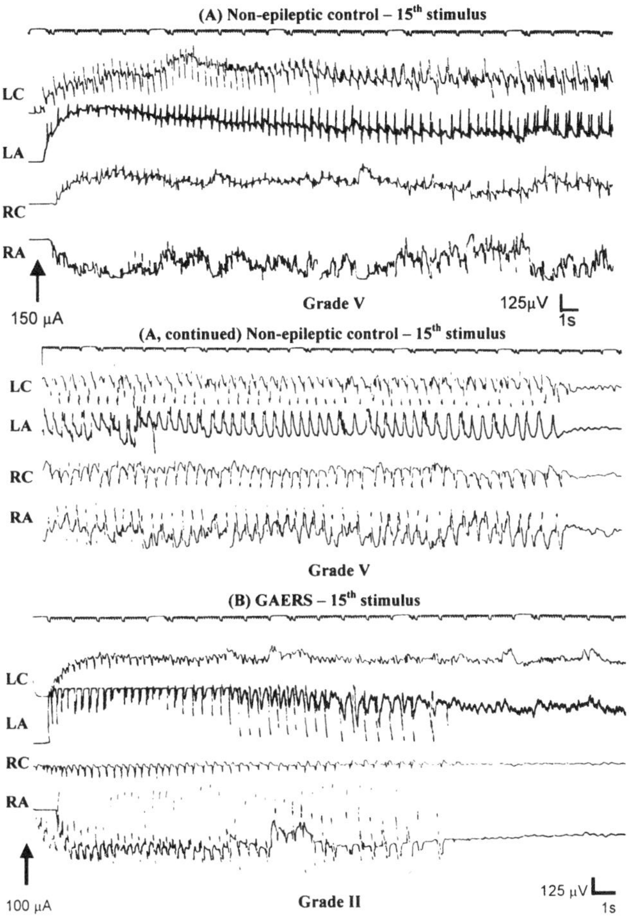

Figure 4 EEG recordings of non-epileptic Wistar control rat (A) and GAERS (B). LC, left cortex, LA; left amygdala, RC; right cortex, RA; right amygdala (with permission from Eşkazan et al[14]).

emission tomography scan in humans and thalamic stimulation-induced epileptiform responses in the hippocampus.[20] Additionally, a recent report of changes specifically in the human dorsal medial thalamus indicated glucose and [^{11}C]flumazenil positron emission tomography abnormalities in thalamic nuclei in mesial temporal lobe epilepsy.[21] Interestingly, Nanobashvili et al[22] reported that costimulation of the thalamic reticular nucleus during hippocampal kindling stimulation reduced the number and the duration of generalized convulsions. They concluded that thalamic reticular nucleus stimulations suppressed limbic motor seizures in hippocampal kindling and provided a new approach for seizure control in temporal lobe epilepsy. Moreover, in patients with partial complex seizures, a significant decrease in generalized tonic-clonic convulsions was observed in a long-term study with electrical stimulation of the centromedian thalamic nucleus, suggesting the propagation of epileptic activities using the thalamic nuclei and/or diffuse thalamic projection system.[23] Even though the mechanism by which electrical stimulation of the centromedian thalamic nucleus induced seizure control is still a matter of speculation, the centromedian thalamic nucleus seems to be involved in the tonic-clonic component of primary generalized seizures. Further studies are required to determine the mechanisms for prevention of generalized convulsions in humans with partial complex seizures and for prevention of generalization of limbic focal activity in rats receiving thalamic reticular nucleus stimulation during the kindling process. Therefore, kindling in GAERS is highly suitable to study the mechanisms involved in the propagation and generalization of seizure activity for the convulsive and non-convulsive components of epilepsy as well as thalamo-limbic interaction.

In our study, failure in provoking convulsive seizures in GAERS might be related to strain differences in the ability of the electrical stimulus to initiate generalization. Likewise, a study of rats with fast or slow amygdala kindling rates resulted in selective breeding of seizure-prone (Fast) vs. seizure-resistant (Slow) rat strains based on their rates of kindling.[24, 25] Moreover, selective breeding of rats to produce Fast vs. Slow kindling revealed differential sensitivity to three different GABAergic antagonists, pentylenetetrazol, bicuculline and picrotoxin.[26] This type of paradoxical response to GABAergic inhibitors has been seen in GAERS when compared to non-epileptic control rats. Injection of the $GABA_A$ receptor antagonists bicuculline, picrotoxin, and inverse agonists of the benzodiazepine site induced myoclonic spike-wave activity followed by clonic or tonic-clonic seizures with paroxysmal activity on the cortical EEG.[27] In contrast, GAERS were less susceptible than non-epileptic animals to the tonic-clonic convulsions induced by other GABAergic inhibitors, including inhibitors of glutamate decarboxylase, isoniazide and 3-mercaptopropionic acid. In fact, rats with genetic absence epilepsy can be accepted as a kindling-resistant strain due to the failure of partial seizures to develop into more complex forms such as the secondarily generalized convulsive form.

In the second part of our study, we hypothesized that suppression of absence seizures may give rise to stage 3-5 seizures, assuming that the neural network involved in absence epilepsy may act as an inhibitor on the generalization of amygdala kindling seizures. Therefore, an agent which is not effective against generalized tonic-clonic and partial seizures but very effective against absence seizures should be chosen for this purpose. Among the therapeutic antiepileptic agents, ethosuximide is the only drug with an indication restricted to absence-type seizures. In addition to its effectiveness as the anti-absence seizure medication, ethosuximide has been reported to be ineffective against kindled amygdaloid seizures in rats, except for neurotoxic doses such as 400 mg/kg.[28, 29] Other agents such as diazepam and valproate are used in absence seizures as well as in generalized tonic-clonic and partial seizures. Our finding that 100 mg/kg of ethosuximide was an effective anti-absence dose confirmed previous studies showing that ethosuximide readily crosses the blood-brain barrier and is evenly distributed within rat brain regions.[30] Interestingly, although systemic administration of ethosuximide effectively suppressed behavioral and spike-wave activities in the EEG, only five animals out of nine reached the stage 3 seizure state, suggesting that absence seizures partially explain the resistance to kindling generalization. Additionally, the observation of no stage 5 seizures in GAERS pretreated with ethosuximide suggested that ethosuximide by itself may be the reason for the resistance in the secondary generalization of limbic seizures during amygdala kindling in GAERS. Therefore, the effect of ethosuximide on kindling needs further investigation in non-epileptic control rats.

In conclusion, the occurrence of only stage 1-2 seizures and no observation of stage 3-5 seizures in GAERS with the maximum number of stimulations would suggest that a mechanism underlying generalized absence epilepsy may be the reason for the resistance in the secondary generalization of limbic seizures during amygdala kindling. The results of the present study demonstrating resistance to propagation of kindled seizures in rats with genetic absence epilepsy are consistent with and expand on previous reports indicating that the neuronal mechanisms of absence epilepsy are distinctly different from convulsive events. Additionally, the failure to progress to generalized convulsions in GAERS pretreated with ethosuximide suggests that the development of kindled seizure stages seems to be independent of spike-wave activity in GAERS. Additional interrelated studies with different doses of ethosuximide and other antiepileptic agents may prove to be of value in studying the antagonistic effect of absence seizure mechanisms on limbic seizure generalization.

5. REFERENCES

1 J. Engel, Excitation and inhibition in epilepsy, *Can. J. Neurol. Sci.* **23**, 167-174 (1996).
2. J. P. A. Manning, D A Richards, and N G Bowery, Pharmacology of absence epilepsy, *TIPS* **24**(10), 542-549 (2004)
3. B. J. Wilder and J. Bruni, in. *Antiepileptic Drugs, fifth edition,* edited by R. H. Levy, R H. Mattson, B S. Meldrum, and E. Perruca (Lippincott Williams and Wilkins, Philadelphia, 2002).
4. W. D. Shieds and E. Saslow, Myoclonic, atonic and absence seizures following institution of carbamazepine therapy in children, *Neurology* **33**, 1487-1489 (1983).
5. S. M. Grant and R. C. Heel, Vigabatrin: a review of its pharmacodynamic and pharmacokinetic properties and therapeutic potential in epilepsy and disorders of motor control, *Drugs* **41**, 889-926 (1991).
6. C. P. Panayiotopoulos, A. Agathonikou, I. Sharoqi, and A. P. Parker, Vigabatrin aggravates absences and absence status, *Neurology* **49**, 1467 (1997)
7. S. L. Moshe and N. Ludvig, in: *Recent Advances in Epilepsy*, edited by T. A. Pedley and B. S. Meldrum (Churchill Livingstone, New York, 1988), pp. 21-44.
8. D. C McIntyre, M. O. Poulter, and K. Gilby, Kindling: some old and some new, *Epilepsy Res.* **50**, 79-92 (2002).
9. R. J. Racine, Modification of seizure activity by electrical stimulation. II. Motor seizure. *Electroencephalogr. Clin. Neurophysiol.* **32**, 281-94 (1972).
10. G.V. Goddard, Development of epileptic seizures through brain stimulation at low intensity, *Nature* **214**, 1020-1021 (1967).
11. M. Vergnes, C. Marescaux, A. Depaulis, G. Micheletti, and J. M. Warter, Spontaneous spike and wave discharges in thalamus and cortex in a rat model of genetic petit mal-like seizures, *Exp Neurol.* Apr; **96**(1), 127-136 (1987).
12. L. Danober, C. Deransart, A. Depaulis, M. Vergnes, and C. Marescaux, Pathophysiological mechanisms of genetic absence epilepsy in the rat. *Progr. Neurobiol.* **55,** 27-57. (1998).
13 G Paxinos and C Watson, in: *The Rat Brain in Stereotaxic Coordinates*, 4th edition, (Academic Press, San Diego, California, 1998).
14. E. Eşkazan, F. Y. Onat, R. Aker, and G Öner, Resistance to propagation of amygdaloid kindling seizures in rats with genetic absence epilepsy, *Epilepsia* **43**(10), 1115-1119 (2002).
15. A. Nehlig, M. Vergnes, S. Boyet, and C. Marescaux, Local cerebral glucose utilization in adult and immature GAERS, *Epilepsy Res.* **32**, 206-212 (1998).
16 S. Sırvancı, C. Meshul, F. Onat, and T. Şan, Immunocytochemical analysis of glutamate and GABA in hippocampus of genetic absence epilepsy rats (GAERS), *Brain Res.* **988**, 180-188 (2003).
17. D. A. Richards, L. A. Morrone, and N.G. Bowery, Hippocampal extracellular amino acids and EEG spectral analysis in a genetic rat model of absence epilepsy, *Neuropharmacology* **39,** 2433-2441 (2000).
18. R. M. Cassidy and K. Gale, Mediodorsal thalamus plays a critical role in the development of limbic motor seizures, *J. Neurosci.* **18**, 9002-9009 (1998).
19. M. R. Sperling, R.C. Gur, and A. Alavi, Subcortical metabolic alterations in partial epilepsy, *Epilepsia* **31**, 145-155 (1990).
20 E. H. Bertram, D. X. Zhang, P. Mangan, N. Fountain and D. Rempe, Functional anatomy of limbic epilepsy: a proposal for central synchronization of a diffusely hyperexcitable network, *Epilepsy Res.* **32**, 194-205 (1998).
21. C. Juhasz, F. Nagy, and C. Watson, Glucose and [C-11]flumazenil positron emission tomography abnormalities thalamic nuclei in temporal lobe epilepsy, *Neurology* **53**, 2037-45 (1999).
22. Z. Nanobashvili, T. Chachua, A. Nanobashvili, I. Bilanishvili, O. Lindvall, and Z. Kokaia, Suppression of limbic motor seizures by electrical stimulation in thalamic reticular nucleus, *Exp. Neurol.* **181**, 224-230 (2003).

23. F Velasco, M. Velasco, F Jimenez, A. L. Velasco, B. Rojas, and M. L. Perez, Centromedian nucleus stimulation for epilepsy· Clinical, electroencephalographic, and behavioral observations, *Thalamus & Related Systems* **1**, 387-398, (2002).
24. R. J. Racine, M. Steingart, and D C. McIntyre, Development of kindling-prone and kindling-resistant rats: selective breeding and electrophysiological studies, *Epilepsy Res.* **35**, 183-195 (1999)
25 D. C. McIntyre, M. E. Kelly, and C Dufresne, FAST and SLOW amygdala kindling rat strains: comparison of amygdala, hippocampal, piriform and perirhinal cortex kindling, Epilepsy *Res* **35**, 197-209, (1999).
26 R. J. Racine, M. Steingart, Y. Bureau, and D. C. McIntyre, Differential sensitivity of genetically Fast vs. Slow kindling rats strains to GABAergic convulsive agents, *Neuropharmacology* **45**, 918-924 (2003)..
27. M. Vergnes, A. Boehrer, S Reibel, S Simler, and M. Marescaux, Selective susceptibility to inhibitors of GABA synthesis and antagonists of $GABA_A$ receptor in rats with genetic absence-like epilepsy, *Exp. Neurol.* **161**, 714-723 (2000).
28. T. E. Albertson, S. L. Peterson, and L. G. Stark, Anticonvulsant drugs and their antagonism of kindled amygdaloid seizures in rats, *Neuropharmacology* **19**, 643-652 (1980).
29. P. S Albright and W. McIntyre Burnham, Development of a new pharmacological seizure model: Effects of anticonvulsants on cortical- and amygdala-kindled seizures in the rat, *Epilepsia* **21**, 681-689 (1980).
30. D. A. Richards, J. P. A. Manning, D. Barnes, L. Rombola, N. G. Bowery, S. Caccia, N. Leresche, and V. Crunelli, Targeting thalamic nuclei is not sufficient for the full anti-absence action of ethosuximide in a rat model of absence epilepsy, *Epilepsy Res.* **54**, 97-107 (2003).

Discussion

Leung: $GABA_B$ receptor antagonists are good for treating absence seizures and in certain models, but they also can promote focal or partial seizures. In the inverse, in the GAERS , we don't know the exact genetic constitution, but I was about to suggest that $GABA_B$ receptors are over-expressed or over-functioning, and if you give a $GABA_B$ antagonist maybe you can have kindling proceed in the GAERS rat.
Onat: That is true, it is another possibility for further study. But you know that there are different types of GABAB antagonists, and their behavior is different than the others. So there is no easy way to study this.
Leung: It also is not easy because the antagonists themselves would provoke partial seizures. You would have to give a dose that is low enough that it would not induce the seizure itself and then kindle.
Buzsaki: Spike induced epilepsy is strain age and sex dependent; so perhaps one way of avoiding the drug effect is looking at animals before 1 month of age when there is no spike-wave activity. Another problem is that in breeding situations many genes co-segregate so we do really know what is happening here, whether the spike wave pattern and its associated pharmacology is the key factor. So it might be worth looking at other strains such as the WAG strain from the Netherlands and to see if you have the same pattern before you do anything else.
Onat: Very interesting comment. We are doing the same study in WAG/Rij rats. As you mentioned, the WAG/Rij strain of rats is derived from the Netherlands. The results are partially different from those in GAERS.
Avanzini: I think that these are very interesting data. I think that you should stress that not only are generalized seizure physiologically different from focal seizure, but nonconvulsive generalized absence seizures may have a different physiology, making interpretation difficult. I think your plan to use euthoseximide also in controls is mandatory. Speaking about euthoseximide, if if is true that its main mechanism of action is on low threshold calcium current, then you can influence other areas where this current is expressed. The interpretation may be complicated, but the demonstration that there is somehow a kind of alternative effect between the

mechanism of generalized, non-convulsion and kindled seizure. I think is a very important demonstration.
Post: In the clinical situation, how does the combination of the two seizure types play out?
Onat: Generalized convulsions coexisting with partial seizures in absence epilepsy patients are rare. But I think that we need more epidemiological data. Recently there is a paper saying that the coexistence of partial seizures in absence epilepsy or absence seizures in partial seizures is very rare.
Moshé: We will have the answer in four or five years. There is a new NINDS-funded study that Tracey Glauser (PI) is doing at this point. We are going to look at all the kids with absence seizures. There are several hospitals involved and we will have the EEGs. My task is to look at the EEGs, and we will find out how often partial seizures may occur, because we have a very detailed approach, trying even to figure whether the classical 3Hz spike wave discharge is indeed the sine que non in absence epilepsy.

PENTYLENETETRAZOL-INDUCED KINDLING AS A MODEL OF ABSENCE AND CONVULSIVE FORMS OF EPILEPSY

Alexey A.Shandra, Leonid S.Godlevsky*

I. INTRODUCTION

The electrically and chemically induced kindling is a model of human epilepsy. Whereas the electrical kindling is regarded as a model of complex partial epilepsy,[1,2,3] the chemical kindling[6,7] is a model of primary generalized epilepsy.[5] Although pentylenetetrazol (PTZ) kindling has often been used to explore the mechanisms of seizure genesis and neurobehavioral and neurophysiological consequences of seizures, limited attention was paid to the contribution of this model in the mechanisms of the generalized nonconvulsive epilepsy- absence model.[8-17] Meanwhile, complex evaluation of this model should be performed as far as divergent and even opposite mechanisms of epileptogenesis might be suspected in the course of generalized absence and convulsive stages of PTZ-kindling. Thus, first one is characterized by absence- like manifestations (spike-wave discharges (SWD)-bursts) and is supported by hyperexcitable state of all cortical neurons, including inhibitory ones, while receptors of GABA are preserved and their hyperactivation is resulted in typical generalized spike- wave generation.[18-19] Further increasing of epileptogenic stimuli is followed by collapse of GABA control, and precipitation of generalized seizures as a result.

That is why the aim of the present work was to perform systematical investigation on the peculiarities of different stages of kindling development. Namely, early precipitated absence- like generalized manifestations and final convulsive stage were in focus of attention.

In accordance to the velocity of kindling development all rats were subdivided into two subpopulations.[20] Thus, more numerous first group (70%) was characterized by the rapid appearance of generalized seizures, which were pronounced already after 10-12 PTZ injections. The second one was characterized as relatively resistant one to kindled motor seizures precipitation. In this group myoclonic jerks and seizure twitches were induced only after 14-15 administrations of epileptogen, and generalized convulsions were not induced despite prolonged periods of PTZ administrations.

* Odessa State Medical University, Odessa, Ukraine 65026

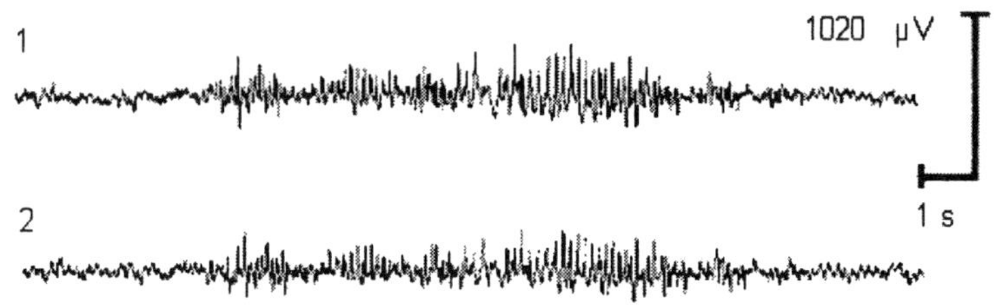

Figure.1. Spike-wave discharges induced by PTZ in left (lead 1) and right (lead 2) hemispheres.
Notes. 30 min from the moment of PTZ (25,0 mg/kg, i.p.) administration. Bipolar fronto- occipital registration was explored.

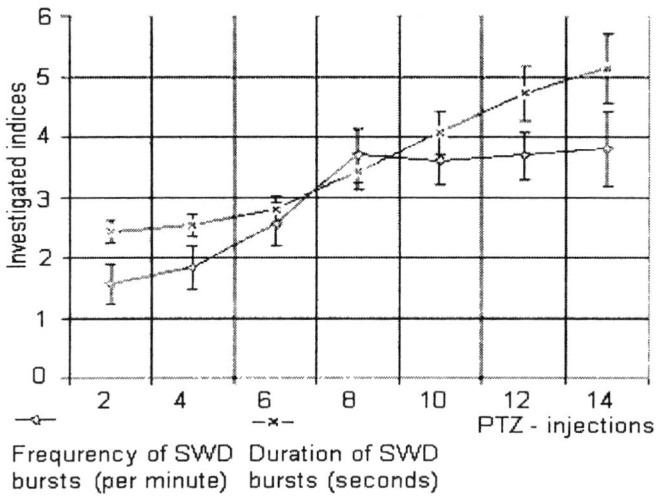

Figure 2. Dynamics of SWD characteristics in the course of repeated PTZ administrations in rats prone to demonstrate absence- like manifestations.

2. KINDLING OF ABSENCE MECHANISMS OF THE BRAIN

Investigations of SWD dynamics in the course of subthreshold PTZ administrations (25,0 mg/rg, i.p.) revealed that first 1-3 injections induced typical SWD with frequency of potentials of 6-9/s, and short duration (from 1,5 up to 3,5 s) (Fig.1). SWD precipitations were followed by subtle manifestations, which compose a typical number of "minor signs"[21-22] of absence epilepsy. Namely, freezing of animals along with tremor of vibrisses, staring, tilting of head and slight breathing acceleration were noted.

Starting from the 8th PTZ administration the duration of SWD bursts significantly exceeded that one registered after second epileptogen injection – by 41,0% (F(1,16)=8,06, P=0,013) (Fig. 2). It is of worth to note that consequent epileptogen administrations were also effective with respect of prolongation of the SWD bursts (Fig. 2), and after 14th administration the investigated index was greater by 49,7% when compared with that one registered after 2^{d} epileptogen injection (F(1,14)=7,87, P=0,014). At the same time, frequency of SWD was "saturated" after

8 injections of epileptogen and maintained at the relatively stable level during consequent PTZ administrations.

The increasing of the number of PTZ administration was followed by the prolongation of the total time of SWD bursts registration (Fig. 3). Thus, duration of SWD registration after 8th PTZ administration was greater by 49,5% when compared with the life-span of bursts after 4th epileptogen injection ($F(1,14)=6,39$, $P=0,024$) (Fig. 3). After 12th administration the investigated index was greater than after 8th administration by 27,1% ($F=(1,14)=4,59$, $P=0,046$).

Obtained data permitted to come to the conclusion that, as a consequence of repeated administrations of subthreshold dosage of PTZ, the significant enlargement of the SWD bursts along with the increasing of their life- span were registered. These facts are in favor for the some sort of kindling of absence mechanisms at the initial stage of kindling in those rats, which were prone to demonstrate postponed precipitation of generalized convulsions.

Heightened seizure susceptibility after drug-free period is quite characteristic feature of the kindling.[23] Renovated seizures are of greater severity, longevity and are resistant to different forms of antiepileptic treatment.[23, 24]

Subthreshold PTZ administration, which was performed in 14 days from the last epileptogen injection drastically – by two times increased the life- span of induced SWD, when compared with the similar index before drug- free period ($H=10,576$, $P=0,001$, Kruskal- Wallis test). The average duration of SWD bursts and frequency of their generation achieved significant differences with the control in 60 and 90 min correspondently from the moment of PTZ administration, and differences were kept till the end of observation. Behavioral disturbances, registered in the course of prolonged bursts of SWD were typical for absence-like manifestations.

Hence, gained data are in favor for the induction of kindling – like process with regard to absence mechanisms of the brain. All principal features of the kindling were identified in the course of the dynamics of SWD bursts induced with PTZ in rats, which were initially prone to demonstrate absence form of kindling. Namely, increasing of duration and frequency of SWD, increasing of life-span of SWD, increasing of EcoG manifestation after period free from epileptogen administrations. Interictal behavior of animals, determined in learning tasks was not differing from that one of intact animals.

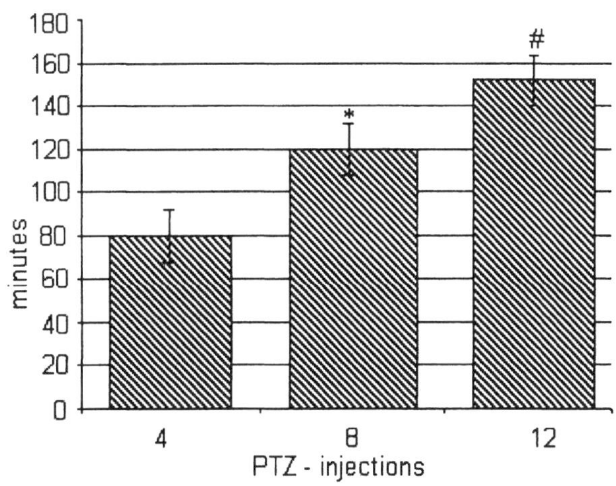

Figure 3. Life-span of SWD bursts registered after corresponded number of PTZ administration (abscissa). Minutes are marked on ordinate.*-P<0,05 vs first and #-P<0,05 vs. second column

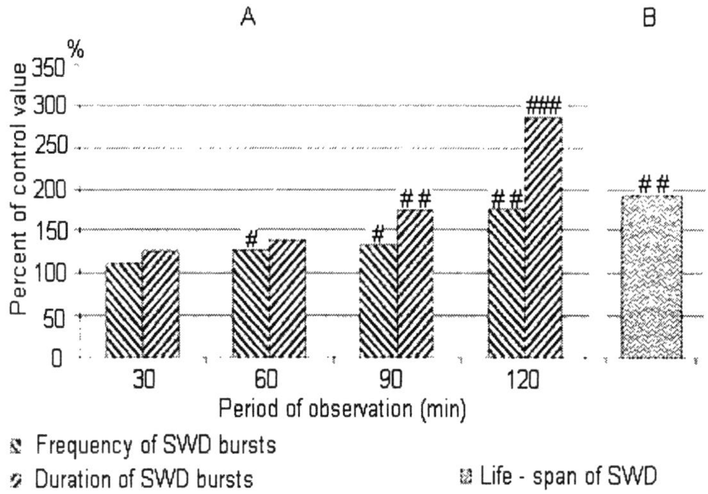

Figure 4. Dynamics of duration and frequency of PTZ- induced SWD bursts after two weeks drug- free period (A), and total period of SWD bursts generation (B).
Notes: ordinate- investigated indices in percents pertained to those ones in proper control groups taken as 100%.
#-P<0,05; ##-P<0,01; ###-P<0,001 in comparison with the control (ANOVA + Newmann- Keuls test).

2.1 Some Neurophysiological Data on Absence Stage of Kindling

The conception on thalamo-cortical genesis of SWD compose the basis for the absence – like mechanisms analysis.[18-19,21,25-27] From this point of view it is of interest to investigate the peculiarities of this system functional state under conditions of modulative influences on thalamic and cortical neurons from distantly located structures. Taking into consideration the intensive connections between nucleus dentatus and motor relay thalamic nuclei, involved into generation of SWD, it might be supposed that some forms of neuroplastic changes in these pathways might be in charge for the modulation of absence- like electrographic manifestations.

We showed that the threshold of locomotor reaction induced through nucleus dentatus ES in rats, which did not demonstrate convulsions after prolonged PTZ administrations was 286,6 μA, (with a range from 80,0 to 400,0 μA). Rats with generalized seizures precipitated after 12-15 PTZ injections had threshold lower when compared with absence kindled animals (172,0 μA, with a range from 40,0 to 400,0 μA) (T=2,060, P=0,039, Mann-Whitney test). Besides, it was established that bilateral destruction of nucleus dentatus was followed by intensive PTZ kindling formation. Namely, more pronounced with regard to frequency and duration SWD were observed during "shortened" first (absence) stage of kindling (Fig. 5).

Intensification of the early stage of PTZ kindling in rats with destructions of nucleus dentatus might be explained by the deafferentation of thalamic nuclei, which resulted in heightened excitation in thalamo- cortical neuronal circuits. Such theory of thalamic deafferentation represents the general concept for absence epilepsy origin.[25, 27]

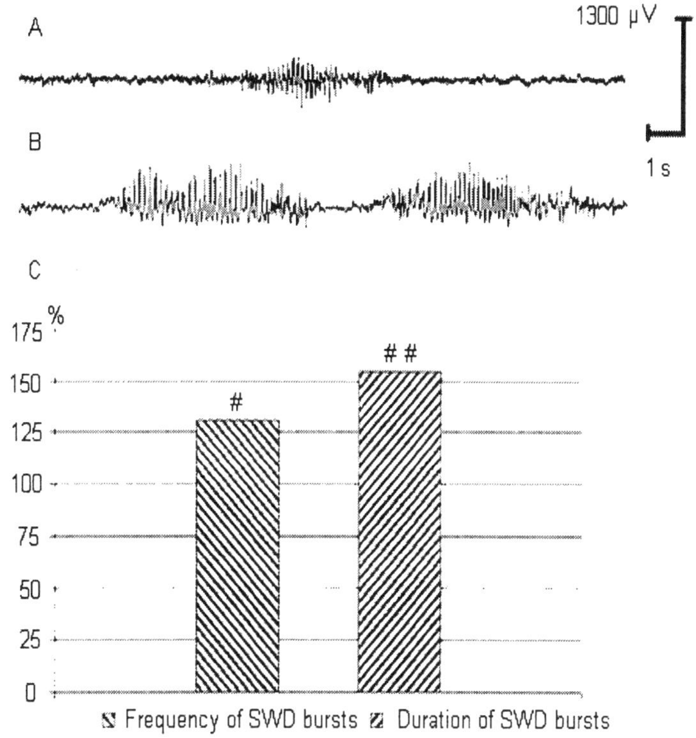

Figure 5. Increased SWD characteristics in "absence kindled" rats after bilateral destruction of nucleus dentatus.
Notes: A- 30 min after 12th PTZ administration, B- 30 min after PTZ administration. Bipolar recordings of EcoG in left frontal cortex was explored.
C: ordinate- investigated indices in percents pertained to those ones in proper control groups taken as 100%
#-P<0,05; ##-P<0,01 in comparison with the control (ANOVA + Newmann- Keuls test).

Hence, gained data permit to assume that the difficulties of achievement of locomotor component in rats, which are not prone to demonstrate convulsive stage of kindling, might be explained by the fact of prevalence of inhibition on the ways of spreading of efferent impulses from nucleus dentatus. It might be assumed that proneness of these rats to display absence seizures manifestations is in great part a result of spinal "gate" control/ filtration of descending motor drives. It also might be supposed that not only cortico-thalamic mechanisms, but spinal ones as well are contributive to precipitation of absence seizures. In both cases role of cerebellar structures is not excluded.

Considering the possible role of cerebellum in absence epilepsy development it should be noted that an increase of the cerebellar activity might be in charge for the arrest of locomotor activity, tremor precipitation, staring, eyeball niystagm-like movement, and autonomic system activation- breathing acceleration as well. The timely developed locking out of cortical function might be induced by hyper functional state of cerebellar structures and precipitation of short-time extraordinary inhibitory influences upon sensorimotor cortex.

3. CONVULSIVE MANIFESTATION IN KINDLED ANIMALS

In those animals, which were prone to demonstrate convulsive kindling the myoclonic seizures were noted in 8-12 days of PTZ administrations. The high-amplitude spike discharges and SWD bursts were registered that time as well (Fig. 6). The appearance in EEG, which were composed in prolonged rhythmic discharges periods with a life-span of 8-25 seconds and mean frequency of 3-6/ s (Fig. 6, D), the fits of generalized seizures were seen, which encompassed the forelimbs and trunk. The complete absence of SWD was noted during such generalized seizures. The further heightening of the frequency of discharges up to 6-10 per second was followed by the increase of the severity of seizure fits. Hence, repeated generalized clonic- tonic fits were registered with a range of longevity from 8-20 up to 300- 360 seconds.[28, 29] The postseizure depression was registered as well.

In 25% of animals with PTZ induced kindling the repeated seizure fits were registered. The intensity and duration of such fits were less as a rule when compared with the previous seizure manifestations. Our data are in favor for the shortening of the latency of first seizures in response to testing dosage of PTZ. Hence, latent period of first seizures was shortened from 8,9± 0,8 min after first epileptogen administration up to 4,1± 0,2 min after 20th administration of PTZ ($F(1,28)=33,88$, $P<0,001$). The latency of generalized seizure fits was also shortened from 12,6± 0,6 up to 7,4± 0,3 min ($F(1,28)=67,22$, $P<0,001$).

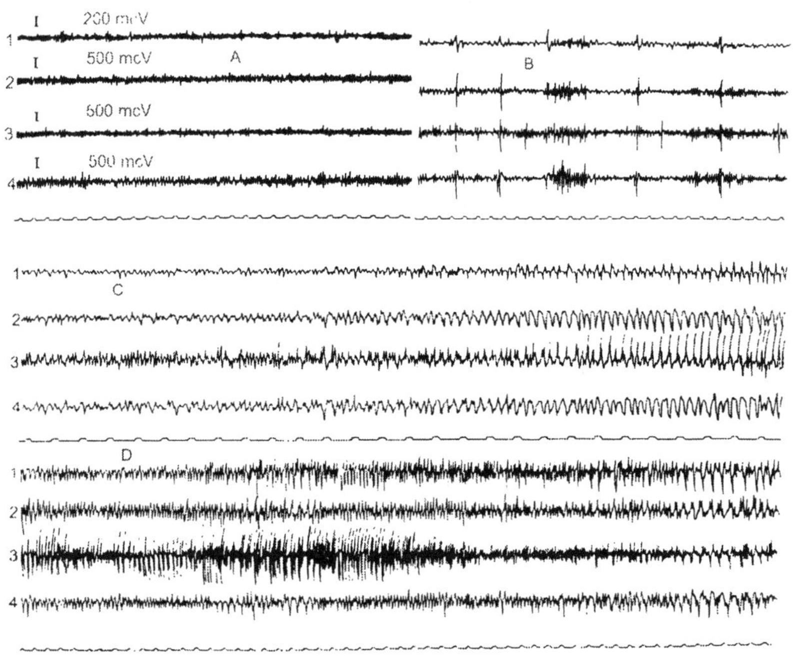

Figure 6. The changes of electrical activity in rat brain at seizure stage of PTZ kindling development. Notes A- before PTZ administration, B-, C- and D- after 10th, 16th, and 20th PTZ administrations (25,0 mg/kg, i.p.). First- time developed generalized convulsions are presented. 1-, and 2- frontal cortex of left and right hemispheres; 3-, and 4- ventral hippocampus of left and right hemispheres.

To determine whether the pattern of onset changed with progressive seizure fits, we evaluated the percentage of focal- and diffuse- onset at each of the seizure fits. For the first seizure fits only 34% of the seizures had diffuse onset, for the tenth seizure fits – however, 65% had a diffuse onset. A regression analysis showed clear tendency for the later seizure fits to have more diffuse onset in comparison with the early stage.

When the stage of full kindling was achieved, widely spread character of epileptiform activity generation was met more frequently, and this fact was in favor for the network nature of observed phenomena. On the ES model of kindling although the focal onset of seizures was more common during the early stages of recovery.[40] That might be due to not fully established connections, whereas diffuse seizure onset was clearly the most common pattern in the fully kindled mature state.[30,40]

Bilateral hippocampal destructions did not significantly affect the early stage of kindling but significantly postponed the precipitation of convulsive manifestations of kindling (Fig. 7). The intensity and duration of seizure fits were reduced in comparison with the corresponded control data. The heightened susceptibility, which was achieved in the course of kindling in animals with destroyed hippocampus was preserved during shorter period of time when compared with the control. Such facts are in favor for the role of hippocampus as a structure involved in maintenance of raised seizure susceptibility once full kindled state was achieved.

Investigations on the hippocampal slices revealed that PTZ-induced kindling is the form of appearance of evoked potentials and extra potentials in the system of pyramidal hippocampal neurons as a response to Shaffer collaterals electrical stimulations. (Fig 8).

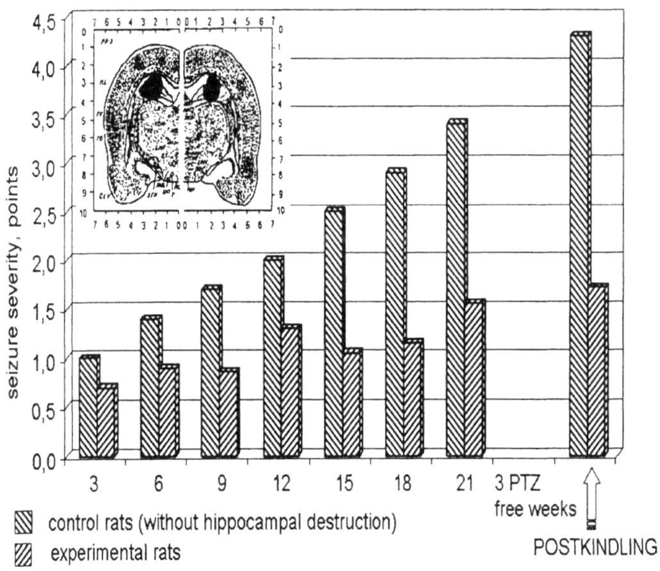

Figure 7 Effects of hippocampal destruction upon PTZ kindling development. Abscissa: number of PTZ injections

The extra PC observed in this work[31] has been reported in literature later both *in vivo* and *in vitro*.[32-37] Stringer[32] showed a double populational spike (PS) in hippocampal CA1 after systemic application of PTZ. Barkel et al.[33] have shown paroxysmal field potential in neocortical slices after PTZ kindling. Fatholahi et al.[34,35] reported the appearance of a PS or two in CA1 of hippocampal slices after PTZ kindling. Hence, the appearance of extra PS is in favor for enhanced excitation, which is developed in CA1 during kindling. It might be that activation of hyperexcitable neuronal population is able for self-maintenance of excitation after cessation of stimulation. Such mechanism is in good correspondence with the reduced threshold for PS appearance in CA1 after Shaffer collaterals stimulation in PTZ kindled rats. Rocha et al.[36] found that extracellular levels of glutamate and aspartate are elevated following a single administration of subconvulsive dose of PTZ. Such evidence suggests that PTZ- induced kindling involves the activation of excitatory aminoacid system from the beginning of the epileptogenic process.

4. CONCLUSION

It should be stressed that if the early stage of PTZ kindling might be regarded as a model of absence – like seizure syndrome, and might be explained, at least partially by plastic changes in thalamo-cortical neuronal loops, the late stage of PTZ kindling is characterized by the

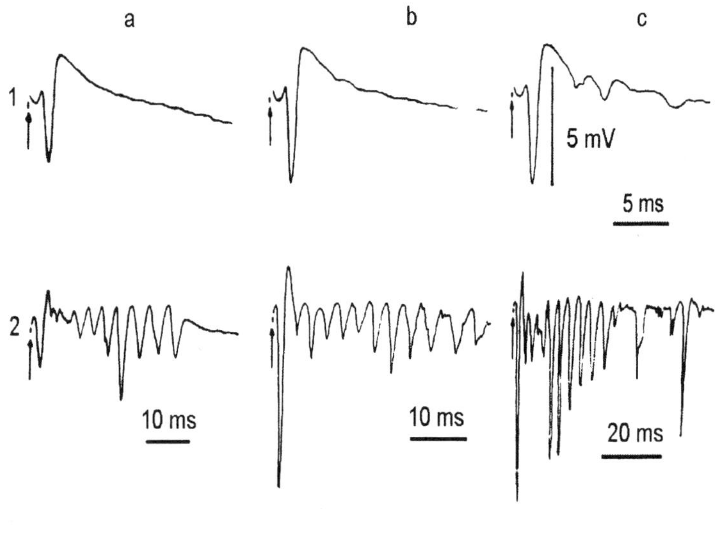

Figure 8. The evoked potentials pattern in the control (1) and experimental (2) hippocampal slices stimulated by the current of the different intensity.[31]
Notes: a- power of stimulation is 9 μA, b- 14 μA; c- 20 μA.

precipitation of generalized clonic- tonic fits and only limbic structures are mentioned as contributive.[1-10,37-39] Hence, “shifting” of responsible structures is regarded as a key element of seizure development. It might be also that such limbic structures involvement is critical for overcoming of all controlling systems on the stage of output of signals and creating of widely spread network system of epileptogenesis. It looks reasonable to discuss “on” and “off” character of thalamo-limbic relationships as a basis for subdivision of animals on prone and not prone to display full- developed kindling.[20]

It should be to mention that high tonic state of endogenous inhibitory mechanisms of the brain are responsible for the behavioral kindled disturbances.[38, 39] The same or similar conception is applicable for the absence- stage of epilepsy, which is characterized by heightened excitability of cortical inhibitory neurons.[18,19] Obtained data, which are in favor for the kindling of absence mechanisms, and which are in correspondence with other authors data[41] suggest that cortical endogenous inhibition might be raised at the early stage of PTZ kindling, representing the substantional mechanism of kindling. Besides, in this direction those facts derived from neurochemistry/neuropharmacology, namely, ability of GABA to induce epileptogenesis in the course of repeated (kindling- like) administrations,[42] might be enriched with neurophysiological sense.

Taking into consideration that cortical inhibition is induced secondarily, in most general perspective kindling serves for the substitution of the concept of "single epileptic neuron" on concept of "neuronal chain" as a basis of brain epileptogenesis.[43]

5. REFERENCES

1. J.O.McNamara, Kindling: an animal model of complex partial epilepsy *Ann Neurol.*, **16**, (suppl.), 72 – 76 (1984).
2. Wada J.A., in: *The Continuing Evolution of the Limbic System Concept*, edited by K.E.Livingston, and H.Hornykiewicz (Plenum Press, New York, 1978), pp. 369 – 388.
3. M.Sato, R.J.Racine, and D.C.McIntyre, Kindling: basic mechanisms and clinical validity, *EEG Clin. Neurophysiol.*, **76**, 459 – 472 (1990).
4. J.Ono, R.F.Vieth, and P.D.Walson, Electrocorticographical observation of seizures induced by pentylenetetrazol (PTZ) injections in rats, *Funct. Neurol.*, **5**, 345 – 352 (1990).
5. R Rossi, Sensitization induced by kindling and kindling-induced phenomena as a model for multiple chemical sensitivity, *Toxicol.*, **111**, 87 – 100 (1996)
6. C G.Wasterlain, and V.Jonec, Chemical kindling by muscarinergic amygdaloid stimulation in the rat, *Brain Res.*, **271**, 311-323 (1983)
7. C.G.Wasterlain, A.M.Morin, D.G.Fujikawa, and J.M.Bronstein, in: *The Clinical Relevance of Kindling*, edited by T G.Bolwig , and M.R Trimble (John Wiley, New York, 1989), pp.35- 53.
8. A.A.Shandra, L.S.Godlevsky, and N.D.Semenyuk, Generalized seizure activity formation in mice after corazol subthrrehol doses daily administration, *Bull. Exp. Biol.*, **195** (4), 527 – 532 (1983), in Russian.
9. C.R.Mason, and R.M.Cooper, A permanent change in convulsive threshold in normal and brain damaged rats, with repeated small doses of pentylenetetrazol, *Epilepsia*, **13**, 663 – 674 (1972).
10. J P.J.Pinel, and K.F.Cheung, Brief communication. Controlled demonstration of metrazol kindling, *Pharmacol. Biochem. Behav.*, **6**, 599 – 600 (1977)
11. T.Ito, M.Hori, K Yoshida, and M.Shimuzu, Effect of anticonvulsants on seizure developing in the course of daily administration of pentetrazol to rats, *Eur.J.Pharmacol.* **45**, 165-172 (1977)
12. W.Pohle, A.Becker, G.Grecksch, A.Juhre, and A Willenberg, Piracetam prevents pentylenetetrazol-induced neuronal loss and learning deficits, *Seizure*, **6**, 467-474 (1997).
13. G.Grecksch, A.Becker, H.Schroeder, and V.Hollt, Involvement of delta-opioid receptors in pentylenetetrazol kindling development and kindling- related processes in rats, *Naunyn-Schmiedeberg's Arch.Pharmacol.* **360**, 151-156 (1999).
14. V.Erakovic, G.Zupan, J.Varljen, J.Laginja, and A.Simonic, Altered activities of rat brain metabolic enzymes caused by pentylenetetrazol kindling and pentylenetetrazol- induced seizure, *Epilepsy Res.* **43**, 165-173 (2001).
15. V.Erakovic, G Zupan, J.Varljen, and A.Simonic, Pentylenetetrazol- induced seizures and kindling: changes in fre fatty acids, superoxide dismutase, and glutathione peroxidase activity, *Neurochem.Internat.* **42**, 173-178 (2003).
16. Y.Hayashi, Y.Morizumi, Y.Hattori, and J.Tanaka, Pentylenetetrazol- induced kindling stimulates the polyamine interconversion pathway in rat brain, *Brain Res.* **828**, 184- 188 (1999)
17. A.Becker, G.Grecksch, and H.Schroeder, Low doses of AMPA exert anticonvulsant effects on pentylenetetrazol – kindled seizures, *Pharmacol.,Biochem. Behav.* **70**, 421- 426 (2001).

18 P.Gloor, Generalized epilepsy with spike and wave discharges: reinterpretation of its electrographic and clinical manifestations, *Epilepsia*, **20**, 571- 588(1979).
19 G K Kostopoulos, Spike-and-wave discharges of absence seizures as a transformation of sleep spindles: the continuing development of a hypothesis, *Clin Neurophys.*, **111**, S27-S38 (2000).
20. R.J.Racine, M.Steingart, and D.C.McIntyre, Development of kindling- prone and kindling- resistant rats:selective breeding and electrophysiological studies, *Epilepsy Res.* **35**, 183- 195 (1999).
21. A.M.L.Coenen, W.H.I.M.Drinkenburg, M Inoue, and E L.J.M.van Luijtelaar, Genetic models of absence epilepsy, with emphasis on the WAG/rij strain of rats, *Epilepsy Res.*, **12**, 75–86 (1992).
22. L Danober, C.Deransart, A Depaulis, M Vergnes, and C.Marescaux, Pathophysiological mechanisms of genetic absence epilepsy in the rat, *Progr. Neurobiol.*, **55**, 27- 57(1998)
23 A A.Shandra, A M Mazarati, L.S Godlevsky, and R.S.Vastyanov, Chemical kindling. implications for antiepileptic drug- sensitive and resistant epilepsy models, *Epilepsia*, **37**(3), 269-274 (1996)
24. W.Losher, Which animal models should be used in the search for new antiepileptic drugs, *Epilepsia*, **34** (suppl.2), 188 (1993).
25. H.K.M.Meeren, E.L.J.M.van Luijtelaar, F.H.Lopes de Silva, R.K.Berdiev, N.E.Chepurnova, S.A.Chepurnov, and A.M.L.Coenen, The cotico-thalamic theory for generalized spike- wave discharges, *Uspehi Fiziologicheskih Nauk*, **35**, (1), 3-19 (2004). In Russian.
26. I.S.Midzianovskaia, G.D.Kuznetsova, A.M.Coenen, A.M.Spiridonov, and E.L.J.M. van Luijtelaar, Electrophysiological and pharmacological characteristics of two types of spike-wave discharges in WAG/Rij rats, *Brain Res.*, **911**, 62-70 (2001).
27. R.Paetau, M.Peltola, E.Liukkonen, M.Granstrom, L.Valanne, G.Blomstedt, and A.Peatau, Continuous spike-wave during sleep: thalamocortical denervation? *Epilepsia*, **43**, S8, 176 (2002).
28. G.N.Kryzhanovsky, A.A.Shandra, R.F.Makulkin, and L.S.Godlevsky, The hippocamp as a determinant stracture generating the epileptiform activity in corazol kindling, *Bull. Exp. Biol.*, **5**, 527 – 532 (1985), in Russian.
29. G.N.Kryzhanovsky, and A.A.Shandra, Corazol-induced changes in the convulsant readiness of mice, *Pharmacol. Toxicol.*, **2**, 16 – 19 (1984), in Russian.
30. G.N.Kryzhanovsky, *Central nervous System Pathology: A New Approach* (Consultant Bureau, Raven Press Publishing Company, New York, 1986).
31 S.L.Buldakova, A.A.Shandra, G.N.Kryzhanovsky, S A.Saakyan, and V.G.Skrebitsky, Characteristics of electrical activity of hippocampal slices in mice with corazol kindling, *Bul.Exp.Biol.*, **3**, 272 – 274 (1985), in Russian
32 J L Stringer, Pentylenetetrazol caused polysynaptic responses to appear in the dentate gyrus, Neurosci., **68**, 407 – 413 (1995).
33. E.Barkel, M.J.Grossman, and M.J.Gutnick, Long-term changes in neocortical activity after chemical kindling with systemic pentylenetetrazol: an in vivo study, *J. Neurophysiol.*, **72**, 72 – 83 (1994).
34. Y.Fatholahi, F.Motamedi, S.Semnanian, M.Zardoshti, Examination of persistent effects of repeated administration of pentylenetetrazol on hippocampalCA1: evidence from in vitro study on hippocampal slice, *Brain Res.*, **758**, 92 – 98 (1997).
35 Y.Fatholahi, F.Motamedi, S.Semnanian, and M.Zardoshti, Reapeted administration of pentylenetetrazol alters susceptibility of rat hoppocampus to primedburst stimulation. evidence from in vivo study on CA1 of hippocampal slice, *Brain Res.*, **738**, 138 – 141 (1996).
36 L.Rocha, M.Briones, R.F.Ackermann, B.Anton, N.T.Maidment, C.Y.Evans, and J.Jr.Engel, Pentylenetetrazol-induced kindling: early involvement of excitatory and inhibitory systems, *Epilepsy Res.*, **26**, 105 – 113 (1996).
37. M.R.Palizvan, Y.Fathollahi, S.Semnanian, S.Hajezadeh, and J.Mirnajafizadh, Differential effects of pentylenetetrazol-kindling on long-term potentiation of population excitatory postsynaptic potentials and population spikes in the CA1 region of rat hippocampus, *Brain Res.*, **898**, 82 – 90 (2001).
38. R.E. Adamec, Evidence that limbic neural plasticity in the right hemisphere mediates partial kindling induced lasting increases in anxiety-like behavior: effects of low frequency stimulation quenching? on long term potentiation of amygdala efferents and behavior following kindling, *Brain Res.* **839**, 133–152 (1999)
39 R Adamec, and B Young Neuroplasticity in specific limbic system circuits may mediate specific kindling induced changes in animal affect—implications for understanding anxiety associated with epilepsy, *Neurosci. Biobehav. Rev.*, **24**, 705–723 (2000).
40. E.Bertram, Functional anatomy of spontaneous seizures in rat model of limbic epilepsy, *Epilepsia*, **38** (1), 95 – 105 (1997).
41. M.Wong, D.F.Wozniak, and K.A.Yamada, An animal model of generalized nonconvulsive status epilepticus: immediate characteristics and long- term effects. *Exp. Neurol.*, **183**(1), 87-99 (2003).
42. R.Q.Hu, M.A.Cortez, H.Y.Man, Y.T.Wang, and O.C Snead, Alteration of GLUR2 expression in the rat brain following absence seizures induced by gamma- hydroxybutiric acid, *Epilepsy Res.*, **44**, 41-51 (2001).
43. L.S.Godlevsky, E.L.J M.van Luijtelaar, A.A Shandra, and A.M.L.Coenen, Cause and effect relations in disease; lessons from epileptic syndromes in animals, *Medical Hypotheses*, **58**, 237- 243 (2002).

Discussion

Schwartzkroin: You showed that lesions of dorsal hippocampus reduce the intensity of kindled seizure activity, but you also said that seizure initiation was likely to be associated with the ventral hippocampus. Have you done lesions of the ventral hippocampus to look at effects on seizure activity?

Shandra: This is an important question, because many studies have shown that the seizure thresholds of ventral and dorsal hippocampus are different. We will continue with lesions of the ventral hippocampus and will study the effect on kindling development.

Onat: My study is parallel to your results. I didn't show our spike and wave discharges results after electrical stimulation in kindling development. As you showed, AD duration after electrical stimulation increased immediately, and it was within 10 minutes after electrical stimulation in our study. This is parallel to your results. Secondly, Pinault (from Strasbourg) showed that substania nigra exerts remote control in the generation of the spike wave discharge. Maybe the substania nigra can be the next possible study for you in the future

Shandra: Indeed, several minutes elapsed between administration of PTZ and precipitation of epileptiform phenomena, and in this sense chemical kindling resembles kindling induced by electrical stimulation. I cannot comment on the absence of SWD during the course of electrical kindling, but, as I suggested in my presentation, the "shifting" from thalamic to limbic mechanisms during seizure precipitation might be of significance. Hence, electrical stimulation of the amygdala might be regarded as direct induction of the "final stage" of kindling. That is why appearance of SWD is hardly to be expected. Concerning the substantia nigra, its role in generalized forms of seizure is quite well known. Modulation of SWD by substantia nigra might have an impact on generalization of both seizure and non-seizure activity in the brain. Hence ascending control, which might be realized via modulation of thalamic nuclei and might be mediated primarily by substantia nigra, deserves further exploration. In any case, such effects should be interpreted along with the primary role played by cortical neurons in genesis of SWD.

Buzsaki: Counting the incidences of spike waves is not very trivial, especially when you would like to compare across lesion groups or pharmacological treatment groups. The reason is spike waves only appear when the animal is immobile. You mentioned, for example, that the lesion of the cerebellum altered the number of spike waves. The question is whether this is a direct effect or is mediated through behavior, because every single treatment that makes the animal move will prevent the occurrence of spike waves. This has to be disassociated; otherwise we don't get closer to the mechanism that we are investigating. Having said that, I would like to ask you, whether development of spike waves and the increased susceptibility to seizures are connected or develop independently.

Shandra: We have discussed the continuously increasing incidence of SWD in the course of PTZ administration from the perspective that deafferentation is responsible for the increase in seizure susceptibility. Hence, for a very short time (comparable to the life span of SWD burst), the net susceptibility of cortical neurons is expected mainly to be decreased, in so far as waves are prevalent over spikes. Also deafferentation per se is happening during this period. Later on, in the interictal state, an increase in susceptibility is expected. Hence the phenomena mentioned are divided with regard to the time course of their precipitation. But. From my point of view, they are mutually interdependent. For a better understanding of this question, measurement of seizure thresholds in a very precise fashion is recommended.

SODIUM CURRENT PROPERTIES IN DIFFERENT MODELS OF EPILEPSY

Wytse J. Wadman, Rogier Min, and Guangchun Sun*

1. INTRODUCTION

The voltage dependent sodium current determines the upstroke of the action potential, and it is therefore one of the determinant factors for neuronal excitability. The steep voltage dependence of activation as well as inactivation in combination with its fast kinetics imply that even relatively small changes in sodium current properties will have a significant influence on, for example, cell firing frequency. It is therefore not surprising that the majority of antiepileptic drugs (AEDs) exert their action by modulating sodium current properties that are relevant for cell firing.[1] The aim of this study is to parameterize the sodium current in CA1 hippocampal pyramidal cells using the classical description originally provided by Hodgkin and Huxley[2, 3] employing voltage clamp data obtained from neurons acutely isolated from the rat hippocampus . Such a description does not comprise every possible detail of the present knowledge of the sodium channel. Nevertheless it appears to be sufficient to describe the changes observed in the inactivation function of the sodium current in neurons isolated from the epileptic focus in rats that were either kindled[4] or in which a status epilepticus model was generated.[5]

The most effective drugs against epileptic seizures exert their effect by specifically shifting the inactivation function in a hyperpolarizing direction in a concentration dependent manner. This modulation can be incorporated into the kinetic scheme of the sodium current as a binding to the inactivated state based on actual experiments in the cells under study. With the description of the sodium current at hand, a model neuron was implemented in the NEURON simulation environment[6] so that its firing properties could be simulated under current clamp conditions. This offers the possibility to predict how carbamazepine binding could affect the firing behavior of these neurons.

* Wytse J Wadman, Center for NeuroScience, Swammerdam Institute for Life Sciences, University of Amsterdam, Kruislaan 320, 1098SM Amsterdam The Netherlands.

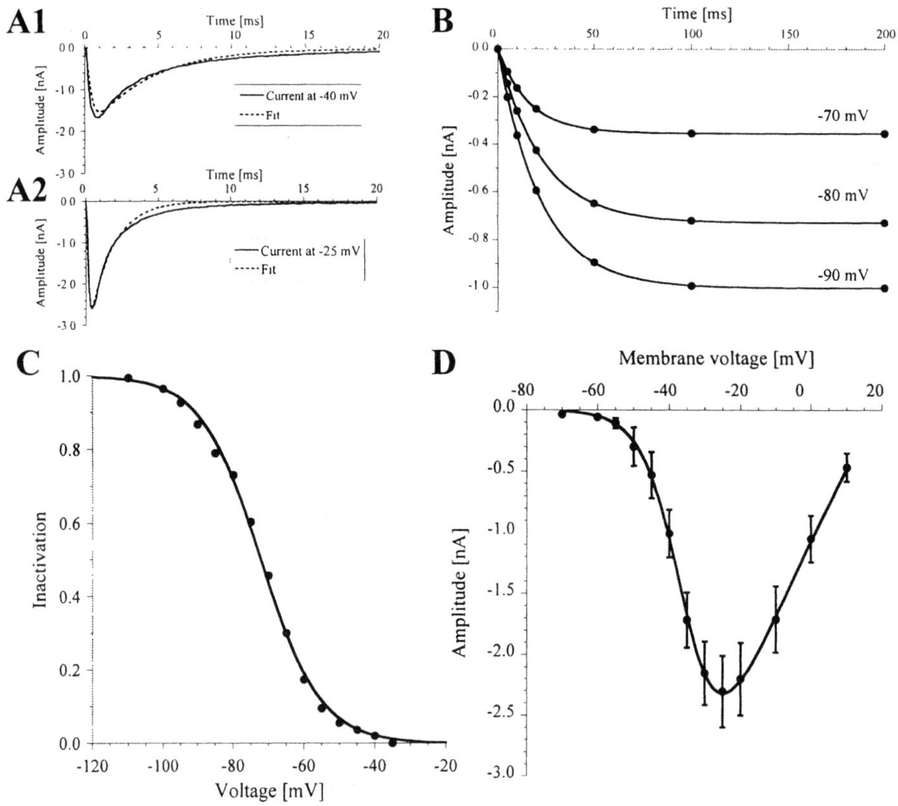

Figure 1. Examples of fits to sodium currents evoked with various protocols in CA1 pyramidal neurons acutely enzymatically isolated from the rat hippocampus (for methods see [7]). **A1** Sodium current evoked by 20 ms depolarization to -40 mV preceded by a 40 ms hyperpolarization to -150 mV. Holding potential is set at -65 mV. The current is fitted with a third order rising exponential and single exponential decay. **A2** Same as A1 but now for a depolarization to -25 mV. **B** Recovery from inactivation measured using a double pulse protocol separated by hyperpolarizations of variable duration and membrane voltage of three different values (-70, -80, and -90 mV). Smooth curves are first order exponential fits to the data points. **C** Inactivation measured by a standard depolarization to -20 mV, preceded by hyperpolarization to levels between -65 and -150 mV. Smooth curve is the fit to a Boltzmann function.[4] **D** Activation function determined as the peak amplitude of the sodium current evoked by a 20 ms depolarization. Smooth curve is the fit to the Goldman-Hodgkin-Katz current equation,[8] using a Boltzmann function to describe the voltage dependence of the permeability.[5]

2. Na-CURRENT IN CA1 PYRAMIDAL NEURONS

The classical description of the sodium channel[2, 3] postulates that the current is controlled by an activation gate (m: the fraction of open gates) and an inactivation gate (h) that together control the voltage and time dependence of the conductance for sodium. The kinetics in the model is determined by the state of the gates, while the voltage dependence is incorporated in the rate constants that define the state transitions:

$$g_{Na}(t) = m(t)^3 \times h(t) \times \bar{g}_{Na} \tag{1}$$

where h defines the fraction of the h gates in the open state

$$h \underset{\beta_h(V)}{\overset{}{\longleftarrow}} \overset{\alpha_h(V)}{\longrightarrow} 1 - h \tag{2}$$

and the change in h is given by

$$\frac{dh}{dt} = \beta_h(V) \times (1 - h) - \alpha_h(V) \times h \tag{3}$$

The rate constants □ and □ determine the steady state (h_∞) at a certain membrane voltage and the voltage dependent time constant of h. A similar set of Eqs. (2)-(3) can be defined for the activation gate m (which stand for the fraction of m gates in the open state). We used a procedure that directly fitted a complete set of currents observed under experimental conditions (examples in Figure 1a and b) in order to estimate the best functions for □ and □ and found:

$$\alpha_m(V) = \frac{23.76}{1 + e^{\frac{-19.91 - V}{8.41}}} \quad \text{and} \quad \beta_m(V) = \frac{27.36}{1 + e^{\frac{V + 121.57}{26.30}}}$$

and

$$\alpha_h(V) = \frac{0.856}{1 + e^{\frac{V + 151.66}{20.50}}} \quad \text{and} \quad \beta_h(V) = \frac{1.917}{1 + e^{\frac{-16.08 - V}{12.71}}}$$

Under these conditions the inactivation function is sufficiently fitted by a Boltzmann equation (Figure 1c):

$$I(V) = \frac{I_{max}}{1 + e^{\frac{V - V_h}{V_c}}} \tag{4}$$

The activation function was fitted with the Goldman-Hodgkin-Katz current equation[8] using a permeability with a Boltzmann type voltage dependence (Figure 1d):

$$I(V) = V \times \frac{g_{max}}{1 + e^{\frac{Vh - V}{Vc}}} \times \frac{\frac{[Na^+]_i}{[Na^+]_o} - e^{-\alpha V}}{1 - e^{-\alpha V}} \tag{5}$$

The sodium current defined by these parameters was implemented in a NEURON simulation model,[6] with a single compartment morphology, which resembles the electrotonic shape of the isolated neuron used for the fitting. Applying the standard voltage clamp protocols to the artificial neuron confirmed the correct implementation of the properties of the sodium current.

3. Na-CURRENT INACTIVATION IN EPILEPSY MODELS

The change in excitability that is associated with the process of epileptogenesis results in part from changes in ionic currents. Here we focus on the role of the sodium current properties.[1] A common observation made in several models of epilepsy is a shift of the inactivation curve in the depolarizing direction. In the epilepsy model of classical kindling, Vh shifted about 3.1 mV in the depolarizing direction.[4] A similar shift (2.5 mV) was found in the SSLSE model of epilepsy,[5] although in those experiments the variance was larger and the difference did not reach significance (Figure 2). The inactivation function is so steep that even a shift of this small magnitude results in a ~25% increase of the number of recruitable sodium channels at resting membrane potential.

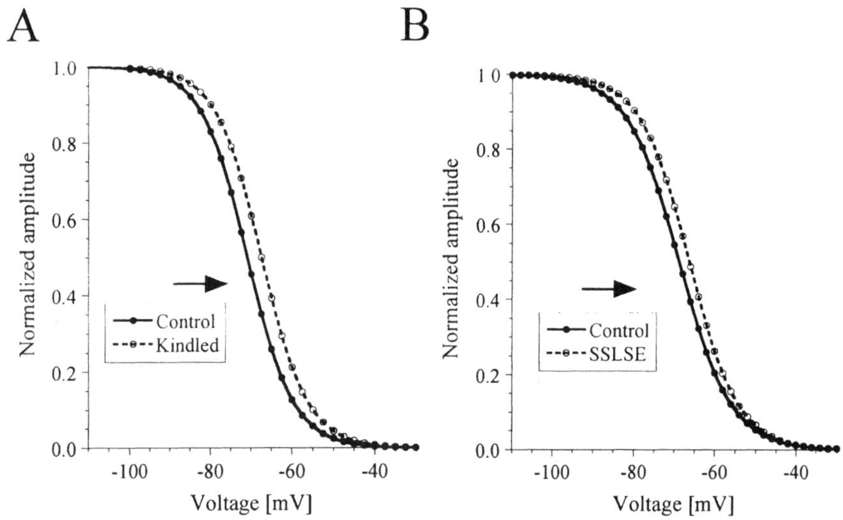

Figure 2: Inactivation properties of sodium current in rat hippocampal neurons. **A.** Shift in depolarizing direction in fully kindled animals.[4] **B.** A similar shift in neurons isolated from rats in the SSLSE model.[5]

4. CBZ MODULATION OF Na-CURRENT INACTIVATION

Carbamazepine (CBZ) is one of the drugs of first choice for the treatment of epileptic patients.[9] Its basic mechanism of action has extensively been studied, in particular under conditions of voltage clamp in isolated neurons. Extrapolating such results to the effects of the drug under current clamp conditions has proven more difficult than expected, and most results are formulated in terms of its effectiveness in the prevention of seizures.

The isolated cell preparation allows us to determine which parameters of the sodium current (as defined in Figure 1) are modulated by CBZ. CBZ does not affect the activation properties of the sodium current (data not shown, but see [10]), but it exerts a highly specific effect on the inactivation of the current (Figure 3). The relation between the CBZ concentration and the shift of the inactivation curve is well described by a Hill relation, with a Hill factor of 1 and an EC_{50} in the range of 40-100 μM (Figure 3b). In a series of studies Kuo and coworkers[11, 12] have shown that CBZ acts by binding to an extracellular site of the sodium channel. The affinity of binding to the inactivated state is much larger than that to the resting state. This mechanism can be investigated under voltage clamp, and it can be added to the kinetic scheme presented by equations 1-3.

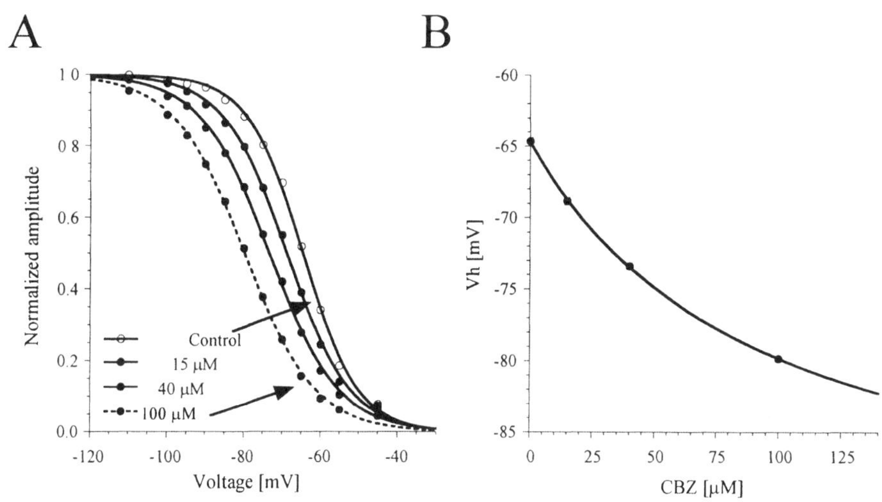

Figure 3: **A.** Application of carbamazepine shifts the inactivation in the hyperpolarizing direction in a dose dependent way. **B**. The relation between Vh and CBZ can be described by a Hill function; the Hill factor for this cell was 1 and the EC_{50} for this cell was 94 μM.

The assumption that binding of CBZ predominantly takes place to the inactivated state of the sodium channel implies the existence of an additional h state (h_{bound}). The distribution of the channels between h_{closed} and h_{bound} is then determined by a CBZ dependent association constant γ(CBZ) and a fixed dissociation constant δ:

$$h_{closed} + CBZ \underset{\delta_h}{\overset{\gamma_h(CBZ)}{\rightleftarrows}} h_{bound} \qquad (6)$$

A protocol was designed in which a depolarizing voltage step of variable duration allowed CBZ to bind to an increasing fraction of inactivated gates.

A standard depolarizing voltage step subsequently applied made it possible to quantify the bound fraction (Figure 5A, voltage protocol given as an inset). The binding rate constant was then obtained from the relation between CBZ and the binding rate. For CBZ in CA1 hippocampal pyramidal cells we obtained a value for γ(CBZ) of 32 ± 3 $mM^{-1}s^{-1}$.

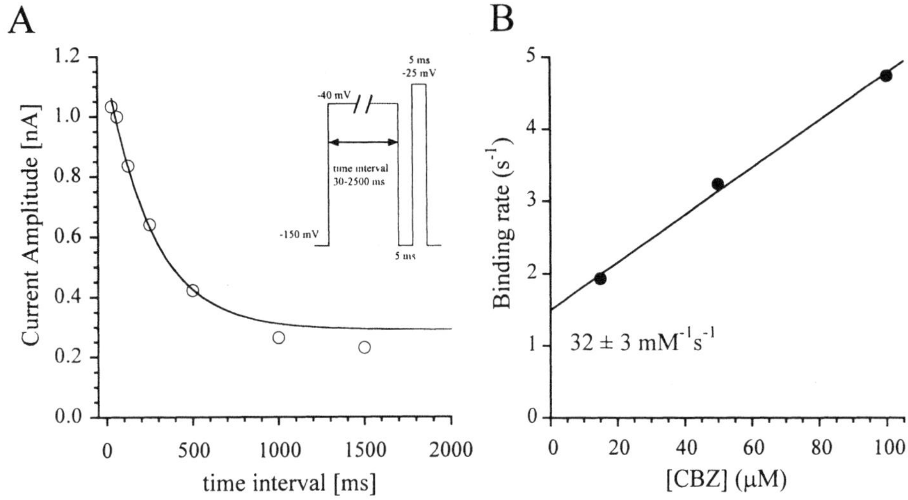

Figure 4. Determining the binding rate constant γ(CBZ) for CBZ to the inactivated state of the sodium channel. **A**. Slow binding to the inactivated state decreases the current evoked by a test pulse after an inactivating voltage step of variable duration. CBZ concentration for this experiment was 50 μM. The smooth curve is the single exponential fit to the data points with a time constant of 270 ms. The voltage protocol is given as an inset. **B.** The binding rate constant can be determined from the relation between binding rate and CBZ concentration, using a least square fit.

5. CBZ MODULATION OF NEURONAL FIRING

The addition of a CBZ dependent "bound" state to the model allows illustration of the equilibrium distribution of the h channels over the three states as a function of CBZ (see Figure 5 for examples of a low and a high concentration of CBZ). It also confirms the observed shift in Vh of the inactivation function and allows investigation of the changes in the time constant of recovery from inactivation that can be anticipated (data not shown). In order to evaluate the possible consequences of CBZ presence on neuronal firing rate, we implemented the sodium current in a single compartment neuron implemented in the simulation environment NEURON.[6] To create realistic firing we had to add at least a voltage dependent potassium current and we implemented the one defined by Hofman and coworkers.[13] Although this oversimplified neuronal model will not demonstrate all the fine variations in firing patterns that are possible (e.g., it will lack properties related to calcium current, calcium accumulation, and calcium dependent potassium channels), it will nevertheless demonstrate how CBZ affects basic firing rate of an isolated neuron (for overview see [14, 15]).

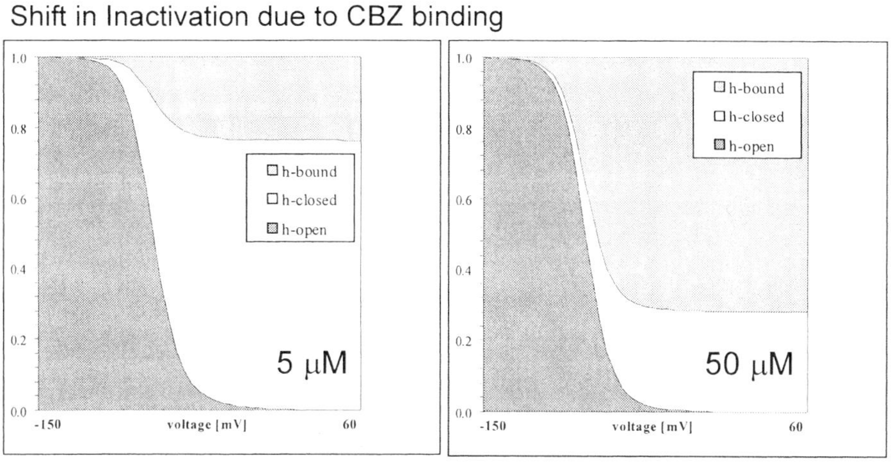

Figure 5. Distribution of the h gates over the three possible states (open, closed, and bound) illustrated for 5 μM (left panel) and 50 μM CBZ (right panel). The Vh of the curve that describes the distribution between h_{open} and h_{closed} shifts from -75.2 mV to -83.7 mV when CBZ is increased from 5 to 50 μM

As is demonstrated in figure 6 the presence of a high concentration of CBZ (100 μM, considerably above the usual adopted therapeutic dose of 15 μM) blocks all action potentials. At a concentration of 50 μM, CBZ effectively reduces firing rate of the neurons, and the difference between figure 6A and B indicates that these properties are use-dependent.

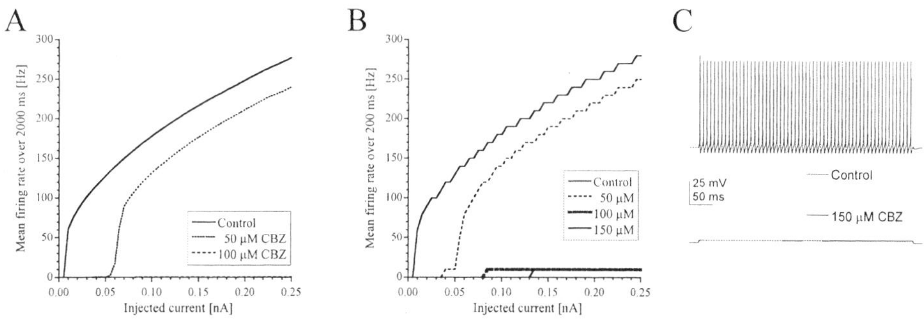

Figure 6. Firing properties of a hippocampal CA1 neuron investigated in a NEURON simulation model. **A.** Relation between current injection and firing rate measured over 2000 ms for control and in the presence of 50 and 100 mM CBZ. The latter concentration completely blocked neuronal firing. **B.** The same relation as illustrated in A, but now firing rate was determined over the first 200 ms. **C.** sample trace of membrane potential for control and in the presence of 150 μM CBZ.

6. CONCLUSIONS

Using a fit procedure that minimizes the collective error of a complete set of current traces describing voltage dependent activation, inactivation, and recovery from inactivation, allowed us to use the classical formalism of Hodgkin and Huxley and model the sodium current with sufficient accuracy to simulate such a set of current traces. Adding one more inactivated state (the one bound to CBZ) also permitted us to incorporate the binding of carbamazepine in a form that catches the essential properties responsible for its anti-epileptic activity.

The changes induced by CBZ counteract the changes observed in sodium current inactivation in the classical kindling model of epilepsy[4, 16] and in the SSLSE model.[5, 17, 18] Of course the effectiveness of CBZ in preventing seizures cannot be predicted from such a modeling study, as many more parameters of currents and neurons might have changed in the epileptic focus. On the other hand, it has provided us with a tool that predicts how an AED will affect firing rate under normal conditions.

7. ACKNOWLEDGMENTS

The help of Martin Vreugdenhil, Susan Ketelaars, Kristel Crommentuijn, Jan Gorter, and Taco Werkman in the experiments is greatly acknowledged. We also thank Hans Kager and Michiel Remme for their help in building and investigating the NEURON model.

8. REFERENCES

1. R. Köhling, Voltage-gated sodium channels in epilepsy, *Epilepsia* **43**, 1278-1295 (2002).
2. A. Hodgkin and A. Huxley, The components of membrane conductance in the giant axon of Loligo, *J. Physiol.* **116,** 473-496 (1952a).
3. A. Hodgkin and A. Huxley, A quantitative description of membrane current and its application to conduction and excitation in nerve, *J. Physiol.* **117,** 500-544 (1952b).
4. M. Vreugdenhil, G.C. Faas, and W.J. Wadman, Sodium currents in isolated rat CA1 neurons after kindling epileptogenesis, *Neuroscience* **86,** 99-107 (1998).
5. S.O.M. Ketelaars, J.A. Gorter, E.A. van Vliet, F.H. Lopes da Silva, and W.J. Wadman, Sodium currents in isolated rat CA1 pyramidal and dentate granule neurones in the post-status epilepticus model of epilepsy, *Neuroscience* **105**, 109-120 (2001).
6. M. Hines and N T Carnevale, The NEURON simulation environment, *Neural Computation* **9**, 1179-1209 (1997).
7. M. Vreugdenhil and W.J. Wadman, Enhancement of calcium currents in rat hippocampal CA1 neurons induced by kindling epileptogenesis, *Neuroscience* **49**, 373-381 (1992).
8. B. Hille, *Ionic channels of excitable membranes* (Sinauer Associates, New York, 2001).
9. R. Macdonald and K. Kelly, Mechanisms of action of currently prescribed and newly developed antiepileptic drugs, *Epilepsia* **35**, S41-S50 (1994).
10. M. Vreugdenhil and W.J. Wadman, Modulation of sodium currents in rat CA1 neurons by carbamazepine and valproate after kindling epileptogenesis, *Epilepsia* **40,** 1512-1522 (1999).
11. C.-C. Kuo, A common anticonvulsant binding site for phenytoin, carbamazepine, and lamotrigine in neuronal Na^+ channels, *Mol. Pharmacol.* **54**, 712-721 (1998).
12. C.-C. Kuo, R.-S. Chen, L. Lu, and R.-C. Chen, Carbamazepine inhibition of neuronal Na+ currents: quantitative distinction from phenytoin and possible therapeutic implications, *Mol. Pharmacol.* **51**, 1077-1083 (1997).
13. D.A. Hofman, J.C. Magee, M.C. Costa, and D. Johnston, K^+ channel regulation os signal propagation in dendrites of hippocampal pyramidal neurons, *Nature* **387**, 869-875 (1997).

14. A. Destexhe and J. Huguenard, Nonlinear thermodynamic models of voltage-dependent currents, *J. Computational Neurosci.* **9**, 259-270 (2000a).
15. A. Destexhe and J. Huguenard, Which formalism to use for modeling voltage-dependent conductances? In: *Computational Neuroscience: Realistic Modeling for Experimentalists*, edited by E. De Schutter E: (Boca Raton, CRC Press., 2000b).
16. G.V. Goddard, D.C. McIntyre, and C.K. Leech, A permanent change in brain functioning resulting from daily electrical stimulation, *Exp. Neurol.* **25**, 295-330 (1969).
17. J.A. Gorter, E.A. van Vliet, E. Aronica, and F.H. Lopes da Silva, Long-lasting increased excitability differs in dentate gyrus vs. CA1 in freely moving chronic epileptic rats after electrically induced status epilepticus, *Hippocampus* **12**, 311-324 (2002).
18. E.W. Lothman, D A. Rempe and P.S. Mangan, Changes in excitatory neurotransmission in the CA1 region and dentate gyrus in a chronic model of temporal lobe epilepsy, *J. Neurophysiol.* **74**, 841-848 (1995).

Discussion

Burnham: Very interesting data, but kindling is a long lasting phenomenon. Is this long lasting? How long after the last seizures were your animals killed.
Wadman: I didn't mention that. The analysis that I showed you was all done directly after the period of 5 seizures. We have another group of animals where we waited for 5 weeks, where there was no seizure in between, and then it appears to be a transient phenomenon. If you go to some of the older epilepsy models like the kaniate model and material from human patients, you see the same changes, but then of course most of the time you have seizures. The changes could well be a function of the seizures.
Heinemann: The removal of inactivation seems to have two time constants, a fast one and a slow one. With Hogkin- Huxley modelling you get only one time constant. How do you deal with that problem?
Wadman: Very pragmatically. If you saw the quality of the fits, you are really not missing out anything, and if you can describe something with one time constant you can have any combination of two time constant to give you the same data. Basically, the data we have provide no statistical support for trying to get out more time constants.
Wasterlain: This is an absolutely beautiful study. Can you tell from the model if the number of channels is affected? We find that seizures cause internalization and endocytosis of $GABA_A$ receptors, but Na+ channels, which technically are not receptors, have the ability to endocytose as well, which has only been recently realized. Can you tell me if the number of channels decreases?
Wadman: The data that I showed will not give you an answer. But of course if we do the initial fit, we get a value for the amplitude and the amplitude is sort of different. If you look at the size of the disassociated cells, there are also a slight difference between the control and the kindled animals. The density was not as different as you might expect. What this probably means is that the density is the same, although the site is different. But you have to be extremely careful, because you can not exclude that we one way or another select neurons of a particular size. From our data, I think that the density of the Na+ channels is not much different.
Buzsaki: In the past we tried to explain everything with synapses. But this reminds me of the epileptic neuron era. It turns out that the intrinsic properties, the biophysical properties of the neuron, may change and the change maybe long lasting. Sodium inactivation affects a lot of things, not only firing rate, but in fact propagation of the action potential, the burst firing of the cell, some of which cannot be measured in a

disassociated culture. What you could do is look at bursting properties to see whether there is any difference between the kindled and non-kindled, whether the action potential rise time or duration is any different, and so on.
Wadman: I could not agree more with you. All the mechanisms are there, and you have to add up all the things you find on the cellular level. We didn't look carefully, but I expect that in the pilocarpine model, on the network scale, you have much more burst firing and things like that.
Engel: It's a beautiful study, but I'm confused, because this is a transient post-ictal phenomenon as opposed to a enduring phenomenon associated with increased epileptogenicity. We know that animals have a refractory period after seizures, when the excitability should be decreased. So I don't understand how this fits that principle.
Wadman: What you need to make it better understandable is to have an idea of the time course of these discharge events. General expression might take days or weeks to work, and we have not tested at many different time points. The other thing is that we first check whether it is related to the expression of other sodium channel subtypes that we have indications for. That could be the case and that would explain it more easily. But maybe it is more an aspect of how the epileptogenic process keeps its self going. After kindled animals have had several stage 5seizures, if you do anything more they will not have seizures. In that sense it could fit. It could be part of a consequence more than a cause, but it is hard to say.

KINDLING THE GABAERGIC PHENOTYPE OF THE GLUTAMATERGIC GRANULE CELLS

Rafael Gutiérrez, Gisela Gómez-Lira, Carmen Vivar, Jasmín Maqueda, Mónica Ramírez, Mónica Lamas*, Mario Treviño, Héctor Romo-Parra*

1. INTRODUCTION

The granule cells (GC) of the dentate gyrus (DG) are very exciting cells, although it seems that they can also be very inhibiting cells. They contain and release the excitatory amino acid glutamate, the metal Zn^{++}, the peptides Y, and dynorphin, brain derived neurotrophic factor (BDNF), and it has recently been well established that they also have the necessary machinery to synthesize, vesiculate and release the inhibitory amino acid GABA.[6,12] Among all these chemical messengers, glutamate is the only one with fast-acting excitatory actions. There is no doubt on the excitatory nature of the GC, but when they fire at high frequencies, peptides Y[14] and dynorphin[35] are released to produce inhibition and glutamate reaches an extracellular concentration that, by spilling over, activates presynaptic mGlu and kainate receptors[21,28] to inhibit further release. To complicate things further, GABA can be released from the mossy fibers (MF) under certain conditions.

It is noteworthy that in all experimental models of epilepsy, as well as in human epileptic patients, a series of plastic changes concur precisely in the GC. Some represent alterations in the effects, metabolism, and content of Zn^{++}, expression of neurotrophins and their receptors, sprouting of the MF, and formation of atypical basal dendrites which underlie recurrent innervatïon, changes in the content and release of opioid peptides, induction of immediate early genes, regulation and transcription of opioid peptides' genes and transcription factors, changes in the homeostasis of Ca^{++} and ionic fluxes[6] and finally, the expression of GAD and GABA.[5,16,24,29,31] Furthermore, some of these plastic phenomena seem not to be exclusively related to epilepsy but can occur after synaptic

* Centro de Investigación y de Estudios Avanzados, Apartado postal 14-7140, México D.F. and * Instituto de Fisiología Celular, Universidad Nacional Autónoma de México.

strengthening by LTP induction,[2,7] and in physiological conditions, such as learning.[23] Thus, some of these adaptive changes may serve to limit the risk of seizure generation while others facilitate this process.

Among the many plastic changes that occur in the GC, we have centered our attention in the regulation of the expression, or rather, up-regulation of the GABAergic phenotype of the GC in the developing and adult hippocampus. The GABAergic phenotype of the granule cells is constitutively expressed during development, i.e., GAD_{67}, GABA and VGAT mRNA are present in these cells.[11,20] Moreover, their stimulation produces $GABA_A$-receptor-mediated monosynaptic responses in pyramidal cells[11,34] and interneurons of CA3.[11,25] Interestingly, during the first week of age AMPA receptors are missing in hippocampal principal neurons, therefore GABA provides the depolarization needed to release NMDA receptors from their Mg^{++} blockade.[17] Although it has been shown that this function is exerted by interneurons driven by the MF, the latter can provide direct GABAergic actions that could underlie fast GABA-glutamate synergism.[11,13] When development is complete, the GABAergic phenotype is shut off and GABAergic transmission in this synapse cannot be detected.[11] In the adult rat, seizures produced *in vivo* or strong synaptic activation *in vitro* leads to an up regulation several peptides with inhibitory actions[19,33] and of all the markers of the GABAergic phenotype: GAD_{67},[7,16,24,29,31] GABA,[3,5,31] VGAT Mrna.[15] When this occurs, GC stimulation provokes again simultaneous glutamatergic and GABAergic monosynaptic responses in their target cells in CA3.[6,7,9,10,25] Therefore, inhibitory actions of the adult GC emerge in an activity-dependent manner and their inhibitory phenotype can thus be kindled. By taking advantage of this property, we sought to characterize the regulation of the GABAergic phenotype in an otherwise glutamatergic system.

2. KINDLING THE GABAERGIC MARKERS IN THE GRANULE CELLS AND THEIR MOSSY FIBERS

The GABAergic makers: GABA, GAD_{67}[3,5,27,31] and VGAT mRNA[15] are normally present, although in trace amounts, in the glutamatergic GC. GAD_{67} and GABA have also been shown to be normally present in monkeys and in humans.[31] Despite this, GABAergic transmission from the MF is not observed in naïve adult rats unless a certain level of hyperexcitability is induced, and for this to occur, protein synthesis is required.[7,25] The overexpression of the GABAergic markers seems to be gradual, as hyperexcitability builds-up, and the emergence of GABAergic transmission depends on the presence of a minimal level of hyperexcitability. After a single seizure GAD_{67} is detected in the MF but not in the GC bodies. However, if several seizures are produced by kindling the amygdala, GAD_{67} is additionally detected in GC bodies.[24] GAD_{67} and GABA in the GC can also be overexpressed with different procedures that lead to seizures *in vivo.*[5,16,19,29,31] Interestingly, by applying an LTP-like stimulation protocol over the perforant path *in vitro*, without producing epileptiform activity, up-regulation of GAD_{67} is observed in the GC and MF[7]. This shows that the presence of hyperexcitability, and not seizures *in vivo* or epileptiform activity *in vitro*, is the mechanism by which the GABAergic markers are up-regulated. The same dependence on excitability has been demonstrated for VGAT mRNA expression, i.e., the more the preparation is kindled (with LTP-inducing high frequency stimuli[7]) the higher is VGAT mRNA expression.[15]

This mechanism is further confirmed by recent data of our laboratory, which show that adult cultured GC can express the GABAergic markers in a manner dependent on glutamate receptor activation (Fig. 1). Again, this requires a certain time of exposure of the cells to the depolarizing stimulus and at least two hours post-stimulus to allow protein synthesis. This and previous evidence[7] show that depolarization and Ca^{++} can be triggering signals for GC differentiation, even in conditions where the normal synaptic circuitry is missing.

In summary, while GAD_{67}, GABA and VGAT mRNA are all present in trace amounts in normal adult rats, GABAergic transmission from the MF does not occur. This suggests that the sole presence of GABA and VGAT is not sufficient for GABA to be released. On the other hand, high levels of activity up-regulate the expression of GAD_{67}, GABA and VGAT mRNA, which coincides with the emergence of GABAergic transmission. Despite this temporal coincidence in the expression of the GABAergic markers and of the GABAergic signaling in and from the MF, it is still not demonstrated if the up-regulation of this markers is enough to trigger GABAergic transmission from the MF.[7,25] Thus, the intracellular signal that triggers, on the one hand, the up-regulation of the GABAergic phenotype in the GC, and on the other, the emergence of GABAergic transmission, is unknown.

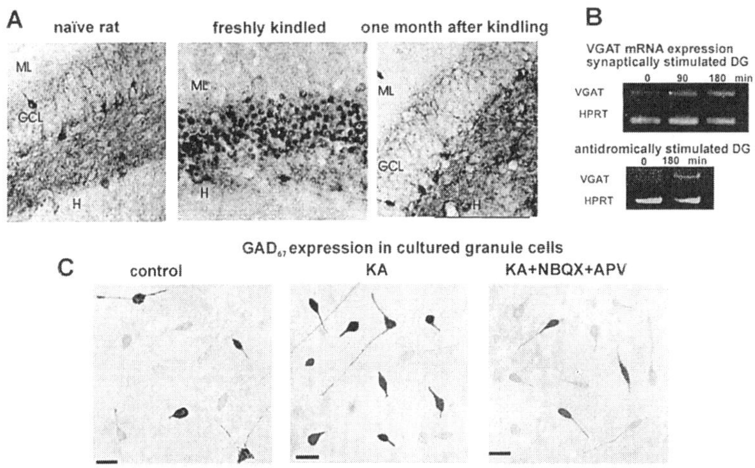

Figure 1. Expresión of GAD_{67} and VGAT mRNA in response to seizures in vivo, and kindling-like stimulation and kainic acid exposure in vitro. A) Immunohistological determination of GAD_{67} in the DG before, and 24 h and one month after the last of 5 kindled seizures. B) RT-PCR experiments show the expression of VGAT mRNA in the DG in an activity-dependent manner. C) Glutamate-receptor-dependent induction of GAD_{67} in granule cells in vitro. (A, from Ramírez and Gutiérrez, 2001; B, from Lamas et al., 2001, with permission; C, from Gómez-Lira et al., unpublished observations).

3. KINDLING GABAERGIC TRANSMISSION FROM THE MOSSY FIBERS

The first electrophysiological evidence of GABAergic transmission from the MFs to CA3 was in total agreement with the immunohistological observations showing that seizures transiently upregulated the expression of GAD_{67}.[29,31] Indeed, we found monosynaptic GABA-mediated transmission from the DG to CA3 in kindled epileptic but not in control healthy rats.[9,10] The MFs form glutamatergic excitatory synapses on pyramidal cells and local inhibitory interneurons, which in turn inhibit pyramidal cells[1,4] (Fig. 2). Thus, activation of the DG provokes monosynaptic excitation and disynaptic inhibition on CA3 neurons that, with intracellular recordings, are evidenced as sequences of excitatory and inhibitory post-synaptic potentials (EPSP/IPSP) in CA3 pyramidal cells (Fig. 2). The perfusion of the NMDA type glutamate receptor antagonist 5APV and the AMPA-Kainate type receptors' antagonist NBQX blocks all synaptic responses (Fig. 2), demonstrating that the inhibitory transmission in the MF-to-CA3 projection is disynaptically mediated. However, in kindled epileptic animals, the blockade of glutamatergic transmission isolated a bicuculline-sensitive IPSP.[9,10] This IPSP had the same latency as the control EPSP (even decreasing the probability of release by lowering extracellular calcium) and could be inhibited by activation of metabotropic glutamate receptors (mGluR).[6] This response was transient because it could be observed 24-48 h after the last kindled seizure but was not present if the experiment was carried out a month after the last seizure (Fig. 2). If seizures are again produced these monosynaptic GABAergic responses reappear.[6] It was also established that the kindled epileptic state was not necessary for this monosynaptic IPSP to be expressed, as a single seizure could provoke its emergence.[6] Due to the difficulty to unequivocally identify responses of MF origin, an experimental design was developed to avoid contamination of the putative MF-evoked responses by the activation of inhibitory fibers different from the MFs, and to record from the same preparation before and after the induction of MF GABAergic transmission in vitro.[7] This stimulation protocol produced hyperexcitability in the absence of epileptic activity by stimulation of the perforant path in a kindling-like manner (three 1-sec trains of 0.1 ms pulses at 100 Hz, 1 min apart from each other every 15 min for 3 hours). After completion of the stimulation protocol, the perfusion of glutamatergic antagonists blocked the EPSP and isolated a fast bicuculline-sensitive IPSP (Fig. 3). The expression of the GABAergic potential was prevented if this stimulation was provided just for one hour or if it was completed in the presence of the protein synthesis inhibitor, cycloheximide. This establishes that a certain level of excitation is required for the granule cells to express their GABAergic phenotype.[7] On the other hand, when the control synaptic responses (EPSP/IPSP sequences) of a given cell were first blocked, the direct kindling stimulation over the same site during perfusion of glutamatergic antagonists resulted in the induction of fast GABAergic potentials as kindling developed. Furthermore, a high spatial specificity of this phenomenon was evidenced by applying test pulse stimulation to an alternative non-kindled parallel MF input, which, contrary to the kindled one, did not evoke GABAergic responses. Altogether, this evidence establishes that the emergence of MF GABAergic transmission does not depend on the activation of the postsynaptic cell and, together with the over-expression of GAD and VGAT mRNA, reveals the presynaptic nature of this phenomenon.

As previously described in pyramidal cells, after seizures or LTP-like stimulation, MF activation in the presence of glutamatergic antagonists also provokes monosynaptic IPSPs in interneurons of CA3.[25] Despite this, the effect of the MF glutamatergic transmission surpasses the inhibitory one. Thus, the interneurons are more readily excited than pyramidal cells, which continue to receive a strong inhibitory control from the interneurons within CA3, besides the MF GABAergic signal.[25] Therefore, the activation of MFs at high frequencies provokes the more pronounced IPSPs in pyramidal cells to be pontentiated and to summate to hyperpolarize them.[10] These data show that, although both pyramidal cells and interneurons receive the same dual MF input, the ratio of released glutamate/GABA onto the target cells is different. This is also confirmed by extracellular recordings.[32] Indeed, MF activation produces GABA-receptor-dependent field responses in stratum lucidum of CA3 and presynaptic inhibition of MF collaterals.[26,32] The MF-GABAergic field potentials are, like the intracellular responses, less pronounced than the glutamatergic ones. The GABA/glutamate ratio is of around 60%. This goes in line with data showing that GABA is less concentrated than glutamate within mossy fiber terminals.[3,5]

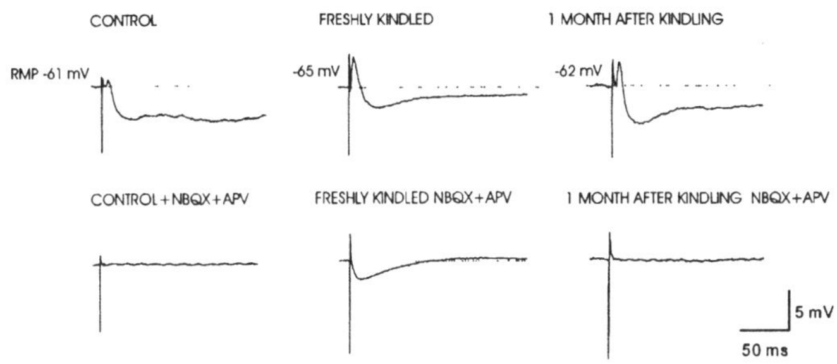

Figure 2. Pyramidal cell responses to MF activation before and during perfusion of glutamate receptors' antagonists. In control rats, NBQX and APV block all synaptic components, whereas in freshly kindled rats, a fast bicuculline-sensitive IPSP can still be elicited. This IPSP has the same onset latency as the control EPSP. A month after the last kindled seizure, perfusion of glutamate receptors' antagonists blocks again all synaptic components. (From Gutiérrez and Heinemann, 2001, with permission).

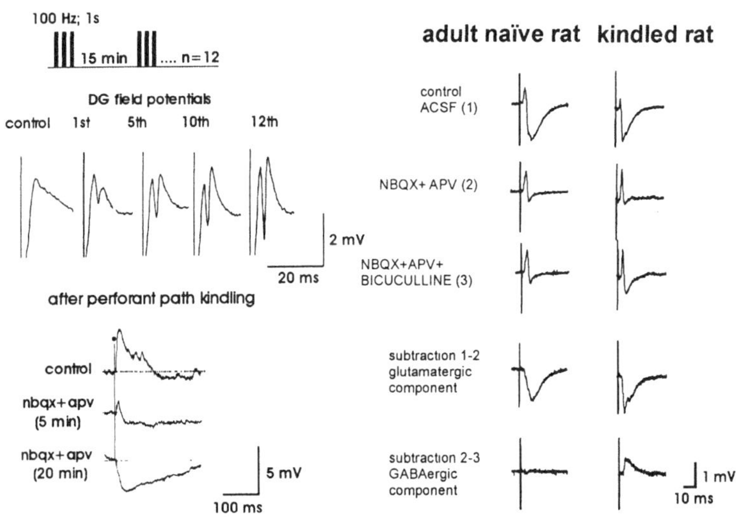

Figure 3. Left panel shows the kindling-like stimulation protocol used to produce synaptic potentiation in the DG in vitro. After 12 stimuli, synaptic responses were recorded in a pyramidal cell before (control) and during perfusion of glutamate receptors' antagonists. MF activation produces a fast EPSP in control condition and, after blockade of glutamate receptors, MF stimulation produces a monosynaptic IPSP that has the same latency as the control EPSP. The right panel depicts field potentials recorded in stratum lucidum of area CA3 in a naïve and a kindled rat. GABA-receptor-dependent field potentials can be digitally isolated in the kindled but not in the naïve rat. (left panel, from Gutiérrez, 2002, with permission; right panel, from Treviño et al., submitted).

It has been repeatedly shown, using intracellular and whole cell recordings, that MF-GABAergic transmission is selectively depressed by the activation of group III mGluR with L-AP4.[6,7,13,25] On the other hand, despite the data demonstrating the presence of groups II and III mGluR in the rat MF,[22,30] it has been considered that MF neurotransmission in the rat is depressed by the activation of group II but not group III mGluR. In accordance with this proposal, we have recently established that, indeed, MF neurotransmission in naïve adult rats is insensitive to L-AP4, but if tested in slices obtained from rats subjected to seizures or kindling-like stimulation in vitro, it has a clear depressing effect. We have suggested that this difference is due to the presence of GABAergic transmission in the latter, which is selectively depressed by L-AP4.[6,32] We have repeatedly shown that MF GABA transmission is preferentially inhibited by the activation of group III mGluR while group II mGluR preferentially inhibits glutamatergic transmission. This has led us to propose that group III mGluR can be linked to the GABA releasing machinery and also that there is a presynaptic segregation of mGluR receptors according to the class of neurotransmitter to be released (for a review see[8]). Recently Kasyanov et al.[13] have provided further data showing a higher depressing effect of the activation of group II than that of group III mGluR on MF-GABAergic transmission. These findings can have clinical implications, since temporal lobe epileptic patients present an up-regulation of III-mGluR precisely in the granule cells[18], whose excessive activation may disrupt this protective MF-GABA release.

4. FUNCTIONAL IMPLICATIONS AND OPEN QUESTIONS

With reasonable confidence, one can state that the function of the up-regulation of the GABAergic markers in the GC is to sustain GABAergic transmission. This phenomenon does not relate exclusively to epilepsy but can occur in conditions in which an enhancement of excitability does not lead to epileptic activity, as we have shown to occur. The transient expression of the GABAergic phenotype in the GC is constitutive during development possibly to foster development until its completion, when it is shut off[11]. In the adult, MF GABA transmission transiently emerges probably as compensatory change in response to enhanced excitability, especially after seizures. It can be again expressed in response to demanding conditions of the system. When this occurs, MF can provoke postsynaptic inhibition in pyramidal cells of CA3 and can also exert presynaptic actions in MF themselves, inhibiting further neurotransmitter release[32]. It is noteworthy that this mechanism can be involved in the anterograde amnesia that appears for some time after seizures, particularly after status epilepticus, so enhanced GABA transmission from DG can hamper storage of information in the hippocampal network.[10]

Despite the strong evidence that has accumulated in the recent years showing that the "glutamatergic" MF have the necessary machinery for the synthesis, vesiculation and release of GABA (for a review see[8]), a direct and undisputable prove of GABA-glutamate corelease is still missing. This can only be solved by simultaneously recording connected granule and pyramidal cells. The recording of autaptic currents, for example, can be an approach suitable to accomplish this. There is no doubt, however, that the GC express a GABAergic phenotype and that GC activation produces GABAergic responses in CA3 during glutamate receptor blockade in hippocampal slices from animals that have presented seizures or if the DG is strongly stimulated. Still, a number of questions arise:

What are the intracellular signals that shut off the GABAergic phenotype in the developing GC? What are those that trigger it again by activity in the adult? Why is this GABAergic transmission transient? What determines its disappearance? What is the contribution of MF GABA to hippocampal network activity?

Solving these and many other open questions will shed light on the role of dentate GC and the function of the DG, especially in epilepsy, when the expression of GABA is increased.

5. REFERENCES

1. L. Acsády, A. Kamondi, A., A. Sik, T. Freund, and G. Buzsáki, GABAergic cells are the major postsynaptic targets of mossy fibers in the hippocampus, *J. Neurosci.* 18,3386-3403 (1998).
2. B. Adams, M. Lee, M. Fahnestock, and R.J. Racine, Long-term potentiation trains induce mossy fiber sprouting, *Brain Res.* 775,193-197 (1997).
3. L. Bergersen, A. Ruiz, J.G. Bjaalie, D.M. Kullmann, V. Gundersen, GABA and GABAA receptors at hippocampal mossy fibre synapses, *Eur. J. Neurosci.* 18,931-941 (2003).
4. I.L. Crawford and J.D. Connor, Localization and release of glutamic acid in relation to the hippocampal mossy fibre pathway, *Nature* 244, 442-443 (1973).
5. G. Gómez-Lira, E. Trillo, M. Ramírez, M. Asai, M. Sitges, R. Gutiérrez, Expression of GABAergic transmission in the mossy fiber synapse coincides with the seizure-induced expression of GABA in mossy fiber synaptosomes, *Exp. Neurol.* 177,276-283 (2002).
6. R. Gutiérrez, Seizures induce simultaneous GABAergic and glutamatergic neurotransmission in the dentate gyrus-CA3 system, *J. Neurophysiol.* 84,3088-3090 (2000).
7. R. Gutiérrez, Activity-dependent expression of simultaneous glutamatergic and GABAergic neurotransmission from the mossy fibers in vitro, *J. Neurophysiol.* 87,2562-2570 (2002).
8. R. Gutiérrez, The GABAergic phenotype of the "glutamatergic" granule cells of the dentate gyrus, *Prog. Neurobiol.* 71,337-358 (2003).
9. R. Gutiérrez and U. Heinemann, Simultaneous release of glutamate and GABA might be induced in mossy fibers after kindling, *Neurosci. Lett.* 48,S23 (1997).
10. R. Gutiérrez and U. Heinemann, Kindling induces transient fast inhibition in the dentate gyrus - CA3 projection, *Eur. J. Neurosci.* 13,1371-1379 (2001).
11. R. Gutiérrez, H. Romo-Parra, J. Maqueda, C. Vivar, M. Ramírez, M.A. Morales and M. Lamas, Plasticity of the GABAergic phenotype of the "glutamatergic" granule cells of the rat dentate gyrus, *J. Neurosci.* 23,5594-5598 (2003).
12. D.A. Henze, N.N. Urban and G. Barrionuevo, The multifarious hippocampal mossy fiber pathway: a review, *Neuroscience* 98,407-427 (2000).
13. A.M. Kasyanov, V.F. Safiulina, L.L. Voronin and E. Cherubini, GABA-mediated giant depolarizing potentials as coincidence detectors for enhancing synaptic efficacy in the developing hippocampus, *Proc. Natl. Acad. Sci. USA* 101,3967–3972 (2004).
14. G.J. Klapstein and W.F. Colmers, On the sites of presynaptic inhibition by neuropeptide Y in rat hippocampus in vitro, *Hippocampus* 3,103-111 (1993).
15. M. Lamas, G. Gómez-Lira and R. Gutiérrez, Vesicular GABA transporter mRNA expression in the dentate gyrus and in mossy fiber synaptosomes, *Mol. Brain Res.* 93,209-214 (2001).
16. H. Lehmann, U. Ebert and W. Löscher, Immunocytochemical localization of GABA immunoreactivity in dentate granule cells of normal and kindled rats, *Neurosci. Lett.* 212, 41-44 (1996).
17. X. Leinekugel, I. Medina, I. Khalilov, Y. Ben-Ari, R. Khazipov, Ca^{2+} oscillations mediated by the synergistic excitatory actions of GABA(A) and NMDA receptors in the neonatal hippocampus, *Neuron* 18,243-255 (1997).
18. A.A. Lie, A. Becker, K. Behle, H. Beck, B. Malitschek, P.J. Conn, R. Kuhn, R. Nitsch, M. Plaschke, J. Schramm, C.E. Elger, O.D. Wiestler, I. Blumcke, Up-regulation of the metabotropic glutamate receptor mGluR4 in hippocampal neurons with reduced seizure vulnerability, *Ann. Neurol.* 47,26-35 (2000).
19. Y. Makiura, F. Suzuki, E. Chevalier and B. Onténiente, Excitatory granule cells of the dentate gyrus exhibit a double inhibitory neurochemical content after intrahippocampal administration of kainate in adult mice, *Exp. Neurol.* 159,73-83 (1999).

20. J. Maqueda, M. Ramírez, M. Lamas and R. Gutiérrez R., Glutamic acid decarboxylase $(GAD)_{67}$, but not GAD_{65}, is constitutively expressed during development and transiently overexpressed by activity in the granule cells of the rat, *Neurosci. Lett.* 353,69–71 (2003).
21. M.Y. Min, D.A. Rusakov and D.M. Kullmann, Activation of AMPA, kainate, and metabotropic receptors at hippocampal mossy fiber synapses: role of glutamate diffusion, *Neuron* 21,561-570 (1998).
22. H. Ohishi, C. Akazawa, R. Shigemoto, S. Nakanishi and N. Mizuno, Distributions of the mRNAs for L-2-amino-4-phosphonobutyrate-sensitive metabotropic glutamate receptors, mGluR4 and mGluR7, in the rat brain, *J. Comp. Neurol.* 360, 555-570 (1995).
23. V. Ramirez-Amaya, M.L. Escobar, V. Chao and F. Bermudez-Rattoni, Synaptogenesis of mossy fibers induced by spatial water maze overtraining, *Hippocampus* 9,631-636 (1999).
24. M. Ramírez and R. Gutiérrez, Activity-dependent expression of GAD_{67} in the granule cells of the rat hippocampus, *Brain Res.* 917,139-146 (2001).
25. H. Romo-Parra, C. Vivar, J. Maqueda, M.A. Morales and R. Gutiérrez R. Activity-dependent induction of multitransmitter signaling onto pyramidal cells and interneurons of area CA3 of the rat hippocampus, *J. Neurophysiol.* 89,3155-3167 (2003).
26. A. Ruiz, R. Fabian-Fine, R. Scott, M.C. Walker, D.A. Rusakov and D.M. Kullmann, $GABA_A$ receptors at hippocampal mossy fibers, *Neuron* 39,961-973 (2003).
27. R. Sandler and A.D. Smith, Coexistence of GABA and glutamate in mossy fiber terminals of the primate hippocampus: an ultrastructural study, *J. Comp. Neurol.* 303,177-192 (1991).
28. D. Schmitz, J. Mellor, M. Frerking and R.A. Nicoll, Presynaptic kainate receptors at hippocampal mossy fiber synapses, *Proc. Natl. Acad. Sci. USA* 98,11003-11008 (2001).
29. C. Schwarzer and G. Sperk, Hippocampal granule cells express glutamic acid decarboxylase-67 after limbic seizures in the rat, *Neuroscience* 69,705-709 (1995).
30. R. Shigemoto, A. Kinoshita, E. Wada, S. Nomura, H. Ohishi, M. Takada, P.J. Flor, A. Neki, T. Abe, S. Nakanishi and N. Mizuno, Differential presynaptic localization of metabotropic glutamate receptor subtypes in the rat hippocampus, *J. Neurosci.* 17, 7503-7522 (1997).
31. R.S. Sloviter, M.A. Dichter, T.L. Rachinsky, E. Dean, J.H. Goodman, A.L. Sollas and D.L. Martin, Basal expression and induction of glutamate decarboxylase and GABA in excitatory granule cells of the rat and monkey hippocampal dentate gyrus, *J. Comp. Neurol.* 373,593-618 (1996).
32. M. Treviño, H. Romo-Parra and R. Gutiérrez, $GABA_A$-receptor-mediated field potentials in CA3 reflect massive GABAergic transmission in the mossy fiber synapse, Submitted.
33. O. Vindrola, R. Briones, M. Asai and A. Fernandez-Guardiola, Amygdaloid kindling enhances the enkephalin content in the rat brain, *Neurosci Lett.* 21,39-43 (1981).
34. M.C. Walker, A. Ruiz and D.M. Kullmann, Monosynaptic GABAergic signaling from dentate to CA3 with a pharmacological and physiological profile typical of mossy fiber synapses, *Neuron* 29,703-715 (2001).
35. M.G. Weisskopf, R.A. Zalutsky and R.A. Nicoll, The opioid peptide dynorphin mediates heterosynaptic depression of hippocampal mossy fibre synapses and modulates long-term potentiation, *Nature* 365,188 (1993).

Discussion

Engel: Istvan Mody likes to talk about the granule cell in our chronic models becoming interneurons, because electrophysiologically they behave like interneurons. Where a granule cell will ordinarily only fire one action potential, once the animal is epileptic it will fire several. If it is true that it is this kind of burst firing that is responsible for activating the release of GABA, what are we doing when we give use-dependent sodium blockers? We are actually reducing the inhibitory mechanisms.

Gutierrez: Yes, probably. If we are reducing the firing rate of these cells, this should reduce the release of whatever they are releasing. The problem is that these cells they are really messy – they release a lot of things.

McNamara: I have three questions. First, what happened to the fast EPSC in the kindled animals, did that change? Second, it is not clear to me that you have unambiguously excluded the possibility you may have unsilenced $GABA_A$ synapses in the CA3 pryamidal cell. While you obviously have some changes presynaptically, in principle the explanation could be post-synaptic. Third, what is the net effect on information flow from granule cells through CA3 pyramidal cells as a consequence of this enhanced $GABA_A$ synaptic function?

Gutierrez: I hope that I have retained the three questions. (Shows slide). This is a pyramidal cell that is being shown under conditions of 10 and 50 Hz stimulation of the mossy fibers. This is a control pyramidal cell from a control animal. The higher the frequency of stimulation, the higher the probability that this cell will fire an action potential. This is a kindled animal, in which we are proposing that GABA is being released from the mossy fibers. And here is summation of IPSC or IPSP that prevents the cell from firing an action potentials. If we go to 50 Hz stimulation, the probability of having action potentials here is lower. We think that the GABA release from the mossy fibers is adding to that GABA release from the interneurons to finally inhibit the pyramidal cells. About the silent synapses, we don't think that is the case, because in order to get the GABAergic responses we need many hours of stimulation in vitro with LTP-like stimulation repeatedly. So if it were a silent synapse, 2 or 3 trains would be enough to activate the silent synapse.

McNamara: That assumes that GABA works like glutamate.

Gutierrez: This is simplistic, yes. We cannot exclude other mechanisms. Of course, the proof for us to say that GABA is released from the mossy fibers would be to do paired recordings. We are aware of that.

Buzsaki: A couple of times you mentioned that in the unkindled control animal there was no change or GABA response. On the other hand, Sloviter, who followed the histology of kindled and control animals, claims that GABA is present in the adult granule cells at a very low rate. So it seems like the machinery is there and you just change quantities.

Gutierrez: It seems that the machinery is there, but there must be some triggering factor for the mossy fibers to release the GABA that is already there. But who knows, we have never seen a control preparations where we can record IPSPs or IPSCs.

HIPPOCAMPAL KINDLING AND $GABA_B$ RECEPTOR FUNCTIONS

L. Stan Leung, Xinhuai Liu, Kevin J. Canning, and Bixia Shen*

1. INTRODUCTION

Inhibition is mediated by two main types of GABA receptors – ionotropic $GABA_A$ receptors that directly open Cl^- channels, and metabotropic $GABA_B$ receptors that act through G-protein coupled second messenger systems.[1, 2, 3] Postsynaptic $GABA_B$ receptors are found mainly on the dendrites of pyramidal cells (Fig. 1) and interneurons in the hippocampus, and they induce a slow K^+-mediated inhibitory postsynaptic current (IPSC).[2] Presynaptic $GABA_B$ receptors at the axon terminals (Fig. 1) mediate presynaptic inhibition of GABAergic terminals (autoreceptors) or glutamatergic terminals (heteroreceptors), likely by decreasing Ca^{2+} influx.[2, 4]

Mice with a $GABA_B$ receptor R1 knockout[5, 6] start to have spontaneous convulsive seizures after 12 days of age, and the seizures could be provoked by handling. The $GABA_B$ receptor antagonists CGP35338 or CGP56999A, administered systematically, induced convulsive seizures.[7] Thus, a lack of $GABA_B$ receptor function results in seizures.

We are interested in two separate but related questions: (1) whether kindling of the hippocampus results in a change in efficacy of $GABA_B$ receptors, and (2) how $GABA_B$ receptor blockade leads to hippocampal seizures. We used the partial hippocampal kindling model, in which repeated hippocampal afterdischarges (ADs) were evoked without accompanying motor convulsions. Partial kindling is known to increase AD duration and make a persistent contribution to epileptogenesis.[8] Partial hippocampal kindling also causes long-term disruption of behavioral and physiological functions.[9] In the following, partial hippocampal kindling was induced by evoking 15 ADs by electrical stimulation of dorsal hippocampal CA1 at hourly intervals, 5 per day and over 3 days.

*L. Stan Leung, Xinhuai Liu, Kevin J. Canning, and Bixia Shen, all at Dept. Physiology-Pharmacology and Clinical Neurological Sciences, University of Western Ontario, London, Canada N6A 5A5

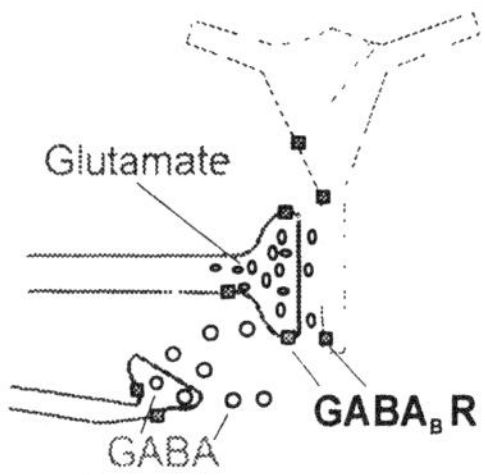

Figure 1. GABA$_B$ receptors (GABA$_B$R; ■) are shown on the postsynaptic membrane of a pyramidal cell, on presynaptic terminals that release glutamate vesicles (heteroreceptors), and on presynaptic GABAergic terminals (autoreceptors). Recordings were made in the pyramidal cell by means of a microelectrode.

2. HIPPOCAMPAL KINDLING DECREASES GABA$_B$ RECEPTOR FUNCTIONS

2.1 GABA$_B$ Autoreceptor Function

Partial hippocampal kindling was shown to decrease GABA$_B$ autoreceptor function *in vitro*. Monosynaptic (GABA$_A$-receptor mediated) IPSCs were evoked by paired-pulse stimulation in the presence of ionotropic glutamate receptor blockers.[10,11] Fig. 2 shows paired-pulse depression (PPD) of the IPSCs, or a smaller 2nd IPSC (IPSC2) as compared to the 1st IPSC (IPSC1). PPD was mediated by GABA$_B$ autoreceptor activation. We found that the IPSC2/IPSC1 ratio was larger (PPD was smaller) in neurons of kindled compared to control rats, although IPSC1 magnitudes were not different between the two groups of rats (Fig. 2). As expected, the IPSC2/IPSC1 ratio increased after adding a GABA$_B$ receptor antagonist CGP35348, but this increase was greater in the control group than in the kindled group of neurons (Fig. 2). Direct activation of GABA$_B$ autoreceptors by applying the GABA$_B$ receptor agonist baclofen (10 μM) in the perfusate suppressed the (single-pulse) IPSC more in neurons from control than kindled rats (Fig. 2). Both synaptic and direct activation indicate that the efficacy of the GABA$_B$ autoreceptors in CA1 was decreased in kindled as compared to control rats. Buhl et al. [12] also reported a smaller PPD of the monosynaptic IPSCs in dentate gyrus (DG) granule cells one day after full kindling of the perforant path. Although pharmacological tests were not reported, it is likely that GABA$_B$ autoreceptor downregulation also occurred in the DG after kindling.

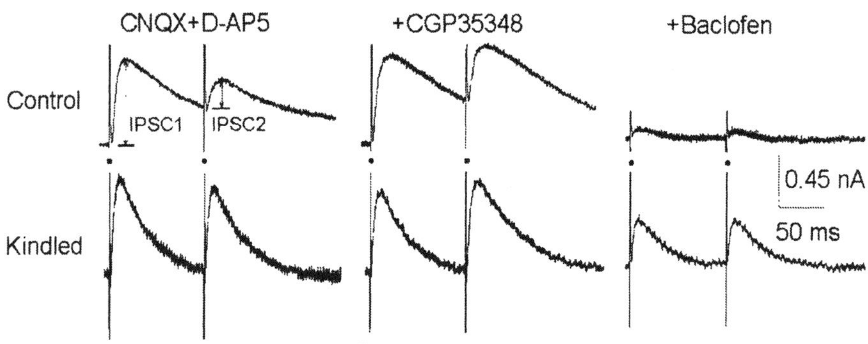

Figure 2. Decrease of GABA$_B$ autoreceptor efficacy in rats after partial hippocampal kindling as compared to control rats. Monosynaptic inhibitory postsynaptic currents (IPSCs) were recorded in the presence of the ionotropic glutamatergic antagonists CNQX and D-AP5 The IPSC2/IPSC1 ratio was increased in kindled as compared to control group of neurons; the IPSC2/IPSC1 ratio also increased with the GABA$_B$ receptor antagonist CGP35348 (1 mM) Baclofen (10 μM) decreased IPSC1 more in the control than the kindled group of neurons (adapted from Wu and Leung[11])

2.2 $GABA_B$ Heteroreceptor Function

The efficacy of $GABA_B$ heteroreceptors on glutamatergic terminals was decreased in CA1 for up to 3 weeks after partial hippocampal kindling, as shown by both direct and synaptic activation of the heteroreceptors *in vitro*[13] (Leung and N. Poon, unpublished). Application of baclofen (10 μM) in the perfusate directly activated $GABA_B$ heteroreceptors, thus decreasing glutamate release and the excitatory postsynaptic potentials (EPSPs) in CA1, recorded either intra- or extracellularly. Baclofen-induced decrease in the EPSP was smaller in slices of kindled as compared to control rats. GABA synaptically released by a brief train of stimulating pulses suppressed the EPSP[14] recorded intracellularly in CA1 neurons, and this was smaller in neurons from kindled rats as compared to those from control rats.

Using microiontophoretic application in urethane-anesthetized animals, Kamphuis et al.[15] reported that GABA was less effective in blocking the glutamate-evoked response in CA1 neurons after full CA1 kindling. The latter result is consistent with a downregulation of the $GABA_B$ heteroreceptors. A similar downregulation of $GABA_B$ heteroreceptors was shown by recording EPSPs from amygdala neurons after amygdala kindling.[16] Since $GABA_B$ heteroreceptors decrease glutamate release, in particular after high neural activity, a reduced heteroreceptor function is expected to increase seizure susceptibility.

2.3 $GABA_B$ Postsynaptic Receptor Function and GABA Release

$GABA_B$ receptor-mediated postsynaptic currents in principal neurons of CA1 and DG were studied using whole-cell recordings *in vitro*, in the presence of ionotropic glutamate receptor antagonists (Fig. 3). Here we present original results on DG neurons, which were patched at the granule cell layer (Liu and Leung, unpublished data). DG neurons from kindled and control rats showed similar $GABA_B$ receptor-mediated IPSCs following a brief train stimulation of 20 pulses at 300 Hz and supramaximal intensity (Fig. 3). The peak of the $GABA_B$-IPSC evoked by the brief train was 105.8 ± 9.2 pA (mean ± standard error of the mean, N=24 cells) in neurons of kindled rats, similar to 97.7 ± 8.3 pA (N=22 cells) in neurons of control rats (P>0.05). Baclofen induced an outward current in DG neurons (Fig. 3C, D) that was blocked by the $GABA_B$ receptor antagonist CGP55845A.

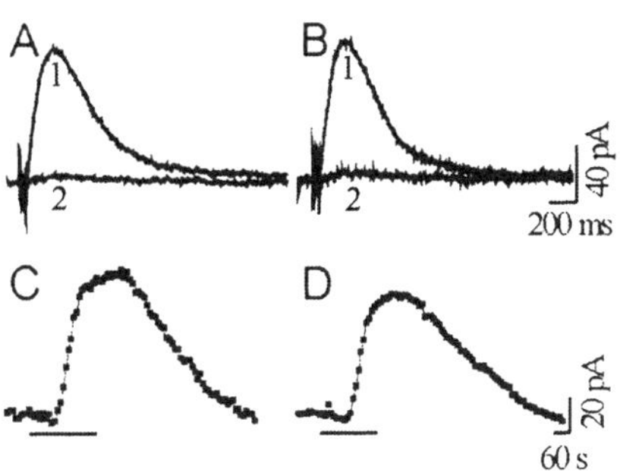

Figure 3. $GABA_B$ postsynaptic currents were not different between kindled and control groups of dentate granule cells. A and B, $GABA_B$-IPSCs from a neuron in kindled (A) or control rat (B), evoked by a brief supramaximal stimulus train applied to the middle molecular layer. $GABA_B$-IPSC (trace 1) was abolished (trace 2) after adding 1 μM $GABA_B$ receptor antagonist CGP55845A. C and D, time courses of outward current induced by 10 μM baclofen (applied during horizontal bar) recorded from representative dentate neurons in kindled (C) or control (D) rats (holding at –62 mV). Each point was the average current over 2.5 s; interval between points was 10 s. All recordings were made in the presence of kynurenic acid (1 mM) and picrotoxin (0.1 mM). Calibration applies to each row.

The magnitude of the baclofen-induced current was not different between kindled (76.6 ± 4.2 pA, N=23) and control (69.8 ± 8.6 pA, N=16) groups of neurons (P>0.05). The $GABA_B$ current reversal potential was also not different between kindled and control groups.

Study of CA1 neurons also showed that the outward $GABA_B$ receptor-mediated current induced by 10 μM baclofen (90 ± 7 pA) did not differ between kindled and control groups of neurons.[17] However, brief train stimulation (20 pulses at 300 Hz) of stratum radiatum induced a larger (P<0.05) $GABA_B$-receptor mediated IPSC in neurons from kindled (65.9 ± 5.2 pA) compared to control rats (45.8 ± 4.8 pA). In addition, the GABA uptake blocker nipecotic acid (1 mM) but not SKF89976A (0.2 mM) induced an outward, $GABA_B$ receptor-mediated current, and this current was larger in neurons of kindled rats than that in control rats. We suggest that the nipecotic acid-induced outward current was caused by non-vesicular GABA release through hetero-exchange, i.e., extracellular nipecotic acid exchanging for cytoplasmic GABA.[18, 19] Since the direct postsynaptic response to baclofen was not different, the increased $GABA_B$-IPSC following brief train stimuli in CA1 in kindled as compared to control rats likely resulted from a higher release of GABA, partly on account of $GABA_B$ autoreceptor downregulation.[17] GABA release in slices[20, 21] was reported to increase after partial or full hipocampal kindling, partly because of $GABA_B$ autoreceptors.[21]

In summary, our results after partial hippocampal kindling indicate that there was no change in postsynaptic $GABA_B$ receptor sensitivity in DG and CA1 *in vitro*. However, electrical train stimulation or nipecotic acid application induced larger $GABA_B$ receptor-mediated currents in CA1 neurons *in vitro* after partial kindling, likely because of enhanced presynaptic GABA release.

2.4. Changes of $GABA_B$ receptor expression after seizures

Various seizures, including kindling, were reported to increase expression of $GABA_B$ receptors. An increase in $GABA_B$-R1a and R2 mRNA was found in the DG at 1 day but not at 28 days after perforant path kindling.[22] An increase in immunoreactivity to $GABA_B$-R1 also suggests an increase in $GABA_B$ receptor protein expression after seizures.[23] Temporal lobe epilepsy is associated with increased $GABA_B$ R1 mRNA and decreased $GABA_B$ receptor density in the hippocampus.[24, 25]

Our electrophysiological studies after hippocampal kindling indicate that there was a decrease in efficacy in presynaptic $GABA_B$ receptors, despite an increase in expression of $GABA_B$ receptors reported after kindling or other seizures. Expression increase may serve to compensate for the loss in function. However, $GABA_B$ receptor function also depends on many signals other than the number of receptors.[1] Furthermore, $GABA_B$ receptors may be pre- and postsynaptic, on principal cells and interneurons, and differential expression of the different receptor subtypes has not been reported.

3. BEHAVIORS AFFECTED BY GABA$_B$ RECEPTOR DYSFUNCTION

3.1 Synaptic plasticity and spatial performance deficits after partial kindling

A GABA$_B$ receptor antagonist blocked long-term potentiation (LTP) induced by theta-frequency primed burst stimulation,[26, 27] presumably by blocking GABA autoreceptors. As expected after a decrease in autoreceptor efficacy (Fig. 2), partial hippocampal kindling also resulted in a decrease in primed burst induced LTP *in vitro*.[28] It has been suggested that LTP is an important mechanism mediating spatial navigation.[29] Thus, disruption of primed-burst LTP may cause poor spatial navigation. Indeed, performance on a radial arm maze (place but not cue task) was disrupted for up to 4 weeks after hippocampal kindling,[9, 30] similar to the time course of GABA$_B$ autoreceptor downregulation. So far, there is no direct evidence that GABA$_B$ autoreceptor function is responsible for the poor spatial performance.

3.2 Wet dog shakes suggest a decrease and GABA$_B$ receptor function after kindling

Wet dog shakes (WDSs) are named after a dog's shaking water out of its fur by means of a rapid rotation of its body along the long axis. WDSs in rats were frequent near the end of a hippocampal AD,[31] but the frequency of WDSs decreased as kindling progressed.[33] WDSs were decreased greatly after the 6th AD in our partial hippocampal kindling. We hypothesize that the decrease in WDS frequency indicates a decrease in efficacy of hippocampal GABA$_B$ receptors. Direct injection of baclofen (0.2 μL of 5 mM) into the hippocampus was used to test this hypothesis. Baclofen injected into stratum radiatum of dorsal CA1 induced WDSs that were blocked by a GABA$_B$ receptor antagonist (CGP55845A) injected through the same cannula. On day 1, 4 and 21 after partial hippocampal kindling, the number of WDSs induced by baclofen was reduced in kindled rats as compared to control rats (Leung and Shen, in preparation). GABA autoreceptor activation can mediate disinhibition of principal neurons, and firing of DG granule cells induced WDSs.[34] We suggest that baclofen was less effective in activating autoreceptors in kindled than control rats, a result already reported *in vitro* (Fig. 2).

3.3 Sensorimotor and auditory gating deficit after kindling

A deficit in sensorimotor gating, measured by prepulse inhibition of an acoustic startle response, was found after hippocampal partial kindling[35] and amygdala kindling.[36] A sensorimotor gating deficit is a characteristic of schizophrenia,[37] and, interestingly, GABA$_B$ receptor knockout mice also showed a deficit in prepulse inhibition.[5] We do not know how a loss of GABA$_B$ receptor function causes the latter deficit. However, paired-pulse auditory potentials recorded in hippocampal CA3, at a 500 ms interpulse interval (IPI), showed reduced paired-pulse inhibition after CGP35348 was injected intraventricularly.[38]

4. NEURAL SYSTEM EFFECTS INDUCED BY $GABA_B$ RECEPTOR BLOCKADE

4.1 Assessment of $GABA_B$ receptor function *in vivo*

Intracerebroventricular (icv) injection of the $GABA_B$ receptor antagonist CGP35348 was used to study $GABA_B$ receptor function in the hippocampus *in vivo*. We infer that $GABA_B$ receptors in the hippocampus are spontaneously active *in vivo*.[39] PPD of the population EPSP (pEPSP) or population spikes in DG and CA1 was sensitive to icv CGP35348. PPD of the pEPSPs (or population spikes) at 200 - 400 ms IPI was decreased by icv CGP35348, consistent with postsynaptic $GABA_B$ receptor blockade. Paradoxically, PPD of the population spikes at <100 ms IPI was increased by CPG35348,[39, 40] suggesting enhanced $GABA_A$ receptor mediated inhibition. In addition, icv CGP35348 also increased trisynaptic transmission from the medial perforant path to CA1 (P. Peloquin, K. Canning, K. Wu & Leung, in preparation), which was also found in partially kindled rats as compared to control rats.[41] $GABA_B$ receptor functions in kindled rats *in vivo* have yet to be studied in detail.

4.2 Hippocampal seizures induced by $GABA_B$ receptor antagonists in behaving rats

Systemic administration of $GABA_B$ receptor antagonists was reported to induce partial (focal) seizures and c-fos expression in the hippocampus.[7] Here we report further details on the hippocampal ADs following icv injection of a $GABA_B$ receptor antagonist.

In a behaving rat, injection of 0.11 mg CGP35348 icv (estimated to give ~160 μM concentration in ~1.5 ml of ventricular fluid) induced a hippocampal AD about 50% of the time. Hippocampal AD was preceded by a small increase in hippocampal gamma (30-70 Hz) activity and sometimes by theta rhythm (5-8 Hz) during immobility. The AD apparently started as 5 - 9 Hz sharp waves in the hippocampus, and it did not initially involve the entorhinal cortex, amygdala or sensorimotor neocortex. A higher dose of CGP35348 (> 0.22 mg icv) induced seizures accompanied by motor convulsions (facial and limb clonus) that appeared to start elsewhere (possibly neocortex) and spread into the hippocampus. Multiple ADs/seizures were induced after a high dose of CGP35348 icv (> 0.22 mg) or after CGP55699A, another $GABA_B$ receptor antagonist. A high icv dose of the $GABA_B$ receptor antagonist also induced seizures that started extra-hippocampally, and seizures could be induced by touch or by an experimenter approaching the injected animal.

Hippocampal ADs after icv CGP35348 can be explained by blockade of postsynaptic $GABA_B$ receptors. Uncontrolled glutamate release after $GABA_B$ heteroreceptor block may also be proconvulsant. Since $GABA_B$ receptor activation peaks at a latency of ~ 200 ms (Fig. 3B), a loss of $GABA_B$ receptor function enhances synaptic facilitation particularly at a frequency of ~ 5 Hz. Perhaps it is no coincidence that hippocampal ADs after icv CGP35348 started at 6 ± 1 Hz (N=20), as distinct from kindled ADs that started at <4 Hz.

5. SUMMARY

$GABA_B$ receptors play an important role in modulating and shaping activity-dependent synaptic activities. $GABA_B$ receptor blockade induced ADs that originated from the hippocampus, possibly by blocking postsynaptic and presynaptic heteroreceptors. Repeated hippocampal ADs induced by partial kindling resulted in a long-lasting decrease in the efficacy of presynaptic but not postsynaptic $GABA_B$ receptors in CA1 and DG. Heteroreceptor downregulation may increase seizure susceptibility after kindling.[42] Autoreceptor downregulation resulted in reduced synaptic plasticity, perhaps a cause for the poor spatial maze performance after kindling. The sensorimotor gating deficit and decrease in frequency of WDSs after kindling may also be attributed to $GABA_B$ receptor downregulation.

6. ACKNOWLEDGEMENTS

This research was supported by grants from the Canadian Institutes of Health MOP-64433 and MOP-36421 and the Natural Sciences and Engineering Research Council.

7. REFERENCES

1. A. Couve, S. J. Moss, and M.N. Pangalos, GABAB receptors: a new paradigm in G protein signaling, *Mol Cell Neurosci.* **16**, 296-312 (2000).
2. U. Misgeld, M. Bijak, and W. Jarolimek, A physiological role for $GABA_B$ receptors and the effects of baclofen in the mammalian central nervous system, *Prog. Neurobiol.* **46**, 423-462 (1995).
3. I. Mody, Y. De Koninck, T. S. Otis, and I. Soltesz, Bridging the cleft at GABA synapses in the brain, *Trends Neurosci.* **17**, 517-525 (1994).
4. N. G. Bowery, B. Bettler, W. Froestl, J. P. Gallagher, F. Marshall, M. Raiteri, T. I. Bonner, and S. J. Enna, International Union of Pharmacology. XXXIII. Mammalian gamma-aminobutyric acid(B) receptors: structure and function, *Pharmacol.Rev.* **54**, 247-264 (2002).
5. H. M. Prosser, C. H. Gill, W. D. Hirst, E. Grau, M. Robbins, A. Calver, E. M. Soffin, C. E. Farmer, C. Lanneau, J. Gray, E. Schenck, B. S. Warmerdam, C. Clapham, C. Reavill, D. C. Rogers, T. Stean, N. Upton, K. Humphreys, A. Randall, M. Geppert, C. H. Davies, and M. N. Pangalos, Epileptogenesis and enhanced prepulse inhibition in GABA(B1)-deficient mice, *Mol. Cell Neurosci.* **17**, 1059-1070 (2001).
6. V. Schuler, C. Luscher, C. Blanchet, N. Klix, G. Sansig, K. Klebs, M. Schmutz, J. Heid, C. Gentry, L. Urban, A. Fox, W. Spooren, A. L. Jaton, J. Vigouret, M. Pozza, P. H. Kelly, J. Mosbacher, W. Froestl, E. Kaslin, R. Korn, S. Bischoff, K. Kaupmann, P. H. van der, and B. Bettler, Epilepsy, hyperalgesia, impaired memory, and loss of pre- and postsynaptic GABA(B) responses in mice lacking GABA(B(1), *Neuron* **31**, 47-58 (2001).
7. M. Vergnes, A. Boehrer, S. Simler, R. Bernasconi, and C. Marescaux, Opposite effects of GABAB receptor antagonists on absences and convulsive seizures, *Eur. J. Pharmacol.* **332**, 245-255 (1997).
8. A. Dennison, G. C. Teskey, and D. P. Cain, Persistence of kindling: effect of partial kindling, retention interval, kindling site, and stimulation parameters, *Epilepsy Res.* **21**, 171-182 (1995).
9. L. S. Leung, C. Wu, K. Wu, B. Shen, R. Sutherland, and D. Zhao, Long-lasting behavioral and electrophysiological effects induced by partial hippocampal kindling, in: *Kindling 5*, edited by M. E. Corcoran, and S. Moshe (Plenum Press, New York, 1998), pp. 395-408.
10. C. H. Davies, S. N. Davies, and G. L. Collingridge, Paired-pulse depression of monosynaptic GABA-mediated inhibitory postsynaptic responses in rat hippocampus, *J. Physiol. (Lond.)* **424**, 513-531 (1990).
11. C. Wu and L. S. Leung, Partial hippocampal kindling decreases efficacy of presynaptic GABAB autoreceptors in CA1, *J. Neurosci.* **17**, 9261-9269 (1997).
12. E. H. Buhl, T. S. Otis, and I. Mody, Zinc-induced collapse of augmented inhibition by GABA in a temporal lobe epilepsy model, *Science* **271**, 369-373 (1996).
13. R. Siushansian, N. Poon, C. Naus, J. F. Bechberger, J. Mao, and L. S. Leung, Up-regulation of mRNA

expression of $GABA_B$ receptor 1B in the rat following hippocampal kindling, *Neurosci. Abstr.* **27**, #558.6. (2001).
15. W. Kamphuis, J. A. Gorter, and F. H. Lopes da Sılva, A long-lasting decrease in the inhibitory effect of GABA on glutamate responses of hippocampal pyramidal neurons induced by kindling epileptogenesis, *Neuroscience* **41**, 425-431 (1991).
14. J. S. Isaacson, J. M. Solis, and R. A. Nıcoll, Local and diffuse synaptic actions of GABA in the hippocampus, *Neuron* **10**, 165-175 (1993).
16. E. K. Asprodini, D. G. Rainnie, and P. Shinnick-Gallagher, Epileptogenesis reduces the sensitivity of presynaptic gamma-aminobutyric acid B receptors on glutamatergic afferents in the amygdala, *J. Pharmacol. Exp. Ther.* **262,** 1011-1021 (1992).
17. X. H. Liu and L. S. Leung, Partial hippocampal kindling increases $GABA_B$ receptor-mediated postsynaptic currents in CA1 pyramidal cells, *Epilepsy Res.* **57**, 33–47 (2003).
18. O. Honmou, J. D. Kocsis, and G. B. Richerson, Gabapentin potentiates the conductance increase induced by nipecotic acid in CA1 pyramidal neurons in vitro, *Epilepsy Res.* **20**, 193-202 (1995).
19. J. M. Solis and R. A. Nicoll, Postsynaptic action of endogenous GABA released by nipecotic acid in the hippocampus, *Neurosci. Lett.* **147**, 16-20 (1992).
20. W. Kamphuis, E. Huisman, M. J. Veerman, and F. H. Lopes da Silva, Development of changes in endogenous GABA release during kindling epileptogenesis in rat hippocampus, *Brain Res.* **545**, 33-40 (1991).
21. M. Zuiderwijk, G. C. Faas, F. H. Lopes da Silva, and W. E. J. M. Ghijsen, Electrically stimulated GABA release in rat hippocampus CA1 region is enhanced in kindling epileptogenesis, *Neurosci. Abstr.* **23**, 2161 (1997).
22. J. Francis, B. P. Jung, G. Zhang, W. Ho, J. Cheng, W. M. Burnham, and J. H. Eubanks, Perforant pathway kindling transiently induces the mRNA expression of GABA-B receptor subtypes R1A and R2 in the adult rat hippocampus, *Brain Res Mol. Brain Res.* **91**, 159-162 (2001).
23. Z. Kokaia and M. Kokaia, Changes in GABA(B) receptor immunoreactivity after recurrent seizures in rats. *Neurosci. Lett.* **315**, 85-88 (2001).
24. A. Billinton, V. H. Baird, M. Thom, J. S. Duncan, N. Upton, and N. G. Bowery, GABA(B(1)) mRNA expression in hippocampal sclerosis associated with human temporal lobe epilepsy, *Braın Res. Mol. Brain Res.* **86**, 84-89 (2001).
25. A. P. Princivalle, J. S. Duncan, M. Thom, and N. G. Bowery, Studies of GABA(B) receptors labelled with [(3)H]-CGP62349 in hippocampus resected from patients with temporal lobe epilepsy, *Br. J. Pharmacol.* **136**, 1099-1106 (2002).
26. C. H. Davies, S. J. Starkey, M. F. Pozza, and G. L. Collingridge, $GABA_B$ autoreceptors regulate the induction of LTP, *Nature* **349**, 609-611 (1991).
27. D. D. Mott and D. V. Lewis, Facilitation of the induction of long-term potentiation by $GABA_B$ receptors, *Science* **252**, 1718-1720 (1991).
28. L. S. Leung and C. Wu, Kindling suppresses primed-burst induced long-term potentiation in hippocampal CA1, *NeuroReport* **14**, 211-214 (2003).
29. R. U. Muller, M. Stead, and J. Pach, The hippocampus as a cognitive graph, *J. Gen. Physiol.* **107**, 663-694 (1996).
30. L. S. Leung, D. Zhao, and B. Shen, Long-lasting effects of partial hippocampal kindling on hippocampal physiology and function, *Hippocampus* **4**, 696-704 (1994).
31. L. S. Leung, Hippocampal electrical activity following local tetanization. I. Afterdischarges, *Brain Res.* **419**, 173-187 (1987).
32. M. Lerner-Natoli, G. Rondouin, and M. Baldy-Moulinier, Evolution of wet dog shakes during kindling in rats: comparison between hippocampal and amygdala kindling, *Exp. Neurol.* **83**, 1-12 (1984).
33. D. P. Frush and J. O. McNamara, Evidence implicating dentate granule cells in wet dog shakes produced by kindling stimulations of entorhinal cortex, *Exp. Neurol.*, **92**, 102-113 (1986).
34. B. P. Damiano and J. D Connor, Hippocampal mediation of shakıng behavior induced by electrical stimulation of the perforant path in the rat, *Brain Res.* **308**, 383-386 (1984).
35. J. Ma and L. S. Leung, Schizophrenia-like behavioral changes after partial hippocampal kindling, *Braın Res.* **997**, 111-118 (2004).
36. M. Koch and U. Ebert, Deficient sensorimotor gating following seizures in amygdala-kindled rats, *Biol. Psychiatry* **44**, 290-297 (1998).

37. D. Braff, C. Stone, E. Callaway, M. Geyer, I. Glick, and L. Bali, Prestimulus effects on human startle reflex in normals and schizophrenics, *Psychophysiology* **15**, 339-343 (1978).
38. K. M. Hershman, R. Freedman, and P. C. Bickford, GABAB antagonists diminish the inhibitory gating of auditory response in the rat hippocampus, *Neurosci. Lett.* **190**(2), 133-6 (1995).
39. K. J. Canning and L. S. Leung, Excitability of rat dentate gyrus granule cells in vivo is controlled by tonic and evoked $GABA_B$ receptor-mediated inhibition, *Brain Res.* **86**, 271-275 (2000).
40. F. H. Brucato, R. A. Morrisett, W. A. Wilson, and H. S. Swartzwelder, The GABAB receptor antagonist, CGP-35348, inhibits paired-pulse disinhibition in the rat dentate gyrus in vivo, *Brain Res.* **588**, 150-153 (1992).
41. K. Wu and L. S. Leung, Enhancement of multisynaptic transmission through the hippocampo-entorhino-hippocampal loop in partially kindled rats in vivo, *Neurosci. Abstr.* **25**, 1351 (1999).
42. M. Sato, R. J. Racine, and D. C. McIntyre, Kindling: basic mechanisms and clinical validity, *Electroencephalogr.Clin.Neurophysiol.* **76**, 459-472 (1990).

Discussion

Gale: First of all, I am thrilled that our microinjections with baclofen found their way to a good use in your studies. My question is, in the kindled animals where you are seeing this reduction in response to the baclofen induced wet-dog shakes, did you try putting bicuculline in the same places producing the wet dog shakes and perhaps seeing that those wet dog shakes were not affected? It would be nice to see that you could separate the postsynaptic and presynaptic mechanisms.
Leung: We have not tried that, but it is a good suggestion.
Gale: It is quite possible that there is a more generalized shift because of the kindling, which may have nothing to do with the circuitry in the hippocampus, but it really could have something to do with the larger network being reorganized. So the whole thing may be shifted, and the ability to get wet dog shakes, which is a very hippocampal phenomenon, maybe be suppressed.
Wasterlain: This is very nice. In addition to those experiments, what do you know about the mechanism of the down regulation? This is a G-protein coupled receptor; so it is expected to be down-regulated a lot, but it could also be internalized. So is it a decreased number of receptors, or is it a change in the kinetics of the receptor, which is presynaptic down regulation? How does the cell down-regulate one and not the other?
Burnham: How long do these changes last?
Leung: Most of the changes we have tested for 21 days. After that I am not sure. There is corroborating evidence from release studies, actually done in Amsterdam, with Dr. Wadman's associates, who indicate the change lasts for about 4 weeks after full kindling.
Gallagher: We published in 1992 that there was a presynaptic change in $GABA_B$ receptors and no change in postsynaptic $GABA_B$ receptors in the amygdala. So my question is, you are getting changes in presynaptic $GABA_B$ receptors that are proepileptogenic and antiepileptogenic - which one wins, and is there a time course change?
Leung: Our study is pretty close to your study with Asprodini in 1992 showing that postsynaptic receptor efficacy doesn't change, although I think that you have shown the heteroreceptors have a decrease in efficacy similar to here. It is hard to know, because of the different types of $GABA_B$ receptors. What we know at the end, what I showed here, is

that they did change, I mean the interneurons. I didn't even look at their responses to different types of $GABA_B$ receptors. So it is a complicated story. I might be over-emphasing the autoreceptor change, but I think in a circuit or network population, it is really hard to tell them apart.

ALTERED INTERACTION BETWEEN THE ENTORHINAL CORTEX AND HIPPOCAMPUS IN AMYGDALA KINDLED RATS

Uwe Heinemann, Doris Albrecht, Achim Behr and Tengis Gloveli*

1. INTRODUCTION

The entorhinal cortex is the main input and output region for the hippocampus. Perirhinal and postrhinal cells send their axons into superficial layers of the EC where they interact with projection cells in layer II and III of the entorhinal cortex but also with apical dendrites of the deep layer entorhinal cortex cells.[1] These receive input from olfactory cortex and subiculum and cholinergic input from the basal ganglia and the nucleus basalis Meynert as well as from the septum. Layer II stellate cells form the major portion of the perforant path, which projects to the dentate gyrus, while layer III cells project to the subiculum and in addition to the stratum moleculare of the CA1 area and perhaps also to area CA3.[2] Studies on the normal interaction between the entorhinal cortex and hippocampus from adult rats have revealed that seizure susceptibility is larger in the EC than in the hippocampus, which tends to develop only interictal discharges and short ictal events.[3,4] In adult rats seizure like events spread from the EC to the subiculum, but rarely fully recruit area CA1 and the DG. This limitation of spread seems to relate to the extensive feedforward and feedback inhibitory network in the DG and the prominent activation of inhibitory interneurons in area CA1.[5] The DG was therefore assumed to play a gating role in transfer of information from layer II of the EC to the DG.[6] We asked the question whether the properties of EC layer II stellate cells also contribute to the limitation of seizure spread from the EC through the DG to area CA3. Indeed layer II stellate cells express a flat input output curve when stimulated at low frequencies. However, they readily propagate action potentials to the DG when activity with frequencies at and above 10 Hz are used to synaptically activate these cells either from deep layers in the medial EC or from superficial layers in the lateral EC. By contrast layer III cells in the medial EC readily produce action potentials when activated with low frequencies, but suspend propagation of action potentials when activated at frequencies above 10 Hz.[7] We studied effects of kindling on patterns of

* Uwe Heinemann, Johannes-Mueller Instituteof Physiology, Humboldt University oBerlin, 10117 Berlin, Germany, uwe.heinemann@charite.de

seizure spread, and on alterations in the entorhinal cortex which might affect interaction between the entorhinal cortex and hippocampus.

2. KINDLING AND SEIZURE SPREAD FROM THE ENTORHINAL CORTEX TO THE HIPPOCAMPUS

Kindling affects the interaction between the entorhinal cortex and hippocampus. In amygdala kindled animals we found that after kindling low Mg^{2+} induced seizure like events propagated more readily through the dentate gyrus to the hippocampus while also readily recruiting the subiculum into seizure activity.[8 9] The facilitated recruitment of the subiculum may in part depend on a transient down regulation of both the early and the late after-hyperpolarization following a train of action potentials[10] also apparent in human subiculum from patients with temporal lobe epilepsy.[11] This is in contrast to pilocarpine treated animals, where seizure spread can occur both through the subiculum to area CA1 and through the dentate gyrus to CA3 Even when seizure activity was locally induced in the entorhinal cortex by focal application of bicuculline and elevated K^+, seizure spread in kindled animals was facilitated through the dentate gyrus.[12]

3. GLUTAMATE RECEPTOR DEPENDENT FASCILITATION OF SEIZURE SPREAD

This facilitated seizure spread points to alterations in the functional organization of the glutamatergic innervation of the dentate gyrus. We tested the relative contribution of NMDA, AMPA, and Kainate receptors on frequency facilitation in transmission from the perforant path to granule cells. This may in part be due to alterations in presynaptic release properties. Indeed we have observed an upregulation of the synaptobrevin-synaptophysin complex following kindling both in cortex and hippocampus.[13] This may provide for a reserve pool of synaptobrevin, promoting enhanced synaptic transmission in the kindled state. However, there was no evidence for altered synaptically evoked AMPA currents in kindled animals. Several lines of evidence indicate a substantial contribution of kainate receptors to temporal lobe seizures. The activation of kainate receptors located on hippocampal inhibitory interneurons was shown to reduce GABA release.[14] A reduced GABA release secondary to kainate receptor activation could contribute to enhanced seizure susceptibility. As the dentate gyrus serves a pivotal gating function in the spread of limbic seizures, we[15] tested the role of kainate receptors in the regulation of GABA release in the dentate gyrus of control and kindled animals. Application of glutamate (100 μM) in the presence of the NMDA receptor antagonist d-APV and the AMPA receptor antagonist, SYM 2206 caused a slight depression of evoked monosynaptic inhibitory postsynaptic currents (IPSCs) in control, but a substantial decrease in kindled dentate granule cells. The observation that kainate receptor activation altered paired-pulse depression and reduced the frequency of TTX-insensitive miniature IPSCs without affecting their amplitude is consistent with a presynaptic action of glutamate via kainate receptors on the inhibitory terminal to reduce GABA release. In kindled preparations, neither glutamate (100 μM) nor kainate (10 μM) applied in a concentration known to depolarize hippocampal interneurons led to an increase of the TTX-sensitive spontaneous IPSC frequency nor to changes in the postsynaptic membrane properties. Consistently, the inhibitory effect on evoked IPSCs was not affected by the presence of the GABA-B receptor antagonist CGP55845A, thus excluding depression by an enhanced release of GABA acting on presynaptic GABA-B receptors. The enhanced inhibition of GABA release following presynaptic kainate receptor activation favours a use-dependent hyperexcitability in the epileptic dentate gyrus.

Apart from this effect on use dependent disinhibition there are two transient effects which may also contribute to fascilitated kindling. One is the transient upregulation of NMDA receptors[16] which contributes to frequency potentiation of EPSP's. We[12] therefore investigated dentate gyrus field potentials and granule cell excitatory postsynaptic potentials (EPSPs) following high-frequency stimulation (10-100 Hz) of the lateral perforant path in kindled rats. Although control slices showed steady EPSP depression at frequencies greater than 20 Hz, slices taken from animals 48 hr after the last seizure presented pronounced EPSP facilitation at 50 and 100 Hz, followed by steady depression. However, 28 days after kindling, the EPSP facilitation was no longer detectable. Using the specific N-methyl-D-aspartate (NMDA) and RS-alpha-amino-3-hydroxy-5-methyl-4-isoxazolepropionic acid (AMPA) receptor antagonists 2-amino-5-phosphonovaleric acid and SYM 2206, we examined the time course of alterations in glutamate receptor-dependent synaptic currents that parallel transient EPSP facilitation. Forty-eight hr after kindling, the fractional AMPA and NMDA receptor-mediated excitatory postsynaptic current (EPSC) components shifted dramatically in favour of the NMDA receptor-mediated response. Four weeks after kindling, however, AMPA and NMDA receptor-mediated EPSCs reverted to control-like values. Although the granule cells of the dentate gyrus contain mRNA-encoding for kainate receptors, neither single nor repetitive perforant path stimuli evoked kainate receptor-mediated EPSCs in control or in kindled rats. The enhanced excitability of the kindled dentate gyrus 48 h after the last seizure, as well as the breakdown of its gating function, appear to result from transiently enhanced NMDA receptor activation that provides significantly slower EPSC kinetics than those observed in control slices and in slices from kindled animals with a 28-day seizure-free interval. Therefore, NMDA receptors seem to play a critical role in the acute throughput of seizure activity and in the induction of the kindled state, but not in the persistence of enhanced seizure susceptibility. Changes in metabotropic glutamate receptor expression may also contribute to facilitated seizure spread.[17]

It is well established that granule cells up regulate GAD, the enzyme synthesizing GABA.[18] The granule cells of the dentate gyrus (DG) send a strong glutamatergic projection, the mossy fibre tract, toward the hippocampal CA3 field, where it excites pyramidal cells and neighbouring inhibitory interneurons. Despite their excitatory nature, granule cells contain small amounts of GAD (glutamate decarboxylase), the main synthetic enzyme for the inhibitory transmitter GABA.[19] Chronic temporal lobe epilepsy results in transient up regulation of GAD and GABA in granule cells, giving rise to the hypothesis that following over excitation, mossy fibres exert an inhibitory effect by release of GABA. We[20] therefore stimulated the DG and recorded synaptic potentials from CA3 pyramidal cells in brain slices from kindled and control rats. In both preparations, DG stimulation caused excitatory postsynaptic potential (EPSP)/inhibitory postsynaptic potential (IPSP) sequences. These potentials could be completely blocked by glutamate receptor antagonists in control rats, while in the kindled rats, a bicuculline-sensitive fast IPSP remained, with onset latency similar to that of the control EPSP. Interestingly, this IPSP disappeared 1 month after the last seizure. When synaptic responses were evoked by high-frequency stimulation, EPSPs in normal rats readily summate to evoke action potentials. In slices from kindled rats, a summation of IPSPs overrides that of the

EPSPs and reduces the probability of evoking action potentials. However at frequencies above 50 Hz this inhibitory effect fades, presumably due to K^+ dependent reduced efficacy of inhibition. Our data show that kindling induces functionally relevant activity-dependent expression of fast inhibition onto pyramidal cells, coming from the DG, that can limit CA3 excitation in a frequency-dependent manner. Interestingly a contribution of GABA to synaptic transmission in this pathway is characteristic for juvenile animals, and it can be provoked by a single seizure and even under in vitro conditions by protocols which induce LTP (see contribution of Gutierrez, this volume). This transient inhibitory effect on CA3 cells may be in contrast to activation of networks in the dentate gyrus. Increased release of GABA from mossy fibre recurrent axon collaterals will lead to disinibition by reducing activation of interneurons and mossy cells.

4. KINDLING INDUCED ALTERATIONS OF ENTORHINAL CORTEX PROPERTIES

Alterations in kindling not only affect the hippocampus but also alter the amygdala and input of the amygdala to the entorhinal cortex. This was found in a recent study where we analysed the neuropathological consequences of kindling with a sensitive silver-staining method for the visualization of damaged neurons and Nissl staining for the estimation of the neuronal densities in different limbic areas.[21] Amygdala-kindled animals had reduced cell density in the amygdala and increased density of fragments of degenerated axons both ipsi- and contralateral to the stimulation site. Reduced neuronal density and the occurrence of degenerated axons in kindled animals were more prominent in the ipsilateral than in the contralateral hemisphere. In addition, more degenerated axons were found in entorhinal and perirhinal cortical structures of kindled than sham-operated animals. These results indicate that kindling induced morphological alterations that were not restricted to either the ipsilateral hemisphere or the stimulated region. With respect to the entorhinal cortex we found a particularly interesting loss of innervation from the amygdala and a somewhat reduced cell number in layer III of the entorhinal cortex which was less marked than the findings of cell loss in the pilocarpine model and kainate model of temporal lobe epilepsy.[22, 23]

We[24] therefore began to study alterations in entorhinal cortex layer III cells which project to the subiculum and to area CA1. We studied the effect of kindling on the frequency-dependent information transfer from the entorhinal cortex layer III to area CA1 in vitro. In control rats repetitive synaptic activation of layer III projection cells resulted in a frequency dependent depression of the synaptic transfer of action potentials to the hippocampus. One-to-two-days after kindling this effect was strongly reduced. Although no substantial change in synaptic inhibition upon a single electrical stimulation was detected in kindled rats, there was a significant depression in the prolonged inhibition[25] following high frequency stimulation. In kindled animals, paired-pulse depression (PPD) of stimulus-evoked IPSCs in layer III neurons was significantly stronger than in control rats. The increase of PPD is most likely caused by an increased presynaptic GABA(B) receptor-mediated auto inhibition. In kindled animals activation of presynaptic GABA(B) receptors by baclofen (10 μM) suppressed monosynaptic IPSCs significantly more than in control rats. In contrast, activation of postsynaptic GABA(B) receptors by baclofen was accompanied by comparable changes in the membrane conductance in both animal groups. Thus, in kindled animals activation of the layer III-CA1 pathway is facilitated by an increased GABA(B) receptor-mediated auto inhibition leading to an enhanced activation of the monosynaptic EC-CA1 pathway.

We therefore investigated further the effects which high frequency activation of the perforant path portion from EC layer III to CA1 would have on the plasticity in this area. We found that stimulation of the PP would induce a potentiation of evoked responses in area CA1. Interestingly high frequency stimulation had a suppressant effect on Schaffer collateral evoked field EPSPs and intracellularly recorded EPSPs. This activation protocol could in part reverse LTP induced by high frequency stimulation of the stratum radiatum. This suggests that memory deficits in patients with temporal lobe epilepsy may have an active component mediated by altered transmission properties of the entorhinal cortex layer III cells to the hippocampal field CA1 (Heinemann et al, in preparation).

5. CONCLUSIONS

Our findings add to the widespread observations that kindling affects a wide spectrum of cellular and synaptic functions within mesial temporal lobe structures. Some of these alterations are lasting, while many of them are transient in nature. On a network level the alterations in the DG favour seizure spread from the entorhinal cortex into the hippocampus. In addition they have effects on synaptic plasticity which might contribute to cognitive impairments characteristic of patients who suffer from temporal lobe epilepsy.

Kindling is thought to model one condition of human temporal lobe epilepsy where hippocampal and extrahippocampal cell loss is minimal. In these patients the functional alterations in cellular properties seem to be more important than the structural reorganisation which is associated with hippocampal sclerosis. However, not all alterations which we know from kindled animals also occur in the tissue of patients suffering from non-Ammon's Horn Sclerosis (AHS). Thus the down regulation of calbindin 28 which is typical for AHS patients and for models of temporal lobe epilepsy induced by a prior status epilepticus and kindling is not observed in tissue from non AHS patients.

Our findings regarding effects of up regulation of GABA B receptors in a model of TLE also explain why GABA B agonists such as baclofen had disappointing effects on the treatment of focal epilepsies. This suggests that preclinical monitoring of potential anticonvulsants should involve studies in chronic epileptic animals and where possible also in human epileptic tissue

6. REFERENCES

1. M. P Witter, Organization of the entorhinal-hippocampal system: A review of current anatomical data, *Hippocampus* **3** Suppl., 33-44 (1993).
2. T Dugladze, U. Heinemann,U , and T. Gloveli, Entorhinal cortex projection cells to the hippocampal formation in vitro, *Brain Res.* **905**, 224-231 (2001).
3. J P. Dreier and U. Heinemann, Regional and time dependent variations of low magnesium induced epileptiform activity in rat temporal cortex, *Exp. Brain Res.* **87**, 581-596 (1991).
4. K. Buchheim, S. Schuchmann, H. Siegmund, H.-J. Gabriel, U Heinemann, and H. Meierkord, Intrinsic optical signal measurement reveal characteristic features during different forms of spontaneous neuronal hyperactivity associated with ECS shrinkage in vitro, *Eur. J. Neurosci.* **11**, 1877-1882 (1999).
5. R. M. Empson and U, Heinemann, The perforant path projection to hippocampal area CA1 in the rat hippocampal-entorhinal cortex combined slice, *J. Physiol. (Lond.)* **484**, 707-729 (1995).
6. U. Heinemann, H Beck, J P Dreier, E Ficker, J. Stabel, and C. L. Zhang, The dentate gyrus as a regulated gate for the propagation of epileptiform activity In: *The Dentate Gyrus and its Role in*

Seizures,.edited by C. E. Ribak, C. M. Gall, and I Mody (Elsevier Science Publishers BV., Amsterdam, 1992), pp. 273-280.

7. T. Gloveli, D. Schmitz, R M Empson, and U Heinemann, Frequency-dependent information flow from the entorhinal cortex to the hippocampus, *J. Neurophysiol.* **78**, 3444-3449 (1997).
8 J Behr, T Gloveli, R Gutiérrez, and U Heinemann, Spread of low Mg2+ induced epileptiform activity from the rat entorhinal cortex to the hippocampus after kindling studied in vitro, *Neurosci. Lett.* **216**, 41-44 (1996).
9. J Behr, K J. Lyson, and I. Mody, Enhanced propagation of epileptiform activity through the kindled dentate gyrus, *J. Neurophysiol.* **79**, 1726-1732 (1998).
10. J.Behr, T. Gloveli, and U. Heinemann, Kindling induces a transient suppression of afterhyperpolarization in rat subicular neurons, *Brain Res.* **867**, 259-264 (2000).
11. C. Wozny, A. Kivi, T. N. Lehmann, C. Dehnicke, U. Heinemann, and J. Behr, Comment on "On the origin of interictal activity in human temporal lobe epilepsy in vitro", *Science* **301** (2003) NIL1-NIL2.
12. J. Behr, U. Heinemann, and I. Mody, Kindling induces transient nmda receptor-mediated facilitation of high- frequency input in the rat dentate gyrus, *J. Neurophysiol.* **85**, 2195-2202 (2001).
13. B Hinz, A. Becher, D. Mitter, K. Schulze, U. Heinemann, A. Draguhn, and G. Ahnert-Hilger, Activity-dependent changes of the presynaptic synaptophysin- synaptobrevin complex in adult rat brain, *Eur. J. Cell Biol.* **80,** 615-619 (2001).
14. D M. Kullmann, Presynaptic kainate receptors in the hippocampus: slowly emerging from obscurity, *Neuron* **32**, 561-564 (2001).
15. J. Behr, C. Gebhardt, U. Heinemann, and I. Mody, Kindling enhances kainate receptor-mediated depression of GABAergic inhibition in rat granule cells, *Eur. J. Neurosci.* **16**, 861-867 (2002).
16. I Mody and U. Heinemann, NMDA receptors of dentate gyrus granule cells participate in synaptic transmission following kindling, *Nature* **326**, 701-704 (1987)
17. M. Friedl, H. Clusmann, T. Kral, D. Dietrich, and J. Schramm, Analysing metabotropic glutamate group III receptor mediated modulation of synaptic transmission in the amygdala-kindled dentate gyrus of the rat, *Brain Res.* **821**, 117-123 (1999).
18. C. Schwarzer, and G. Sperk,, Hippocampal granule cells express glutamic acid decarboxylase- 67 after limbic seizures in the rat, Neuroscience, 69 (1995) 705-709.
19 R. S. Sloviter, M. A Dichter, T. L. Rachinsky, E. Dean, J H. Goodman, A. L. Sollas, and D. L. Martin, Basal expression and induction of glutamate decarboxylase and GABA in excitatory granule cells of the rat and monkey hippocampal dentate gyrus, *J. Comp. Neurol.* **373**, 593-618 (1996).
20. R. Gutierrez and U. Heinemann, Kindling induces transient fast inhibition in the dentate gyrus-CA3 projection, *Eur. J. Neurosci.* **13**, 1371-1379 (2001)
21. H. O. Bohlen und, K. Schulze, and D. Albrecht, Amygdala-kindling induces alterations in neuronal density and in density of degenerated fibers, *Hippocampus* **14**, 311-318 (2004).
22. F. Du, W. O. Whetsell, Jr., B. Abou-Khalil, B. Blumenkopf, E. W. Lothman, and R. Schwarcz, Preferential neuronal loss in layer III of the entorhinal cortex in patients with temporal lobe epilepsy, *Epilepsy Res.* **16**, 223-233 (1993).
23. F Du, T. Eid, E. W. Lothman, C. Köhler, and R. Schwarcz, Preferential neuronal loss in layer III of the medial entorhinal cortex in rat models of temporal lobe epilepsy, *J. Neurosci.* **15**, 6301-6313 (1995).
24. T. Gloveli, J. Behr, T. Dugladze, Z. Kokaia, M. Kokaia, and U. Heinemann, Kindling alters entorhinal cortex-hippocampal interaction by increased efficacy of presynaptic GABA(B) autoreceptors in layer III of the entorhinal cortex, *Neurobiol. Dis.* **13**, 203-212 (2003).
25. T. Gloveli, D. Schmitz, and U. Heinemann, Prolonged inhibitory potentials in layer III projection cells of the rat medial entorhinal cortex induced by synaptic stimulation in vitro, *Neuroscience* **80**, 119-131 (1997).

Discussion

Leung Very nice data. I'd like to ask you about the autoreceptors that mediate depression on the layer 3 cells. Its seems like in your control neuron the depression was not very large. And have you looked at the difference after $GABA_B$ antagonist, to see what is the component not $GABA_B$ receptor mediated?

Heinemann: Most of the depression, the paired-pulse depression after application of the $GABA_B$ agonist, is removed. Second, you saw a very strong depression, but certainly the paired-pulse depression in the inhibitory IPSC is not very remarkable in these cells, about 20% lower than in your case and in studies of the dentate gyrus. So regarding epilepsy in these networks, if you look from synapse to synapse, you will differences quantitative but also qualitative differences.

Buzsaki: I am just curious to know how reliable is the identification of layer 2 and layer 3 neurons by back firing. When you are stimulating the outer molecular layer obviously some of the current spreads to the molecular layer also. How do we know which cells we are recording from?

Heinemann: Projection cells in layer II and III are recognized based on a number of criteria. First is their location in layer II or III. Second, we identify them by antidromic activation by deep layer stimulation or angular bundle stimulation. Third, stellate cells are easily recognized due to their depolarizing and hyperpolarizing sag potentials and occurrence of membrane potential oscillations when depolarized to near threshold. Fourth, they can be identified by filling of cells with biocytin and subsequent reconstruction.

Schwartzkroin: Let me ask both you and Stan Leung, how should we think about comparing the data that both of you have presented given the different location of the kindling process and the different degree to which the kindling has proceeded?

Heinemann: I have learned from studies on dopamine, 5-HT, and so forth, that what you observe in the entorhinal cortex is often opposite to what happens in the hippocampus. In these kind of things you have to see this not as a sign of an isolated event, but you really have to put it into the network component. What we need is really realistic modelling of the whole of the network. Considering that this is a network, which unlike the neocortex may have only 2 or 3 working modes, here have very very different working modes. I think we have to broaden our view in this kind of change.

Leung: I agree with what you said. I have nothing really to add with respect to why one region may have an autoreceptor down-regulation and another have an increase. Could I ask a question since you mentioned network? I just want to clarify your view or your data on the entorhinal-CA1 or entorhinal-hippocampal transmission. Is there a difference between kindling and your status-epilepticus?

Heinemann: With respect to seizure spread was if we looked at low magnesium or focal application of bicuculine, high potassium, what we see was a marked difference in seizure spread. First off, the seizure susceptibility is more increased in the entorhinal cortex than the hippocampus proper. Second, if you induce in the middle and deep layers of the entorhinal cortex a seizure on the whole preparation, in the kindled tissue it always propagates through the dentate gyrus; so the kindling process removes this gating function that we originally proposed, and Lothman proposed. In contrast, what you have in pilocarpine and kainate status animals is that you have seizures spread through the dentate gyrus and also very often through the subiculum-CA1, and then you have occluding seizure waves which hit it in the CA3 region and then something very interesting happens to the seizures that collide.

NATURE AND CONSEQUENCES OF SEIZURES ORIGINATING IN THE BRAINSTEM

Shigeru Chiba, Takahiro Ishimoto, Yuji Ishimaru, Yoshiyuki Tamura, Akihiko Nunomura*

1. INTRODUCTION

Experimental studies suggest that the brainstem, particularly the brainstem reticular formation (RF), is involved in the secondary generalization of partial seizures originating in the forebrain. Electrolytic lesioning of the midbrain involving the medial part of the mesencephalic RF (MRF) prior to amygdala (AM) kindling retards the development of AM kindling.[10] Extensive midline bisection of the brainstem (the midbrain to the pons) can inhibit the development of a stage 5 AM kindled seizure.[8] In feline AM kindling, the emergence of afterdischarge (AD) in the ipsilateral and then contralateral midbrain reticular formation precedes generalization and synchronization of AD leading to the final stage 6 generalized convulsion.[45] In addition, a unilateral lesion (ipsilateral to the kindled AM) in the midbrain reticular formation not only elevates the established generalized seizure triggering threshold but also prevents recalling the kindled stage 6 seizures in cats.[45, 47] However, bisection of the forebrain commissures does not prevent stage 6 convulsive development.[46] In prefrontal kindling in epileptic Senegalese baboons, *Papio papio*, bisynchronous electroencephalographic discharges in the bulbar RF are associated with photically induced generalized convulsions and AM kindled generalized seizures.[44]

Further evidence has suggested that the brainstem RF is itself associated with the generation of primary generalized seizures. Electrical stimulation applied to the brainstem RF triggered tonic convulsions in rats,[5, 25] rabbits,[2] and cats,[25] whereas lesions within this area can attenuate the maximal electroshock seizure or pentylenetetrazol-induced seizure in rats.[3] In genetically epilepsy-prone rats, the propagation of seizure discharges from the inferior colliculus to the brainstem RF is considered to be important in the development of audiogenic seizures.[4, 15-17, 32] Furthermore, a brain chimera study on 12-somite stage embryos of Fayoumi chickens with hereditary reflex epilepsy in response

* Department of Psychiatry and Neurology, Asahikawa Medical College. Midorigaoka higashi 2-1-1-1, Asahikawa 078-8510, Japan

to either light or sound stimulation demonstrated that the mesencephalon contains the generator of the epileptic manifestations of running and generalized convulsions.[1, 31] To gain further insight into the role of the MRF in the generation of epileptic seizures, we focused on the nature and consequences of seizures originating in the brainstem.

2. CHEMICAL PROCONVULSANT MICROINJECTIONS TO MRF

An imbalance between inhibitory and excitatory neurotransmission is considered to be associated with the generation and expression of human[13, 14] and animal epileptic seizures[7, 11, 12, 18, 20-23, 28-30, 34, 38-41] in the forebrain.

In inhibitory neurotransmission, repeated administration of the γ-aminobutyric acid (GABA) A receptor antagonists picrotoxin[7] or bicuculline methiodide (BIC),[41] into the unilateral amygdala induces AM seizure in rats. Systemic administration of the GABA receptor agonist muscimol strongly suppresses AM kindled seizures[29] in rats.

Excitatory amino acids, such as N-methyl-D-aspartate (NMDA), Alpha-amino-3-hydroxy-5-methyl-4-isoxazolepropionic acid (AMPA), and kainic acid, are associated with excitatory neurotransmission. Systemic administration of NMDA,[27] or the focal administration of NMDA to the AM,[11, 12] massa intermedia,[21, 22] or substantia nigra pars reticulate,[39, 40] produces convulsive seizures. In addition, a competitive NMDA receptor antagonist, 3-(2-carboxypiperazine-4-yl)-propyl-1-phosphonic acid (CPP),[34] and a non-competitive NMDA receptor antagonist, dizocilpine (MK-801),[38] have potent inhibitory effects on the development of AM kindling in rats. AMPA-sensitive quisqualate receptor binding[23] and AMPA receptor mRNA[18] are both reported to increase in the epileptogenic hippocampus in the human brain. A selective AMPA receptor antagonist, 2,3-dihydroxy-6-amino-7-sulfamoyl-benzo(F)-quinoxaline,[29] has also been shown to have potent inhibitory effects on the development of AM kindling in rats. The focal injection of kainic acid into limbic structures[41] or into the MRF in rats and cats[20] also produces acute generalized convulsive seizures.

In order to examine whether diminution of inhibitory neurotransmission or potentiation of excitatory neurotransmission in the brainstem RF can produce epileptic seizures patterns, we microinjected the proconvulsants BIC, AMPA, and NMDA into the unilateral MRF and observed the behavioral and EEG changes in rats. BIC is a selective GABAA antagonist which was used to diminish GABAergic inhibitory neurotransmission, while AMPA and NMDA were used to potentiate excitatory amino acid neurotransmission.

2.1. BICUCULLINE

Sixteen male adult Sprague-Dawley rats (2 to 3 months of age) were used in the BIC study. Under pentobarbital anesthesia, bipolar electrodes made of twisted stainless steel wire (200 μm in diameter) were stereotaxically[35] inserted into the left AM (2.8 mm posterior, 5.0 mm lateral, and 9.0 mm ventral from the bregma). Three surface electrodes (stainless steel screws) were driven into the skull, two for recording from the motor cortices bilaterally and the remaining one for the reference electrode over the olfactory bulb unilaterally. In addition, chemitrodes, i.e., 24 G guide cannulas with bipolar electrodes made of twisted stainless steel wire (200 μm in diameter), were implanted into the left MRF (the deep mesencephalic nucleus; 5.8 mm posterior, 1.7 mm lateral, and 6.6 mm ventral from the bregma: n = 6) or the left pontine RF (PRF) (the pontine reticular nucleus, oral part; 7.8 mm posterior, 1.2 mm leteral, and 8.4 mm ventral from the bregma: n = 6). The tips of the bipolar electrodes extended 1.0 mm beyond the ends of the guide cannulas. Seven days after the operation, a single 20 nmol dose of BIC (Sigma St. Louis, U.S.A.) was administered into the MRF (the MRF group; n = 6) or the PRF (the PRF group; n = 6). The dose of BIC was dissolved in saline and delivered in a volume of 1.0 μl at a rate of 0.5 μl/minute by a microsyringe pump (Eicom corp., Kyoto, Japan, EP-60). The microinjections were performed with injection needles (30 G) which extended 1.0 mm beyond the ends of the guide cannulas. The remaining 4 of 16 rats received a saline injection (1.0 μl) in a manner identical to that of the MRF/PRF groups and were used as the control group (2 for the MRF and 2 for the PRF group). On completion of the experiment, the rats were deeply anesthetized and their brains were prepared for histological verification of BIC injection needle and depth electrode placements.

Histological examination showed that the BIC injection needles and depth electrodes were located in the intended area (within 0.5 mm of the target site).

The control group did not show any behavioral or electrographic changes during the 45 minutes from the beginning of the saline injection.The MRF group showed seizure patterns in the following order: (I) a running/bouncing seizure, and (II) a generalized tonic-clonic seizure (GTCS) which lasted longer than 10 min and finally involved both forelimb and partial/complete hindlimb. The seizure II tonic patterns were in the following order (from mild to severe symptoms): tonic flexion of neck and trunk, tonic extension of forelimb, and partial/complete extension of hindlimb. The mean (range) latencies from the beginning of the BIC injection to the seizures I and II were 1.88 (1.17-3.50) and 4.40 (2.08-7.33) min, respectively. The PRF group showed convulsive seizures that were identical to those of the MRF group, although their mean (range) latencies from the beginning of the BIC injection to the seizures I and II were slightly shorter than the respective latencies in the MRF group [1.33 (0.83-2.00) and 3.85 (1.67-8.83) min, respectively]. However, there was no significant difference in the latencies between the two groups (Mann-Whitney U test). There were no differences in the features of the electroencephalographic seizure discharges between the two groups. One rat of the MRF group and 3 rats of the PRF group died at 19 to 23 min after the beginning of the BIC injection while showing seizure II. The remaining rats (5 of 6 in the MRF and 3 of 6 in the PRF group) received treatment for seizure II with pentobarbital (40 mg/kg, i.p.) and survived thereafter. Representative electroencephalograms of the seizures I and II in the MRF group are shown in Figure 1. Electroencephalographic seizure discharges were observed in the subcortical structures (the MRF and the AM) during the seizure I. However, after the seizure II appeared, cortical involvement in the seizure discharges was apparent.

A

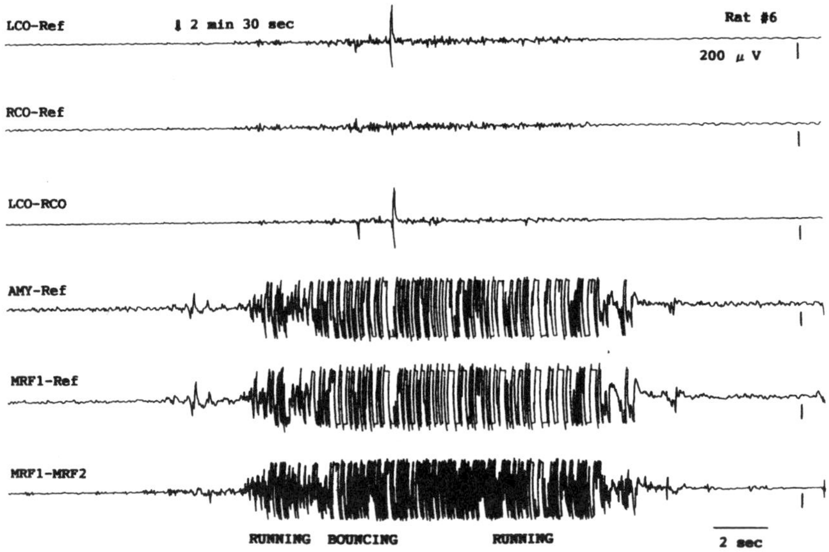

B

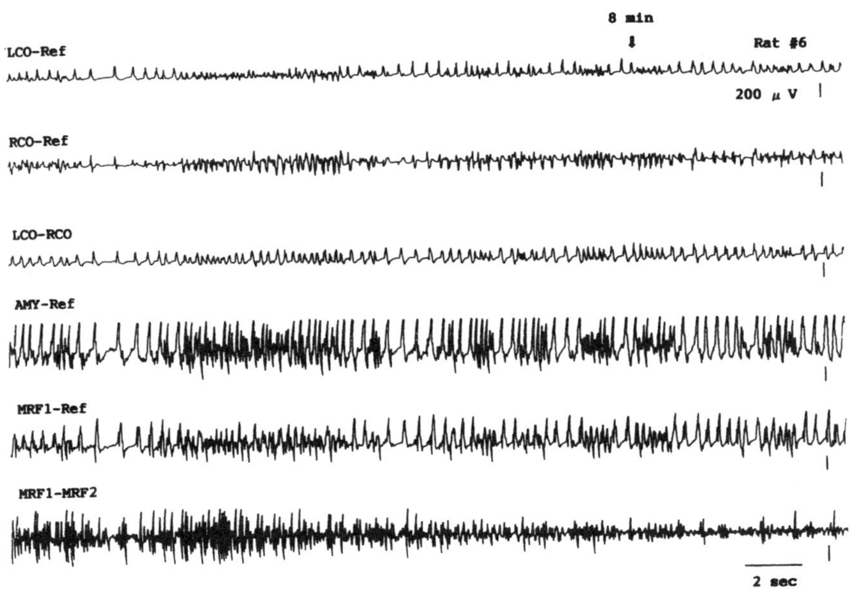

Figure 1. Electroencephalograms of a running/bouncing seizure (A) and a GTCS (B) in a MRF group rat. LCO, left motor cortex; RCO, right motor cortex; AMY, amygdala; MRF, mesencephalic reticular formation; Ref, reference electrode.

2.2. AMPA

Forty-five male adult Sprague-Dawley rats were used in the AMPA study. The rats were randomly assigned to Group A (n = 15), B (n = 15), or C (n = 15). The methods of electrode (in the motor cortices, AM, MRF) and MRF chemitrode implantation for the MRF were identical to those of the BIC study described above.

Seven days after the operation, a single 10 nmol dose of AMPA (Sigma, St Louis, MO, USA) was administered into the MRF in Group A, while Group B received only a 2 nmol dose. The AMPA dose was dissolved in saline and delivered in a volume of 1.0 μl at a rate of 1.0 μl/min by a microsyringe pump (Eicom corp., Kyoto, Japan, EP-60). The microinjections were performed with 30 G needles extending 1.0 mm beyond the ends of the guide cannulas. Group C received a saline injection (1.0 μl) into the left MRF in a manner identical to that carried out in Groups A and B. Behavioral and EEG changes were recorded for 45 min from the end of the AMPA/saline microinjections.

We applied 60 sec of sound stimulation to the rats of Groups A, B, and C at 15, 30, and 45 min from the end of the AMPA or saline microinjections. Although there are several sound stimulation methods using key jingling, in Exp. 2 we used a short manual shake of a bunch of keys (6 metal door keys on a metal key-ring) held at 50 cm above the floor of an open-topped observation box (35 cm × 35 cm × 35 cm);[24] the frequency and intensity of the sound is reported elsewhere.[24] On completion of Exp. 2, the animals were deeply anesthetized and their brains were subjected to perfusion fixation with 10% formalin, and subsequently cut into 10 μm thick frozen sections to histologically confirm the position of the depth electrode. Statistical comparisons were made using Fisher's exact probability test.

Histological examination revealed that the chemitrodes and the depth electrodes were located in the intended area and within 0.5 mm of the target site in all the rats used.

Microinjection of the 2 nmol dose of AMPA induced only hyperactivity and running/circling. Microinjection of the 10 nmol dose of AMPA induced seizure patterns in the following order: (1) hyperactivity (a state in which a rat is frequently moving or restlessly walking around in the observation box); (2) a running/circling seizure which consisted of sudden running forward and circling in the open-topped observation box without any regular direction; (3) a GTCS which finally involved both forelimb and partial/complete hindlimb extension; and (4) amygdala kindling-like seizures (AMKS) which consisted of facial movements, head nodding, bilateral forelimb clonus, and rearing. Electroencephalographic seizure discharge was observed in the MRF and AM during hyperactivity and running/circling seizures, and in the MRF, AM, and cortices during a GTCS. In contrast, seizure discharge was observed predominantly in the AM during AMKS, although discharge was also seen in the MRF and cortices (Figure 2). AMKS was observed in 8 rats that exhibited hyperactivity and running/circling in Group A, but was not seen earlier than hyperactivity, running/circling, or GTCS. Additionally, AMKS appeared without exception within 15 min of the rats receiving AMPA microinjection. Group C did not show any behavioral or electroencephalographic changes during the 15 min following the end of the saline injection. The incidence of hyperactivity and running/circling was significantly higher in Groups A and B than in Group C: A, 15/15 and 13/15; B, 15/15 and 10/15; C, 0/15 and 0/15, respectively: ($p<0.01$ by Fisher's probability exact test). The incidence of GTCS and AMKS was

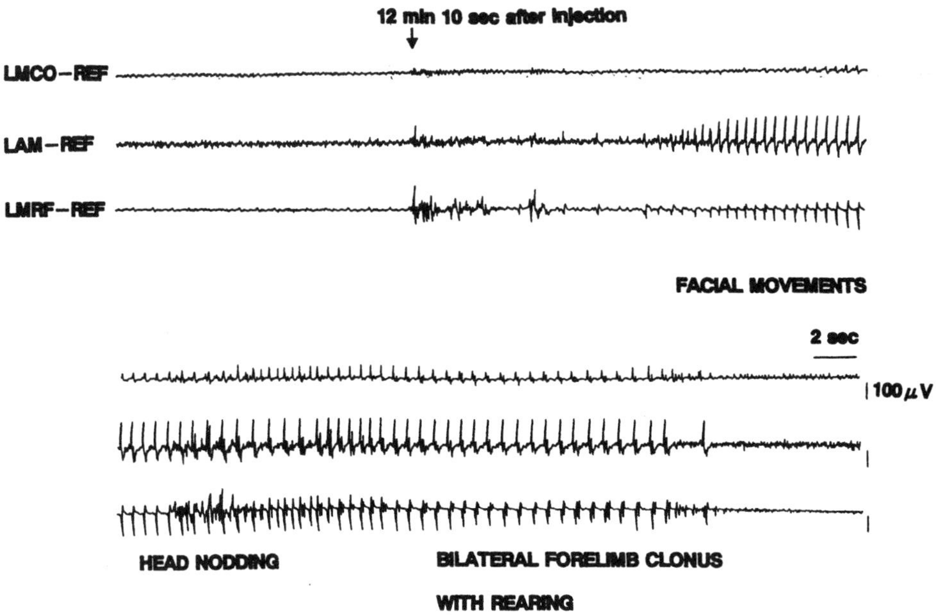

Figure 2. Electroencephalograms of AMKS in a Group A rat. AMKS, i.e., facial movements, head nodding and bilateral forelimb clonus with rearing appeared initially at 12 min 10 s after AMPA microinjection.

significantly higher in Group A than in Groups B and C: A, 4/15 and 8/15; B, 0/15 and 0/15; C, 0/15 and 0/15, respectively ($p<0.05$ and 0.01 by Fisher's probability exact test, respectively). AMPA-induced seizures were not observed during the period from 15 to 45 min after the end of AMPA injections.

Sound stimulation applied at 15, 30, and 45 min after AMPA injections induced hyperactivity (Groups A and B), running/circling (only Group A), and GTCS (only Group A), but did not produce AMKS. The EEG findings observed during sound stimulations in Exp. 2 were similar to those observed for 15 min after AMPA injection. In Group C, no behavioral or EEG changes were elicited by sound stimulation. The incidence of hyperactivity, running/circling and GTCS was significantly higher in Group A (15/15, 14/15, and 6/15, respectively) than in Groups B (5/15, 0/15, and 0/15, respectively) and C (0/15, 0/15, and 0/15, respectively).

2.3. NMDA

The effects of microinjections of a single 2 or 10 nmol dose of NMDA into the unilateral MRF (the deep mesencephalic nucleus) on behavior and the EEG were examined in male adult Sprague-Dawley rats (n = 18) during a 15 min period (Exp. 1), and subsequent effects of sound stimulation with key jingling (described above) applied at 15, 30, and 45 min after the injections were observed (Exp. 2).[24] The methods of implantation of electrodes and MRF chemitrodes were identical to those of our BIC study described above.

Histological examination revealed that the chemitrodes and the depth electrodes were located in the intended area (within 0.5 mm of the target site) in all the rats used.

Microinjection of the 2 nmol dose of NMDA (n = 10) induced hyperactivity (9/10 rats) and running/circling (8/10 rats) in Exp. 1, and hyperactivity (3/10 rats) in Exp. 2. Moreover, microinjection of the 10 nmol dose of NMDA (n = 8) induced not only hyperactivity (8/8 rats) and running/circling (7/8 rats) but also a GTCS which finally involved both forelimb and partial/complete hindlimb extension (5/8 rats) in Exp. 1; these seizure patterns were also elicited by sound stimulation in Exp. 2. The seizure patterns were accompanied by electroencephalographic seizure discharge in the MRF and the motor cortex (in the MRF predominantly). In contrast, the control group (n = 10), which received a single dose of saline microinjection into the unilateral MRF, showed no behavioral or electroencephalographic changes in both Exp. 1 and 2. Additionally, in Exp. 1 and 2, no rats in Groups A or B showed the AMKS that was observed in our AMPA study described above[24].

3. KINDLING OF MRF AND SUBSEQUENT AM KINDLING

Twelve male adult Sprague-Dawley rats were used in the kindling study (MRF stimulation group, n = 6; MRF non-stimulation group, n = 6). Under pentobarbital anesthesia, bipolar electrodes made of twisted stainless steel wire, 200 μA in diameter, were stereotaxically[35] implanted into the unilateral MRF (the deep mesencephalic nucleus; 5.8 mm posterior, 1.7 mm lateral, and 6.6 mm ventral from the bregma) and the AM (2.8 mm posterior, 5.0 mm lateral, and 9.0 mm ventral from the bregma) ipsilateral to the MRF electrode. Two surface electrodes (stainless steel screws) were inserted into the skull over the ipsilateral motor cortex and cerebellar cortex (the latter was used as the reference electrode).

Two weeks after the operation, the AD threshold of the MRF was determined by applying 1-sec stimulation beginning at 100 μA (60 Hz biphasic square pulses). Every 24 hours, stimulation was increased by 100 μA until AD was induced. The first stimulus producing an AD was arbitrarily designated as AD threshold. The MRF was stimulated once a day at the AD threshold until 3 consecutive generalized tonic-clonic seizures of at least 50 sec in duration were induced. Twenty-four hours after the last MRF stimulation, 3 of 6 rats in the MRF stimulation group received ipsilateral AM stimulation once a day at the AD threshold until 3 consecutive stage 5 seizures[37] were induced. The AD threshold of the AM was determined by applying 1-sec stimulation beginning at 100 μA (60 Hz biphasic square pulses). Every 24 hours, stimulation was increased by 100 μA until AD was induced. Rats without MRF stimulation were used as the MRF non-stimulation group. MRF restimulation at the AD threshold was performed 24 hours after AM kindling in 3 rats and 30 days after kindling in 2 rats from the MRF stimulation group. On completion of the experiments, all animals were deeply anesthetized and their brains were perfused, serially sectioned, and stained by hematoxylin and eosin.

Histological examination confirmed that the depth electrodes were located in the intended area. The mean AD threshold at the MRF was 317 μA (range 200-500 μA) and the mean number of ADs required to complete MRF kindling was 12.8 (range 5-22). As shown in Table 1, daily MRF stimulations at the AD threshold produced a progressive change in the seizure pattern, from generalized tonic to generalized tonic-clonic seizure, and progressive AD growth involving the motor cortex. A representative example of seizure development during MRF kindling is shown in Figure 3.

The AM kindling profile of the MRF stimulation and non-stimulation groups is shown in Table 2. All the rats in each group showed AD at the intensity of 100 μA, with no statistical difference between the two groups. The MRF stimulation group reached

Table 1. MRF kindling profile

Clinical seizures	AD MRF	AD AMY	AD MCO	AD duration (sec, mean with range)
the first seizure: generalized tonic	+	+	–	17.8 (8-27)
the last seizure: generalized tonic-clonic	+	+	+	66.8 (50-88)

AD, afterdischarge; MRF, mesencephalic reticular formation; AMY, amygdala; MCO, motor cortex.
+, AD is present; –, AD is not present. The mean (range) number of MRF stimulations: 12.8 (5-22).

Table 2. AM kindling profile of the MRF stimulation and non-stimulation groups.

Groups	Number of stimulations to reach stage 5	AD duration (sec) at the first stage 5	No. of rats with a PLPC
MRF stimulation (n = 3)	4.0 (3-5) *	42.3 (32-50)	3/3 **
MRF non-stimulation (n = 6)	9.2 (6-12)	65.5 (42-86)	0/6

AD, afterdischarge; PLPC, a prolonged loss of postural control; MRF, mesencephalic reticular formation.
*p<0.05 by Mann-Whitney U test, **p<0.05 by chi-square test.

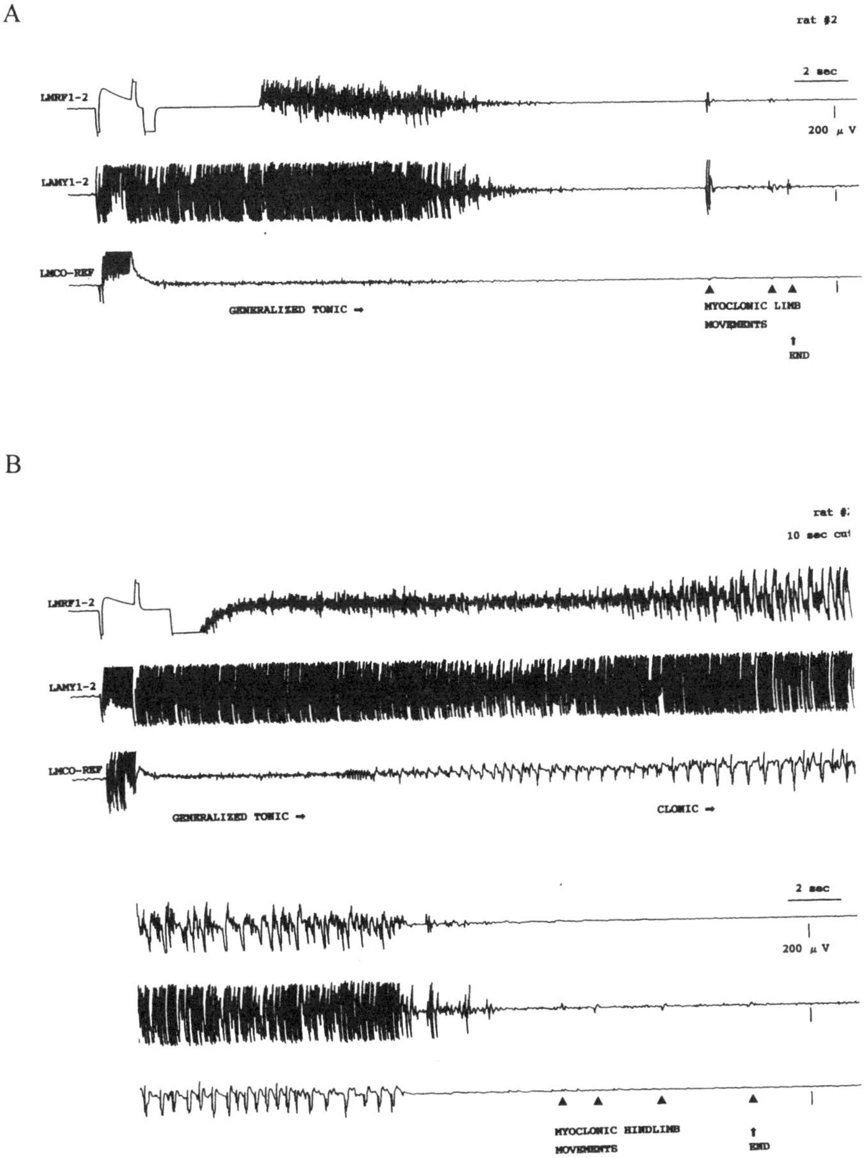

Figure 3. A representative seizure development of MRF kindling. The rat displayed AD 17 times.

A: The first stimulation (AD threshold 500 μA). The rat showed a brief generalized tonic seizure with myoclonic limb movement (AD = 25 sec).

B: The last (17th) stimulation (AD threshold 500 μA). The rat showed a prolonged tonic-clonic seizure with myoclonic limb movement (AD = 55 sec).

LMRF, left mesencephalic reticular formation; LAMY, left amygdala; LMCO, left motor cortex; REF, reference electrode.

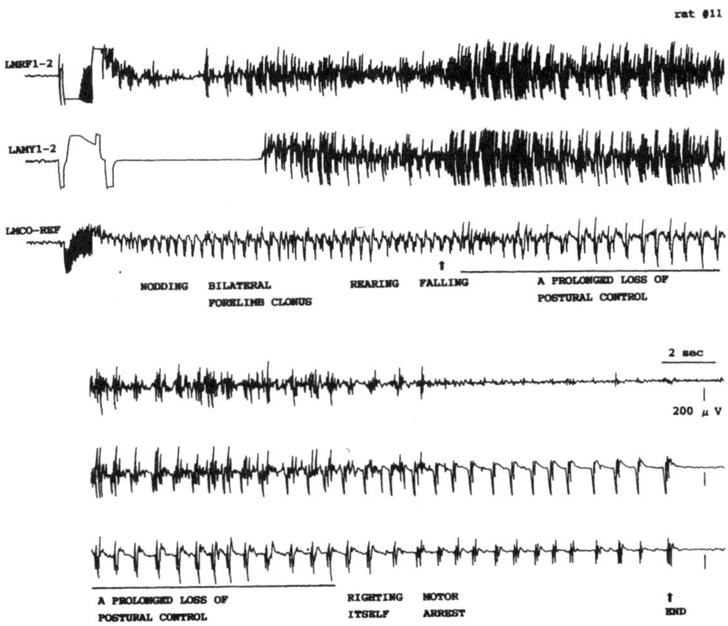

Figure 4. Stage 5 Amygdala (AM)-kindled seizure in an MRF kindled rat (the third stimulation of left AM). The rat developed a stage 5 seizure at the third stimulation, with a generalized tonic seizure and a prolonged loss of postural control (20 sec).

stage 5 significantly faster than the non-stimulation group. There was no significant difference between the two groups in the AD duration during the first stage 5 seizure. However, all rats in the MRF group exhibited a prolonged loss of postural control (>4 sec, range 5-18 sec) with generalized tonic seizures during the stage 5 seizure, whereas none of the no-stimulation group rats did.

Both at 24 hours and 30 days after AM kindling, MRF restimulation induced a generalized tonic-clonic seizure that was similar to the final generalized tonic-clonic seizure of the MRF kindling session in all rats in the MRF kindling group.

4. DISCUSSION AND CONCLUSIONS

4.1. Microinjection study

In our study, microinjections of the GABAA antagonist BIC (20 nmol) into the unilateral MRF/PRF induced seizure patterns in the following order: (I) a running/bouncing seizure, and (II) a GTCS lasting longer than 10 min. There were no significant differences in the features of the seizure patterns between the MRF and PRF groups. The final form of behavioral seizures was accompanied by EEG seizure discharge in the motor cortex as well as the MRF and AM. These findings suggest that the brainstem RF has an important role in the development of a running/bouncing seizure and GTCS, and that blocking of GABAergic inhibitory neurotransmission in the brainstem RF participates in seizure development.

In addition, microinjections of the excitatory amino acids AMPA or NMDA induced seizure patterns in the following order: (1) hyperactivity, (2) a running/circling seizure, (3) a GTCS which finally involved both forelimb and partial/complete hindlimb extension, and (4) AMKS which consisted of facial movements, head nodding, bilateral forelimb clonus, and rearing (only by AMPA injection). Sound stimulation applied at 15, 30, and 45 min after AMPA or NMDA injections could induce hyperactivity, running/circling, and GTCS, but did not produce AMKS.

Tonic seizures are considered to depend on the brainstem but not the forebrain. Although the specific substrates within the brainstem responsible for producing tonic seizures have not been elucidated, the brainstem RF (the midbrain to the pons) is likely to be involved in the generation of tonic seizures.[2-5, 8, 10, 24, 25] It is assumed that the propagation of seizure discharge from the inferior colliculus to the brainstem RF is crucial for the development of audiogenic seizures (running or generalized tonic seizure) and that the brainstem RF neurons play a major role in generation of audiogenic seizure in genetically epilepsy-prone rats.[15-17] It is therefore suggested that the MRF has an important role in the development of GTCS, which follows hyperactivity and running/circling, and that potentiation of excitatory neurotransmission in the MRF participates in the development of audiogenic seizures as well as GTCS.

The development of GTCS, in which the cortico-reticular system is considered to be involved, seems to be more easily produced by excitation of NMDA rather than AMPA neurotransmission. This speculation is not inconsistent with the fact[47] that in rat audiogenic seizures the development of GTCS is mainly controlled by excitation of NMDA rather than AMPA neurotransmission. In contrast, AMKS was induced by microinjection of AMPA, and not NMDA, into the MRF, probably due to excitation of the limbic-brainstem connection.

4.2. Kindling Study

Goddard et al.[19] reported that 200 daily electrical stimulations (50 μA, 60 Hz biphasic square pulses, 60 sec) of the brainstem (midbrain periaqueductal gray, MRF, substantia nigra, etc.) neither produced a convulsive response nor established kindling. However, recent studies have shown that the inferior colliculus,[27] the interpeduncular nucleus,[9] the dorsal tegmentum,[26] and the midbrain periaquaductal gray[33] can be kindled in rats.

A previous study[5] indicates that electrical stimulation (1100-1500 μA, 10 sec) to the MRF produce generalized tonic seizure without EEG seizure discharge in rats. However, we were able to induce not only the seizure but also EEG seizure discharge. The reason for this difference is not clear, but it is possible that the stimulated site in our study is different from that in the previous study.[5]

Our study revealed that MRF stimulated rats showed a progressive change in the behavioral seizure pattern and progressive AD growth involving the motor cortex. Furthermore, the MRF restimulation experiment demonstrated that the increased seizure susceptibility established by MRF kindling was persistent. Therefore, it is indicated that the MRF can be kindled effectively. In the AM kindling experiment, MRF-kindled rats reached stage 5 significantly faster than the MRF non-stimulation group, suggesting that MRF kindling has a facilitatory influence on subsequent AM kindling. This finding supports the hypothesis that the vertical (limbic-brainstem) rather than horizontal pathway (interlimbic pathway through the forebrain commissures) is crucial for the development of AM kindling.[6, 7, 36, 43, 45] Moreover, the MRF-kindled group had a significantly higher incidence of tonic seizures associated with a prolonged loss of postural control as compared with the non-stimulation group during stage 5 seizures. This fact suggests that AM kindling utilizes the increased seizure susceptibility of the brainstem induced by previous MRF kindling.

4.3. Conclusion

On the basis of our study, it is assumed that nature of seizures originating in the MRF consists of varied patterns in the following order (from mild to severe symptoms): hyperactivity, a running/circling/bouncing seizure, tonic seizure, a tonic-clonic seizure, and amygdala kindling-like seizures with EEG seizure discharge involving forebrain structures. The MRF can be kindled, and subsequent AM kindling utilizes not only the strengthened vertical (limbic-brainstem) connection but also the proconvulsant neuroplastic changes that have been already established by MRF kindling.

5. REFERENCES

1. C. Batini, M-A. Teillet, R. Naquet, and N. M. Le Douarin, Brain chimeras in birds: application to the study of a genetic form of reflex epilepsy, *Trends Neurosci.* **19**, 246-252 (1996).
2. F. Bergmann, A. Costin, and J. Gutman, A low threshold convulsive area in the rabbit's mesencephalon, *Electroencephalogr. Clin. Neurophysiol.* **15**, 683-690 (1963).
3. R. A. Browning and D. K. Nelson, Modification of electroshock and pentylenetetrazol seizure patterns in rats after precollicular transactions, *Exp. Neurol.* **93**, 546-556 (1986).
4. R. A. Browning, C. Wang, and C. L. Faingold, Effect of norepinephrine depletion on audiogenic-like seizures elicited by microinfusion of an excitant amino acid into the inferior colliculus of normal rats, *Exp. Neurol.* **112**, 200-205 (1991).
5. W. M. Burnham, P. Albright, J. Schneiderman, P. Chiu, and T. Ninchoji, "Centrencephalic"mechanisms in the kindling model, in: *Kindling 2*, edited by J. A. Wada (Raven Press, New York,1981), pp. 161-178.
6. D. P. Cain, Transfer of kindling: clinical relevance and a hypothesis of its mechanism, *Prog. Neuropsychopharmacol. Biol. Psychiatry* **9**, 467-472 (1985).
7. D. P. Cain, Kindling by repeated intraperitoneal or intracerebral injection of picrotoxin transfers to electrical kindling, *Exp. Neurol.* **97**, 243-254 (1987).
8. S. Chiba and J. A. Wada, Amygdala kindling in rats with brainstem bisection, *Brain Res.* **682**, 50-54 (1995).
9. Chiba, S. and Wada, J.A., Kindling of the interpeduncular nucleus and its influence on subsequent amygdala kindling in rats, *Epilepsia* **36**, 410-415 (1995).
10. S. Chiba and J. A. Wada, The effect of electolytic lesioning of the midbrain prior to amygdala kindling in rats, *Neurosci. Lett.* **227**, 83-86 (1997).
11. M. J. Croucher, K. L. Cotterell, and H. F. Bradford, Amygdaloid kindling by repeated focal N-methyl-D-aspartate administration: comparison with electrical kindling, *Eur. J. Pharmacol.* **286**, 265-271 (1995).
12. M. J. Croucher, K. L. Ruffle, and H. F. Bradford, The effects of focal N-methyl-D-aspartate pretreatment on the parameters of amygdaloid electrical kindling, *Eur. J. Pharmacol.* **319**, 207- 213 (1997).

13. M. J. During, K. M. Ryder, and D. D. Spencer, Hippocampal GABA transporter function in temporal-lobe epilepsy, *Nature* **376**, 174-177 (1995).
14. M. J. During and D. D. Spencer, Extracellular hippocampal glutamate and spontaneous seizure in the conscious human brain, *Lancet* **341**, 1607-1611 (1993).
15. C. L. Faingold, Neuronal network in the genetically epilepsy-prone rat, in: *Advances in Neurology*, vol. 79, edited by A. V. Delgado-Escueta, W. A. Wilson, R. W. Olsen, and R. J. Porter (Lipppincott Willams & Wilkins, Philadelphia, 1999), pp. 311-321.
16. C. L. Faingold, M. J. Marcinczyk, D. J. Casebeer, M. E. Randall, S. P. Arneric, and R. A. Browning, GABA in the inferior colliculus plays a critical role in control of audiogenic seizures, *Brain Res.* **640,** 40-47 (1994).
17. C. L. Faingold and M. E. Randall, Pontine reticular formation neurons exhibit a premature and precipitous increase in acoustic responses prior to audiogenic seizures in genetically epilepsy-prone rats, *Brain Res.* **704**, 218-226 (1995).
18. F. J. Garcia-Ladona, J. M. Palacios, A. Probst, H. G. Wieser, and G. Mengod, Exitatory amino acid AMPA receptor mRNA localization in several resions of normal and neurological disease affected human brain. An in situ hybridization histochemistry study, *.Mol.. Brain Res.* **21**, 75-84 (1994).
19. G. V. Goddard, D. C. Mclntyre, and C. K. Leech, A permanent change in brain function resulting from daily electrical stimulation, *Exp. Neurol.* **25**, 295-330 (1969).
20. K. Hashizume, T. Tanaka, T. Fujita, and S. Tanaka, Generalized seizures induced by an epileptic focus in the mesencephalic reticular formation: Impact on the Understanding of the generalizing mechanism, *Stereotact. Funct. Neurosurg.* **74**, 153-160 (2000).
21. Y. Hirayasu and J. A. Wada, Convulsive seizures in rats induced by N-methyl-D-aspartate injection into the massa intermedia, *Brain Res.* **57**, 36-40 (1992).
22. Y. Hirayasu and J. A. Wada, N-methyl-D-aspartate injection into the massa intermedia facilitates development of limbic kindling in rats, *Epilepsia* **33**, 965-970 (1992).
23. D. A. Hosford, B. J. Crain, Z. Cao, D. W. Bonhaus, A. H. Friedman, M. M. Okazaki, J. V. Nadler, Jand J. O. McNamara, Increase AMPA-sensitive quisqualate receptor binding and reduced NMDA receptor binding in epileptic human hippocampus, *J. Neurosci.* **11**, 428-34 (1991).
24. T. Ishimoto, N. Omori, F. Mutoh, and S. Chiba, Convulsive seizures induced by N-methyl-D-aspartate microinjection into the mesencephalic reticular formation in rats, *Brain Res.* **881**, 152-158 (2000).
25. A. Kreindler, E. Zuckermann, M. Steriade, and D. Chimion, Electroclinical features of convulsions induced by stimulation of brainstem, *J. Neurophysiol* **21**, 430-436 (1958).
26. B. Maton, E. Hirsch, M. Vergnes, A. Depaulis, and C. Marescaux, Dorsal tegmentum kindling in rats, *Neurosci. Lett.* **134**, 284-287 (1992).
27. T. J. McCown, R. S. Greenwood, G. R. Frye, and G. R. Breese, Electrical elicited seizure from the inferior colliculus: a potential site for the genesis of epilepsy?, *Exp. Neurol.* **86**, 527-542 (1984).
28. N. Mori and J. A. Wada, Bidirectional transfer between kindling induced by excitatory amino acids and electrical stimulation, *Brain Res.* **425**, 45-48 (1987).
29. K. Morimoto, S. E. Mason, and G. V. Goddard, Kindling-induced changes in the EEG recorded during stimulation from the site of stimulation. III. Direct pharmacological manipulations of the kindled amygdala, *Exp. Neurol.* **97**, 17-34 (1987).
30. T. Namba, K. Morimoto, K. Sato, N. Yamada, and S. Kuroda, Antiepileptogenic and anticonvulsant effects of NBQX, a selective AMPA receptor antagonist, in the rat kindling model of epilepsy, *Brain Res.* **638**, 36-44 (1994).
31. R. Naquet and A. Valin, Experimental models of reflex epilepsy, in: *Advances in Neurology*, vol.75, edited by B. G. Zifkin, F. Andermann, A. Beaumanoir, and A. J. Rowan (Lippincott Willams & Wilkins, Philadelphia, 1998), pp. 311-321.
32. P. N'Gouemo and C. L. Faingold, Periaqueductal gray neurons exhibit increased responsiveness associated with audiogenic seizures in the genetically epilepsy-prone rat, *Neuroscience* **84**, 619-625 (1998).
33. N. Omori, T. Ishimoto, F. Mutoh, and C. Chiba, Kindling of the midbrain periaqueductal gray in rats, *Brain Res.* **903**, 162-167 (2001).
34. M. Osonoe, N. Mori, S. Hoshino, Y. Yamada, K. Osonoe, H. Kittaka, and Y. Iwata, The effects of N-methyl-D-aspartate (NMDA) and its competitive antagonist,3-(2-carboxypiperazine-4-yl)-propyl-

1-phosphonic acid (CPP),injected into caudate-putamen on kindled amygdaloid seizures in rats, *Brain Res.* **728**, 242-246 (1996).
35. G. Paxinos and C. Watson, *The Rat Brain in Stereotaxic Coordinates*, 2nd edition (Academic Press,Sydney,1986).
36. J. P. J. Pinel, Spontaneous kindled motor seizures in rats, in: *Kindling 2*, edited by J. A. Wada (Raven Press, New York, 1981), pp. 179-192.
37. R. J. Racine, Modification of seizure activity by electrical stimulation: motor seizures, *Electroencephalogr. Clin. Neurophysiol.* **32**, 281-293 (1972).
38. K. Sato, M. Morimoto, and M. Okamoto, Anticonvulsant action of a non-competitive antagonist of NMDA receptors(MK-801)in the kindling model of epilepsy, *Brain Res.* **463**, 12-20 (1988).
39. C. Shin, F. A. Scialabba, and J. O. McNamara, Stimulation of substantia nigra pars reticulata enhances dentate granule cell excitability, *Brain Res.* **411**, 21-7 (1987).
40. C. Shin, J. M. Silver, D. W. Bonhaus, and J. O. McNamara, The role of substantia nigra in the development of kindling: pharmacologic and lesion studies, *Brain Res.* **412**, 311-317 (1987).
41. T. Tanaka, S. Tanaka, T. Fujita, K. Takano, H. Fukuda, K. Sako, and Y. Yonemasu, Experimental complex partial seizures induced by a microinjection of kainic acid into limbic structures, *Prog.. Neurobiol.* **38**, 317-334 (1992).
42. S. Uemura and H. Kimura, Amygdaloid kindling with bicuculline methiodide in rats. *Exp. Neurol.* **102**, 346-353 (1998).
43. J. A. Wada and T. Osawa, Spontaneous recurrent seizure state induced by daily electric amygdaloid stimulation in Senegalese baboons (Papio papio), *Neurology* **26**, 273-286 (1974).
44. J. A. Wada, T. Osawa, and T. Mizoguchi, Recurrent Spontaneous Seizure State Induced by Prefrontal Kindling in Senegalese Baboons, Papio Papio, in: *Kindling*, edited by J. A. Wada (Raven Press, New York, 1976), pp. 173-202.
45 J A. Wada and M. Sato, Generalized convulsive seizures induced by daily electrical stimulation of the amygdala in cats: Correlative electrographic and behavioral features, *Neurology* **24**, 565-574 (1974).
46. J. A. Wada and M. Sato, The generalized convulsive seizure state induced by daily electrical stimulation of the amygdala in split brain cats, *Epilepsia* **16**, 417-430 (1975).
47. J. A. Wada and M. Sato, Effects of unilateral lesion in the midbrain reticular formation on kindled amygdaloid convulsion in cats, *Epilepsia* **16**, 693-697 (1975).
48. S. Yasuda, N. Ishida, A. Higashiyama, and N. Kato, Running fit and generalized tonic-clonic seizure are differently controlled by different subtype receptors in the brainstem, *Jpn. J. Psychopharmacol.* **18**, 123-126 (1998), in Japanese.

Discussion

Wada: It is a beautiful study. It reminds me of a an observation we made many years ago in cats, when Dr. Sato was with us, that when cat was kindled you could precipitate a kindled seizure without any problem following primary site kindling. Then we placed a lesion in the lateral MRF, and then we could not evoke a kindled seizure at all. We continued to stimulate for something like 16 or 20 days, and no kindled seizure. Secondly, we did concurrent MRF stimulation and amygdaloid kindling, and it accelerated amygdala kindling tremendously. If I remember correctly, something like 5 to 8 amygdala stimulations kindled the animal and practically all concurrent stimulated animals developed recurrent spontaneous seizures subsequently, which is most unusual in my experience. Have you noticed any spontaneous seizures in your MRF-AM kindled animals?

Chiba: We expected this, but did not observe it. So in MRF-kindled rats spontaneous seizure hardly occurred. We did not carry out the simultaneous stimulation of the MRF and AM.

Wada: Did you check for positive transfer in the rats?

Chiba: No not yet.

Burnham: Congratulations on a lovely study. I couldn't see from the back of the room – were you getting AD in the brainstem, and were you recording bipolar or monopolar?

Chiba: Bipolar, and yes I saw AD in the brain stem. I read your paper from 1981, where you reported that you cannot induce AD in the MRF, but the difference is the stimulated sites. So, we can induce AD in the brain sites as you see. The deep mesencephalic nucleus can produce the AD.

Velisek: One comment, two questions. The description of generalized tonic-clonic seizures you provide on your slides reminded me about the seizure pattern we have seen after systemic NMDA injection in baby rats. The questions is, when you are showing the transfer from generalized tonic-clonic seizures to amygdala-like kindling seizures for AMPA, you were showing it for 10nmol dose, but could it be a function of the dose? What was the result with 2nmol?

Chiba: What kind of function?

Velisek: The transfer from GTCS to amygdala-like seizures. My question is how is it with 2nmol dose?

Chiba: Not a very difficult question. If you kindle the inter-peduncular nucleus you can see the GTCS is followed by amygdala kindling-like seizure. So this pattern, namely GTCS converted to amygdala kindled-like seizures, is not so rare in the brain stem seizures.

Velisek: The last question is, you were infusing 1microliter of the substances, quite a huge amount. Do you have a control for the spread?

Chiba: don't think that 1 microliter is too large.

Shouse: I just wanted to remind you that Augusto Fernandez-Guardiola showed sometime ago that another site in the brain stem that you can get AD is the dorsal raphe nucleus.

Chiba: I have not tried the raphe yet.

INVOLVEMENT OF THE CLAUSTRUM AND VENTROMEDIAL THALAMUS IN EPILEPTOGENESIS

Aaron Sheerin, Xia Zhang, Deb Saucier and Michael Corcoran*

1. INTRODUCTION

Characterizing the differences in kindling rates and associated convulsive behaviors following kindling of various structures may aid in understanding the structural and functional mechanisms of epileptogenesis. Overall, the structures and pathways that are critical to kindling are largely unknown. The results of recent research suggest a role for the CLA, namely that the CLA may be part of a network of structures involved in seizure generalization. The anterior CLA is highly susceptible to kindling, involving a pattern of dual-phase seizure development.[1]

In the same study, Zhang et. al.[1] revealed a strong projection to the nucleus submedius (NSUB) from the anterior CLA. The anatomical connections of the anterior CLA with various regions thought to play an important role in epileptogenesis, such as the amygdala (AM), motor cortex, frontal cortex, limbic cortex, and midbrain structures[1] are consistent with its high susceptibility to kindling and support the view that it is part of a network of structures participating in the generalization of seizures. However, the involvement of the NSUB in kindling remains to be examined.

If the anterior CLA and NSUB play a role in a seizure generalization network, several predictions are generated. First, the state of kindled seizure susceptibility produced by kindling should persist over an interval without stimulation. Second, lesions restricted to the anterior CLA or NSUB should retard the development of generalized seizures from other sites. Third, one might expect that kindling from the anterior CLA or NSUB would result in transfer. In the following experiments, we tested these predictions.

---

* Aaron Sheerin and Deb Saucier, Department of Psychology, University of Saskatchewan, Saskatoon, SK, Canada. Xia Zhang, Neuropsychiatry Research Unit, University of Saskatchewan, Saskatoon, SK, Canada. Michael Corcoran, Department of Anatomy and Cell Biology, University of Saskatchewan, Saskatoon, SK, Canada.

2. METHODS

2.1 General Methodology

Male Long-Evans rats (Charles River, Quebec), weighing 275-375 g at the time of surgery, were used. Food and water were freely available, and procedures were conducted during the light portion of the 12-hr light/dark cycle. All procedures were conducted in strict accordance with the guidelines established by the Canadian Council on Animal Care as approved by the University of Saskatchewan Animal Care Committee.

Rats received stereotaxic implantation of two kindling electrodes with a level skull. A reference wire and four additional dental screws were anchored to the skull, and the electrode assembly was affixed to the skull with dental acrylic. Bilateral radiofrequency (RF) lesions were produced by brief application of a train of RF current (Radionics, Model RFG-4A, Burlington, MA) at approximately 5 mA for 90 sec via two lesion electrodes during the implantation surgery.

One week after surgery, afterdischarge thresholds (ADT) were determined at one of the two kindling electrodes. In preliminary experiments, electrical stimulation was generated on a Grass S88 stimulator and consisted of a 1-sec train of balanced biphasic square-wave pulses 1 msec in duration and delivered at 60 pps to the AM at an initial current of 20 μA (base to peak) and increasing in increments of 10 μA at 1 min intervals, alternating between left and right kindling electrodes, until an AD was evoked from one. Following the completion of kindling, rats were killed with CO_2. Brains were removed and fixed in 10% formalin for at least 1 wk before sectioning. Frozen coronal sections 40 μm thick were taken from the regions of the electrode tracks and lesions and stained with cresyl violet.

2.2. Contribution of the anterior CLA to epileptogenesis

2.2.1. Persistence of anterior CLA kindling

In order to test persistence of kindling in the anterior CLA, daily stimulation was suspended after five generalized stage 5 seizures were evoked. After an arbitrarily chosen interval of 14 days, kindling resumed at the initial ADT, using the parameters described above.

2.2.2 The effect of lesions of the anterior CLA on kindling and kindled seizures

Rats received stereotaxic implantation of four electrodes: two kindling electrodes implanted bilaterally in the AM (2.6 mm posterior to bregma; 4.5 mm lateral to the midline; 9.1 mm ventral from dura), and two lesion electrodes implanted in the anterior CLA (2.8 mm anterior to bregma; 2.1 mm lateral to the midline; 4.5 mm ventral from dura). Bilateral RF lesions were produced as described above. Ten rats received RF current via the lesioning electrodes during the implantation surgery (i.e., before AM kindling), 10 rats received RF current in a separate surgery after 5 stage 5 seizures had been evoked from the AM and received stimulation until five additional stage 5 seizures were evoked, and 9 control rats received sham lesions.

2.2.3 Transfer of kindling between anterior CLA and AM

Rats received stereotaxic implantation of two kindling electrodes: one directed to the AM and one directed to the contralateral anterior CLA. One week after surgery, ADT was determined at one of the two kindling electrodes in order to begin kindling the primary site in either the CLA or the AM; the initial site kindled was counterbalanced across rats. Electrical stimulation and the procedure for determining the ADT were as described above. Twenty-four hr after completion of primary site kindling, ADT at the secondary site was determined. Once-daily kindling sessions of the secondary site began twenty-four hr later, with stimulation applied at the ADT. Daily stimulation of the secondary site continued until five generalized stage 5 seizures were evoked.

2.3. Contribution of the NSUB to epileptogenesis

We hypothesized that given its anatomical location and proximity to proposed substrates of seizure generalization, the NSUB should display rapid kindling. To test this idea, we kindled the NSUB in short-term (5 ADs) and long-term (30 ADs) kindling preparations. As well, we wished to determine the persistence of kindled seizure susceptibility produced by kindling in the NSUB, because kindled susceptibility to seizures should persist over an interval without stimulation, a characteristic of kindling from other sites in the brain.[2]

2.3.1. Characterization of kindling of the NSUB

Rats displayed a motor response to conventional high-frequency kindling stimulation (see Results). We therefore applied low-frequency stimulation,[3] which consisted of a 5-sec train of constant current biphasic square wave pulses at 3 pps/sec delivered unilaterally to the NSUB at an initial current of 20uA and increasing at 1-min intervals until AD was evoked. AD threshold (ADT) was arbitrarily defined as the minimum stimulation intensity sufficient to trigger an AD lasting approximately 5 sec. Kindling sessions began 24 hr later, and stimulation was applied once daily at the ADT and continued until five generalized seizures were evoked (in the short-term group). After five ADs were evoked, kindling was suspended for an arbitrarily determined period of 14 days, at which time ADT was remeasured using the parameters described above. For 30 ADs (long-term group), surgical and kindling procedures were as described above, except daily low-frequency stimulation continued until 30 ADs were evoked.

2.3.2. The effect of RF lesions of the ventral midline thalamus on kindling

Bilateral RF lesions were produced as described above. Twenty rats received RF current via the lesioning electrodes during the implantation surgery. Twenty control rats received sham lesions. In the same surgery, rats received stereotaxic implantation of two kindling electrodes bilaterally in the AM or in the anterior CLA. One week after surgery, ADT was determined at one of the two kindling electrodes (i.e., CLA or AM). Conventional high-frequency electrical stimulation (60pps) and the procedure for determining the ADT were as described in Section 2.1.

2.3.3. Transfer of kindling from the ventromedial thalamus to the anterior CLA and AM

Rats received stereotaxic implantation of two kindling electrodes: one in the NSUB (2.7 mm A/P, 0.6 mm lateral to midline and 7.0 mm ventral from dura), and one directed to the contralateral AM or the contralateral anterior CLA. One week after surgery, ADT was determined at the kindling electrode implanted in the NSUB in half of the rats in order to begin kindling. Low-frequency stimulation, the procedure for determining the ADT, and the method for assessing transfer were as described above, except stimulation was applied at the ADT until 30 ADs were evoked and daily stimulation of the secondary site continued until stage 5 seizures were evoked.

3. RESULTS

3.1. Contribution of the anterior CLA to epileptogenesis

3.1.1. Persistence of anterior CLA kindling

The mean ADT was 390 ± 89.4 μA and the mean duration of the initial AD was 11.8 ± 2.1 sec. A mean of 7.6 ± 1.9 ADs was required for the development of the first stage 5 seizure. The mean AD duration (ADD) of the initial stage 5 seizure was 30 ± 20.5 sec, and in the final stage 5 seizure before suspending kindling it was 44.4 ± 27.2 sec. Following the 14-day suspension of kindling, all rats required only one stimulation to evoke a stage 5 seizure.

3.1.2. The effect of lesions of the anterior CLA on kindling and kindled seizures

3.1.2a. Anterior CLA lesions before amygdaloid kindling.

There were no significant differences in the mean initial ADT (p=0.259) or mean duration of initial AD (p=0.383). However, rats with lesions in the anterior CLA required significantly more ADs to develop the first stage 5 seizure (lesioned: 16.6 ± 6.58; nonlesioned: 10.1 ± 6.44; p=0.047), suggesting that damage to the CLA impairs kindling without altering the other characteristics of AD development and expression. Lesioned rats required 5.4 ± 2.3 ADs to progress from stage 0 to stage 1 seizures, whereas nonlesioned rats required significantly fewer ADs (3.0 ± 1.6 ADs; p= 0.037) to make the same transition.

3.1.2b. Anterior CLA lesions after amygdaloid kindling.

Control and lesioned rats all required only one stimulation to evoke the first postlesion stage 5 seizure. There was no significant difference between lesioned and control rats in the duration of AD in the first two prelesion stage 5 seizures (p=0.299). However, lesions of the anterior CLA shortened the duration of ADs associated with stage 5 seizures (lesioned: 84.4 ± 8.3 sec; nonlesioned: 109.2 ± 6.4 sec; p=0 .036).

3.1.3. Transfer of kindling between anterior CLA and AM

Transfer was measured by comparing the speed of primary-site kindling of a given site to the speed of secondary-site kindling of the same site. Primary kindling of the AM resulted in significant transfer to the secondary site CLA (secondary-site CLA: 5.86 ± 1.24 ADs; primary-site CLA: 10.3 ± 1.24 ADs; p=0.038; savings of 43.3 percent). In contrast, primary kindling of the CLA failed to result in significant transfer to the secondary-site AM (secondary-site AM, 13.8 ± 1.67 ADs; primary-site AM, 11.9 ± 1.89 ADs; p= 0.524; no savings).

3.2. Contribution of the NSUB to epileptogenesis

3.2.1. Characterization of kindling of the NSUB

Kindling with high-frequency stimulation applied to the midline thalamic region was unsuccessful. Attempts to determine (at 60pps, 1-sec duration) an initial stimulation threshold resulted in diffuse and dramatic motor responses consisting of jumping and running behavior and intermittent observations of the rat rolling along its longitudinal axis. Subsequent attempts to kindle through the motor response and below the 'motor-response' threshold failed, as we were unable to obtain AD, an indication that kindling was not progressing.

3.2.2. Low-frequency kindling.

3.2.2a. 5ADs. The initial ADT was typically very high (3802.1 ± 1692.3 μA), and fairly well-developed seizures tended to appear suddenly, involving rhythmic hopping with bilateral forelimb clonus, and rolling on the side and displaying bilateral fore/hindlimb tonic-clonic seizures. These symptoms appeared in the absence of myoclonus of the head and facial musculature typical of early stages of limbic kindling. The seizures tended to be brief in duration (18.9 ± 7.7 sec). Surprisingly, we observed a significant attenuation of ADD/seizure duration between the 1st and 5th seizures (13.3 ± 6.1 sec; p= 0.001).

After 14 days without stimulation the first stimulation triggered an immediate generalized seizure, indicating persistence of the seizure state. However, following the 14-day interval, we observed a significant decrease between the initial ADT (3802.1 ± 1692.3 μA) and the ADT determined in the rekindling sessions (1103.6 ± 1023.3 μA; p< 0.001). Furthermore, an examination of the ADDs for the 1st initial seizure and the 1st seizure following the resumption of kindling indicated that ADD failed to recover to initial levels (18.9 ± 7.7 sec vs 14.7 ± 5.2 sec, respectively; p= 0.047).

3.2.2b. 30ADs. ADTs were very high (4500.0 ± 1087.1μA), and the initial seizures tended to be brief (15.6 ± 5.2 sec). Beyond the fifth AD, a significant increase in AD duration and subsequent generalization to late-stage limbic-type seizures occurred, which included bilateral forelimb clonus, rearing, and repeated falling. It was not until the tenth AD that we saw a significant increase in AD duration (20.9 ± 10.5 sec) compared to the first AD (15.6 ± 5.22; p=0.027).

3.3. Lesions of the ventromedial thalamus on kindling

3.3.1a. Lesions of the ventromedial thalamus and AM kindling. There were no significant differences in the mean initial ADT (p=0.329) or mean duration of initial AD (p=0.332) between lesioned and nonlesioned rats. Lesions altered the progression of kindling, in that lesioned rats required significantly more ADs to exhibit the first stage 3 seizure (12.4 ± 4.7 ADs), than control rats (9.9 ± 3.0 ADs; p= 0.048).

Both the lesioned and control rats progressed to stage 5 seizures at the same rate (p=0.402); however, lesioned rats exhibited a significantly shorter stage 5 seizure duration (lesioned: 50.6 ± 26.6 sec; control: 77.5 ± 22.7 sec; p=0.025). As well, although there were nonsignificant differences in the latency to clonus (p=0.371), the duration of the clonic component of the generalized seizures was significantly shorter in the lesioned rats (28.4 ± 12.5 sec; control 51.0 ± 16.7 sec; p=0.007). In proportionate terms, lesioned rats spent 56.1% of a generalized seizure in clonus, whereas clonus accounted for 65.8% of the total generalized seizure duration in nonlesioned rats.

3.3.1b. Lesions of the ventromedial thalamus and CLA kindling. There were no significant differences in the mean initial ADT (p=0.106) or mean duration of initial AD (p=0.532) between lesioned and control rats. Lesioned rats required significantly more ADs to exhibit the first stage 3 seizure (lesioned: 11.5 ± 4.3 ADs; control: 8.1 ± 2.0 ADs, p= 0.039). As well, the control rats progressed to stage 5 seizures at a significantly faster rate than lesioned rats (9.5 ± 2.7 ADs vs. 13.0 ± 4.5 ADs, respectively; p=0.041). No other significant differences were observed between lesioned and control rats in the duration of stage 5 seizures (p=0.175) or the duration of and latency to clonus within stage 5 seizures (p's = 0.072 and 0.066, respectively).

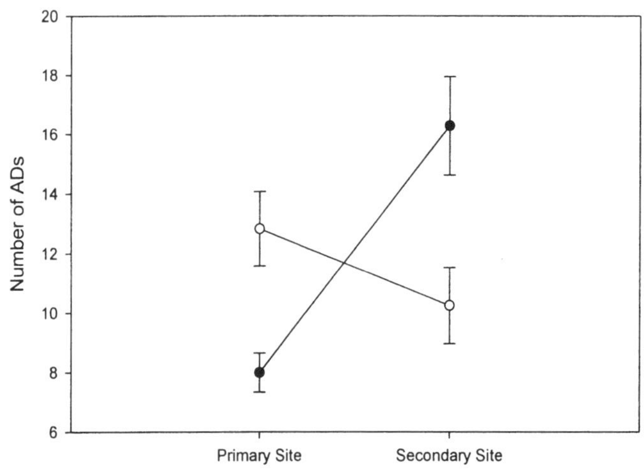

Figure 1: Transfer of kindling, assessed by comparing the rate of kindling of AM (○) and CLA (●) in naïve rats (primary site) and in rats previously stimulated in the NSUB

3.4. Transfer of kindling from the ventromedial thalamus to the CLA and AM

Transfer was measured by comparing the speed of kindling in previously stimulated rats to the speed of kindling of the same site in naive rats (i.e., yoked controls). Prior stimulation of the ventromedial thalamus failed to produce significant transfer to the AM (previous stimulation: 10.3 ± 3.62 ADs; controls: 12.8 ± 3.06 ADs; $p=0.184$). In contrast, stimulation of the NSUB prior to AM kindling produced a significant anti-transfer or antagonistic effect as the anterior CLA required nearly double the number of ADs to elicit a generalized seizure (previous stimulation, 16.3 ± 4.39 ADs; controls, 8.0 ± 1.73 ADs; $p= 0.001$; Figure 1).

4. DISCUSSION

Several predictions deriving from the hypothesis that the anterior CLA and NSUB play a role in seizure generalization were tested. The predictions relate to the characterisitics of kindling, to the effects of lesions on limbic kindling, and to transfer of kindled seizure susceptibility between various sites.

Following a 14-day period without stimulation, rekindling of the CLA resulted in immediate triggering of generalized seizures, associated with small increases in the duration of seizures suggestive of increased seizure intensity. We found that small lesions of the anterior CLA do indeed retard amygdaloid kindling, adding support to the notion that the CLA contributes to the development of generalized limbic seizures. On the other hand, post-kindling lesions of the anterior CLA failed to affect established seizures. This may indicate that alternate routes of propagation are available in the kindled brain.

The asymmetrical transfer of kindling between the anterior CLA and AM was unexpected, particularly given that AD propagated to the AM from the earliest stimulations. One possible explanation is that, as with anterior neocortical kindling,[4] anterior CLA kindling would facilitate subsequent amygdaloid kindling only if stimulation were repeatedly applied to the CLA until limbic-type generalized seizures have been kindled. An alternate possibility is that structures more distant from the substrates of generalization require more kindling than sites closer to the substrates and that these sites would be resistant to transfer from the closer sites. Thus AD elicited from the CLA and other structures in the generalization network may not drive plastic changes in limbic sites sufficient to produce transfer.

The ventromedial thalamus, and the NSUB in particular, are able to drive seizures. That is, low frequency stimulation of this area produces immediate seizures that accompany the first AD, a result consistent with earlier reports of kindling in sensory regions of the lateral thalamus in rats[5] and dorsomedial and centromedial thalamic nuclei in cats,[6] with associated significant decreases in the intensity of the stimulation needed to elicit AD. As seen in previous studies[3, 7, 8] with limbic stimulation, the application of low-frequency kindling stimulation commonly elicits a rapid progression of kindling. One unanticipated finding of this study is the attenuation of AD duration during the first five stimulations, an event rarely if ever observed when kindling other sites.

In the current experiment, application of 30 daily stimulations resulted in the development of prolonged limbic-type seizures, a result that points to a possible plastic mechanism that operates in two phases. The mechanics of such a progression have yet to be fully examined; however, there likely is activation of more complex brainstem mechanisms that are involved with limbic convulsions. This particular idea is consistent with a hypothesis that candidate brainstem structures may be involved in convulsive seizure generation by acting to accept synchronous activity in a circuitry-dependent manner and generate a convulsive motor program associated with activation of specific circuitry.

Lesions of the ventromedial thalamus failed to completely abolish seizures but delayed the development of generalized seizures with CLA but not AM kindling. Given the prominent connection between the anterior CLA and the NSUB, this result is not particularly surprising. However, the fact that amygdaloid kindling progressed without noticeable delay indicates that the AD produced in the AM may be accessing alternate routes of propagation and generalization that do not necessarily require ventromedial thalamus.

Interestingly, prestimulation of the NSUB significantly delayed kindling of the anterior CLA, thus producing delayed or negative transfer. The first, and perhaps most obvious explanation, is that stimulation of the NSUB results in distinct plastic changes (e.g. in the CLA) that imply some sort of directionality to the propagation of epileptiform activity. In terms of the gating hypothesis put forth by Burchfiel and Applegate,[9] stimulation of the NSUB may close the forebrain gate until the AD elicited from anterior CLA can become organized enough to overcome the gate and access the substrates responsible for subsequent seizure generalization. The delayed transfer may be due to the need for stimulation of the CLA to remove any inhibitory influence, or to modify circuitry supporting epileptogenesis. The data presented here can neither confirm nor refute these suppositions, and they are pure speculation at this point.

5. ACKNOWLEDGEMENTS

This research was supported by a Discovery Grant from NSERC awarded to M.E. Corcoran.

6. REFERENCES

1. X. Zhang, D. K. Hannesson, D M. Saucier, A.E. Wallace, J. Howland and M.E. Corcoran, Susceptibility to kindling and neuronal connections of the anterior claustrum, *J. Neurosci.* **21**, 3674-3687 (2001).
2. M.E. Corcoran, "Characteristics and mechanisms of kindling", in Sensitization of the nervous system, edited by p. Kalivas and C. Barnes (The Telford Press, Caldwell, NJ, 1988), pp. 81-116.
3. D. P. Cain and M. E. Corcoran, Kindling with low-frequency stimulation: generality, transfer, and recruiting effects, *Exp. Neurol.* **73**, 219-32 (1981).
4. W. T. Seidel and M. E. Corcoran, Relations between amygdaloid and anterior neocortical kindling, *Brain Res.* **385**, 375-378 (1986).
5. D. P. Cain, Kindling in sensory systems: thalamus, *Exp. Neurol.*, **66,** 319-329 (1979).
6. Y. Wada, Y. Minabe, H. Okuda, I. Jibiki, K. Yoshida and N. Yamaguchi, Lateral geniculate kindling and long-lasting photosensitivity in cats, *Exp. Neurol.* **91**, 343-354 (1986).
7 M. E. Corcoran and D. P. Cain, Kindling of seizures with low-frequency electrical stimulation, *Brain Res.* **196**, 262-265 (1980).
8. D. P. Cain, D. J. McKitrick and F. Boon, Rapid and reliable induction of partial status epilepticus in naive rats by low-frequency (3-Hz) stimulation of the amygdale, *Epilepsy. Res.* **12**, 51-55 (1992).
9. J. L. Burchfiel and C. D. Applegate, Stepwise progression of kindling: perspectives from the kindling antagonism model, *Neurosci. Biobehav. Rev.* **13**, 289-99 (1989).

DISCUSSION

Gale: I found your transfer experiments very interesting. Can you say something about how you to interpret the fact that you got unidirectional transfer, amygdala to claustrum but not claustrum to amygdala?
Sheerin: I suggest that the claustrum is closer to the substrates that support generalization of motor seizures. It predominates and produces the asymmetrical transfer.
Gale: But if that were true and you kindle the claustrum, if that is downstream from the amygdala, then that should be included in the response to the amygdala and you should get accelerated kindling with the amygdala. But that is not happening. The only way that I can see accounting for it would to say that the claustrum has an alternative or preferential route of propagation when you kindle the claustrum, independent of the circuitry that the amygdala works on. Thus there would be two ways that claustrum kindling could express itself, one independent of the amygdala and one that uses the amygdala circuitry. So if you kindle the amygdala, then you have facilitated claustrum response by an unusual mechanism, meaning that it is not the normal mechanism - in other words, it would not be the normal mechanism that is kindled when you kindle with the primary site, the claustrum.
Sheerin: Right, thank you.
Wada: Beautiful study. We planned almost an identical study many years ago. We hypothesized that if we kindled claustrum, we should kindle the amygdala very quickly. Unfortunately our cats did not cooperate, and we found out that the electrode placements were wrong; so we could not accomplish is. But you have done it beautifully. Now as for the thalamus, we didn't go where you did, but we tried to play with medial thalamus. You can induce generalized convulsion in one trial, once you do that you can kindle from amygdala very rapidly. If you destroy the medial thalamus you don't get to convulsive state. Another thing about transfer you mentioned, you were talking about claustrum and the kindled ipsilateral amygdala. For the positive transfer effect, we do know medial thalamus down to brainstem is involved. Have you ever checked after kindling claustrum or amygdala manipulating your thalamic system? What happens to the secondary hemisphere site? Have you tried to see if there was any transfer or interference?
Sheerin: No, not yet.
Avanzini: When you kindled the thalamus with 3 per second, did you find any electrical correlate like spike wave or things like that? During the stimulation that may have some influence in preventing or interfering with it.
Sheerin: During the stimulation, what I saw was rhythmic hopping with the stimulation. Using our system I am not able to record EEG during the stimulation; so I wasn't able to look at that time frame.
Onat: My question is related to Dr. Avanzini's. After thalamic stimulation, have you done histological microscopic examination of what happened in thalamus at the microscopic level?
Sheerin: No, I have only done microscopic evaluation of electrode tracks. I have not yet performed cell counts or looked for cell death.
Goodman: How long did you stimulate for? You are using very high currents. How long was the stimulus?
Sheerin: Five seconds.
Goodman: In previous work by Corcoran and Cain, using low frequency stimulation to kindle, they stimulated for very long durations. So there probably was no chance for damage

from the stimulus itself?
Sheerin: I don't yet know.
McIntyre: What was the time window between your kindling primary site and kindling your secondary site? Did you just start it the next day?
Sheerin: I just started the next day.
McIntrye: Then it would be interesting to see whether your negative transfer in fact was an inhibitory thing, whether with time it would just go away.
Sheerin: Yes, I agree.

FROM ULTRASTRUCTURE TO NETWORKS: Kindling-induced changes in neocortex

G. Campbell Teskey*, Corey Flynn, Marie-H. Monfils, Crystal D. Goertzen, Luke C. Henry, and Pamela A. Valentine*

1. INTRODUCTION

The tendency for chemical, mechanical, electrical, biological and artificial neural network systems to undergo oscillatory behaviour has been extensively studied. At least two general principles appear to emerge from these studies. Systems have a greater tendency to oscillate if they become more strongly interconnected and simpler. We have chosen to investigate whether these two principles apply to seizure phenomenon within the vertebrate brain. We postulate that during progression to a more seizure prone state, the involved neuronal system will become more strongly interconnected and simpler. This chapter contains a series of observations from the synaptic and cellular levels in the neocortex of rats that support this postulate. Furthermore, the consequence of greater synaptic efficacy of excitatory connections is manifested at the systems level as a series of predictable changes in network behaviour. We present an overview of a variety of studies that we have conducted in the neocortex of rats and cats that examined evoked and spontaneous network properties. We argue that the networks show greater interconnection and reduced functional specificity as a consequence of kindling-induced alterations.

We have chosen the forelimb area of sensorimotor cortex of the rat and the primary auditory (AI) cortex of the cat, in which to conduct our studies. The rat sensorimotor cortex kindles, shows synaptic potentiation and depression phenomena, as well as plastic changes at the cellular and system levels to a variety of conditions and treatments.[1-7] Cat AI cortex exhibits a tonotopic organization that is dynamically regulated and continually altered by ongoing experience even into adulthood.[8] Cat AI also presents an ideal

---

* G. Campbell Teskey, Behavioural Neuroscience Research Group, Department of Psychology, University of Calgary, Calgary, Alberta, Canada, T2N 1N4

structure for examining unit firing characteristics following kindling-induced reorganization because the naïve system has previously been described in detail, and a great deal is known about the underlying mechanisms.[9]

2. SYNAPTIC AND CELLULAR LEVEL OF ANALYSIS

If repeated seizures cause a brain to become more strongly interconnected and simpler, then we predict an increase in the presence of highly efficacious synapses, as well as a decrease in the number of neurons following kindling. We tested those predictions in the rat neocortex using electron and light microscopic techniques.

2.1. Density of excitatory synapses

We examined the dendritic area close to the cell bodies in layer V of rat sensorimotor neocortex following 3 weeks of kindling stimulation applied to the corpus callosum. Brains were extracted either 2 days or 3 weeks following the last seizure, and 70 nm serial sections were prepared for quantitative ultrastructural analysis using the physical dissector method. We categorized and measured two types of excitatory post-synaptic densities (perforated and non-perforated), as well as whether the synaptic connections were onto dendritic spines (axospinous) or directly onto the dendrite (axodendritic). Overall there was not a significant difference in the density of synapses in the kindled tissue as compared to tissue from non-kindled rats. However, there was a significant *increase* in the total number of synapses per mm^3 of the more efficacious perforated subtype that made axodendritic connections in layer V from kindled as compared to non-kindled rats.[10]

Previous work at the synaptic ultrastructural level in the hippocampus has shown that kindling leads to an overall decrease in the number of synapses.[11-13] Geinisman[12] argued that the proportion of perforated to non-perforated synapses shifted toward the larger, and highly efficacious perforated synaptic type, and that this accounts for some of the strengthened electrophysiological responses following kindling. Our work in the neocortex demonstrated that following kindling there was an overall *increase* in the perforated synapses *and* a positive shift in the proportion of perforated to non-perforated synapses. Furthermore, we saw an increase in the axodendritic type of synaptic connections. Synapses made onto spines vary in efficacy because the properties of the axospinous connections are regulated by the conformational properties of the spine. In this way spines can act as modulators of efficacy. Neocortical kindling results in increased numbers of axodendritic synapses that are not subject to this type of regulation.

2.2. Number of neurons

We examined 1µm toluidine blue stained sections from layer V of rat sensorimotor cortex from the same 300µm block of tissue in which we took our synaptic density measurements. We first examined whether kindling stimulation altered neocortical thickness and found that it did not. We then used a computer-assisted microscope, a stereology software package and the physical dissector method to obtain unbiased estimates of neuron density. We found a ~30% reduction in neuron number at 3 weeks following the cessation of kindling.[10] Other researchers have found neuronal loss in the amygdala and hippocampus following kindling of those structures. Thus, neuronal loss may be a common response to the repeated elicitation of electrically kindled seizures.

It is important to note that when the number and type of synapses are expressed on a per neuron basis, there is an increase in the number of perforated axodendritic and axospinous synapses on the remaining neurons in layer V from brains that were kindled relative to sham controls[10]. We do, however, acknowledge that while it is likely that most of the synapses examined in layer V are on layer V pyramidal cell neurons, some small proportion of the synapses may be on dendrites of neurons from layers VI.

2.3. Pyramidal cell dendritic spines

Our ultrastructural analysis revealed a greater number of synaptic contacts on the remaining pyramidal cells in layer V. We confirmed the per neuron increase in axospinous synapses by examining spine density with the light microscope. Golgi impregnated pyramidal cells in both layers III and V of sensorimotor neocortex from rats that were kindled for 3 weeks were examined. There was an increase in spine density at both 2 days and 3 weeks post kindling in layer V basilar dendrites (Teskey et al., unpublished observations). However, within layer III, we observed that there was a decrease in spine density in kindled rats, 2 days following the last seizure. At 3 weeks post-kindling this relationship had reversed and there was a significant increase in spine density relative to sham-kindled controls.[14, 15] Dendritic spines have also been shown to regress and then regrow in response to the onset and cessation of epileptic activity in the hippocampus and amygdala. Our results in the neocortex are generally consistent with those from the hippocampus and amygdala and support the per neuron increase in axospinous synaptic connections onto remaining layer V pyramidal cells.

2.4. Pyramidal cell dendritic length and branches

It is difficult to make specific predictions regarding the direction in which dendrites should change with kindling. For instance one could hypothesize that larger dendritic arborizations would have more potential space for synaptic contacts whereas smaller dendritic trees might provide a simpler and be more efficient at integrating excitatory postsynaptic potentials. As with the dendritic spines, we also examined dendritic length and branching of Golgi impregnated pyramidal cells in both layer III and V of sensorimotor neocortex from rats that were kindled for 3 weeks. Within layer III, we observed that there was a decrease in dendritic length and branching in kindled rats 2 days following the last seizure. At 3 weeks post-kindling the dendritic length and branching measures were slightly, but not significantly, larger than control.[14, 15] In layer V, the pyramidal cells showed the opposite profile with respect to dendritic length and branching. Kindling resulted in an increase in length and branches at 2 days post kindling but a decreased length and branches was observed 3 weeks post kindling. The fact that dendritic regression and growth differs in layers III and V with kindling and time suggests that kindling of the corpus callosum does not have equivalent effects on the different layers of the neocortex.

3. SYSTEMS LEVEL OF ANALYSIS

Increasing synaptic efficacy in structures with multiple layers of synaptically connected units should result in broader evoked responses that reflect the recruitment of more downstream cells into participating in the response. We examined evoked responses in the rat by applying an ascending series of square wave pulses to the corpus callosum and recording the response in the sensorimotor neocortex. We also examined movement representation maps in the sensorimotor cortex of the rat using intracortical microstimulation and predicted that movements should be elicited from a larger cortical area. In the cat we evoked responses with brief pure sounds of varying intensity and frequency while recording the unit responses in primary auditory cortex. We also explored the tonotopic organization of multiple single unit responses and the inter-correlation of spontaneous unit activity from multiple sites within cat primary auditory cortex. We predicted that more strongly interconnected networks should result in broader evoked responses, degraded tonotopic organization and higher inter-correlated unit activity.

3.1. Evoked potentials

The callosal-neocortical pathway shows kindling-induced potentiation when kindling stimulation is applied either to the corpus callosum or amygdale.[1, 2] A study from Racine's laboratory, using a current source density analysis, revealed that the evoked callosal-neocortical late component response has a sink in upper layer V and a source located in deep layer V.[16] The late component of the evoked response is likely polysynaptic in nature because it failed to be expressed when stimulations were delivered at high frequencies. The kindling-induced potentiation of the late component of the evoked potential likely reflects polysynaptic activity mediated by horizontal fibers, and the recruitment of neurons into participating in the response.[17] Thus, the enhanced polysynaptic response is a manifestation of strengthened neural communication.

Kindling-induced potentiation has been reported in multiple forebrain structures.[17, 18] However, it doesn't appear to be essential for kindling because it is not always observed with kindling[2, 19, 20] and prior induction of long-term potentiation of the evoked responses only partially facilitates subsequent kindling.[21, 22]

3.2. Movement representations

The presence of a highly ordered somatotopic map of movement representations was intuited by John Hughlings Jackson based on his observations of the ordered spread of tonic clonic activity to adjacent muscle groups during a single ictal event. Penfield later confirmed Jackson's notion and provided the classic descriptions of movement representations within the human precentral gyrus.[23] Much of the initial characterization of these motor maps came from intracortical stimulation experiments conducted on human patients prior to surgery intended to alleviate epileptic seizures. Later work showed that stimulation-evoked movements in some people with epilepsy could be elicited from regions well outside the classic motor strip located along the Rolandic fissure.[24] Perhaps changes in the topography of movement representations within the primary motor cortex may occur in response to recurrent seizure activity. This suggests that the classic depictions of cortical representations that have been found to contain an error[25] may be due to the fact that they were obtained from tissue exposed to chronic epileptiform activity.

We decided to examine whether repeated seizure activity could reorganize the topography of movement representations, if there was a correlation between reorganization and synaptic potentiation, if the reorganization was persistent and if it could be driven by seizures from either cortical or limbic sites. Once again we asked these questions in the rat sensorimotor cortex, specifically the caudal forelimb area. The caudal forelimb area is defined as that area of rat sensorimotor cortex in which intracortical microstimulation (ICMS) applied to layer V elicits movements of the digits, wrist, elbow or shoulder of the contralateral forelimb.[26]

The primary finding of the study was that under identical stimulation parameters, kindled animals exhibited a profound increase in the area of cortex from which forelimb movements could be elicited.[2] This is not to say that the expanded areas necessarily represent "new" areas of forelimb motor cortex. Rather, these areas have undergone some functional change due to the kindling experience. This functional change facilitates the ability to elicit forelimb movements from regions of cortex that would not previously produce forelimb movements in response to similar stimulation. At higher stimulation intensities, kindled rats also showed a greater number of multiple forelimb muscle responses compared to controls. Taken together, our observations of changes in the topography of cortical movement representations appear to be mediated by synaptic enhancement of horizontal connections.

Kindled rats displayed expanded caudal forelimb movement representations that went well beyond the boundaries determined in the non-kindled control rats. The dramatically expanded forelimb movement representations in the kindled rats were not a result of alterations in anesthetic levels, stimulation sensitivity (thresholds) or a simple expansion into the rostral forelimb region. We also observed that the kindling-induced enhanced synaptic potentiation and expanded caudal forelimb motor map were expressed both 1 and 21 days after the last seizure. In neither case was there a significant decrement in either the size of the synaptic potentiation or size of movement representations in the 3 weeks following the last seizure. Our observation of persistent evoked potential and movement representations also indicates that these effects were not directly seizure related and transient, but were related to the more long-term nature of kindling[27]. Because ICMS appears to elicit movement by stimulating a pool of neurons interconnected by horizontal fibers within the cortex rather than via direct stimulation of corticospinal neurons,[26] the enhanced polysynaptic response may be responsible for the expanded motor representations. Stimulation-induced potentiation of horizontal fibers has been demonstrated *in vitro* and has been proposed as the mechanism underlying functional reorganization within motor cortex.[28] ICMS-induced movements are most easily elicited when the stimulating electrode is located within layer V, and it is possible that expansion of the movement representations are primarily mediated by potentiated layer V currents. While we have shown that polysynaptic potentiation correlates with expanded motor maps, it is possible that there are other contributions, such as increased monosynaptic drive onto layer V neurons, other cells that polysynaptically activate layer V and potentiation of other intracortical horizontal connections that may contribute to the observed changes. It may also be the case that the neocortex is reorganized as a compensatory response to epileptiform activity so that functional movement representations are not lost as layer V cells die.

In order to more deeply explore the relationship between epileptogenesis, evoked potentials and the size of movement representations, we used the epileptogenic-prone (FAST) and epileptogenic-resistant (SLOW) rat strains. These strains have become a valuable tool for investigating the neurochemical and neurophysiological basis of epilepsy. FAST and SLOW rats received high-frequency stimulation of the corpus callosum in order to induce long-term polysynaptic potentiation and then underwent high-resolution ICMS in order to assess functional movement representations of the left caudal forelimb area of the sensorimotor cortex. We observed that high-frequency stimulation induced similar increases in polysynaptic potentiation in both rat strains. These results are consistent with Racine et al.[30] who used high-frequency stimulation to induce long-term potentiation of the perforant path to the dentate gyrus, and the amygdala to the entorhinal cortex pathway, and found no differences between the FAST and SLOW strains.[30] Thus, synaptic potentiation likely does not underlie the inherent differences in epileptogenesis between the two strains. Despite the similarities in the magnitude of potentiation, we found that only the FAST potentiated rats showed an increase in the size of their movement representations. SLOW potentiated rats did not manifest an increase in the size of their movement representations compared to SLOW control rats, whereas the FAST potentiated rats had movement representations that were twice the area of FAST control rats. Thus, our results suggest that there is dissociation between polysynaptic potentiation and the size of movement representations.

3.3. Evoked frequency responses

Neurons in primary auditory cortex of the cat respond most vigorously to simple sounds (tone pips) of a particular frequency range. The tuning curve provides a three dimensional plot of the firing rate of auditory cortical neurons over a restricted range of stimulus frequencies and intensities, from which the characteristic frequency may be defined as the frequency to which the neurons are maximally responsive. Examination of tuning curves provides a useful way of examining the response properties of single cells within larger neural networks. We examined the response properties of neurons in the primary auditory cortex from cats that were either kindled, sham kindled or non-implanted controls and observed that the tuning properties of neurons in the primary auditory cortex were altered in the kindled animals. In primary auditory cortex there is normally a high percentage of neurons that express narrow V-shaped tuning curves. In the kindled animals there was a greater tendency for double-tuned multiple single-units (29%) than for the normal controls where the percentage of double tuned neurons in cat AI was nominally only about 8% for the non-implanted controls and 14% for the sham controls. This may represents less specification between thalamocortical afferents that provide the normal tuning, and/or the stronger interconnection of cortical horizontal fibers that convey the characteristic frequency range. Thus, single auditory neurons from kindled cats respond to a broader range of sound frequencies and this is consistent with the neurons from kindled cats being more strongly interconnected.[9]

3.4. Tonotopy and Inter-correlations

Cat auditory cortex normally exhibits a tonotopic organization with a gradual progression from low characteristic frequencies at the posterior ectosylvian sulcus to high characteristic frequencies towards the anterior ectosylvian sulcus. We recorded multiple single-unit activity under light ketamine anesthesia with a rectangular array of eight (4 x 2) microelectrodes positioned across the frequency map in AI. Substantial changes in the tonotopic maps in AI were observed in all kindled cats. In the kindled cats, the maps in AI typically exhibited the lack of a tonotopic gradient, with large portions of AI tuned to a relatively narrow range of CFs. Due to the vastly increased size of the blood vessels, a finding common in all fully kindled cats and likely indicating an increased cortical energy demand during the seizures, the mapping was less dense than in the sham controls. During spontaneous neural activity, we also observed that neurons recorded from different electrodes within the array show higher inter-correlations in the kindled cats[9]. The enhanced spontaneous neural synchrony was not only observed between neurons, but extended over a greater area of primary auditory cortex. Both of these observations can be explained by a greater amount of interconnections mediated by horizontal cortico-cortical synaptic connections.

4. SUMMARY OF THE MODEL AND FUTURE DIRECTIONS

Seizures are hyper synchronous and hyper excitatory oscillatory discharges that can occur in the vertebrate central nervous system as the result of a wide variety of intrinsic (i.e., ion channel mutations, neurotransmitter imbalance) and extrinsic (i.e., drugs, infections) factors. Seizures appear to induce a Hebb-like increase in synaptic efficacy: cells that fire together during seizures subsequently wire together. We believe that a system is more likely to go into oscillation if it is more strongly interconnected and simpler. We have presented data from the kindled neocortex, that show the presence of more highly efficacious synapses, fewer neurons and increased numbers of synapses per neuron. These anatomical changes that occur with kindling likely account for a corollary to Hebb's rule: cells that are wired together, fire together *and* recruit more cells into the response. We observed broadened evoked potentials and tuning curves as well as expanded movement representations, loss of tonotopic organization and higher inter-correlations among active neurons that likely reflect enhanced interconnection of remaining excitatory units.

Kindling is an important technique and phenomenon that models certain aspects of epilepsy.[31] Repeatedly elicited seizures affect brain anatomy, physiology and function as well as lower thresholds for subsequent seizures. Future research should investigate, in more depth, our observations that seizures result in a simpler and more strongly interconnected network that in turn is more likely to go into pathological oscillation. We intend to design experiments to delineate the relative contributions of enhanced synaptic strength versus cell loss to kindling. There is also clearly a need to examine the neurotransmitter and receptor systems associated with the additional synaptic contacts. The generality of more perforated synapses and cell loss to other structures and other seizure/epilepsy models should also be explored.

5. ACKNOWLEDGEMENTS

Sincere thanks to colleagues and collaborators Richard Dyck, Jos Eggermont, Jeff Kleim, Bryan Kolb and Dan McIntyre.

6. REFERENCES

1. R. J. Racine, C. A. Chapman, G. C. Teskey, and N. W. Milgram, Post-activation potentiation in the neocortex: III. Kindling-induced potentiation in the chronic preparation, *Brain Res.* **702**, 77-86 (1995).
2. G. C. Teskey, M.-H. Monfils, P. VandenBerg, and J. A. Kleim, Motor map expansion following repeated cortical and limbic seizures is related to synaptic potentiation, *Cereb. Cortex* **12**, 98-105 (2002).
3. C. Trepel and R. J. Racine, Long-term potentiation in the neocortex of the adult freely moving rat, *Cereb. Cortex* **8**, 719-729 (1998).
4. M.-H. Monfils and G. C. Teskey, Skilled-learning-induced potentiation in rat sensorimotor cortex: a transient form of behavioral LTP. *Neurosci.* **125**, 329-336 (2004).
5. M.-H. Monfils, P. M. VandenBerg, J. A. Kleim and G. C. Teskey. Long-term potentiation induces expanded movement representations and dendritic hypertrophy in layer V of rat sensorimotor neocortex, *Cereb. Cortex* **14**: 586-593 (2004).
6. D. J. Froc, C. A. Chapman, C. Trepel, and R. J. Racine, Long-term depression and depotentiation in the sensorimotor cortex of the freely moving rat, *J. Neurosci.* **20**, 438-445 (2000).
7. M.-H. Monfils and G.C. Teskey, Induction of long-term depression is associated with decreased dendritic length and spine density in layers III and V of sensorimotor neocortex, *Synapse* **53**, 114-121 (2004).
8. S. Seki and J. J. Eggermont, Changes in cat primary auditory cortex after minor-to-moderate pure-tone induced hearing loss, *Hearing Res.* **173**, 172-186 (2002).
9. P. A. Valentine, G. C. Teskey, and J. J. Eggermont, Kindling changes burst firing, neural synchrony and tonotopic organization of cat primary auditory cortex, *Cereb. Cortex* **14**, 823-835 (2004).
10. C. D. Goertzen and G. C. Teskey, Kindling-induced alterations in synaptic morphology of rat neocortical neurons in layer V of the caudal forelimb area, *Soc. Neurosci. Abstr.* **28**, 477.6 (2003).
11. Y. Geinisman, F. Morrell, and L. deToledo-Morrell, Remodelling of synaptic architecture during hippocampal "kindling", *Proc. Natl. Acad. Sci. USA* **85**, 3260-3264 (1988).
12. Y. Geinisman, F. Morrell, and L. deToledo-Morrell, Increase in the relative proportion of perforated axospinous synapses following hippocampal kindling is specific for the synaptic field of stimulated axons, *Brain Res.* **507**, 325-331 (1990).
13. Y. Geinisman, F. Morrell, and L. deToledo-Morrell, Increase in the number of axospinous synapses with segmented postsynaptic densities following hippocampal kindling, *Brain Res.* **569**, 341-357 (1992).
14. G. C. Teskey, J. E. Hutchinson, and B. Kolb, Sex differences in cortical plasticity and behavior following anterior cortical kindling in rats, *Cereb. Cortex* **9**, 675-682 (1999).
15. G. C. Teskey, J. E. Hutchinson and B. Kolb. Cortical layer III dendritic morphology normalizes within three weeks after kindling and is dissociated from kindling-induced potentiation, *Brain Res.* **911**, 125-133 (2001).
16. C. A. Chapman, C. Trepel, T. L. Ivanco, D. J. Froc, K. Wilson, and R. J. Racine, Changes in field potentials and membrane currents in rat sensorimotor cortex following repeated tetanization of the corpus callosum *in vivo*. *Cereb. Cortex* **8**, 730-742 (1998).
17. G. C. Teskey and P. A. Valentine, Post-activation potentiation in the neocortex of awake freely moving rats, *Neurosci. Biobehav. Rev.* **22**, 195-207 (1998).
18. D. P. Cain, Long-term potentiation and kindling: how similar are the mechanisms? *Trends Neurosci.* **12**, 6-10 (1989).
19. J. L. Giacchino, G. G. Somjen, D. P. Frush, and J. O. McNamara, Lateral entorhinal cortical kindling can be established without potentiation of the entorhinal-granule cell synapse, *Exp. Neurol.* **86**, 483-492 (1984).
20. E. Maru, Is synaptic potentiation necessary for the development of kindling? *Jpn. J. Psychiatry Neurol.* **45**, 291-296 (1991).
21. R. J. Racine and W. M. Burnham, The kindling model. in: *The Electrophysiology of Epilepsy*, edited by P. Schwartzkroin and H.Wheal (Academic Press, New York, 1984), pp. 153-171.
22. R. J. Racine, F. Newberry and W. M. Burnham, Post-activation potentiation and the kindling phenomenon, *EEG Clin. Neurophysiol.*, **39**, 261-271 (1975).
23. W. Penfield and T. Rasmussen, *The Cerebral Cortex of Man.* (McMillian, New York, 1950).
24. S. Uematsu, R. Lesser, R. S. Fisher, B. Gordon, K. Hara, G. L. Krauss, E. P. Vining, and R. W. Webber, Motor and sensory cortex in humans: topography studied with chronic subdural stimulation. *Neurosurgery* **31**, 59-72 (1992).
25. P. Servos, S. A. Engel, J. Gati, and R. Menon, fMRI evidence for an inverted face representation in human somatosensory cortex, *Neuroreport* **10**, 1393-1395 (1999).
26. J A. Kleim, S. Barbay, and R. J. Nudo, Functional reorganization of the rat motor cortex following motor skill learning, *J. Neurophysiol.* **80**, 3321-3325 (1998).

27. Z. Dennison, G. C. Teskey, and D. P. Cain, Persistence of kindling: Effect of partial kindling, retention interval, kindling site and stimulation parameters, *Epilepsy Res.* **21**, 171-182 (1995).
28. G. Hess and J. P. Donoghue, Long-term potentiation of horizontal connections provides a mechanism to reorganize cortical motor maps, *J. Neurophysiol.* **71**, 2543-2547 (1994).
29. C. Flynn, M.-H. Monfils, J. A. Kleim, B. Kolb, D. C. McIntyre, and G. C. Teskey, Differential neuroplastic changes in cortical movement representations and dendritic morphology in epilepsy-prone and epilepsy-resistant rat strains following high-frequency stimulation, *Euro. J. Neurosci.* **19**, 2319-2328 (2004).
30. R. J. Racine, M. Steingart and D. C. McIntyre, Development of kindling-prone and kindling-resistant rats: selective breeding and electrophysiological studies, *Epilepsy Res.* **35**, 183-195 (1999).
31. G. C. Teskey, Using kindling to model the neuroplastic changes associated with learning and memory, neuropsychiatric disorders, and epilepsy, in: *Toward a theory of neuroplasticity*, edited by C.A. Shaw & J.C. McEachern, (Taylor and Francis, Philadelphia, 2001), pp. 347-358.

Discussion

Post: Have you done any work or theorized about what it would take to move these back towards normal?
Teskey: Well, I can't make the cells come back to life, that's a problem. If cell loss is a function of kindling and epilepsy, then it really might be unidirectional. But I have thought about the perforated synapses and how can we go after the perforated synapses and try to reverse that. I think that we need to understand what receptor systems and receptors are involved with those perforated synapses, to target them.
Engel: I am really glad to see that you are pursuing the perforated synapses and that is a beautiful demonstration of the functional corelate. I have 2 questions. First, do you have a feeling when looking at the areas of the cortex with perforated synapses whether these are homogeneously distributed or whether they exist in small islands?
Teskey: They are patchy.
Engel: Second, you mentioned that at the end you have an increase in spines. Most work shows a decrease in spines, and the assumption is with a decrease in spines you get increased connectivity because then the terminals are directly on the post-synaptic membrane – spines actually interpose a problem for synapses. I am puzzled by that.
Teskey: We found that there was not an increase in the spines, that is to just the synapses on the spines, when you just look at synaptic density. When you look at synapses per neuron, that is when we did found the increase in spines. What Geinisman and Morrell had shown was that there was a overall decrease in synapses, but so much of the perforated type. We didn't find that. We found that there was an overall increase, especially when you express it as the number of synapses per neuron. You have to be careful when you interpret data on the number of synapses per neuron, because what we do is take the tissue just apical to the cell bodies in layer 5 and it is possible that there are some layer 6 dendrites up there, and so some of those synapses that we could be counting my not be on layer 5 dendrites. That is an important caveat.
Avanzini: What would be the relationship between excitatory and inhibitory neurons and synapses? There are several observations that when the number of GABAergic neurons is reduced, the number of synapses is increased, and this is a very powerful system of synchronization. How much do you think would be involved?
Teskey: We need to look at the inhibitory systems, synapses, and neurons. We haven't done that yet. We had to make a choice on where to look for our synaptic contacts, and the choice we made did not allow us to record that many inhibitory synapses. That is a

very excellent comment.
Schwartzkroin: Along those same lines, there's a whole bunch of old literature suggesting that if you decrease GABAergic inhibition in the cortex, the specificity of receptor field expands and you get bigger fields. I think Dr. Avanzini's question is really important, and also regarding the question of whether your cell loss, whether you can say anything about whether your cell loss is preferentially inhibitory versus excitatory neurons.
Teskey: That is an excellent point, and we should look at that. In respect to the cells, we are only classifiying layer 5 pyramidal cells when we look at cell loss.
Wada: Beautiful study, congratulations. Have you looked at those changes that you so elegantly displayed over a period of time? Is it right after the experiment was over, or some length of time after this experiment was done? I ask this particular question knowing that cortical kindling is quite different from limbic kindling in that the time course and quality of kindling affects is quite different in terms of time. If you haven't done that I strongly recommend that you look at some time data later on after your procedure is over.
Teskey: I have been at enough conferences with Mac Burnham, Dan McIntyre, Ron Racine, and Peter Cain to know that you have to look a multiple time points. So I look at 2 days after the last seizure and 3 weeks after the last seizure and the changes in maps are persistent, and the changes in the synapses are persistent. So they are related to the long-term effects of kindling rather than just the transient effects of the seizure.
Wada: This is in cat?
Teskey: This is in cat and in rat, in both.
Wada: What I am not to sure about is whether just 3 weeks justifies the conclusion that there is no change. I think one might consider a bit longer time span perhaps. I say this because in cat cortical kindling, kindled epileptogenesis may get degraded up to 3 months. Second is just a question, you showed us your elegant way of mapping of this area. One of the most intriguing observations we have in clinical epileptology is that there are a condition called epilepsy partialis continua. The seizure begins most often in the face area and never spreads no matter how many times it fires, and indeed when we try to kindle in primate, we could not kindle the face area. We can produce seizure, localized through the contralateral face. After 200 hundred stimulations there was no growth. This is strikingly different from the hand area or leg area and what we often talk about, the Jacksonia march, which often begins hand to face going down this way. I think that it must have a different mechanism instead of going face to arm to leg and so on. This intrigues me a great deal, and I hope your future study would shed some light.
Burnham: Chantal Gravelin in our group has studied sensory motor ability in kids with moderate epilepsy and found that it is impaired. She said the parents told her that the kids were clumsy, and they are. It would be interesting to study auditory ability in kids with epilepsy. A question: You are painting a picture of a very excitable network of neurons. Have you checked your threshold, you predicted that threshold would drop?
Teskey: The threshold for the input-output curve in LTP did not drop, whereas the ADT does drop.
Velisek: On the sub-synaptic level there's an old study from the late 80's maybe early 90's, showing that during kindling there is a redistribution of synaptic vesicles from basically cytoplasm to the synaptic zone, which actually adds to your synaptic strengthening the synaptic activity.

Teskey: Thank you. It is an interesting paper, but unfortunately, if I remember the methodology, they sacrificed the animal pretty soon after the last seizure. It is probably seizure related, and we need to look at a longer time frame.

KINDLING AS A TOOL FOR STUDYING THE ROLE OF SUBCORTICAL STRUCTURES IN LIMBIC SEIZURES

Edward H. Bertram and John Williamson*

1. INTRODUCTION AND DISCUSSION

Determining the circuitry of a seizure has been of interest for many years. The primary goals have been to define the regions that are involved in the seizures and to determine the roles that these different regions play in the course of a seizure. Kindling has been used over the years to examine this issue, and it has many advantages for this purpose. Some of the advantages include the ability to determine exactly where to stimulate, when to stimulate and to time the stimulation so that potentially modifying treatments can be given at predetermined points around the seizure inducing stimulation. Over the years a number of observations have been made regarding the potential roles of the different regions of the brain based on seizure thresholds, speed of kindling, ability to alter established seizure patterns or to slow the rate of kindling.[9, 15, 16, 22] These observations have been interpreted in terms of presumed or known anatomical connections, and our understanding of the significance of findings has evolved with our improving understanding of the anatomical relationships among the different regions. In this paper we will re-examine the issue of seizure circuitry through a series of experiments that were designed around past observations and evolving information about neural networks in the limbic system.

Understanding the circuitry of epilepsy is important for several reasons. From a neurobiological perspective this information can teach us the relative role of the pathways that exist in a single region or between regions. There are multiple connections from multiple sources in any given region of the brain, some of which may be far more critical to supporting seizure spread, but it is not possible, based on anatomical relationships alone, to determine what this role might be. Understanding the important pathways and the structures that are involved in the seizures can then

* Edward H. Bertram, Department of Neurology, University of Virginia, Charlottesville, Virginia 22901 and SEIN, Epilepsy Institute of the Netherlands, Heemstede, 2103 SW, The Netherlands

lead to the second benefit of understanding seizure circuitry: determining targets for therapeutic interventions.

Kindling can be an especially useful tool for the purpose of understanding seizure circuitry, and there have been many such studies over the years. These studies have included examinations of the relative sensitivity of different brain regions to the induction of a seizure or the speed at which the stable kindled state is reached.[9, 19] Other studies have included the effect of lesions or pharmacological manipulations in one region on the rate of kindling or the severity of a kindled seizure induced in another region.[11, 15] One of the great advantages of the kindling model is that the site of seizure induction can be chosen and the timing of an intervention can be consistently linked to the seizure induction by the electrical stimulation. Knowing the site of seizure initiation can allow the investigator to determine the evolution of seizure activity through the circuitry through all of the possible sites. This approach can allow for an examination of which regions are involved in the initiation of the seizure, which regions are targets for seizure spread and which are not involved or only occasionally involved in the seizure activity.

In the course of a seizure a region can serve one or more critical roles. In the role of seizure trigger a region acts as the point of seizure initiation by providing the critical drive that switches the system into a seizure. In the real situation of epilepsy, this region can be a single area or multiple areas acting together or independently. Without these regions, a seizure would never begin. Another role in seizure circuitry is distributor or amplifier. These areas will be an initial target for the seizure activity from the triggers. The distributors pick up the activity and amplify it, and finally spread the seizure to other regions. Another potential role is modulator. These regions may not participate directly in seizure activity, but instead they influence the excitability of the system through a variety of receptors and channels that alter neuronal excitability. Depending on where the modulator acts in the system, it can alter (enhance or reduce) the chance of a seizure (if it acts at the trigger regions) or it may affect the spread of the seizure by acting at a target or distributor site. By approaching seizures with a concept of role and pattern of activation and spread, we can develop an overall picture of the actual circuitry.

Kindling is regarded as an excellent model for the limbic or mesial temporal lobe epilepsy syndrome.[25] It involves many of the same regions of the brain such as the hippocampus and amygdala, and the electrographic seizure activity is similar in the two. Over the years, based in part on our concept of the local and regional circuitry of the mesial temporal structures (hippocampus, entorhinal cortex and amygdala), and in part from the experience in temporal lobe surgery for this syndrome, our interpretation of the data frequently focused on the circuitry within, to and from the hippocampus, which is a central structure in most models of limbic epilepsy. However, in the last 10 to 15 years there has been a significant expansion in our knowledge about the circuitry of the limbic system and the possible changes that may be present in limbic epilepsy. Although the majority of the changes are seen in the traditional limbic structures such as the hippocampus and amygdala, there is some long standing but now increasing evidence that there are changes in subcortical structures as well, including the thalamus, with reported primary changes in the medial dorsal nucleus.[3, 5, 12, 14] This observation is of interest because the medial dorsal nucleus has significant, and largely reciprocal, connections with the limbic structures that are often associated with mesial temporal lobe epilepsy.[1, 10, 23, 27] As a result of these observed structural alterations and regional connections, the questions arise, does this subcortical structure have a role in limbic seizures, and if so, what is that role? There have been a number of studies over theyears that have suggested that this region of the thalamus participates in limbic seizures. In general these

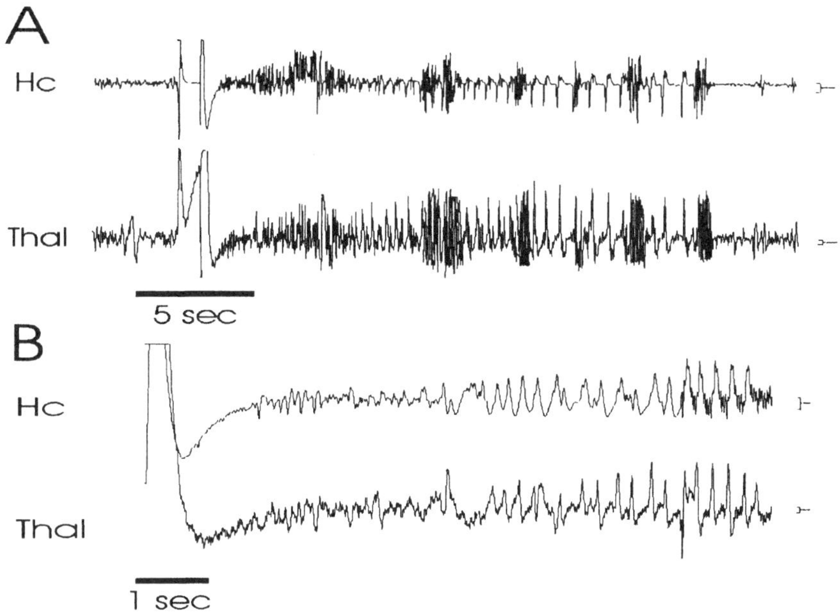

Figure 1. First stimulated seizure in a rat with the stimulating electrode in the hippocampus and a second recording electrode in the medial dorsal thalamic nucleus. B is the same seizure displayed at faster time base. Tracings show early involvement of this thalamic region with synchrony of activity throughout the seizure.(From Bertram et al., 2001.)

studies have examined the effect of drug infusions into this thalamic area on the behavioral expression of limbic seizures, and, over all, the studies have consistently shown that when compounds that are inhibitory or diminish excitation are infused into the region of the medial dorsal nucleus, there is a significantly reduced behavioral seizure severity.[8, 11, 17, 20] These observations led to the conclusion that the medial dorsal nucleus was a key path to seizure spread or generalization. However, these studies were examinations of the behavioral effects of pharmacological manipulation of the medial dorsal nucleus. There was no examination of the potential effect of this manipulation on the electrographic seizure activity. If it were found that, in addition to reducing seizure behavior, intervention directed at the medial dorsal nucleus also reduced seizure duration, the conclusions regarding its role in limbic seizures would expand to include playing a role in the primary circuits that initiated seizures, participate in the triggering of the seizure. The anatomic connections between this thalamic nucleus and the hippocampus, amygdala and entorhinal cortex suggest this possibility, and recent physiological studies demonstrate that this midline thalamic region can induce significant excitatory responses in these limbic sites.[4, 29]

In some initial work our laboratory had examined the potential effect that inactivation of the midline region of the thalamus would have on seizures elicited by hippocampal stimulation.[6] To perform these experiments we used a variation of the Lothman rapid kindling model in which suprathreshold stimulations were given ever 5 minutes.[13] The animals were under urethane anesthesia which causes the afterdischarges to remain short (but stable in duration) over a 6 hour period. There is no behavioral accompaniment to these seizures. Because the rats were in a stereotactic frame, it was possible to obtain consistent placement of electrodes and infusion cannulas. By infusing lidocaine into the medial dorsal nucleus we were able to attenuate the seizure activity in the hippocampus significantly, without affecting the evoked responses in CA1 (suggesting that there was no direct effect of the drug infusion on the hippocampus). We then tried a series of experiments in which, instead of blocking neuronal activity with a sodium channel blocker, we used synaptic agonists and antagonists. When using bicuculline, AMPA and NMDA, all proexcitatory, we found a significant increase in electrographic seizure duration that lasted for several stimulations immediately following the infusion.[28] These observations strongly suggested that this region of the thalamus played an important role in primary limbic seizure activity.

With this observation of modulating seizure duration from infusions into the medial dorsal nucleus, we wished to know if this region was also involved in the seizure activity that was induced in awake animals. We had previously recorded spontaneous limbic seizures in which there appeared to be synchronized onset of seizure activity in the hippocampus and the midline thalamic region, and the seizure activity evolved and stopped together in the two sites.[6] These findings suggested that these two regions were operating in a primary seizure circuit. However, these recordings were obtained when the seizures were fully matured. It was unclear whether this relationship was present from the very beginning, or whether it evolved over time. To examine that question we kindled animals with hippocampal stimulation while recording simultaneously from the hippocampus and medial dorsal region. From the very first, non motor, stimulated seizure, this thalamic region was involved and was synchronized with the hippocampus, and, as the kindling proceeded, there was an well synchronized evolution of seizure activity at the two sites, including cessation (Figure 1).[6]

In addition to determining the role of the medial dorsal nucleus in the seizure circuit, we wished to know if it played a role in the process of generalization to convulsive seizures. Previous work by others had strongly suggested that it might be a critical control point for generalization. To study this issue, we used rats that had been fully kindled to consistent class 5 seizures. A cannula was placed into the midline thalamic region. Five minutes before a planned stimulation the rats received 2μl of 1mM muscimol. Following the infusion there was a significant reduction in seizure duration and behavioral seizure score if the cannula was in the dorsal midline of the thalamus, within the medial dorsal nucleus (Figure 2). If the cannula was outside this small area, the infusion had no effect on the afterdischarge and a slight but non significant reduction in the behavioral seizure score. These observations, together with the previous observations by others, that pharmacological manipulation of the thalamus in the region of the medial dorsal nucleus of the thalamus, suggests that this structure has an important role in the primary circuitry of seizure as well as in the spread of seizure activity to other areas during generalization.[26]

These experiments have clearly suggested that the medial dorsal thalamic nucleus is involved in the circuitry of limbic seizures, but they have not defined the role of this structure in the initiation, synchronization or spread of seizures. For this question we studied two factors: the relative rates of kindling and the afterdischarge thresholds of the several traditional kindling sites (amygdala and hippocampus) with several thalamic nuclei (medial dorsal and the ventro-medial or primary motor relay nucleus).

Stimuli (10 seconds) were given 6 times daily, every other day. The points for study included time to first full Stage 5 motor seizure, progression of afterdischarge duration and the relationship of seizure activity among the different regions (each rat had a bipolar electrode in either the hippocampus or amygdala and in one of the thalamic nuclei).

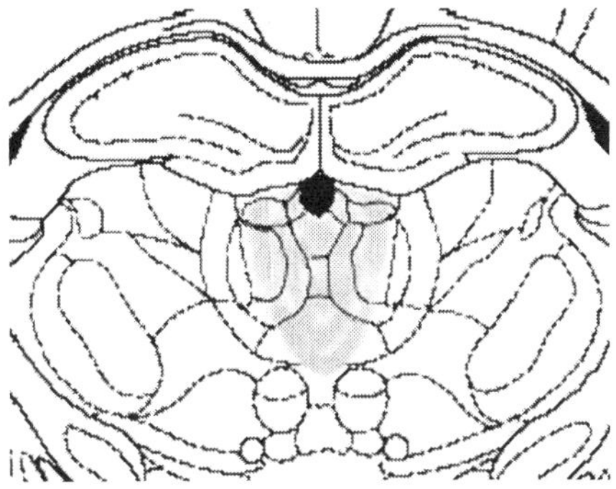

Figure 2. Region of the thalamus in which infusions of muscimol successfully shortened the duration of the afterdischarge and reduced the severity of the behavioural seizure. Infusions into regions adjacent to but outside of the shaded area were uniformly unsuccessful at changing the afterdischarge and had variable but, in general, minimal effects on the seizure behaviour. Shaded area is located approximately 2.3 mm posterior to gregma, below the dorsal hippocampus. It covers the paraventricular nucleus, the medial dorsal nucleus and the most medial aspects of the centromedian nucleus. (Brain section outline from Paxinos and Watson)

Afterdischarges following stimulation in the motor nucleus were intermittent and inconsistent, but, when present, usually lasted only a short time following cessation of stimulation. Seizures induced at this site only rarely involved the limbic sites, and then only after a number of stimulations. Finally it should be noted that the medial dorsal nucleus was always synchronzied with seizure activity in the amygdala or hippocampus, but the ventral medial motor nucleus was only involved following a stimulation in the hippocampus or amygdala when the rat had a motor seizure (Figure 3). The following outline summarizes the relation among the four sites with regard to these kindling attributes. In interpreting the findings from these kindling experiments one must put the data in the context of our understanding of the connectivity of the brain, with emphasis on the limbic system. In addition one must also consider the structural changes associated with limbic epilepsy as well as what previous work has suggested about the relative roles of the different structures in a seizure circuit. The studies presented here have shown that the medial dorsal nucleus is always involved in limbic seizures and involved very early in the course of the seizures. In addition direct stimulation of this midline structure was the fastest route to generalization of the seizure. On the other hand it required a much less intense stimulus to induce seizures from the amygdala and hippocampus. The motor nucleus of the thalamus was involved later in the seizures (if involved at all), it was extremely difficult to induce seizures from this site, and these seizures only rarely involved limbic structures.

Afterdischarge Thresholds
(highest first)
VENTROMEDIA L>> MEDIAL DORSAL >> AMYGDALA = HIPPOCAMPUS

Progression to Motor Seizures
(fastest first)
MEDIAL DORSAL > AMYGDALA > HIPPOCAMPUS >>> VENTROMEDIAL

Progression of Afterdischarge Duration
(fastest first)
AMYGDALA = HIPPOCAMPUS >MEDIAL DORSA L>>> VENTROMEDIAL

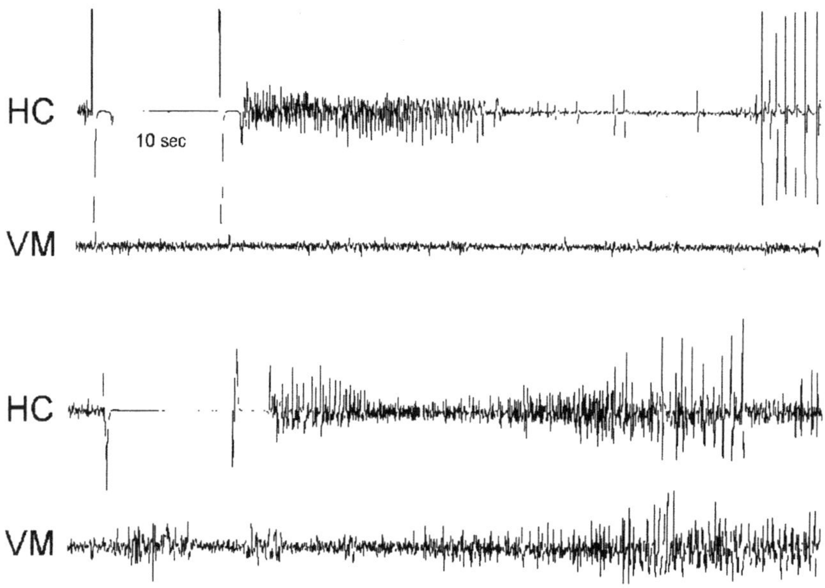

Figure 3. Involvement of the ventromedial nucleus during the evolution of kindling from the hippocampus. The top tracing shows the very first stimulated seizure from the hippocampus. There is a reasonably well developed afterdischarge in the hippocampus that is associated with with a non motor seizure, but there is no involvement of the ventromedial nucleus (second line of tracing). The second tracing shows the changes associated with a motor seizure. The afterdischarge initially involves the hippocampus (top line in tracing), but as the seizure proceeds to the motor stages, there is a gradual recruitment of the motor nucleus (beginning of bottom pair of lines). Contrast this observation to the one in Figure 1, in which the medial dorsal nucleus was involved from the beginning

Previous work from our laboratory and from others had shown that it is possible to attenuate the behavioral severity of the seizures as well as the electrographic duration by infusing agents that inhibit neuronal activity into this region. Anatomical studies have shown that the medial dorsal nucleus has widespread and frequently reciprocal connections with much of the limbic system as well as with a number of regions of the neocortex. Also of importance is the observation from the generalized seizure models that the cortex and thalamus play complementary roles that are of equal importance in spike and wave discharges. The cortex supplies the excitatory drive and the thalamus organizes and distributes the activity into the recognizable ictal pattern.[2] Something similar could also occur in limbic epilepsy. Based on the lower afterdischarge thresholds, the hippocampus and amygdala (and perhaps the entorhinal and piriform cortices) provide the excitatory drive, whereas the medial dorsal nucleus, based on its anatomic connections, consistent involvement and rapid path to generalization may serve to organize the ictal activity and pass it on.

In this chapter we have described an initial examination of the role of several regions in the thalamus in limbic seizures using the classical kindling techniques of afterdischarge threshold, progression of afterdischarge duration and progression of behavioral seizures. In addition we described the effects of pharmacological manipulation of the midline region of the thalamus on afterdischarge duration and seizure behavior. Overall these findings suggest that the medial dorsal nucleus is involved in the earliest stages of a limbic seizure and that it plays a critical role in the initiation, maintenance and spread of limbic seizures. These observations support and expand those made by other investigators in earlier studies which also demonstrated that this region played a role in seizure generalization. Kindling has a number of advantages for studying the functional anatomy of seizures, most important of which is, in this work, the ability to determine precisely where the seizure will be initiated. This advantage allows us to define the relationships among the different regions. How do these findings translate to true mesial temporal or limbic epilepsy? It is unclear, but earlier work had shown that this region is involved early in spontaneous seizures. Kindling has given us targets and potential approaches to examine this question.

2. REFERENCES

1. J. P. Aggleton, and M. Mishkin, Projections of the amygdala to the thalamus in the Cynomolgus monkey, *J Comp. Neurol.* **222,** 56-68 (1984)
2. M. Avoli, and P. Gloor, Interaction of cortex and thalamus in spike and wave discharges of feline generalized penicillin epilepsy, *Exp. Neurol.* **76,** 196-217 (1982).
3. Y Ben-Ari, E Tremblay, and O. P Ottersen, Injections of kainic acid into the amygdaloid complex of the rat: an electrographic, clinical and histological study in relation to the pathology of epilepsy, *Neuroscience* 5, 515-528 (1980).
4. E. H. Bertram, and D. X. Zhang, Thalamic excitation of hippocampal CA1 neurons: a comparison with the effects of CA3 stimulation, *Neuroscience* **92**, 15-26 (1999).
5. E. H. Bertram, and C. Scott, Pathological substrate for limbic epilepsy: is the medial dorsal thalamic nucleus the unifying abnormality, *Epilepsia* (Suppl. 6) **41**, S3-S9 (2000).
6. E H. Bertram, J. J Williamson, C. Scott, P. S. Mangan, and D. X Zhang, The midline thalamus: alterations and a potential role in limbic epilepsy, *Epilepsia* **42**, 967-978 (2001).
7. E H. Bertram, D X. Zhang, and J. M. Williamson, M., Midline thalamic region has a significant modulatory influence over hippocampal seizure activity but cannot induce seizures in the kindling model of epilepsy, *Epilepsia* **41**, (Suppl 7): 10 (2000).
8. R. M. Cassidy, and K. Gale, Mediodorsal thalamus plays a critical role in the development of limbic motor seizures, *J. Neurosci.* **18**, 9002-9009 (1998).
9. G. V. Goddard, D. C. McIntyre, and C. K. Leech, A permanent change in brain function resulting from daily electrical stimulation, *Exp. Neurol.* **25**, 295-330 (1969).
10. M. Herkenham, The connections of the nucleus reuniens thalami: evidence for a direct thalamo-hippocampal pathway in the rat, *J Comp. Neurol.* **177**, 589-610 (1978).
11. Y. Hirayasu, and J. A. Wada, N-methyl-D-aspartate injection into the massa intermedia facilitates development of limbic kindling in rats, *Epilepsia* **33**, 965-970 (1992).
12. C. Juhász, F Nagy, C. Watson, E. A. da Silva, O. Muzik, D. C. Chugani, J. Shah, and H. T. Chugani, Glucose and [C-11]flumazenil positron emission tomography abnormalities of thalamic nuclei in temporal lobe epilepsy, *Neurology* **53**, 2037-2045 (1999).
13. E. W. Lothman, J. B. Perlin, and R. A. Salerno, Response properties of rapidly recurring hippocampal seizures in rats, *Epilepsy Res.* **2**, 356-366 (1986).
14. J. H. Margerison and J. A. N. Corsellis, Epilepsy and the temporal lobes. A clinical, electroencephalographic and neuropathological study of the brain in epilepsy, with particular reference to the temporal lobes, *Brain.* **89**, 499-530 (1966).
15. J. A. McCaughran, Jr., M. E. Corcoran, and J. A. Wada, Role of the nonspecific thalamus in amygdaloid kindling, *Exp. Neurol.* **58**, 471-485 (1978).

16. D. C. McIntyre, M. E. Kelly, and J. N. Armstrong, Kindling in the perirhinal cortex, Brain Res. **615**, 1-6 (1993).
17 J. W. Miller, and J. A. Ferrendelli, Characterization of GABAergic seizure regulation in the midline thalamus, Neuropharmacology **29**, 649-655 (1990).
18 N. Mori, and J. A. Wada, Kindlıng of the massa intermedia of the thalamus in rat,. *Brain Res.* **575**, 148-50 (1992).
19. K. Morimoto, M Dragunow, and G.. V Goddard, Deep prepyriform cortex kindling and its relation to amygdala kindling in the rat, *Exp. Neurol.* **94**, 637-648 (1986)
20. S. Patel, M. H. Millan, and B. S. Meldrum, Decrease in excitatory transmıssion wıthin the lateral habenula and the mediodorsal thalamus protects against limbic seizures in rats, *Exp. Neurol.* **101**, 63-74 (1988).
21. G. Paxinos, and C. Watson, *Compact Third Edition: The Rat Brain in Stereotaxic Coordinates* (Academic Press, San Diego, 1997).
22. S. Piredda and K. Gale, A crucial epileptogenic site in the deep prepiriform cortex, *Nature* **31**, 71:623-625.
23. F. T. Russchen, D. G. Amaral, and J. L. Price, The afferent input to the magnocellular division of the mediodorsal thalamic nucleus in the monkey, *Macaca fascıcularis*, *J. Comp. Neurol.* **256**, 175-210 (1987).
24. B. H. Turner and M Herkenham, Thalamoamygdaloid projections in the rat: a test of the amygdala's role in sensory processing, *J. Comp. Neurol.* **313**, 295-325 (1991).
25. J. A. Wada, Kindling as a model of epilepsy, *Electroencephalogr. Clin. Neurophysiol. - Suppl.* **34**, 309-316 (1978).
26. J. M. Williamson and E. H. Bertram, Kindled limbic seizures and the thalamus: specificity to the midline region, *Epilepsia* **42** (Suppl. 7), 129 (2001)..
27. F. G. Wouterlood, E. Saldana, and M. P. Witter, from the nucleus reuniens thalami to the hippocampal region: light and electron microscopic tracing study in the rat with the anterograde tracer *Phaseolus vulgaris*-leucoagglutınin, *J. Comp. Neurol.* **296**, 179-203 (1990).
28 D. X. Zhang, and E H. Bertram, Modulation of CA1 seizure duration by ınfusion of glutamatergıc agonısts and GABAergic antagonısts into the midline thalamus, *Epılepsıa* **42** (Suppl. 7), 124 (2001).
29. D. X. Zhang, and E. H. Bertram, Midline thalamic region has widespread excitatory input to the entorhinal cortex and amygdala, *J. Neurosci.* **22**, 3277-3284 (2002).

Discussion

Leung: You indicated that your muscimol injection in the medial thalamus suppressed something like a 90-second AD, while it had something like stage 5 seizures, and I think earlier you had lidocaine slides saying that it actually suppressed CA1 seizure or AD, which are of very short duration. I just want you to clarify, is there any relation as to the AD duration and the participation of the medial thalamus?
Bertram: One of the differences is that in the study using lidocaine the animals are anesthetized, and these the seizures are very short. We did this so that the seizures are short and relatively easy to suppress, and I think the spread of seizure activity is relatively limited. So it was nice compact circuitry, if you will, to look at. The study using muscimol with the longer seizures, those animals were awake, where everything seemed to be working.
Leung: So I suppose that I am not as surprised at the later ADs being truncated, because it had spread out the hippocampus. But if it did affect the hippocampal seizure or ADT, that would be somewhat surprising.
Bertram: It didn't. We have no idea what effect it had on ADT; the duration is all we measured.
McNamara: I was wondering whether, when you stimulated to induce seizures in the mediodorsal nucleus, you did that 5 times or whatever, was there a persistence of that? In other words, if you let them go for 3 weeks and then you went back and stimulated them, what happened?
Bertram: We never looked at that. One of the dirty little secrets of rapid kindling is that often, after a little rest, you have to ramp them back up again, even if you are

doing it on an every other day basis. We have done this for 30 to 40 stimulations, but we have never looked at if we stopped it and go later.

McNamara: Have you ever done it in the hippocampus or amygdala?

Bertram: Not formally. So we don't see how long it persists, but the thing is that the generalization happens almost immediately, even in the unkindled animal.

McNamara: It is easy for me to imagine that you might have persistence of this if you stimulated amygdala or hippocampus, but not mediodorsal thalamus, and if you did I think that it would be very interesting.

Post: This reminds me of the concept of multiple thalamo-cortical-striatal loops and the different systems. My question is, are these data consistent with that, and could one setup the study to look to see whether those different loops could be kindled somewhat separately?

Bertram: I am sure you could. The issue to struggle with is to just look at the connectivity of say the mediodorsal thalamus. Although its the connectivity to the limbic system is pointed out, it has so many projections to the motor cortex, the frontal cortex, to the orbital frontal, it basically projects to lots of places and also to some subcortical nuclei. And so, when you start talking about circuitry, it really is a question of where do you jump on the loop? After that point it gets really complex. But I am sure with enough animals and stimulation you could do just that to tease it out. We tried to get into the circuitry from one direction, and doing that you may run into a lot of resistance or walk away from it, whereas if you came in to the same point from a different direction it might work beautifully.

Wasterlain: But if you give a fair amount of current and produce a stage three motor seizure with the first stimulation, is that truly kindling?

Bertram: It's not kindling, you are inducing a seizure and getting a response, but kindling is a process. What I think the implications were, how rapidly did you get to the motor circuit, and that seemed to be a very quick way to get to it.

Wada: What we have done is mediodorsal stimulation, which can cause one-stimulation generalized convulsion with appropriate ADT, threshold. After having given say 10 MD general convulsions in one month, then go to amygdala, and you can just kindle like this (snapped his fingers). Futhermore, if you gave make an ibotenic-acid lesion in the midline thalamus, of course initially it produces a number of seizures that would cause subsequent hippocampal development, in cats. As you know that hippocampal seizure development is slow, and this is tremendously accelerated after the lesion. The fact is that by then the midline thalamus is gone, yet that cat can be kindled. But if mediodorsal nucleus is destroyed, you cannot kindle from that hemisphere. So although both of them facilitate kindling, the loss of that particular thalamic portion has different consequences.

Bertram: I've seem something similar, or we had tried to lesion the mediodorsal thalamus in rats. We've had mixed results, and one of the problems we've run into is that even with reasonable lesions that animals don't do very well for very long, they just seem to become very listless, and don't eat after a while. So we've had transient effects from lesioning, but nothing quite like what you saw.

KINDLING, EPILEPSY, AND THE PLASTICITY OF NETWORK SYNCHRONIZATION

Thomas P. Sutula*

1. INTRODUCTION

When Graham Goddard and colleagues were performing experiments in 1968 that led to the observation of kindling, there was only limited appreciation that neurons and neural circuits were capable of the remarkable capacity for adaptation, alteration, and reorganization of function now referred to as "plasticity". At the time of Goddard's experiments,[1,2] the adult brain was still generally regarded as "hard-wired". In the developing brain, neural activity was recognized to influence synapse formation and patterning of neural connections, but the potent effects of neural activity on neurons and circuitry throughout the lifespan, now appreciated as fundamental processes of neurobiology, were not recognized.

At the time of Goddard's pioneering observation of the kindling phenomena, it was not yet appreciated that synapses in intact animals were capable of long-lasting activity-dependent modification of synaptic transmission such as long-term potentiation (LTP) or long-term depression (LTD),[3, 4] that injury and deafferentation in neural circuits were followed by axonal sprouting and synaptic reorganization,[5, 6] and that cortical maps could be functionally and structurally modified by the patterning of neural activity as well as deafferentation.[7] Goddard's ground-breaking experiments, which investigated the effects of electrical stimulation of neural pathways on learning paradigms in rats, unexpectedly revealed that repeated periodic electrical stimulation induced progressive seizures and permanent susceptibility to additional seizures, a process that he referred to as "kindling". The recognition of kindling, along with synaptic plasticity (LTP and LTD), lesion-induced plasticity, and cortical map plasticity, were major contributions to the contemporary view that neurons and neural circuits are capable of robust short-term as well as permanent structural and functional modification across every level of biological

* Department of Neurology H6/570 CSC, University of Wisconsin, 600 Highland Avenue, Madison, WI 53792, USA; email address: sutula@neurology.wisc.edu

organization, from genes and molecules to neurons, networks, and behaviors, in species ranging from amphibians to primates.

As Goddard's recognition of kindling preceded the descriptions of LTP and the phenomena of cortical map and lesion-induced plasticity, it is not surprising that the significance of kindling as a distinctive form of seizure-induced plasticity was not initially fully appreciated. The permanence of the kindling phenomenon was regarded as an example of information storage in neural pathways, and Goddard speculated that the mechanisms of kindling induction might be relevant to learning and memory formation. Although he emphasized that kindling was potentially important for studying the development of epilepsy and speculated that kindling might produce structural alterations "within or between neurons"[1], the difficulty in detecting anatomical alterations in kindled animals created an initial impression that the kindled brain might be structurally normal, and contributed to the view that underlying mechanisms and alterations associated with kindling might be components of normal developmental or learning processes.

In retrospect, Goddard's emphasis on the pathological outcomes of kindling, specifically the induction of permanently increased susceptibility and progression to epilepsy as a consequence of repeated seizures, is the distinctive feature that distinguishes kindling from other forms of activity-dependent plasticity such as LTP and cortical map plasticity. Kindling is among the most commonly used experimental models for chronic epilepsy, and as originally emphasized by Goddard and others, is a form of plasticity that is potentially relevant to the pathogenesis of epilepsy. Remarkably, however, there is continuing uncertainty and skepticism about the relevance of kindling to human epilepsy and even as a model of temporal lobe epilepsy. This skepticism persists despite the evidence that repeated seizures can be evoked by kindling in species ranging from amphibians to rodents to non-human primates,[1,8-10] are eventually followed by spontaneous seizures,[10-12] and induce alterations including mossy fiber sprouting, neuronal loss resembling hippocampal sclerosis, and memory dysfunction, as observed in intractable human temporal lobe epilepsy.[13]

This analysis will summarize current information about the sequence of alterations induced in hippocampal circuits by repeated seizures in adult rats, the differences in effects of seizures occurring at different developmental ages, and will critically address the relevance of the seizure-induced hippocampal alterations evoked by kindling to human epilepsy and other forms of plasticity.

2. THE SEQUENCE OF NEURONAL AND CIRCUIT ALTERATIONS INDUCED BY KINDLING IN ADULTHOOD

Goddard speculated that kindling might induce structural alterations "within or between neurons"[1], but more than 20 years elapsed before evidence emerged that kindling systematically induces a predictable sequence of long-term functional and structural modifications initiated by alterations in synaptic transmission which eventually evolve into morphological reorganization of neurons and neural circuits of the hippocampus.

2.1 Single and repeated seizures induce increases in NMDA dependent synaptic transmission and gene transcription

The initial episode of network synchronization during a series of repeated kindled seizures induces an acute increase in the NMDA receptor dependent component of excitatory synaptic transmission that is relatively long-lasting (~3 months), but not permanent.[14] The NMDA receptor dependent current increases intracellular Ca^{2+}, which activates other second messengers and signal transduction pathways and eventually results in transcription of early immediate genes,[15, 16] initiating a cascade of gene transcription contributing to long-term and potentially permanent neuronal and circuit alterations.[17] Evidence is accumulating that the neurotrophins including BDNF, NT3, NT4, and the their tyrosine kinase receptors in the trkA,B family are among the genes that may be contributing to long-term effects.[18-22]

2.2 Apoptosis and neuron loss induced by repeated kindled seizures

A single seizure evoked by kindling induces apoptotic neuronal death[23], and repeated seizures eventually result in cumulative neuronal loss.[24, 25] The neuronal loss caused by kindled seizures has been confirmed in multiple laboratories by different techniques, including TUNEL staining for apoptotic neurons, unbiased estimates of neuronal numbers, and silver impregnation for degenerating cells.[23-29] With repeated seizures, apoptosis can be detected in CA1, subiculum, and neocortex.[28, 29] Cumulative neuronal loss can eventually be detected by stereological counting methods, presumably as a consequence of repeated seizure-induced apoptosis, and has been observed in subfields of the hippocampus including the hilus of the dentate gyrus, CA1, and CA3, in a pattern resembling classical human hippocampal sclerosis.[24, 25] The distribution of seizure-induced neuronal loss depends on the site of stimulation.[24, 25]

2. 3 Neurogenesis induced by repeated kindled seizures

Repeated seizures evoked by kindling induce proliferation of neural progenitor cells in the dentate gyrus.[23, 30] The majority of newly born cells differentiate into neuronal phenotypes, but some also differentiate into glia,[23] and in other models appear to integrate into local networks and become activated during recurring seizures.[31] Neurogenesis has been detected after only partial seizures induced by a kindling stimulus.[32]

2. 4 Axon sprouting induced by repeated kindled seizures

Goddard's prediction that kindling might produce structural alterations "within or between neurons"[1, 33] was fulfilled by the observation that repeated seizures evoked by kindling induced mossy fiber sprouting in the dentate gyrus.[34] Mossy fiber sprouting develops after only a few brief seizures, progresses with repeated seizures, and is permanent.[35, 36] The mossy fiber axons of granule cells undergo sprouting in the inner molecular layer and the hilus of the dentate gyrus,[37] and in their terminal zone in CA3.[38] Mossy fiber sprouting has been detected in numerous chronic models of epilepsy[39] and in the human epileptic temporal lobe.[40-42] Seizures also induce sprouting in CA1.[43, 44]

Because mossy fiber sprouting is progressively induced by kindling and is also observed in the dentate gyrus and hippocampus of humans with temporal lobe epilepsy, poorly controlled seizures in humans may induce progressive sprouting and synaptic reorganization.

Sprouted axon terminals in the dentate gyrus of kindled rats form recurrent excitatory circuits,[45] and more than 98% of the synapses formed by sprouted mossy fiber axons in pilocarpine-treated rats are with other granule cells.[46] Although physiological evidence for recurrent excitation in circuits reorganized by mossy fiber sprouting is difficult to detect in normal physiological conditions,[47, 48] evidence for recurrent excitation reliably emerges *in vitro* when inhibition is reduced or in conditions when the extracellular environment is altered.[49, 50] The functional effects of mossy fiber sprouting and seizure-induced recurrent excitatory circuits may be conditionally expressed or unmasked during episodic synchronization when inhibition is reduced and the extracellular environment is altered by activity-dependent increases in K^+_o, thereby contributing to the progressive features of kindling.

2. 5 Glial and astrocytic proliferation induced by repeated seizures

Repeated seizures evoked by kindling upregulate GFAP (glial fibrillary acidic protein) mRNA and protein levels in a time-dependent manner and cause glial cell hypertrophy and proliferation.[51, 52] Gliosis, astrocytic proliferation, and activated microglia are also observed in chronic epileptic tissue.[53] Alterations in glial functions (such as K^+ buffering) could contribute to activity-dependent changes in the extracellular environment that unmask excitatory circuits formed by mossy fiber sprouting and could modify inhibition in neural circuits reorganized be repeated episodes of network synchronization.

2. 6 Interneuron loss induced by repeated seizures

Repeated brief seizures evoked by kindling reduce interneurons labeled by CCK and the neuronal GABA transporter GAT-1 in association with emergence of spontaneous seizures.[12] The reduction of these interneuron subclasses, which provide axo-somatic and axo-axonic inhibition that powerfully regulates propagation of activity into axons,[56] is accompanied by alterations in the kinetics of evoked inhibitory postsynaptic currents and overall reduction in inhibition[12]. Interneurons labeled by parvalbumin and the GABA transporter GAT-1 are also reduced in the human epileptic hippocampus.[54, 55]

3. IMPLICATIONS OF KINDLING FOR THE CUMULATIVE EFFECTS OF REPEATED SEIZURES IN HUMAN EPILEPSY

The relevance of kindling to human epilepsy has been controversial and continues to be debated in contemporary literature. It has been suggested that kindling-like processes could play a primary role in the emergence of seizures following an initial precipitating injury (ie., epileptogenesis), but it is perhaps more likely that the cumulative functional effects of repeated seizures, as revealed by kindling, contribute to consequences of repeated seizures in poorly controlled human epilepsy and development of intractability *after* an initial inciting epileptogenic event. The progressive circuit alterations induced by kindling resemble pathological features of the human epileptic hippocampus, and are accompanied by increasing seizure susceptibility,[1] gradually developing memory dysfunction,[57-60] and with the emergence of spontaneous seizures, by alterations in inhibitory circuitry that are also observed in the human epileptic hippocampus.[12]

3.1 Kindling-induced memory dysfunction: implications for long-term cognitive declines in human epilepsy

Kindled rats develop prominent long-lasting deficits in memory and impairments in a variety of behavioral domains.[57-60] Memory deficits assessed in a radial arm maze gradually develop in kindled rats as a function of repeated seizures,[60] and are associated with cumulative hippocampal neuronal loss resembling hippocampal sclerosis.[25] Kindled rats also show acquisition and retention deficits in the Morris water maze, another behavioral task for spatial memory which is sensitive to hippocampal damage.[57-59]

The induction of seizure-induced behavioral deficits in kindled rats demonstrates that cumulative cellular and circuit alterations induced by repeated network synchronization and seizures have long-term adverse consequences on cognitive function. Recent long-term retrospective and longitudinal neuropsychological studies in patients with temporal lobe epilepsy have documented declines in intellectual functions after 15-30 years compared to age-matched controls in memory tasks and global measures such as IQ.[61, 62] In a recent prospective study of surgically treated patients, memory deficits developed during a period of 10 years in patients experiencing repeated seizures, but were not observed in patients whose seizures were effectively controlled by medications or surgery.[63]

The memory and behavioral dysfunction in kindled rats are most likely caused by cumulative alterations in hippocampal circuits undergoing progressive reorganization as a result of repeated episodes of network synchronization. In people with inadequately controlled temporal lobe epilepsy, cumulative alterations in hippocampal circuits undergoing reorganization as a consequence of repeated episodes of network synchronization may similarly cause progressive evolving memory and cognitive impairments.

3.2 Kindling-induced increases in seizure susceptibility, alterations in inhibitory circuitry, and emergence of spontaneous seizures: implications for progressive features of intractable human epilepsy

Recurring spontaneous seizures are the defining feature of epilepsy, and progression of kindling to spontaneous seizures has been demonstrated in species including rodents and primates.[1, 8-10] Spontaneous seizures develop in kindled rats after about 90-100 evoked Class V seizures, in association with loss of the CCK and GAT-1interneuron subclasses and reduction in functional inhibition.[12] As interneurons labeled by CCK, GAT-1, and parvalbumin are also reduced in association with hippocampal sclerosis in human epilepsy,[54, 55] seizure-induced loss of the critical CCK and GAT-1 could contribute to increasing susceptibility to seizures and some of the progressive features of intractable temporal lobe epilepsy, even if kindling-like processes in humans are not sufficient to induce spontaneous seizures as in rodents or non-human primates.

While it is clear that not all cases of human epilepsy are progressive, MRI studies in human epilepsy, nearly 40 years after Goddard's recognition of progressive features of chronic epilepsy in animals, have demonstrated progressive hippocampal atrophy in a subset of patients with epilepsy.[64-66] Atrophy detected by volumetric MRI analysis is correlated with neuronal loss and circuit reorganization,[67] and in prospective longitudinal MRI studies of patients with new onset seizures without initial precipitating injury, progressive structural atrophy in human hippocampal and neocortical regions has been observed in a subset of patients.[64-66]

As the phenomenon of kindling is observed in species ranging from amphibians to non-human primates, it seems unlikely that human hippocampal neurons and circuits would be unaffected by the cumulative structural and functional effects of repeated seizures, which in rodents eventually contribute to memory dysfunction, increased seizure susceptibility, and spontaneous seizures. Progressive volume loss in prospective imaging studies and neuropsychological declines appearing after 15-20 years of epilepsy suggest that adverse effects of human epilepsy and poorly controlled recurring seizures, like kindling, may become apparent only after long periods of observation.

4. KINDLING AND THE PLASTICITY OF NETWORK SYNCHRONIZATION: EFFECTS OF PERIODICITY OF SYNCHRONOUS EVENTS, AGE, GENETIC BACKGROUND, AND THE PREVIOUS HISTORY OF NEURAL ACTIVITY

Repeated episodes of neuronal and network synchronization during afterdischarges are an absolute requirement for kindling[68] and progressive epileptic modification of neural circuits. The plasticity induced by repeated episodes of network synchronization is prominently influenced by the timing and periodicity of repeated synchronous events, developmental age, and the previous history of synchrony activity in neural circuits.

4.1 Effects of timing and periodicity of synchronous events

Goddard's original description of kindling included systematic assessment of the effect of interstimulus interval on the acquisition of kindling and rate of progression to secondary generalized tonic-clonic (Class V) seizures. Stimulus trains delivered at ~ 24 hour intervals appeared to be optimal for progression of kindling. Stimulus trains delivered at shorter intervals reduced or impaired the rate of progression, which was referred to as the "massed stimulation effect."[1]

There are distinct differences in the outcomes and adverse consequences of repeated seizures as a function of the interval between recurring seizures. Rapidly recurring or "massed" seizures, as during status epilepticus, induce obvious macroscopic damage and adverse long-term effects in neural circuits. Neuronal loss appears to be maximal after an initial episode of status epilepticus, an extreme example of "massed seizures", with little or no continuing loss observed in association with subsequent spontaneous seizures.[69, 70] In contrast, brief seizures evoked by conventional kindling stimulation with longer interseizure intervals (eg., once or twice daily) induce apoptosis which results in cumulative but progressive neuronal loss resembling hippocampal sclerosis.[23-25] Although Goddard's original observations demonstrated that 24 hour interseizure intervals are optimal for inducing gradually progressive effects, permanent increases in seizure susceptibility can be induced when longer stimulus trains are delivered at shorter intervals (eg. 10 sec trains repeated every minute for 6 hours, or trains at 30 minute intervals), which has been referred to as "rapid kindling."[71] These systematic observations indicate that functional and structural consequences of seizures vary depending on the interval between repeated episodes of network synchronization.

4.2 Effects of developmental age

The progressive effects of kindling were initially observed in adult rats, cats, and primates, but it was subsequently recognized that repeated evoked seizures in pre-weanling rat pups also induced long-term increases in seizure susceptibility.[72-74] While long-term increases in seizure susceptibility are induced by repeated network synchronization in both adults and pre-weanling rat pups, notable age-dependent differences include reduced massed stimulation effect in pre-weanling pups.[72-74] Moreover, systematic studies of the effects of seizures occurring during early postnatal development have demonstrated that seizures induce robust systematic long-term functional alterations in hippocampal circuitry which are age-dependent[75]. In contrast to the progressive increases in seizure susceptibility induced by repeated seizures in adult and pre-weanling rats, seizures during the earliest postnatal period induce a long-term decrease in seizure susceptibility accompanied by reduction in capacity for LTP, enhanced inhibition, and memory disturbances in adulthood.[75] It thus appears that the progressive phenomenon of kindling in adulthood is only one example of a range of systematic age-dependent consequences of repeated network synchronization and seizures.

4.3 Effects of genetic background

Genetic background has a powerful influence on the consequences of repeated seizures and status epilepticus, and on the development of kindling. The effects of genetic background on seizure-induced plasticity are evident in strains of rats which have been selected for slow and rapid development of kindling.[76, 77] C57BL/6 mice are remarkably resistant to status induced damage compared to the FVB/N strain, and the phenotypes of transgenic mice are highly influenced by the genetic background.[78, 79] It is highly likely that genetic background in humans also influences structural and functional responses to repeated seizures and plasticity associated with network synchronization.

4.4 Effects of the previous history of neural activity in neural circuits

The consequences of repeated seizures and the phenomena of seizure-induced plasticity are highly dependent on the previous history of activity in neural circuits. Prior kindling reduces the extent of damage from subsequent status epilepticus,[80] and repeated seizures evoked by electroconvulsive shock reduce apoptotic damage from kainic acid.[81] Priming episodes of status epilepticus also reduce damage from a second episode of status epilepticus evoked by kainic acid.[82] Kindling and previous seizures thus appear to have a "neuroprotective" effect against status-induced damage.[80] As noted in section 2.2, however, multiple studies have demonstrated that kindling by brief repeated seizures also induces apoptosis with progressive cumulative neuronal loss revealed by stereological methods,[23-29] but stereological methods have not detected additional loss from repeated spontaneous seizures following status epilepticus evoked by electrical stimulation.[69, 70] The initial seizures evoked either by kindling or status epilepticus may damage vulnerable population of neurons leaving surviving neurons which are relatively resistant to additional seizure-induced loss. Alternatively, these observations could be consistent with a neuroprotective effect of previous seizures evoked by either kindling or status epilepticus against additional damage, perhaps by inducing relative resistance to additional damage as a homeostatic adaptive mechanism to protect circuits from continuing activity-dependent remodeling. Both processes may be occurring simultaneously, suggesting that seizure-induced plasticity may have "bidirectional" features, inducing initial cumulative damage and increased seizure susceptibility while also inducing resistance to additional damage, as a function of timing or inter-seizure interval. In either case, the structural and functional consequences of seizures depend on the previous history of neural activity in neural circuits

4. CONCLUSIONS

The structural and functional modifications induced in neural circuits by repeated episodes of network synchronization during kindling can be phenomenologically distinguished from other forms of activity-dependent plasticity such as LTP/LTD and cortical map plasticity, as kindling reliably induces a predictable sequence of neuronal and circuit alterations in limbic pathways which eventually progress to widespread but specific transsynaptic alterations in pathways remote from the site of initial stimulation. These progressive transynaptic alterations systematically vary depending on age, the timing and periodicity of seizures, genetic background, and the previous history of neural activity. From a contemporary perspective of the variety of activity-dependent processes that influence the structure and function of neurons and neural circuits during development and adulthood, progressive kindling represents only one example of systematic seizure-induced plasticity associated with network synchronization. The molecular, cellular, and circuit alterations which underlie the systematic effects of repeated network synchronization during development and adulthood are manifestations of the neurobiology of the "plasticity of network synchronization". The continuing opportunities for systematic experimental analysis of the plasticity of network synchronization, of which kindling is but one example, are not only likely to improve understanding of epilepsy, but to further advance understanding about the remarkable capacity of neurons and brain circuits to undergo modification and adaptation.

5. REFERENCES

1. G. V. Goddard, D. McIntyre, and C. Leech, A permanent change in brain function resulting from daily electrical stimulation, *Exp. Neurol.* **25,** 295-330 (1969).
2. G. V. Goddard, G., The development of epileptic seizures through brain stimulation at low intensity, *Nature* **214,** 1020-1021 (1967).
3. T. V. Bliss and A.R. Gardner-Medwin, Long-lasting potentiation of synaptic transmission in the dentate area of the unanaestetized rabbit following stimulation of the perforant path, *J. Physiol.* **232**(2), 357-374 (1973).
4. T. V. Bliss and T. Lomo, Long-lasting potentiation of synaptic transmission in the dentate area of the anaesthetized rabbit following stimulation of the perforant path, *J. Physiol.* **232**(2), 331-356 (1973).
5. O. Steward, Reinnervation of dentate gyrus by homologous afferents following entorhinal cortical lesions in adult rats, *Science* **194** (4263). 426-428 (1976).

6. O. Steward, C. Cotman, and G. Lynch, A quantitative autoradiographic and electrophysiological study of the reinnervation of the dentate gyrus by the contralateral entorhinal cortex following ipsilateral entorhinal lesions, *Brain Res.* **114**(2), 181-200 (1976).
7. D. V. Buonomano and M.M. Merzenich, Cortical plasticity: from synapses to maps, *Ann. Rev. Neurosci.* **21,** 149-186 (1998).
8. F. Morrell and N. Tsuru, Kindling in the frog: development of spontaneous epileptiform activity, *Electroencephalogr. Clin. Neurophysiol.* **40**(1), 1-11 (1976).
9. J. A. Wada, J.A., T. Mizoguchi, and T. Osawa, Secondarily generalized convulsive seizures induced by daily amygdaloid stimulation in rhesus monkeys, *Neurology* 8. **28**(10), 1026-36 (1978).
10. J. A. Wada, T. Osawa, and T. Mizoguchi, Recurrent spontaneous seizure state induced by prefrontal kindling in senegalese baboons, Papio papio, *Can. J. Neurol. Sci.* **2**(4), 477-92 (1975).
11. J. P. Pinel and L.I. Rovner, Experimental epileptogenesis: kindling-induced epilepsy in rats, *Exp. Neurol.* **58**(2), 190-202 (1978).
12. U. Sayin, U., et al., Spontaneous seizures and loss of axo-axonic and axo-somatic inhibition induced by repeated brief seizures in kindled rats, *J. Neurosci.***23**(7), 2759-2768 (2003).
13. T. Sutula, Secondary epileptogenesis, kindling, and intractable epilepsy: a reappraisal from the perspective of neural plasticity, in *Brain Plasticity and Epilepsy*, edited by J. Engel, P. Schwartzkroin, S. Moshe, and D. Lowenstein (Academic Press, New York, 2001), pp. 355-379.
14. U. Sayin, P. Rutecki, and T. Sutula, NMDA-dependent currents in granule cells of the dentate gyrus contribute to induction but not permanence of kindling, *J. Neurophysiol.* **81**(2), 564-574 (1999).
15. P. E. Hughes, et al., Differential regulation by MK801 of immediate-early genes, brain-derived neurotrophic factor and trk receptor mRNA induced by a kindling afterdischarge, *Mol. Brain Res.* **53**(1-2), 138-51 (1998).
16. C. Shin et al., Induction of c-fos mRNA expression by afterdischarge in the hippocampus of naive and kindled rats, *J. Neurochem.* **55**(3), 1050-1055 (1990).
17. T. Sutula, et al., NMDA receptor dependence of kindling and mossy fiber sprouting: evidence that the NMDA receptor regulates patterning of hippocampal circuits in the adult brain, *J. Neurosci.* **16**(22), 7398-7406 (1996).
18. P. Osehobo et al., Brain-derived neurotrophic factor infusion delays amygdala and perforant path kindling without affecting paired-pulse measures of neuronal inhibition in adult rats, *Neuroscience* **92**(4), 1367-75 (1999)
19. B Adams et al., Nerve growth factor accelerates seizure development, enhances mossy fiber sprouting, and attenuates seizure-induced decreases in neuronal density in the kindling model of epilepsy *J. Neurosci.* **17**(14). 5288-5296 (1997).
20. K. Rashid et al., A nerve growth factor peptide retards seizure development and inhibits neuronal sprouting in a rat model of epilepsy, *Proc. Natl. Acad. Sci. (USA)* **92**(21), 9495-9499 (1995).
21. H. K. R. Xiao-Ping, N. Serge, B. Luikart, L. Parada, and J. McNamara, Conditional deletion of trkB but not BDNF prevents epileptogenesis in the kindling model, *Neuron* **43**: 31-42 (2004).
22. C. E. Van der Zee et al., Intraventricular administration of antibodies to nerve growth factor retards kindling and blocks mossy fiber sprouting in adult rats. *J. Neurosci.* **15**(7 Pt 2), 5316-5323 (1995).
23. J. Bengzon et al., Apoptosis and proliferation of dentate gyrus neurons after single and intermittent limbic seizures, *Proc. Natl. Acad. Sci. (USA)* **94**(19), 10432-10437 (1997).
24. J. E. Cavazos et al., Neuronal loss induced in limbic pathways by kindling: evidence for induction of hippocampal sclerosis by repeated brief seizures, *J. Neurosci.* **14**(5 Pt 2), 3106-3121 (1994).
25. R. L. M. Kotloski, S. Lauersdorf S, and T. Sutula, Repeated brief seizures induce progressive hippocampal neuron loss and memory deficits, *Prog. Brain Res.* **135,** 95-110 (2002).
26. N. O. Dalby, M. West, and B. Finsen B, Hilar somatostatin-mRNA containing neurons are preserved after perforant path kindling in the rat, *Neurosci. Lett.* **255,** 45-48 (1998).
27. K. Z. Haas et al., Resistance of immature hippocampus to morphologic and physiologic alterations following status epilepticus or kindling, *Hippocampus* **11**(6), 615-625 (2001).
28. S. Pretel, C. D. Applegate, and D. Piekut. Apoptotic and necrotic cell death following kindling induced seizures. *Acta Histochem.* **99,** 71-79 (1997).
29. L. X. Zhang , Apoptosis of hippocampal neurons after amygdala kindled seizures, *Mol. Brain Res.* **55**(2), 198-208 (1998).
30. J. M. Parent et al., Increased dentate granule cell neurogenesis following amygdala kindling in the adult rat, *Neurosci. Lett.* **247**(1), 9-12 (1998).
31. H. E. Scharfman, A.L. Sollas, and J.H. Goodman, Spontaneous recurrent seizures after pilocarpine-induced status epilepticus activate calbindin-immunoreactive hilar cells of the rat dentate gyrus, *Neuroscience* **111**(1), 71-81 (2002).

32. E. Nakagawa et al., Enhancement of progenitor cell division in the dentate gyrus triggered by initial limbic seizures in rat models of epilepsy, *Epilepsia* **41**(1), 10-18 (2000).
33. J. A. Messenheimer, E.W. Harris, and O. Steward, Sprouting fibers gain access to circuitry transsynaptically altered by kindling. *Exp. Neurol.* **64**(3), 469-481 (1979).
34. T. Sutula et al., Synaptic reorganization in the hippocampus induced by abnormal functional activity, *Science* **239**(4844), 1147-50 (1998).
35. A Represa, G. Le Gal La Salle, and Y. Ben-Ari, Hippocampal plasticity in the kindling model of epilepsy in rats, *Neurosci. Lett.* **99**(3), 345-350 (1989).
36. J. E. Cavazos, G. Golarai, and T.P. Sutula, Mossy fiber synaptic reorganization induced by kindling: time course of development, progression, and permanence, *J. Neurosci.* **11**(9), 2795-2803 (1991).
37. T. Sutula et al., Synaptic and axonal remodeling of mossy fibers in the hilus and supragranular region of the dentate gyrus in kainate-treated rats. *J. Comp. Neurol.* **390**(4), 578-594 (1998).
38. A. Represa and Y. Ben-Ari, Kindling is associated with the formation of novel mossy fibre synapses in the CA3 region, *Exp. Brain Res.* **92**(1), 69-78 (1992).
39. M. M. Okazaki, D.A. Evenson, and J.V. Nadler, Hippocampal mossy fiber sprouting and synapse formation after status epilepticus in rats: visualization after retrograde transport of biocytin, *J. Comp. Neurol.* **352**(4), 515-534 (1995).
40. Represa, A., et al., Hippocampal plasticity in childhood epilepsy. *Neurosci Lett*, 1989. **99**(3): 351-355.
41. T. Sutula et al., Mossy fiber synaptic reorganization in the epileptic human temporal lobe, *Ann. Neurol.* **26**(3), 321-330 (1989).
42. C. R. Houser et al., Altered patterns of dynorphin immunoreactivity suggest mossy fiber reorganization in human hippocampal epilepsy, *J. Neurosci.* **10**(1), 267-82 (1990).
43. M. Esclapez et al., Newly formed excitatory pathways provide a substrate for hyperexcitability in experimental temporal lobe epilepsy, *J. Comp. Neurol.* **408**(4), 449-460 (1999).
44. B. N. Smith and F.E. Dudek, Network interactions mediated by new excitatory connections between CA1 pyramidal cells in rats with kainate-induced epilepsy, *J. Neurophysiol.* **87**(3), 1655-1658 (2002).
45. J. E. Cavazos et al., Ultrastructural features of sprouted mossy fiber synapses in kindled and kainic acid-treated rats, *J. Comp. Neurol.* **458**(3), 272-292 (2003).
46. P. S. Buckmaster et al., Axon sprouting in a model of temporal lobe epilepsy creates a predominantly excitatory feedback circuit, *J. Neurosci.* **22**(15), 6650-6658 (2002).
47. J. Cronin and F.E. Dudek, Chronic seizures and collateral sprouting of dentate mossy fibers after kainic acid treatment in rats, *Brain Res.* **474**(1), 181-184 (1988).
48. J. Cronin et al., Electrophysiology of dentate granule cells after kainate-induced synaptic reorganization of the mossy fibers, *Brain Res.* **573**(2), 305-310 (1992).
49. J. P. Wuarin and F.E. Dudek, Electrographic seizures and new recurrent excitatory circuits in the dentate gyrus of hippocampal slices from kainate-treated epileptic rats, *J. Neurosci.* **16**(14), 4438-48 (1996).
50. M. Lynch and T. Sutula, Recurrent excitatory connectivity in the dentate gyrus of kindled and kainic acid-treated rats, *J. Neurophysiol.* **83**(2), 693-704 (2000).
51. M. Khurgel and G.O. Ivy, Astrocytes in kindling: relevance to epileptogenesis, *Epilepsy Res.* **26**(1), 163-175 (1996).
52. A. Hansen et al., Hippocampal kindling alters the concentration of glial fibrillary acidic protein and other marker proteins in rat brain, *Brain Res.* **531**, 307-311 (1990).
53. K. Borges et al., Neuronal and glial pathological changes during epileptogenesis in the mouse pilocarpine model, *Exp. Neurol.* **182**(1), 21-34 (2003).
54. M. J. During, K.M. Ryder, and D.D. Spencer, Hippocampal GABA transporter function in temporal-lobe epilepsy, *Nature* **376**(6536), 174-177 (1995).
55. Z. Q. Zhu et al., Disproportionate loss of CA4 parvalbumin-immunoreactive interneurons in patients with Ammon's horn sclerosis, *J. Neuropath. Exp. Neurol.* **56**(9), 988-998 (1997).
56. T. F. Freund and G. Buzsaki, Interneurons of the hippocampus, *Hippocampus* **6**(4), 347-470 (1996).
57. D. K. Hannesson, P. Mohapel, and M.E. Corcoran, Dorsal hippocampal kindling selectively impairs spatial learning/short-term memory, *Hippocampus* **11**(3), 275-286 (2001).
58. D. K. Hannesson et al., Dorsal hippocampal kindling produces a selective and enduring disruption of hippocampally mediated behavior, *J. Neurosci*, **21**(12), 4443-4450 (2001).
59. T. H. Gilbert, D.K. Hannesson, et al., Hippocampal kindled seizures impair spatial cognition in the Morris water maze, *Epilepsy Res.* **38**(2-3), 115-125 (2000).
60. T. Sutula et al., Deficits in radial arm maze performance in kindled rats: evidence for long-lasting memory dysfunction induced by repeated brief seizures, *J. Neurosci.* **15**(12), 8295-8301 (1995).
61. H. Jokeit and A. Ebner, Effects of chronic epilepsy on intellectual functions, *Prog. Brain Res.* **135**, 455-463 (2002).

62. T. D. M. Oyegbile, J. Jones, B. Bell, P. Rutecki, R. Sheteh, M. Seidenberg, and B. Hermann, The nature and course of neuropsychological morbidity in chronic temporal lobe epilepsy, *Neurology* **62**, 1736-1742 (2004).
63. C. Helmstaedter et al., Chronic epilepsy and cognition: a longitudinal study in temporal lobe epilepsy, *Ann. Neurol.* **54**(4), 425-32 (2003).
64. R. S. Briellmann et al., Seizure-associated hippocampal volume loss: a longitudinal magnetic resonance study of temporal lobe epilepsy, *Ann. Neurol.* **51**(5), 641-644 (2002).
65. D. S. Fuerst, A. Shah, and C. Watson, Hippocampal sclerosis is a progressive disorder: A longitudinal volumetric MRI study, *Ann. Neurol.* **53**(3), 413-416 (2003).
66. R. S. Liu et al., Progressive neocortical damage in epilepsy. *Ann Neurol.* **53**(3), 312-324 (2003).
67. G. D. Cascino et al., Magnetic resonance imaging-based volume studies in temporal lobe epilepsy: pathological correlations, *Ann. Neurol.* **30**(1). 31-36 (1991).
68. R. J. Racine, Modification of seizure activity by electrical stimulation. I. After-discharge threshold, *Electroencephalogr. Clin. Neurophysiol.* **32**(3), 269-279 (1972).
69. J. A. Gorter et al., Neuronal cell death in a rat model for mesial temporal lobe epilepsy is induced by the initial status epilepticus and not by later repeated spontaneous seizures, *Epilepsia* **44**(5), 647-658 (2003).
70. A. Pitkanen et al., Progression of neuronal damage after status epilepticus and during spontaneous seizures in a rat model of temporal lobe epilepsy, *Prog. Brain Res.* **135**, 67-83 (2002).
71. E. W. Lothman and J. M. Williamson, Closely spaced recurrent hippocampal seizures elicit two types of heightened epileptogenesis: a rapidly developing, transient kindling and a slowly developing, enduring kindling, *Brain Res.* **649**(1-2), 71-84 (1994).
72. S. L. Moshe, N.S. Sharpless, and J. Kaplan, Kindling in developing rats: variability of afterdischarge thresholds with age, *Brain Res.* **211**(1), 190-195 (1981).
73. S. L. Moshe et al., The effects of age on the kindling phenomenon, *Dev. Psychobiol.* **14**(1), 75-81 (1981).
74. S. L. Moshe and B.J. Albala, Kindling in developing rats: persistence of seizures into adulthood. *Brain Res.* **256**(1), 67-71 (1982).
75. M. Lynch et al., Long-term consequences of of early postnatal seizures on hippocampal learning and plasticity, *Eur. J. Neurosci.* **12**, 2252-2264 (2000).
76. R. J. Racine et al., Differential sensitivity of genetically Fast vs. Slow kindling rats strains to GABAergic convulsive agents, *Neuropharmacology* **45**(7), 918-924 (2003).
77. R. J. Racine, M. Steingart, and D.C. McIntyre, Development of kindling-prone and kindling-resistant rats: selective breeding and electrophysiological studies, *Epilepsy Res.* **35**(3), 183-195 (1999).
78 P. E. Schauwecker and O. Steward, Genetic determinants of susceptibility to excitotoxic cell death: implications for gene targeting approaches, *Proc. Natl. Acad. Sci. (USA)* **94**(8), 4103-4108 (1997).
79. P. E. Schauwecker, Complications associated with genetic background effects in models of experimental epilepsy, *Prog. Brain Res.* **135**, 139-148 (2002).
80. M. E. Kelly and D.C. McIntyre, Hippocampal kindling protects several structures from the neuronal damage resulting from kainic acid-induced status epilepticus, *Brain Res.* **634**(2), 245-256 (1994).
81. A. Kondratyev, N. Sahibzada, and K. Gale, Electroconvulsive shock exposure prevents neuronal apoptosis after kainic acid-evoked status epilepticus, *Mol. Brain Res.* **91**(1-2), 1-13 (2001).
82. X. Zhang et al., Relations between brain pathology and temporal lobe epilepsy, *J. Neurosci.* **22**(14), 6052-6061 (2002).

Discussion

Moshé: This is a very nice and interesting talk. There is one beautiful paper, actually Goddard's 1969 paper that Dan McIntyre was coauthor of, that said that you could stimulate every hour to every 24 hours to every week. And I think it said that if you stimulate every week that you get better kindling rates than if you stimulate every 24 hours, but if you stimulate every hour it's less so, and 20 minutes has no effect. The problem was that none of us would have a grant or anything that would allow us to stimulate every week; so we sat with the stimulation once a day, which does not necessarily mean that's the best way. So when I start doing kindling I scratched my head and I thought if I stimulated every 24 hours by the time my rat pups finished their stimulations the animal is going to be too old I will lose the electrode and I could not do the study. I decided to take the smallest amount of time that kindling occurs in the adult and use it in the pup. So I stimulated every hour the whole week, that meant I was up to 36 hours, a great achievement. And kindling occured and difference in the development and the first stage five was the same in both the adults and pups. the difference being that the pups did not easily get to stage five. Actually you have to get above the P16 in this strain of rats, you have to be about P17 to get to stage five – before that, the frontal lobes are not developed and the pups don't develop stage five, but stage three and four very quickly. To show that this was kindling, we allowed some of the rats to grow for three months and the rats remembered the kindling. So if you stimulate every hour it's the same response as you see in the adults, but you can stimulate every 20 minutes in the pups, which suggests a propensity of developing of seizures much faster. And what it means, in my mind, is that kindling in early life persists to adulthood in the same side that you stimulated or in the contralateral side. It does not have synaptic reorganization, it may have some minimal loss, but once it is established, it is very very powerful, the same way it stays in the adult. Just to point out, the febrile model that Tallie Baram is using, and she has described that she doesn't get one seizure at a time, there are many seizures that occur in a certain period, suggesting that the kindling-like effect persists. This effect may not include anything of the other features that you see in the adult. Just for the record.

Sutula: I think that I am in agreement with that. This is the kind of perspective that I think is worth paying attention to. Even though the outcome is the same, how it got there may be different. By the way, unless Dan McIntyre remembers otherwise, the original 1969 paper mentioned that when you spaced it beyond the stimulation greater than once every 24 hrs it was not apparent that there was an additional progressive effect. But the closer they got together, less than 24 hours, then it diminished.

Gale: Tom, I am really pleased that you mentioned the Zhang study from Mike Corcoran's group, because one of the things that study shows is that you can completely block the damage and a lot of the plasticity that's related to the damage and not block the epileptogenesis in the kainic acid and pilocarpine models. This is really something that should give everyone cause for thinking about how we go forth and design experiments, because right now, and this is true even for the kindling, we are looking at sequela that could account for epileptogenesis. So much of that sequela is probably related to what will turn out to be epiphenomena, such as the damage and the damage-induced changes that look like they are not necessary for the eventual epileptogenesis. So the question is, why aren't people using models such as what was used by Zhang to block the damage, to block a lot of the things that appear to be unnecessary, and then say maybe it is trophic factors or more subtle changes that are responsible for the epileptogenesis. You bring up this issue of timing for kindling, and I was struck by the fact that very few studies are really saying, okay let's take a kindling situation and then change the parameters so that we make the kindling very weak by, say, giving it every 20 minutes and compare the outcome. Again, we can find out what is related to the experience of the seizures and stimulation and so forth, and what is really related to the epileptogenesis. I don't know why that isn't happening more, but maybe we are stuck in a rut. Thank you for mentioning that.

Gutierrez: What do you think of adding to you list of forms of plasticity, phenotypic plasticity?

Sutula: No problem with that at all. I am not in anyway claiming that the list of things that I put is exclusive. I think the list is going to expand. We were talking earlier about complex systems, I think that noone argues that a neuron and a neural circuit isn't a complex system. That is one of the features of it. You are struggling to design experiments in which you have control over the variables, and I think there's going to be many others things that they add to list.

Wada: Your conceptualization of plasticity of network synchronization including those items you outlined and you are speaking about, within the context of interactive epilepsy: How would your system be applicable to the partial onset epilepsy that is benign and then eventually disappears?

Sutula: I think that it is a great question, and it also occurred to me this morning listening to absence discussions that there are obviously forms of networks synchronization that may not fit in to this at all. The way that I intend this is not to be focused entirely on intractable epilepsy, but in fact to include the kinds of disorders that don't turn out to be intractable. I would say that if there is an underlying neurobiology for this that has common principles, then we should be able to use these principles to explain conditions that don't progress to intractability. The insight provided by fast and slow kindling rats suggest that there really are genes and potentially those kinds of influences that determine why a syndrome is severe, why it isn't, why one episode of head trauma looks comparable but to another person produces a very different outcome. So I think that this kind of logical structure might provide the framework to address the conditions that don't turn out to be intractable too.

GENERAL DISCUSSION 1

Schwartzkroin: A couple of comments related to Dr. Sutula's presentation: First, I really applaud you on what I think is a provocative presentation. So let me be provocative in response and say that I think that it is dangerous to be searching too hard for common mechanisms, that that's the way the field got in trouble at the very beginning when we started looking for a silver bullet that would explain everything, and that the field really exploded when we understood that epilepsy wasn't epilepsy, it was the epilepsies. As an outsider to a kindling conference, I would say that kindling isn't kindling, that these are in fact multiple and potentially very different kinds of phenomena that people are dealing with, and that there is kind of a common tool that they use to study them. I would draw the analogy to the advent of slice technology: slice is a technique. Is there a entity called kindling, or in fact is as, I perceive it to be today, that all the things that we are talking about at this kindling conference are exactly the same kind of issues that we deal with in any epilepsy or synaptic plasticity meeting, that they are in fact a set of overlapping problems that we need to deal with and kindling is a way, a technical way, of approaching that kind of question?

Gallagher: We have performed studies that provide some insight into the commonality of epilepsy mechanisms. In those experiments we used electrical kindling in the amygdala (3 to 5 days after the last of 5 stage 5 seizures) and compared it to chronic cocaine administration (24 to 48 hours after 15mg/kg ip injections, twice a day for 14 days). The low concentration of cocaine in this model did not cause kindling, because cocaine-kindled seizures have high lethality. So in these studies we were not really comparing two models of kindling, but we reasoned that the two models of plasticity could in fact have similar underlying mechanisms. We analyzed amygdala responses to group II and III metabotropic glutamate receptors agonists and it turned out that with cocaine plasticity, the responses completely disappeared whereas, in electrical kindling, the responses were increased. Those data suggest that plasticity in those models is really different. My view, based on our experience, is that there may be common ground in models of plasticity, but also major differences.

Schwartzkroin: That's exactly the point that I was trying to make, that there is not a common or may not be a common underlying basis. I have heard people talking about amygdala kindling, and normal protocols with 24 hrs, and rapid kindling, and mass stimulation, and then people talking about hippocampal kindling, brainstem kindling, thalamic kindling, and chemical kindling. So why do we think that this is a phenomenon or a model for a unitary kind of thing that must have a common underlying basis?

Engel: I would be surprised if anyone in this room thought that it did. That's why somebody asked Ed Bertram whether the mediodorsal kindling was persistent, because if it isn't persistent even though it looks like kindling, it is a different kind of plasticity. I just got up here because I wanted to make a semantic comment on Tom Sutula's presentation that is relevant to your issue, and it is the word "intractability." Tom, you were using intractability to mean chronic epilepsy but not necessarily intractable epilepsy, because you made a jump from kindling to intractable epilepsy. You get kindling, and then you get chronic seizures, and you have benign seizures that go away. That also happens, but there is another step between chronic seizures and seizures that are intractable to antiepileptic drugs – in human, mesial temporal lobe epilepsy that may be a progressive phenomenon. As Ann Berg showed so nicely last year, in her paper in *Neurology*, initially seizures are easily treated and then over time in some patients, and we don't know how many, they come back, and at that point you can't treat them. I think that there is another step, you can call it kindling if you want, but it is another plasticity that goes on that makes these seizures so refractory to medication.
Buzsaki: I think it was Feynman who said that people who like complexity but are not willing to simplify are artists, but people who like complexity and are willing to simplify are scientists. So I think that it is very important that we pin-point those things that are different. One of the most important things about synchronization is it produces plasticity, but that's happening every night in my brain. What we would like to understand is what is the difference between network synchronization on a global scale and another type of synchronization that produces epilepsy. There is a big difference, and I think that this is exactly what we would like to pin-point, very specifically what is the difference between a train that produces LTP and a train that produces an AD and produces a very long-lasting change. As a outsider, I would like to understand where is the progress. Where can we show there's a particular mechanism that makes it so different from everything else? I don't see it yet.
Wasterlain: Obviously, kindling is an incredibly complex phenomenon because it is one of the most intriguing phenomena ever described and after so many years and so many people working on it, we have no understanding of what it really is. However, I think that to say that is as diverse as the epilepsies is very difficult to defend. Obviously the complexity of the phenomenon implies tremendous adversity. There is evidence that part of kindling is a Hebbian phenomenon, which is how Graham Goddard, Dan McIntyre, and the people who worked in kindling early conceptualized it. The link is so much stronger than it is for epilepsy in general. Probably the part that is Hebbian, by which I mean simply a phenomenon of synaptic plasticity induced by experience, is enormous compared to the evidence up to here that this is a really very important part of the epilepsies. In that sense it is a unitary phenomenon – part of it looks like it reflects a single process, which varies with each region, but still a Hebbian phenomenon of experience-dependent synaptic plasticity.
Wada: Kindling has provided at least a window through which to view how seizure and eventually epilepsy might develop, and indeed a number of people have shown by kindling, spontaneous recurrent seizures states evolve. I am just limiting my consideration of epilepsy to partial onset epilepsy, recurrence or partial seizures, that's epilepsy clinically. So we have at least the methodology to produce developmental epilepsy, and we can indeed capture those developed recurrent spontaneous electroclinical seizures. The difficulty is at the main enigma of epilepsy: What ignites those spontaneous seizures? We have no answer

whatsoever, but at least we have opportunity of exploring it through the kindling model. That is where I see the major benefit of this model.

McNamara: I agree with you completely. In response to Phil Schwartzkroin, there are a lot of different questions that one can begin to ask with what is broadly defined here today as kindling. There are many questions, and the process is useful for addressing a variety of these questions. The question that I am personally most interested in, which Juhn Wada articulated just now, is whether it is a model of partial or limbic epileptogenesis. What I would very much like to understand is the pathways that are critical for the development or for limbic epileptogenesis, because as a practical matter we could then intervene pharmacologically with an once of prevention in patients at high risk. Then the question arises, why study kindling as opposed to these many other models, and the answer to that is the rigorous control that one has allows one to address questions that are far more difficult in pilocarpine status epilepticus. It is not that one shouldn't study pilocarpine or that one shouldn't study kindling, but that there are advantages and disadvantages to each. Having said all that, I will make a bold prediction, and you guys can quote me in this darn book, that there will be some signaling cascades in common to multiple forms of limbic epileptogenesis. If you can discover those it might lead to the magic bullet that would prevent limbic epileptogenesis, whether it is cause by a vascular malformation or Ammon's horn sclerosis or some inflammatory process. Now it is easy for me to say that because by the time that anyone figures this out, I will have been 10 feet under for about 3 decades, but that is my prediction.

Schwartzkroin: I don't want to be misunderstood here about what I have said. I don't mean to by any means suggest that kindling is not an incredibly powerful and useful tool. It has helped us understand a whole lot about synaptic plasticity and seizure genesis. I don't have any issue with that. What I am saying, from my perspective, is that the experiments you kindlers are doing sound often like very different kinds of experiments. The repeated stimulation protocols may have very different underlying mechanisms, and therefore talking about kindling as a unitary model is kind of a disservice. It is like our simplistic view that the pilocarpine and the kainate models are equivalent because they produces temporal lobe sclerosis. We know that is not the case. I think that it is important that we understand and try to identify the specific and different mechanisms as well as the similarities. Talking about synaptic plasticity and Hebbian plasticity doesn't do much for me, because all of us pretty much believe at this point that that is the way our brains work, that it is going on all the time. The key as somebody said is to identify what the differences are between normal function and epileptogenesis, and between this kind of process and epileptogenesis, and that kind of process that produces a slightly different form of seizure.

Burnham: I would like to say something about this and then perhaps take the discussion in a different direction. When I am talking about kindling to my students, I define it as all the long term changes that seizures make in the brain. Now some of those may be cell loss, some may be glutamatergic enhancement. It seems unlikely to me that they all have the same mechanism, but they may be important. I teach that some lead to generalized seizures, some lead to threshold drop, some lead to changes in behavior, and some lead to changes in sleep patterns. But to go on a little bit, why are we studying these mechanisms of change? I think in my old age I am getting a little more clinical in my emphasis, and I would say that there probably are two big problems in epilepsy therapy today: One is intractable seizures; the second is probably comorbidities, and no one pays much attention to it at all. If you begin to study the lives of people with uncontrollable seizures, you find that their IQs average around 80, you find that at least 1/3 of them have psychiatric problems, you find that children have ADHD, you'll find that infertility and hormonal problems are common. We can mimic all of these in kindled animals. Now, I believe it was at Kindling 3 that I said seizures hurt your brain, and Nico Moshé disagreed with me. Here we have heard Dr. Shouse talk about behavior problems in her kittens, Dr. Saucier talking about even subtle febrile seizures leading to long term changes, we have heard from Dr. Teskey that we are losing cells and there is reorganization of the cortex, we have heard a nice summary by Dr. Sutula. If these changes occur in animals, they must occur in humans. We act like they don't occur in humans, but they do.

Avanzini: I would like to go back to the previous discussion. If we consider the essential definition for kindling as experience-dependent synaptic plasticity, then we can talk about some unitary concept. If we try to be more specific in this, because this a very general concept, the experience-dependent synaptic plasticity is the basis for many types of plastic changes in the brain including visual rearrangement and so on. I don't know how it can be dealt with as a unitary phenomenon. First of all, not all the types of seizure have the same tendency to create a process, which means that at least one of the basic mechanisms of the different types of epilepsy can be different from each other in this sense. Also from the theoretical point of view, how can one expect that different areas of the brain, which have very different types of organization, may share a common mechanism for synaptic plasticity? For instance, we heard about a thalamic type of kindling and it is done in the rat. But the thalamus of the rat has no interneurons. So that cannot be in anyway comparable to the structure of the limbic system or of the cortex. So I would suggest as an outsider to the field that it would be much more productive to be specific for the different mechanisms of kindling, rather than to try to find a common mechanism, expect that we go to some common basis, which is actually very generic in respect to the phenomenon.

Moshé: I have two comments to make, one going back to Mac Burnham. In clinical practice you know that not all the seizures are intractable, not all the seizures produce brain injury, not all the seizures produce hormonal deficits. And the answer is that some of them do. And the best example that I can give you is if I see a child with a simple febrile convulsion, I would like to have at least a way of predicting what this child is going to do, because the majority are going to be all right in terms of seizures, some are going to develop learning disabilities, some are going to develop intractable epilepsy, and I have no idea why this is going to happen. But the majority of them are going to do all right. And that means that there is not always the bad outcome that you can predict about the change. Which takes me back to the question that I asked you "What was the title of Goddard's 1969 paper," and you told me it was "A permanent change in brain function..." The question is, what makes us believe every permanent change that we see is detrimental to brain function? Sometimes it may have some effects that are not bad, and may even be beneficial. Dr. Teskey, could you please comment on whether you think the seizure-driven increase in the size of the sensorimotor maps is positive or negative.

Teskey: My colleague Jeff Kleim has shown that there is variation in the size of the representation of the caudal forelimb area in sensorimotor cortex (maps) with a strong positive correlation between size and proficiency of forelimb usage on a skilled reaching task. In other words, rats with larger forelimb maps are better on skilled tasks that require use of the forelimbs. We have shown that kindling results in huge maps (~200% increase in map size). Preliminary data from my laboratory also show that kindled rats do not obtain the same level of skill proficiency as non-kindled rats (i.e., that they make more errors) and the way in which kindled rats perform the skilled behaviors is different from control rats. Thus I conclude that seizures that result in extremely large maps lead to functionally maladaptive behaviors. I would add parenthetically that we also have preliminary data that suggest that the kindling-induced interference in forelimb skill proficiency only appears to occur when kindled rats are required to perform a new skilled behavior. When rats were pre-trained on a task that required skilled use of the forelimbs and then kindled, we didn't observe a deficit in their post-kindling abilities.

Goodman: What happens to the map if you train the rat first?

Teskey: The map I showed in my presentation illustrated a distribution of proximal and distal forelimb muscle groups. When rats are trained to use a forelimb there is an internal reorganizing of the map. It is only with kindling or LTP that you actually get an expansion of the map to other areas.

I would like to make a comment, just a sociological observation, with respect to kindling. It has been said that there are two kinds of people; lumpers and splitters. Schwartzkroin is clearly a splitter and Sutula is a lumper. My position is that we have to be both. We have to mindful that we are looking for general mechanisms that cut across species and structures. We also have to be very mindful of the differences of the particular systems we are working in.

Goodman: It is too simple to say you are creating a broader map. What you are showing us is that the map comprises elements. So in reaching, obviously if your shoulder is frozen and yet you have marvelous finger dexterity, you are still going to be impaired, and then by getting the imbalance of the map, that is how you are going to get a functional disturbance of behavior. It is a quality issue.

Teskey: You are absolutely right. A normal animal will bring its elbow in and reach through a slit in a particular way. What we find in the kindled animals is that they just swipe at the pellet. They don't behave the same way as non-kindled rats, and they certainly don't have the same level of accuracy. In fact they behave like rats with small lesions of the sensorimotor cortex. But when we look at the maps in the auditory cortex of the cat we find those neurons to be very responsive to auditory stimuli, they are still working properly, but they've lost that specification for a narrow band of sound frequency. So the behavioral question that we would like to ask is: Are those kindled cats able to make sound discriminations in the same way, as non-kindled animals?

Phillips: Let me go back to the point the Dr. McNamara is making about coming up with important common mechanisms. A couple of the papers that I really enjoyed today were the electrophysiological studies in which people take kindled preparations and say exactly how the sodium channel properties were changed as a consequence of kindling. If that turns out to be a generalizable observation it would be incredibly important. The other point that we heard today is the importance of the induction of GABA release in mossy fiber systems that were not previously doing such a thing. So these seemed to be the general points of convergence, and if we found that those sorts of mechanisms were implicated in numerous forms of plasticity, there may be a common denominator. There is a paper that has stayed with me now for 18 months, a paper in *Science* looking at the properties of dissected tissue from a human epileptic patients, where they did an analysis of the properties of neurons, and the take home message is that the ventral subiculum is a particularly important area in epileptogenesis, and in that region the neurons were responding as though GABA were excitatory. The inference was that perhaps there is some developmental delay, because that's a process that happens during the first week or two of life, the GABA neurons are excitatory and then become inhibitory as the K+/Cl- co-exchanger comes into being. If epilepsy is reverting GABA neurons back into an earlier phenotype, in various forms of epilepsy, that would be a point of common denominator. (To Dr. Gutierrez): I really like the approach that you are taking as well as the other approaches that are using sophisticated electrophysiology to try to give us a signature of what is happening in this pathological tissue. And I think that is where the field is really going to make a lot of advances.

Burnham: If we can return to the question of intractable seizures, I do hope your patients do well. Sixty percent of people put on drugs benefit enormously, but about 1 in 250 still have some seizures and about 1 in 500 have nearly constant seizures. Lisa Kalynchuk and John Pinel have been talking about long-term kindling, going up to about 100 or 200 seizures. One of my friends had a child with epilepsy, and she counted seizures for 2 years and then stopped because she had reached 3000. I have known people who have had multiple seizures everyday. Now these people have got to be kindling, and you may argue that some of the changes are neutral or beneficial, but we get sprouting and cell loss. These cannot be helping. If you look at a any group of people with intractable epilepsy, they have real problems in life. I don't believe it is all social stigma or economic deprivation. I think that their brains are deteriorating, and I think we should face it and try to stop it.

Heinemann: The debate is in part on plasticity. I think that we have to remember that there are 4 types of plasticity. One is developmental plasticity, and it was once predicted that if we understand developmental plasticity we also understand learning and memory. This has not become fully true. So the mechanisms of developmental plasticity are not fully explaining memory and learning. There is repair plasticity. Then we have a phenomenon I would call pathogenic plasticity, and it is not restricted to epilepsy. So people who get chronic pain, who get tinitus, who get spasticity of the throat, that is not an immediate event. A reorganization of the network underlies these phenomena. Part of what we study in kindling is pathogenic plasticity. The issue here is not so much to find the magic bullet, but really to understand where we have the separation from repair plasticity to pathogenic plasticity. What are the key elements that make this difference? For example, in epilepsy we have cell loss and a lot of what we see is repair plasticity. Some of that is really antiepiletogenesis, yet there is a point somewhere when we repeatedly stimulate without making lesions that we do not necessarily cause repair plasticity, but perhaps invoke developmental plasticity. On each level, on the molecular level, on the cellular level, on the network level, we have to understand where this condition becomes pathogenic and pathogenic plasticity occurs. If we could define that more precisely, then I also think we would have the so-called magic bullet that you are looking for.

Gale: I would like to make a comment about that, because I don't think that it is either or. I think that you can have adaptive plasticity that can be maladaptive, and it depends on the context. If you have some sort of abnormal development in the brain, then there may be other things that are abnormal that actually serve a compensatory role, which by themselves out of that context could be bad. So fever, for example, is something that is not good, unless you are sick, in which case the fever is activating immune functions to help fight off the bacteria. We can't take these things in isolation and say this is good, or this is bad; it really has to be looked at in context. Uwe Heinemann is absolutely right that some of these things are adaptive and maladaptive, but we really have more to be splitters in the sense of asking of every phenomenon whether, in fact, it serves some sort of adaptive function in context, or whether taking it away may in fact make the situation worse. We really haven't gotten there yet. I would actually like to ask a slightly different question because we had so many wonderful talks this morning and this afternoon that examined circuitry and really got at issues of interesting interplays between situations where you have a propensity to switch from something that is excitatory to inhibitory, or from something that looks like it is antiepileptogenic to something that is epileptogenic. There are all these dynamics. I also heard a lot of things that had to do with brainstem phenomena versus forebrain phenomena and would like to go back to that in just a second. Perhaps Dr. Gutierrez will say something about the switch from excitatory to inhibitory.

Gutierrez: I would like to thank you for the previous comment on my work. I forgot to say earlier that one of the things that Augusto Fernandez Guardiola always said was that kindling is a model of plasticity, and the seizures, once they begin to appear, constitute the noise of the system – that's the storm and the artifact obscuring the underlying plasticity. What is really interesting about kindling is the progression of the synaptic changes, and all the types of plasticity imply progression of changes, and it may be that seizure is the ultimate plasticity. But, of course, there are many types of plasticity going on before the seizures appear. In that sense, one of the things that I very much wanted to say was that I use kindling as a model of plasticity of phenotypic changes. One of the things I see when I read papers on kindling is that many people think that kindling is established once the seizures appear. Many times one forgets that the seizures are evoked after many stimulations, and many of the phenomena that we see after these seizures can be obtained after just one seizure. You don't need to kindle the system to get the same plastic changes that you see after one seizure. One the other hand, kindling itself is making a lot of changes that do not necessarily come after seizures. One of the things that I wanted to do in my work was to differentiate the effects of kindling a phenomenon, and I called it kindling of the phenotype and I noticed that this effect that I was seeing after 5 seizures could be evoked after just 1 seizure. To get a seizure you need hyperexcitability, and this hyperexcitability can be produced without yielding a seizure, and this hyperexcitability also produces a phenotypic change. I just wanted to call attention to this big difference of classically kindling seizures, which produces specific effects, before and after the seizures appear, and producing a single seizure, that perhaps produces the same effects as the many full-blown kindled seizures. Kinding "a change" does not necessarily imply provoking seizures.

Engel: I wanted to just see if I could get some consensus and some agreement among the lumpers and the splitters here. We could agree that kindling is not a phenomenon, it is a technique, and that the unitary concept for us all who have been doing this is in the technique, not in the phenomenology that it produces. Kindling is of interest and of great value because it provides us with a mechanism to push the brain to change functionally and structurally in some enduring fashion in a highly selective way, and to bring it under laboratory control utilizing mechanism that are available to the normal brain. It allows us to study a tremendous number of plastic mechanisms. To try to put all the effects of kindling into a single concept is ridiculous, but to put it into a technique is not.

Wada: This is what I was referring to, that kindling is not a unitary process, that there is a tremendous complexity going on. But using kindling as a technique, you can induce in the animal an entrenched susceptible state. Then the animal can graduate from that to the spontaneous seizure state. As a physician, clinician, neurologist, the most important thing to understand is, what is the last spark that evokes spontaneous seizure. Kindling allows us to have the opportunity to explore this.

Gale: I want to ask a question about the issue of brainstem. As indicated today, you can do things by way of brainstem stimulation that will augment the possibility and may be necessary for kindling in the limbic system. I heard several things this morning that reiterated or echoed this. When she talked about the kittens, Nicki Shouse was talking about this incredibly rapid kindling phenomenon and rapid progression to spontaneous seizures, and then she showed us a video of a kitten having a seizure. Those seizures were very much brainstem seizures, in that they went into the clonic-tonic seizure that is typical of brainstem involvement. I asked her about the brainstem, and she said it gets involved even before you see the stage 6 seizure, even when they are at stage 3; so there is brainstem involvement very early on. One wonders whether that may have something to do with the rapidity of kindling, which would be consistent with facilitating the kindling as Dr. Chiba had described this morning. I also noticed that Dr. Onat mentioned ethosuximide and the possibility that ethosuximide was actually interfering with the kindling you want to test. Ethosuximide is notorious for being very good for blocking brainstem seizures. When you give pentylenetetrazole, for example, you can block the brainstem seizures without blocking the forebrain seizures with ethosuximide. That is another situation where maybe you are interfering by shutting down brainstem mechanisms. I was intrigued by what Jana Veliskova presented about the differences in seizure susceptibility in different rat strains, in fast and slow kindling animals and Sprague-Dawley rats. In those videos there was hindlimb backing up behavior that has a very brainstem look to it . In adult rats minimal electroshock causes beautiful limbic type seizures, stage 5 seizures, but when you do that in neonatal rats, the seizures look very much like brainstem seizures. As animals mature they lose that brainstem phenotype and look much more like a forebrain phenotype. Again this may have something to do with the rapidity with which they were kindled. Different mouse strains are wildly different in the extent to which they will respond to a drug. For example, kainic acid in some mouse strains will only give you a brainstem seizure. You will not get the forebrain seizure, because they will die before that happens. In other mouse strains you can see a mixture of both. That's another thing, the difference of kindling in the amygdala, where you are targeting the forebrain, and giving kainic acid, where you targeting both the hindbrain and forebrain because the drug is going everywhere. Uwe Heinemann mentioned that you get a different sequela after using kainic acid as opposed to the focal stimulation, which could also be because of the presence or absence of hindbrain components. I wanted to ask Jana Veliskova whether you think that the different animals are showing different phenotypes either because of maturity or because they are from different strains.

Veliskova: In rats I really believe that it is maturity. The immature brain is more sensitive to the development of brainstem seizure. You can see it also with the NMDA model, which is very typical of brainstem seizures, and these NMDA seizures are very pronounced in immature rats. I really believe that it is more a function of maturity than the difference between the strains.

Gale: Do you think that plays a role in terms of the ease at which you can kindle the animal?

Veliskova: Yes. Also in the flurothyl model, there is no separation between the clonic and tonic-clonic seizure almost at all. Sometimes the rats go directly to the tonic-clonic seizures without any clonic component. This is specific to immature brains. In contrast, in adult rats there is a separation between the first clonic seizure and the progression into tonic-clonic seizure several minutes in flurothyl seizures.

Chiba: In rat pups the peak of myelination is two weeks postnatal; so the brainstem and limbic system involvement is probably key in rat pups. But there is almost no involvement of the motor cortex. And during sleep, probably the susceptibility to synchronization is very likely to occur. We can see in the sleep record a spindle or slow wave. So in sleep, there is an increased tendency to the synchronization, and so the brain seizure tends to occur. This is my hypothesis.

Burnham: It was a long time ago, I think I was influenced by Juhn Wada's work, that I framed the idea that the kindled convulsion is driven by the brainstem. Then Dan McIntyre got geared up, performing a series of studies making it look almost inevitable that it was a motor cortex that was doing it. But from the work of Drs. Chiba and Bertram, it is looking as though the force is coming back with us, and I would say probably somewhere in heaven Penfield is smiling at us.

Gale: What a great note to end on, but first I would like to make the observation that I was sitting here today thinking why today has been so much fun, I mean really really fun. And the reason that it was really fun for me is that everyone was talking about networks and systems. For a whole day we have heard about networks and systems, not to ignore channels and the molecules and the genes and the mechanistic fundamental reductionistic pieces. But it's all in the context of networks and systems. I don't think that I sat through an entire day of talks anywhere at any meeting that have been so network and system and functionally oriented, and I thank everyone for that.

ELECTROPHYSIOLOGICAL AND ANATOMICAL DIFFERENCES, BEHAVIORAL COMORBIDITIES AND GENE EXPRESSION IN FAST AND SLOW KINDLING RAT STRAINS

Dan C. McIntyre and Krista L. Gilby[*]

1. INTRODUCTION

The view that genetic mechanisms underlie many forms of epilepsy is widely accepted.[1] Thus, genetically-derived models of epilepsy can serve as important vehicles to study the mechanisms of epileptogenesis,[2] particularly if they can isolate underlying factors that are uncontaminated by a recent history of seizure activity.

While kindling has been used frequently to model complex partial seizures and temporal lobe epilepsy (TLE) in man,[3,4] there has been a long-standing need for a genetic model of TLE. To meet this need, we selectively bred two new strains of rats for their vulnerability versus resistance to amygdala kindling.[5] The selection was based on the *rate* of amygdala kindling development to stage-5 convulsions,[6] but did not take into account afterdischarge (AD) thresholds or other focal attributes. The original parent population used for the selection was a Long-Evans Hooded and Wistar cross. Within 6 generations of kindling selection, there was no evidence of overlap in kindling rates between the two new strains (Fast vs Slow kindlers). Thus, after the 11th generation, selection was relaxed to involve second cousin breeding within each strain, and no apparent drift in the kindling attributes of the strains has occurred since that time.[5,7] It is also important to note that while AD thresholds were found to be similar in the amygdalae of the newly developed strains, they were different in the adjacent piriform and perirhinal cortices, and dorsal hippocampus (Figure 1). In these cases, the seizure-prone Fast rats exhibited significantly lower AD thresholds. Moreover, in all four structures the initial AD duration was significantly longer in Fast compared to Slow rats, particularly in both the basal and

[*] Department of Psychology, Institute of Neuroscience, Carleton University, Ottawa, Ontario, Canada, K1S 5B6

medial nuclei of the amygdale,[7] where discharges in Fast rats were more than twice the duration of Slow rats. This finding strongly suggests that seizure offset and/or recruitment mechanisms are quite different in efficacy between the amygdalae of the strains. In addition, because AD thresholds in Fast rats were much lower in cortical areas surrounding the amygdala, areas believed to be important in the convulsive generalization process,[8,9] augmented recruitment following focal amygdala seizures likely contributes substantially to the ability of the seizure to synchronize those adjacent cortical areas and produce the rapid kindling profile observed in Fast rats.

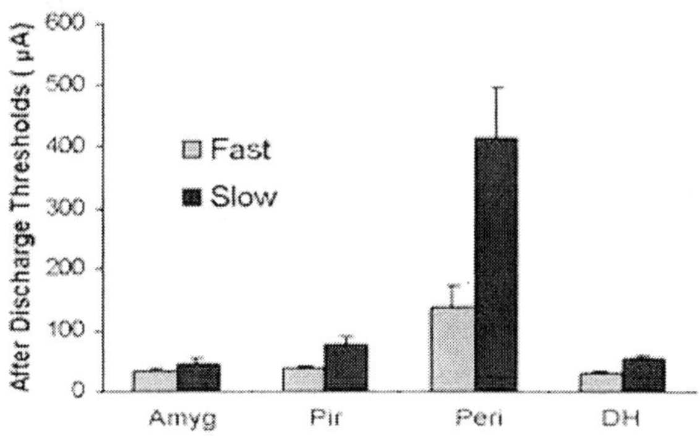

Figure 1. After-discharge thresholds (µA) in the amygdala, piriform (Pir) and perirhinal (Peri) cortices and dorsal hippocampus (DH) of Fast and Slow rats before kindling.

2. MECHANISMS

Mechanisms underwriting strain differences in seizure proclivity are potentially enormous. Examination of neurotransmitter systems in baseline conditions showed relatively small differences in several systems, including excitatory and inhibitory amino acids,[10] monoamines[10,11] and others. However, when challenged through various means, such as stress, kindling or systemic drug application, significant strain differences in transmitter responses were observed (e.g., 12). Interestingly, innate differences in transmitter activity were paralleled by differences in behavior between the strains. For example, intraperitoneal (i.p.) injection of the excitatory amino acid, kainic acid, resulted in faster onset and more severe expression of status epilepticus in Fast compared to Slow rats.[13] Similarly, manipulation of the GABA system with i.p. injection of antagonists, including picrotoxin, bicuculline and pentylenetetrazol, induced convulsive seizures in Fast rats at much lower doses than in Slow rats.[14] This increased sensitivity of Fast rats to negative GABA modulators was matched by an increased sensitivity of Slow rats to positive GABA modulators like pentobarbital and diazepam.[7]

Due to the obvious difference in sensitivity to GABAergic modulators between the strains, we examined expression profiles of $GABA_A$ subunits in various limbic structures. Results of that study revealed no strain differences between many of the subunits (δ was not examined), with the exception of four of the six alpha subunits, including α1,2,3and 5. Specifically, the α2,3 and 5 subunits were over-expressed in the amygdala, piriform and perirhinal cortex of Fast rats, yet were under-expressed in Slow rats compared to Long-Evans Hooded controls (LEH; one of the two original parent populations). By contrast, Slow rats over-expressed the α1 subunit in comparison to LEH rats, while it was under-expressed in Fast rats.[15] Interestingly, α1 is the dominant adult subunit, while α2,3 and 5 subunits are expressed most strongly during development. As the latter subunits are over-expressed in Fast rats, we interpreted this to mean that they perhaps have experienced a stall in their development, which maintains them in a more juvenile developmental stage in relation to Slow rats.[15] This proposal is consistent with observations that juvenile rats and humans are more seizure-prone than adults.[16]

We then questioned whether the differences in pharmacology and receptor subunit expression also might be associated with differences in GABAergic physiology. Here we recorded the spontaneous miniature inhibitory postsynaptic currents (mIPSCs) in the perirhinal cortex of Fast and Slow rats compared to LEH controls. The perirhinal cortex was selected because of its proposed involvement in convulsive seizure generalization,[9] and its large difference in focal AD threshold expression between the strains.[7] We performed patch clamp recordings from pyramidal neurons and interneurons (identified by biocytin fills) from layers 3 and 5 in coronal slices of the perirhinal cortex.[17] Although there were small strain differences in the recordings from pyramidal cells, the largest differences were observed between various interneuron populations. As can be seen in Figure 2, the mIPSCs in Fast rats were relatively small in amplitude and slow in deactivation, while the mIPSCs in Slow rats were very large in amplitude and more rapidly deactivating.[17] Interestingly, the negative charge transfer associated with these two different mIPSC profiles was similar. Since there was no difference in the charge transfer, yet a huge difference in mIPSC amplitude, one might predict that the GABAergic contribution to network activity would be associated with timing. Here the high amplitude, rapidly deactivating responses in Slow rats might promote synchrony of interneurons in the higher frequency range, while the smaller, delayed deactivation in Fast rats could favor lower frequency oscillations. Certainly, power in the lower frequency range is more commonly associated with the kindled state and epilepsy,[18] and might result in better recruitment.

Differences in oscillatory behavior have been speculated to differentially impact learning and memory. For example, there is a clear benefit for memorial activity that is tied to higher frequency oscillatory behavior, particularly in the gamma range.[19] Preliminary evidence in EEG recordings from the strains indicate that Slow rats show increased power in the higher frequency range compared to Fast rats during awake immobility. If such a disposition is predictive of better learning and memory, Slow rats should be the better learners of the two strains. Indeed this is the case, as indicated below.

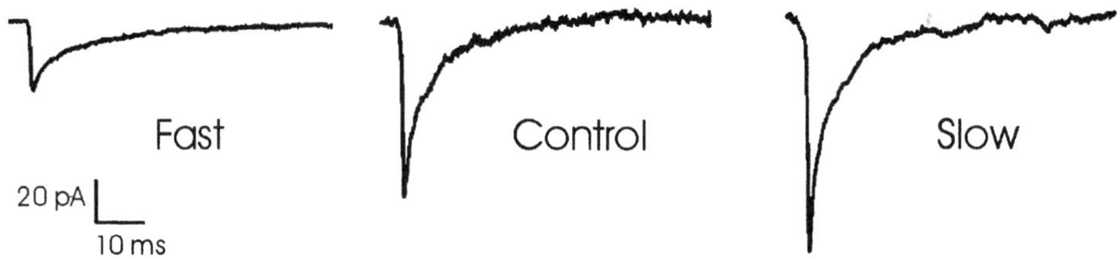

Figure 2. Examples of averaged spontaneous mIPSCs from non-pyramidal neurons in layers 3-5 of the perirhinal cortex in *in vitro* slices taken from adult Fast, Control (LEH) and Slow rats.

3. BEHAVIORAL COMORBIDITIES

Being seizure-prone or -resistant is not without consequence for normal behavior. In a variety of behavioral tests, Fast and Slow rats have shown themselves to be quite different. Considering basic levels of activity, Fast rats express a pattern of high activity that uniquely is non-habituating over days in an open field test compared to lower activity with typical habituation in Slow rats.[20] The heightened activity levels in Fast rats were also apparent when placed in a stressful situation involving restraint. When tightly confined in a triangular shaped plastic bag, with their nose in the air hole at its apex, Fast rats struggled about 50% of the restraint time, while Slow rats mostly remained immobile and only struggled about 10% of the time.[12] One observes similar response differences between the strains when trying to connect the leads of the rats to the stimulator for daily kindling trials. The Fast rats are constantly on the move, while Slow rats are usually rigidly immobile.

There is a sense of fear in the immobility expressed by Slow rats, which translates well to reality when tested in various learning paradigms. For example, in the inhibitory avoidance test, although both strains show similar *initial* step-down latencies onto the grid floor from the safety of the wooden platform immediately before footshock, the next day Slow rats show complete immobility on the platform for nearly the entire 5 min session while Fast rats exhibit relatively high activity.[20] In other aversively motivated tests of learning, Slow rats retain their superior status. Across numerous versions of the Morris Water Maze, a spatial learning test that involves swimming to a submerged platform, Slow rats perform much better than Fast rats. This facility includes variations involving the simplest of conditions, i.e., using a fixed position for the submerged platform over days, to much more difficult tasks where the platform is located in a new position each day.[21] In both cases, Slow rats are better in acquisition. Interestingly, during acquisition, Fast rats appear more distractible than Slow rats yet once the task is well learned by both strains, Fast rats are seldom distracted by extraneous stimuli and appear highly focused compared to Slow rats. Several of these differences in behavioral attributes associated with learning suggest that Fast rats might be expressing a common condition known in humans as attention deficit hyperactivity disorder.(ADHD)[21] Accordingly, it is rarely reported but routinely acknowledged by both therapists and patients that children and adults with ADHD can 'hyper-focus' on tasks they know well, or in behaviors that are easy and/or enjoyable for them. Thus, like Fast rats, they are only distractible during early acquisition of a task before they become skilled in performance. Nonetheless, Fast rats rarely approach the level of performance of Slow rats in learning tasks without receiving special training.[21] It is also important to note that up to 20% of children with ADHD are epileptic compared to ~1% of the normal population.[22] Thus, a comorbid relationship between epilepsy and ADHD clearly exists.

4. NEUROANATOMICAL DIFFERENCES

Considering the number of differences between the two strains, including electrophysiological and behavioral outcomes, one might not be surprised to find some associated differences in neuroanatomical features. In routine histological preparation for electrode identification in kindling studies, we noticed that the lateral ventricles of Fast rats were considerably larger than Slow rats. This observation precipitated a more extensive evaluation of their brains by comparing every 5th coronal section at comparable levels from the anterior commissure through the ventral hippocampus. It was immediately apparent using volumetric analysis that the ventricles were much larger in Fast rats. This outcome is likely a direct consequence of the dorsal hippocampus being much smaller in Fast compared to Slow rats (Figure 3). Since the dorsal hippocampus is believed to be critical for the normal acquisition of spatial behaviors, its reduced volume in Fast rats may lie at the root of their spatial learning deficits in the Morris Water maze[21] compared to Slow rats as described earlier. Our preliminary evidence reveals that these strain differences in neuroanatomy are evident by post-natal day (PND) 21.

An additional gross neuroanatomical difference between the strains is evident in the large myelinated fiber bundles. For example, the corpus callosum is clearly smaller/thinner in Fast compared to Slow rats (Figure 3). The reason for this difference is not yet known. However, it is unlikely due to fewer axons, since the overall 'size' of the Fast brain at the cortical level is at least as large as that of Slow rats; thus, the number of cortical cells contributing to the commissural axon bundles should be similar. More than likely the difference lies in the myelination process, where the quality or quantity of the myelin is inferior in Fast rats. Evidence for this suggestion has been forthcoming from our molecular studies, which are designed to detect differences in constitutive gene expression within individual brain structures of the strains in their naïve state.

5. GENETIC DIFFERENCES

We anticipated that many of the observed strain differences are under genetic control and are likely polygenic in origin. Thus, in the past 4 years we have used a variety of molecular tools to expose putative candidate genes that may underwrite some or all of these differences, including their seizure proclivity. The technique used most frequently in our laboratory has been differential display (DD). In this procedure, expressed mRNA species from specific cellular populations are reversed transcribed to create cDNAs that are then amplified via PCR and fractionated on an acrylamide gel to enable visualization of expression differences.[23,24] As such, DD serves as a powerful random gene screening tool capable of screening several RNA populations simultaneously and targeting both known and novel genes.

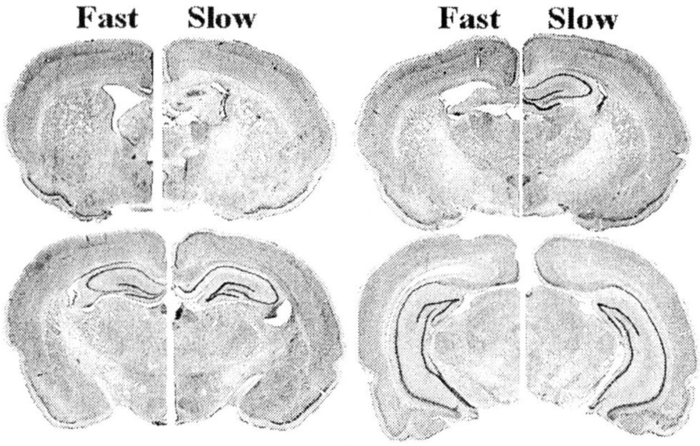

Figure 3. Illustration of brain morphology differences between adult Fast and Slow rats. The left half of each brain is from a Fast rat taken from the same coronal level as its paired Slow rat. Brains were initially prepared in a brain blocker, which presents each brain in the same coronal plane. Note that although both brain halves are from the same anterior-posterior level (indicated by the optic tracts, temporal lobe features and thalamic nuclei), the dorsal hippocampus begins posteriorly in Fast rats and shows larger ventricles and a thinner corpus callosum than Slow rats.

Using DD, we have asked two fundamental questions regarding how differences between the strains might arise. First, we asked whether the strains exhibit differences in constitutive gene expression within limbic structures while in their naïve state, and whether one strain thereby diverges from the LEH parental strain to make them more or less seizure prone. Clearly, genes that hold the most interest for us are those that are expressed in all members of one strain, but in no members of the other strain. Indeed, many such differences in constitutive gene expression have been identified within limbic structures of our naïve strains suggesting that differences exist in basal cellular processing. Interestingly, the majority of expression differences observed in limbic structures of the strains involve genes linked to lipid metabolism, transport and/or synthesis. It is also interesting, and methodologically relevant to studies that use outbred rat strains, that while our strains exhibit considerable homogeneity with respect to similarities and differences in constitutive gene expression, several examples of within strain variability were observed in LEH rats (Figure 4). This finding may serve as a putative explanation for higher variability in seizure susceptibilities and behavior within outbred animals, and will certainly introduce noise into molecular experiments that use these animals. The second use of DD in our laboratory was designed to investigate whether strain differences are predetermined during embryogenesis via genetic or epigenetic mechanisms. Thus, comparisons of the level and timing of developmental gene expression across numerous embryonic timepoints between the strains are underway. To date, we have identified one gene in thousands screened that showed a differential timing of expression between the strains such that it is expressed in Fast embryos by embryonic day 11 (E11), but is not yet evident in Slow embryos at that

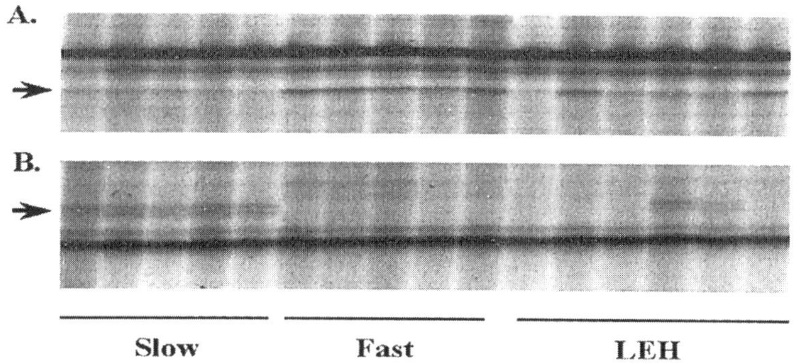

Figure 4. Portions of two different differential display autoradiograms showing a gene (indicated by the arrows) that is expressed (A) in all 5.Fast rats, no Slow rats, and variably in 6 LEH control rats, or (B) two genes (doublet) in all 5 Slow rats, no Fast rats and variably in 6 LEH control rats.

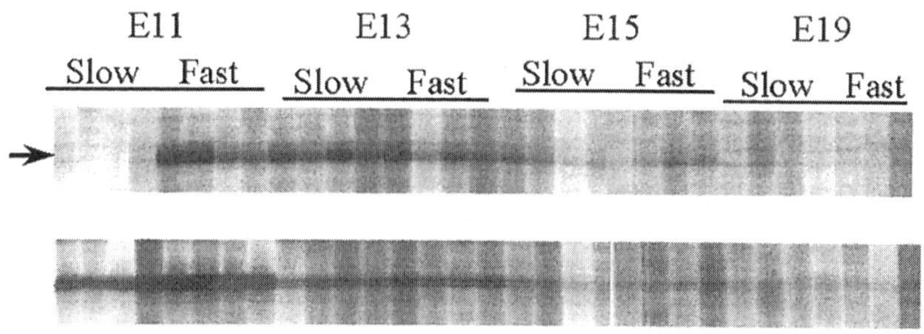

Figure 5. Portions of a differential display autoradiogram demonstrating the alpha 2 macroglobulin (A2M) expression profile during embryogenesis in the Fast and Slow rats. Note that A2M is missing in all 4 Slow embryos (from two litters) at embryonic day 11 (indicated by an arrow), but is expressed in all 4 Fast embryos (two litters). Similar strain expression profiles are evident at E13-19. Thus while Slow embryos can express the gene they do not do so at E11. The lower row illustrates that at E11 all Slow embryos have similar expression levels to Fast embryos for a different gene.

timepoint. However, by E13 and onward, the strains showed near identical expression levels for this cDNA fragment (Figure 5). We have identified this fragment to be alpha-2 macroglobulin, a gene known to play a role in lipid transport, and are currently in the process of determining how early that differential expression pattern appears in embryonic development. In future experiments, we will attempt to manipulate the timing of expression of this gene during embryogenesis in Fast rats with the hopes of altering the cascade of events and behaviors that it controls, perhaps even the rate of epileptogenesis as adults.

6. REFERENCES

1. S. F. Berkovic, Genetics of epilepsy syndromes, in: *Epilepsy: A Comprehensive Textbook*, edited by J. Engel, Jr. and T.A. Pedley (Lippincott-Raven, Philadelphia, 1997), pp. 217-224.
2. J. L. Noebels, Targeting epilepsy genes, Neuron **16**, 241-244 (1996).
3. J. A. Wada, The clinical relevance of kindling: species, brain sites and seizure susceptibility, in: *The continuing Evolution of the Limbic System Concept*, edited by K.E. Livingston and H. Hornykiewicz (Plenum Press, New York, 1978), pp. 369-388.
4. M. Sato, R. J. Racine, and D. C. McIntyre, Kindling: basic mechanisms and clinical validity, *Electroencephalogr. Clin. Neurophysiol.* **76**, 459-472 (1990).
5. R. J. Racine, M. Steingart, and D. C. McIntyre, Development of kindling-prone and kindling-resistant rats. selective breeding and electrophysiological studies, *Epilepsy Res.* **35**, 183-195 (1999).
6. R. J. Racine, Modification of seizure activity by electrical stimulation: Motor seizure, *Electroencephalogr. Clin. Neurophysiol.* **32**, 281-294 (1972).
7. D. C. McIntyre, C. Dufresne, and M. E. Kelly, FAST and SLOW amygdala kindling rats strains: comparison of amygdala, hippocampal, piriform and perirhinal cortex kindling, *Epilepsy Res.* **35**, 197-209 (1999).
8. D. C. McIntyre, and R. J. Racine, Kindling mechanisms: current progress on an experimental epilepsy model, *Prog. Neurobiol.* **27**, 1-12 (1986).
9. Kelly, M. E., and D. C. McIntyre, Perirhinal cortex involvement in limbic kindled seizures, *Epilepsy Res.* **26**, 233-243 (1996).
10. R. S. Shin, H. Anisman, Z. Merali, and D. C. McIntyre, Changes in extracellular levels of amygdala amino acids in genetically fast and slow kindling rat strains, *Brain Res.* **946**, 31-42 (2002).
11. R. S. Shin, H. Anisman, Z. Merali, and D. C. McIntyre, Amygdala amino acid and monoamine levels in genetically Fast and Slow kindling rat strains during massed amygdala kindling: A microdialysis study, *Eur. J. Neurosci.* (in press).
12. D. C. McIntyre, P. Kent, S. Hayley, Z. Merali, and H. Anisman, Influence of psychogenic and neurogenic stressors on neuroendrocrine and central monoamine activity in fast and slow kindling rats, *Brain Res.* **840**, 65-74 (1999).
13. K. L. Gilby, A. Goncalves Da Silva, and D. C. McIntyre, Differential $GABA_A$ subunit expressionfollowing status epilepticus in seizure-prone and seizure-resistant rats: A possible mechanism for refractory drug response, *Epilepsy Res.* (in press).
14. R.J. Racine, M. Steingart, Y. Bureau, and D. C. McIntyre, Differential sensitivity of genetically Fast vs. Slow kindling rat strains to GABAergic convulsive agents, *Neuropharmacol.* **45**, 918-924 (2003).
15. M. O. Poulter, L. A. Brown, S. Tynan, G. Willick, R. Williams, and D. C. McIntyre, Differential expression of α1, α2, α3 and α5 $GABA_A$ receptor subunits in seizure-prone and seizure-resistant rat models of temporal lobe epilepsy, *J. Neurosci.* **19**, 4654-4661 (1999).
16. S. L. Moshé, The effects of age on the kindling phenomenon, *Dev. Psychobiol.* **14**, 75-81 (1981).
17. D. C. McIntyre, B. Hutcheon, K. Schwabe, and M. O. Poulter, Divergent $GABA_A$ receptor-mediated synaptic transmission in genetically seizure-prone and seizure-resistant rats, *J. Neurosci.* **22**, 9922-9931 (2002).
18. D. C. McIntyre, and G. L. Chew, Power spectral analysis of electro-encephalographic activity in kindled rats. *Exp. Neurol.* **92**, 261-266 (1986).
19. A. Bragin, J. Engel Jr, C.L. Wilson, I. Fried, G. Buzsaki, High-frequency oscillations in human brain, *Hippocampus* **9**, 137-42 (1999).
20. P. Mohapel, and D. C. McIntyre, Amygdala kindling-resistant (SLOW) or prone (FAST) rat strains show differential fear, *Behav. Neurosci.* **112**, 1402-1413 (1998).
21. H. Anisman, and D. C. McIntyre, Conceptual, spatial and cue learning in the Morris Water Maze in Fast and Slow kindling rats: Attention deficit comorbidity, *J. Neurosci.* **22**, 7809-7817 (2002).
22. S. M. Wolf, and A. Forsythe, Behavioral disturbance, phenobarbital, and febrile seizures, *Pediatrics* **61**, 728-731 (1978).
23. J.S. Zhang, E. L. Duncan, C. M. Chang, R. R. Reddel, Differential display of mRNA, *Mol. Biotechnol.* **10**, 155-165 (1998).
24. E.M. Denovan-Wright, K.L. Gilby, S.E. Howlett, and H.A. Robertson, Cloning of differentially expressed brain cDNAs. in: *PCR5: Differential Display: 4 Practical Approach*, edited by R. Leslie and H.A. Robertson (Oxford Press, New York, 1998), pp.131-148.

Discussion

Kalynchuk: I am wondering whether the memory deficit that you see in the fast animals is really a memory or learning impairment. Or, given that those animals are so impulsive, is it just that they are moving and making their choice so quickly that they are not really paying attention to the task?
McIntyre: I think that's really it more than anything else. I should bite my tongue every time I say learning, because they have an attention problem. This is like kids who have ADHD – it's not that they are dumber than anyone else, they just have trouble paying attention. One of the beautiful things about this model is that when we would train those animals they were distracted and gave us this poor learning curve. But once they learned it they were actually very good at it, and then if you tried to distract them after they learned it they were not distractable. I asked some people in this business to see if they get this with humans, and they said yes, we call that hyper-focusing. Once kids with ADHD find something they like or are good at, you can't distract them. So it probably isn't an intelligence thing at all, it probably deals with distractability. And we know that if we pre-train them we can eliminate virtually all of that difference; so that is what they do with ADHD kids to, they teach them a new way to do a job.
Kalynchuk: Given the impulsivity and what you were just saying about ADHD, have you looked at all at the dopamine system in these rats?
McIntyre: No.
Schwartzkroin: I wondered if you have looked at ripples, the waves occurring between 140 and 230 Hz, because that would be an obvious candidate for finding something in the systems level.
McIntyre: We did not, and you are the guys that should do that.

GENE EXPRESSION CHANGES IN KINDLING

Dong Liang and Thomas N. Seyfried*

1. INTRODUCTION

Kindling is a phenomenon by which the periodic application of an initially subconvulsive electrical stimulus results in a progressive development of epileptiform activity, culminating in the manifestation of generalized seizures in response to the stimulus. Once the animal is kindled, the enhanced state of excitability is permanent[1-5] Kindling stimulation can produce a great number of effects, any of which could contribute to the development of an epileptogenic state. Many studies have shown interactions between kindling and neurotransmitter systems,[6-11] second messengers,[12-14] and ion channels.[15-17] Histological studies with light and electron microscopy also reveal neural structure changes and mossy fiber sprouting in kindled animals.[18-20] Thus, kindling produces both structural and chemical changes, many of which are associated with or result from changes in gene expression. This chapter focuses on changes in regulator of G-protein signaling (RGS), Ca^{2+}/calmodulin-dependent protein kinase II (CaMK II) gene expression, and long-lasting changes in gene expression in kindling.

2. GENE EXPRESSION CHANGES IN KINLDING

2.1. Regulator of G-Protein

RGS is a key regulator of G-protein signaling. G-protein signaling pathways are essential for all aspects of cell and organ physiology. In this signaling pathway, G protein (Gαβγ subunits) plays a transducing role between a signal - receiving G protein-coupled

* Dong Liang, Qiagen Sciences, Inc., RNAi application development, 19300 Germantown Road, Germantown, MD 209874, Department of Neurology, Johns Hopkins University, 733 North Broadway Street, BRB732, Baltimore, MD 21205. Thomas N. Seyfried, Department of Biology, Boston College, Chestnut Hill, MA 02467.

receptor (GPCR) and the signal-generating downstream target effectors. A neurotransmitter or hormone activated GPCR stimulates the exchange of GDP for GTP on Gα to initiate heterotrimer dissociation and activation of effector proteins that, in turn, initiate a cascade of cellular signaling events.[21, 22] Many proteins including RGS regulate this signaling pathway. The RGS proteins bind directly to the activated Gα-GTP to serve as GTPase-activating proteins (GAPs), which limit the lifetime of Gα-GTP and terminate signaling event(s). So far, more than 30 mammalian family members of RGS have been identified.[23-25]

In hippocampal kindled mouse brain, RGS4 is significantly increased in hippocampus 0.5 - 24 h after kindling and the elevated expression returns to basal level one week later. However, the expression of RGS4 is decreased in forebrain and the down-regulation lasts more than one month after kindling (Fig. 1).[26] Differential alterations of RGS4 in different brain regions are also observed in chronic stress animals.[27] Another study using a chronic maximal electroshock seizures (MES) paradigm also shows a decreased expression of RGS4 in forebrain 24 h after generalized seizure.[28] However, there is no down-regulation of RGS4 in forebrain when using acute seizure models such as MES, audiogenic seizure, or PTZ seizure (Liang & Seyfried, unpublished data). In contrast to RGS4, RGS2 expression is strongly increased after kindling as well as after acute seizures in hippocampus and forebrain. The change in RGS2 expression shows up as early as 0.5 h after last kindling stimulation and returns to basal level 8 hr later (Liang & Seyfried, unpublished data).

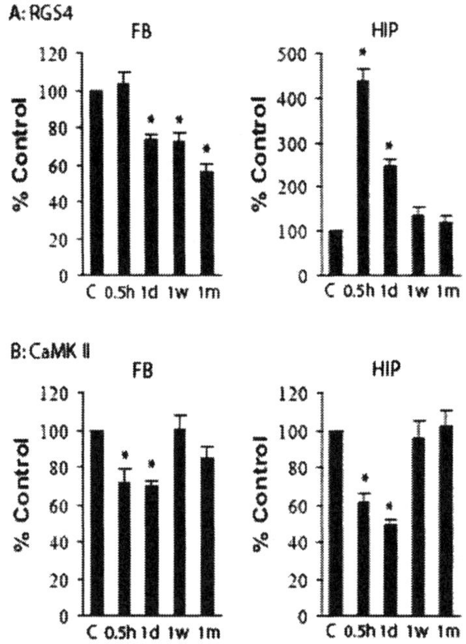

Fig. 1. Influence of rapid kindling on RGS4 (A) and CaMK II (B) gene expression in the forebrain (FB) and the hippocampus (HIP). The mRNA levels were normalized against β-actin and are represented as a percentage of the level in sham-operated controls. The asterisk indicates that the value for the kindled mice is significantly different from that of the control at $P<0.01$ (two-tail Student's *t*-test). The expression of genes was analyzed in four mice at each time-point in hippocampus, in 6 mice for CaMK II and in 8 mice for RGS4 at each time-point in forebrain and the values are expressed as mean ± S.E.M. This figure is from the data of a portion of Figure 4 of the paper published in *Molecular Brain Research* (2001, **96**: page 99), with permission from the publisher.

Although RGS2 and RGS4 belong to same gene family, each has unique functional characteristics. For example, RGS4 binds both Gqα and Giα family members of G-protein.[29-31] While RGS2 selectively binds Gqα.[32] In fact, RGS2 was 5-fold more potent than RGS4 as an inhibitor of Gqα-stimulated phosphoinositide hydrolysis in vivo.[33] Whereas RGS4 was 8-fold more potent than RGS2 as an inhibitor of Giα-mediated signaling.[33] Furthermore, RGS2 expression is transiently increased in hippocampus after kindling and acute seizure. In contract, RGS4 expression is down-regulated in forebrain for more than one month after kindling, but not after acute seizure.[26]

Although the endogenous expression of RGS2 and RGS4 is low in hippocampus[34, 35] as well as in forebrain for RGS2,[36] the expression of both genes increased in hippocampus for a short period in the epilepsy models. These findings may reflect a role for RGS2 and RGS4 in blocking seizure associated excessive G-protein signaling during and immediately after a seizure. In contrast to RGS2, the endogenous RGS4 expression is high in the forebrain.[34, 35] The high levels of constitutive expression suggest that RGS4 might be readily available for acute desensitization of signaling events. The kindling-induced down-regulation of RGS4 may enhance the G-protein signaling in response to its initiator including neurotransmitters and thus augment synaptic transmission and seizure spread. The long-term down-regulation of RGS4 may play a role in maintaining the kindling-induced epileptogenesis.

2.2. CaMK II

CaMK II is a multifunctional protein kinase. After binding to the Ca^{2+}/calmodulin complex, CaMK II α and β subunits are autophosphorylated resulting in autonomous activity.[37, 38] CaMKII is abundant in brain and is widely distributed in cell bodies, dendrites, and synaptic membranes.[39] Its unique properties, abundance, and location contribute to the role of CaMKII as a key transducer of extracellular signals.

Several lines of evidence suggest the involvement of CaMK II in the mechanisms that underlie kindling.[14, 26, 40-43] Studies showed epileptic-like electrical activity in cultured neurons following inhibition of the α subunit of CaMK II[44] and enhanced hyperexcitability in the α subunit of CaMK II knock-out mice.[45] However, the expression change of CaMK II in epilepsy model is controversial.[14, 26, 40-43, 46] Morimoto et al. (1997) reported up-regulation of the β subunit of CaMK II in hippocampus at 4 - 24 h after kindling, but no expression change of the α subunit of CaMK II following kindling. Generally, most of other studies report down-regulation of CaMKII expression and enzymatic activity following kindling.[14, 40, 41]

The changes in gene expression in rapidly kindled mouse brain screened with an RT-PCR differential display method identified the α subunit of CaMK II as down-regulated. Northern blot analysis further confirms this down-regulation in hippocampus and forebrain at 0.5 – 24 h after kindling with a return to basal level one week later (Fig. 1).[26] Furthermore, other studies have shown decreases in CaMK II immunoreactivity.[14] This down-regulation of CaMK II is long lasting.[14, 40] In fact, an decrease of CaMKII activity was observed eight weeks after septal kindling.[42] These results suggest that CaMK II may play a role in the maintenance of epileptogenesis.

2.3. Long-lasting changes in gene expression

One of the major challenges is to delineate the molecular mechanisms by which epileptogenesis is induced by kindling stimulation. Although much has been learned about the rapid and short-term changes that occur following kindling, relatively little is known about how the brief stimulation used in kindling can induce epileptogenesis that can persist for the lifetime of the animal. Covalent modifications of the sort used in short-term brain function change have been proposed to become self-reinforcing for certain instances of long-term brain function change.[47] However, many studies indicate that brain function changes lasting more than 1 day require expression of specific genes that may not be involved in for short-term change of brain function.[48]

Epileptogenesis induced by kindling stimulation lasting whole life of the animals clearly persist longer than the half-life of most proteins. In these cases, it is attractive to propose mechanisms of retention that are independent of protein and mRNA turnover, depending more on persistent changes in transcription states that are maintained by cooperative or self-reinforcing mechanisms. The induction of a gene cascade for the change lasts weeks or months, in which early regulatory gene expression leads to maintain alterations in the expression of later effector genes. By this way, long-lasting epileptogenesis could be induced and maintenanced as a form of a cascade of gene expression change.

The possibility that long-lasting changes in gene expression underlie kindling is supported by the findings that electrical seizure afterdischarge induces the transient expression of *c-fos*, *c-jun*, and other immediate-early genes (IEGs).[49, 50] The products of these genes are usually nuclear proteins that are thought to affect transcription or mRNA processing.[51] In kindled animals, *c-fos* mRNA levels are rapidly induced and then quickly return to baseline levels after the conclusion of the last kindling stimulation.[50, 52-55]. Additionally, *c-jun*, *jun-B*, and *Krox*-24 are reportedly up-regulated briefly following kindling.[49,54,55] Since these changes in IEG expression are temporary, long-term plasticity changes associated with kindling cannot be explained solely by IEG induction. However, IEGs are transcripitional factors that regulate the expression of other genes.[51] Thus, it is possible that these brief changes in IEGs may help to initiate events that elicit a cascade of changes in gene products that ultimately produce long-term changes in gene expression and long-lasting increase of epileptogenesis in kindling.

Long-term changes in gene expression represent an ideal mechanism to produce the enduring changes in the neuronal plasticity associated with kindling. In addition to the reports of long-lasting change in RGS4 and CaMK II, there are studies showing similar long-lasting expression changes in other genes. Iwasa *et al*[56] observed a remarkable increase in Gsα mRNA in bilateral cerebral cortex at 24 h after the last generalized seizure and the expression changes persisted three weeks in some brain regions of kindled rats. $G\alpha i_2$ mRNA level was also increased on the stimulated side of cerebral cortex 24 hr after kindling and persisted for three weeks more.[56] Another study reported that immunoreactivity levels of Gαi2 were significantly reduced in ipsilateral amygdala-pyriform cortex and dorsal hippocampus at 24 h and one month after last seizure.[57] These result suggest that dysfunction of Gs and $G\alpha i_2$ might relate to the mechanisms of seizure generation and the maintenance of epileptogenesis.

Attempts to identify long-lasting changes in glutamate receptors in hippocampus of kindled animals have yielded varied results. Kraus *et al.*[58] noted a substantial increase in the binding of the NMDA receptor channel ligand [^{3}H]CPP in the CA3 region 28 days after kindling. Other studies reported acute changes in NR1, NR2A and NR2B expression[59, 60] or a prolonged increase in the hippocampal CA3 kainate receptor subunit KA-1 mRNA 28 days after amygdala kindling.[61] In other brain regions, a significant increase in NR1 mRNA level was observed in the ipsilateral frontal and temporal cortices at four weeks after kindling.[62]

Increases in [^{3}H]AMPA receptor binding was also demonstrated in PTZ-kindled mice one month after kindling.[63] The long-lasting changes were region-specific and restricted to the motor cortex and basal ganglia. Concomittely, GluR-B mRNA expression levels were altered in the motor and somatosensory cortices one month after the last seizure.[63]. For metabotropic glutamate receptors (mGluR), although mGluR1 mRNA in most hippocampal neurons show transient up-regulation, mGluR5 in most hippocampal neurons show a down-regulation in mRNA for at least 28 days after kindling.[64]

Furthermore, mRNA encoding the membrane-bound protein ligatin was significantly reduced in kindled brains for over four months after kindling.[46] The change in ligatin gene expression was not produced by single or multiple seizures that did not induce kindling, but was blocked by MK801.[46] Additional findings showed an elevated expression level of type II adenylyl cyclase mRNA in hippocampus persisting for four weeks after amygdala kindling.[65] Moreover, in the nucleus accumbens and striatum, expression of D2 receptor mRNA increases more than 30 days after kindling.[66]. Tα1-tubulin mRNA level also increased significantly in the dentate gyrus and CA3 of hippocampus ipsilateral to stimulation four weeks after amygdala kindled seizures.[67] Taken together, these findings provide evidence that kindling can cause persistent changes in the expression of specific genes in neurons and suggests that long-lasting changes in gene expression may be a basic molecular mechanism underlying kindling-induced epileptogenesis.

3. ACKNOWLEDGEMENTS

We thank John Lumpkin for critical reading and helpful comments of the manuscript. This research was supported by PHS grant (NS23355) and the Boston College Research Fund.

4. REFERENCES

1. G. V. Goddard, Development of epileptic seizures through brain stimulation at low intensity, *Nature* **214,** 1020-1021 (1967).
2. G. V. Goddard, D. C. McIntyre, and C. K. Leech, A permanent change in brain function resulting from daily electrical stimulation, *Exp. Neurol.* **25,** 295-330 (1969).
3. R. Racine, V. Okujava, and S. Chipashvili, Modification of seizure activity by electrical stimulation. 3. Mechanisms, *Electroencephalogr. Clin. Neurophysiol.* **32,** 295-299 (1972).
4. R. J. Racine, Modification of seizure activity by electrical stimulation. II. Motor seizure, *Electroencephalogr. Clin. Neurophysiol.* **32,** 281-294 (1972).
5. R. J. Racine, Modification of seizure activity by electrical stimulation. I. After-discharge threshold, *Electroencephalogr. Clin. Neurophysiol.* **32,** 269-279 (1972).
6. M. E. Corcoran, H. C. Fibiger, J. A. McCaughran, Jr., and J. A. Wada, Potentiation of amygdaloid kindling and metrazol-induced seizures by 6-hydroxydopamine in rats, *Exp. Neurol.* **45,** 118-133 (1974).

7. J. Engel, Jr. and N. S. Sharpless, Long-lasting depletion of dopamine in the rat amygdala induced by kindling stimulation, *Brain Res.* **136,** 381-386 (1977).
8. M. S. Myslobodsky and E. S. Valenstein, Amygdaloid kindling and the GABA system, *Epilepsia* **21,** 163-175 (1980).
9. E. Mohr and M. E. Corcoran, Depletion of noradrenaline and amygdaloid kindling, *Exp. Neurol.* **72,** 507-511 (1981).
10. I. Mody and U. Heinemann, NMDA receptors of dentate gyrus granule cells participate in synaptic transmission following kindling, *Nature* **326,** 701-704 (1987).
11. C. Shin, J. M. Silver, D. W. Bonhaus, and J. O. McNamara, The role of substantia nigra in the development of kindling: pharmacologic and lesion studies, *Brain Res.* **412,** 311-317 (1987).
12. P. Gloor, Epilepsy: relationships between electrophysiology and intracellular mechanisms involving second messengers and gene expression, *Can. J. Neurol. Sci.* **16,** 8-21 (1989).
13. D. L. Yourick, M. C. LaPlaca, and J. L. Meyerhoff, Norepinephrine-stimulated phosphatidylinositol metabolism in genetically epilepsy-prone and kindled rats, *Brain Res.* **551,** 315-318 (1991).
14. J. M. Bronstein, P. Micevych, P. Popper, G. Huez, D. B. Farber, and C. G. Wasterlain, Long-lasting decreases of type II calmodulin kinase expression in kindled rat brains, *Brain Res.* **584,** 257-260 (1992).
15. M. E. Moneta, M. de la Fuente, J. L. Liberona, and E. Jaimovich, Sodium pathway markers in normal and kindled frog brains, *Neurosci. Lett.* **65,** 331-335 (1986).
16. I. Mody, J. N. Reynolds, M. W. Salter, P. L. Carlen, and J. F. MacDonald, Kindling-induced epilepsy alters calcium currents in granule cells of rat hippocampal slices, *Brain Res.* **531,** 88-94 (1990).
17. W. B. Sneddon, W. M. Burnham, S. Prowse, J. Cheng, and S. H. Kish, Reduced GABA-mediated chloride flux in entorhinal cortex-kindled rat brains, *Epilepsy Res.* **6,** 18-22 (1990).
18. M. Langmeier, J. Fischer, and J. Mares, Ultrastructural changes in cortical synapses shortly after termination of a seizure during kindling, *Physiol. Bohemoslov.* **31,** 213-216 (1982).
19. Y. Geinisman, F. Morrell, and L. deToledo-Morrell, Remodeling of synaptic architecture during hippocampal "kindling", *Proc. Natl. Acad. Sci. U S A* **85,** 3260-3264 (1988).
20. J. E. Cavazos, G. Golarai, and T. P. Sutula, Mossy fiber synaptic reorganization induced by kindling: time course of development, progression, and permanence, *J. Neurosci.* **11,** 2795-2803 (1991).
21. Y. Landry and J. P. Gies, Heterotrimeric G proteins control diverse pathways of transmembrane signaling, a base for drug discovery, *Mini Rev. Med. Chem.* **2,** 361-372 (2002).
22. C. M. Tan, A. E. Brady, H. H. Nickols, Q. Wang, and L. E. Limbird, Membrane trafficking of G protein-coupled receptors, *Annu. Rev. Pharmacol. Toxicol.* **44,** 559-609 (2004).
23. R. R. Neubig and D. P. Siderovski, Regulators of G-protein signalling as new central nervous system drug targets, *Nat. Rev. Drug Discov.* **1,** 187-197 (2002).
24. J. R. Hepler, RGS protein and G protein interactions: a little help from their friends, *Mol. Pharmacol.* **64,** 547-549 (2003).
25. M. Ishii and Y. Kurachi, Physiological actions of regulators of G-protein signaling (RGS) proteins, *Life Sci.* **74,** 163-171 (2003).
26. D. Liang and T. N. Seyfried, Genes differentially expressed in the kindled mouse brain, *Mol. Brain Res.* **96,** 94-102 (2001).
27. Y. G. Ni, S. J. Gold, P. A. Iredale, R. Z. Terwilliger, R. S. Duman, and E. J. Nestler, Region-specific regulation of RGS4 (Regulator of G-protein-signaling protein type 4) in brain by stress and glucocorticoids: in vivo and in vitro studies, *J. Neurosci/* **19,** 3674-3680 (1999).
28. S. J. Gold, B. D. Heifets, C. M. Pudiak, B. W. Potts, and E. J. Nestler, Regulation of regulators of G protein signaling mRNA expression in rat brain by acute and chronic electroconvulsive seizures, *J/ Neurochem.* **82,** 828-838 (2002).
29. D. M. Berman, T. Kozasa, and A. G. Gilman, The GTPase-activating protein RGS4 stabilizes the transition state for nucleotide hydrolysis, *J. Biol. Chem.* **271,** 27209-27212 (1996).
30. C. Huang, J. R. Hepler, A. G. Gilman, and S. M. Mumby, Attenuation of Gi- and Gq-mediated signaling by expression of RGS4 or GAIP in mammalian cells, *Proc. Natl. Acad. Sci. U S A* **94,** 6159-6163 (1997).
31. Y. Yan, P. P. Chi, and H. R. Bourne, RGS4 inhibits Gq-mediated activation of mitogen-activated protein kinase and phosphoinositide synthesis, *J. Biol. Chem.* **272,** 11924-11927 (1997).
32. S. P. Heximer, N. Watson, M. E. Linder, K. J. Blumer, and J. R. Hepler, RGS2/G0S8 is a selective inhibitor of Gqalpha function, *Proc. Natl. Acad. Sci. U S A* **94,** 14389-14393 (1997).
33. S. P. Heximer, S. P. Srinivasa, L. S. Bernstein, J. L. Bernard, M. E. Linder, J. R. Hepler, and K. J. Blumer, G protein selectivity is a determinant of RGS2 function, *J. Biol. Chem.* **274,** 34253-34259 (1999).
34. S. J. Gold, Y. G. Ni, H. G. Dohlman, and E. J. Nestler, Regulators of G-protein signaling (RGS) proteins: region-specific expression of nine subtypes in rat brain, *J. Neurosci.* **17,** 8024-8037 (1997).

35. S. Nomoto, K. Adachi, L. X. Yang, Y. Hirata, S. Muraguchi, and K. Kiuchi, Distribution of RGS4 mRNA in mouse brain shown by in situ hybridization, *Biochem. Biophys. Res. Commun.* **241,** 281-287 (1997).
36. T. Ingi, A. M. Krumins, P. Chidiac, G. M. Brothers, S. Chung, B. E. Snow, C. A. Barnes, A. A. Lanahan, D. P. Siderovski, E. M. Ross, A. G. Gilman, and P. F. Worley, Dynamic regulation of RGS2 suggests a novel mechanism in G-protein signaling and neuronal plasticity., *J. Neurosci.* **18,** 7178-7188 (1998).
37. R. J. Colbran, Regulation and role of brain calcium/calmodulin-dependent protein kinase II, *Neurochem. Int.* **21,** 469-497 (1992).
38. A. Hudmon and H. Schulman, Neuronal CA2+/calmodulin-dependent protein kinase II: the role of structure and autoregulation in cellular function, *Annu. Rev. Biochem.* **71,** 473-510 (2002).
39. P. I. Hanson and H. Schulman, Neuronal Ca2+/calmodulin-dependent protein kinases, *Annu. Rev. Biochem.* **61,** 559-601 (1992).
40. J. M. Bronstein, D. B. Farber, P. E. Micevych, R. Lasher, and C. G. Wasterlain, Kindling induced changes in calmodulin kinase II immunoreactivity, *Brain Res.* **524,** 49-53 (1990).
41. K. Wu, C. Wasterlain, L. Sachs, and P. Siekevitz, Effect of septal kindling on glutamate binding and calcium/calmodulin-dependent phosphorylation in a postsynaptic density fraction isolated from rat cerebral cortex, *Proc. Natl. Acad. Sci. U S A* **87,** 5298-5302 (1990).
42. C. G. Wasterlain and D. B. Farber, Kindling alters the calcium/calmodulin-dependent phosphorylation of synaptic plasma membrane proteins in rat hippocampus, *Proc. Natl. Acad. Sci. U S A* **81,** 1253-1257 (1984).
43. K. Morimoto, K. Sato, K. Kashihara, and T. Hayabara, Increased levels of mRNA for beta- but not alpha-subunit of calmodulin kinase II following kindled seizures, *Brain Res. Bull.* **43,** 375-380 (1997).
44. S. B. Churn, S. Sombati, E. R. Jakoi, L. Severt, and R. J. DeLorenzo, Inhibition of calcium/calmodulin kinase II alpha subunit expression results in epileptiform activity in cultured hippocampal neurons, *Proc. Natl. Acad. Sci. U S A* **97,** 5604-5609 (2000).
45. L. S. Butler, A. J. Silva, A. Abeliovich, Y. Watanabe, S. Tonegawa, and J. O. McNamara, Limbic epilepsy in transgenic mice carrying a Ca2+/calmodulin-dependent kinase II alpha-subunit mutation, *Proc. Natl. Acad. Sci. U S A* **92,** 6852-6855 (1995).
46. J. B. Perlin, C. M. Gerwin, D. M. Panchision, R. S. Vick, E. R. Jakoi, and R. J. DeLorenzo, Kindling produces long-lasting and selective changes in gene expression of hippocampal neurons, *Proc. Natl. Acad. Sci. U S A* **90,** 1741-1745 (1993).
47. S. G Miller and M. B. Kennedy, Regulation of brain type II Ca2+/calmodulin-dependent protein kinase by autophosphorylation: a Ca2+-triggered molecular switch, *Cell* **44,** 861-870 (1986).
48. T. Tully, T. Preat, S. C. Boynton, and M. Del-Vecchio, Genetic dissection of consolidated memory in Drosophila, *Cell* **79,** 35-47 (1994).
49. M. Simonato, D. A. Hosford, D. M. Labiner, C. Shin, H. H. Mansbach, and J. O. McNamara, Differential expression of immediate early genes in the hippocampus in the kindling model of epilepsy, *Mol. Brain Res.* **11,** 115-124. (1991).
50. C. Shin, J. O. McNamara, J. I. Morgan, T. Curran, and D. R. Cohen, Induction of c-fos mRNA expression by afterdischarge in the hippocampus of naive and kindled rats, *J. Neurochem.* **55,** 1050-1055 (1990).
51 J. I. Morgan and T. Curran, Stimulus-transcription coupling in the nervous system: involvement of the inducible proto-oncogenes fos and jun, *Annu. Rev. Neurosci.* **14,** 421-451 (1991).
52. G. C. Teskey, B. G. Atkinson, and D. P. Cain, Expression of the proto-oncogene c-fos following electrical kindling in the rat, *Mol. Brain Res.* **11,** 1-10 (1991).
53. M. Dragunow and H. A. Robertson, Kindling stimulation induces c-fos protein(s) in granule cells of the rat dentate gyrus, *Nature* **329,** 441-442 (1987).
54. P. Hughes, K. Singleton, and M. Dragunow, MK-801 does not attenuate immediate-early gene expression following an amygdala afterdischarge, *Exp. Neurol.* **128,** 276-283 (1994).
55. P. E. Hughes, D. Young, K. M. Preston, Q. Yan, and M. Dragunow, Differential regulation by MK801 of immediate-early genes, brain-derived neurotrophic factor and trk receptor mRNA induced by a kindling after-discharge, *Mol. Brain Res.* **53,** 138-151 (1998).
56. H. Iwasa, S. Kikuchi, H. Miyagishima, S. Mine, K. Koseki, and S. Hasegawa, *in* "Brain Res", Vol. 818, p. 570-574, 1999.
57. J. C. Cutz, P. P. Li, W. M. Burnham, and J. J. Warsh, Brain region-specific and time-dependent decreases of G alpha i2 in amygdala-kindled rats, *Synapse* **22,** 226-231 (1996).
58. J. E. Kraus, G. C. Yeh, D. W. Bonhaus, J. V. Nadler, and J. O. McNamara, Kindling induces the long-lasting expression of a novel population of NMDA receptors in hippocampal region CA3, *J. Neurosci.* **14,** 4196-4205 (1994).

59. W. Kamphuis, H. Hendriksen, P. C. Diegenbach, and F. H. Lopes da Silva, N-methyl-D-aspartate and kainate receptor gene expression in hippocampal pyramidal and granular neurons in the kindling model of epileptogenesis, *Neuroscience* **67,** 551-559 (1995).
60. G. D. Pratt, M. Kokaia, J. Bengzon, Z. Kokaia, J. M. Fritschy, H. Mohler, and O. Lindvall, Differential regulation of N-methyl-D-aspartate receptor subunit messenger RNAs in kindling-induced epileptogenesis, *Neuroscience* **57,** 307-318 (1993).
61. M. Hikiji, H. Tomita, M. Ono, Y. Fujiwara, and K. Akiyama, Increase of kainate receptor mRNA in the hippocampal CA3 of amygdala-kindled rats detected by in situ hybridization, *Life Sci* **53,** 857-864 (1993).
62. S. Kikuchi, H. Iwasa, and T. Sato, Lasting changes in NMDAR1 mRNA level in various regions of cerebral cortex in epileptogenesis of amygdaloid-kindled rat, *Psychiatry Clin. Neurosci.* **54,** 573-577 (2000).
63. A. Ekonomou, A. L. Smith, and F. Angelatou, Changes in AMPA receptor binding and subunit messenger RNA expression in hippocampus and cortex in the pentylenetetrazole-induced 'kindling' model of epilepsy, *Mol. Brain Res.* **95,** 27-35 (2001).
64. M. T. Akbar, M. Rattray, J. F. Powell, and B. S. Meldrum, Altered expression of group I metabotropic glutamate receptors in the hippocampus of amygdala-kindled rats, *Mol. Brain Res.* **43,** 105-116 (1996).
65. H. Iwasa, S. Kikuchi, S. Mine, H. Miyagishima, K. Sugita, T. Sato, and S. Hasegawa, Up-regulation of type II adenylyl cyclase mRNA in kindling model of epilepsy in rats, *Neurosci. Lett.* **282,** 173-176 (2000).
66. H. A. Gelbard and C. D. Applegate, Persistent increases in dopamine D2 receptor mRNA expression in basal ganglia following kindling, *Epilepsy Res.* **17,** 23-29 (1994).
67. K. Sato and K. Abe, Increases in mRNA levels for Talpha1-tubulin in the rat kindling model of epilepsy, *Brain Res.* **904,** 157-160 (2001).

Discussion

Gale: Its very nice that you have made the effort to compare kindling with seizures, since there is more to kindling than seizures. I suggest, however, that the seizures you are using to compare have some other complications and that's because most of them are seizures that are not focused on forebrain seizures, they're not the minimal type of seizures, the facial and forelimb clonus limbic-type seizures. We have found in a number of situations that if you either user ear-clip electroshock as you have, or if you use drug doses that produce seizures that involve more than just the limbic seizure, you can actually down-regulate genes that would be up-regulated if they were just limbic seizures. So adding additional components can actually cause a suppression of gene expression. My suggestion to you is that the electroshock seizures that would be very appropriate as controls or your reference group would be seizures that are corneal electroshock or minimal as opposed to your maximal electroshock. Then you may actual see more specific differences that would be subtracting out the limbic seizures selectively rather than having more generalized kind of seizure.
Liang: That is correct
McNamara: I think you did a nice job here. Is C3H an inbred strain or an outbred strain?
Liang: Inbred.

A PROTEOMIC APPROACH TO THE MOLECULAR ANALYSIS OF KINDLING

John T. Slevin, Sidney W. Whiteheart, Thomas C. Vanaman*

1. INTRODUCTION

Application of the newly developed tools of proteomics analysis to the brain is still at an early stage of development. Primary application of 2D gel electrophoresis coupled with mass spectrometry based protein identification has been primarily directed at analyzing whole brain proteomes. As has been recently reviewed,[1] this has led to the identification of a number of the more abundant brain proteins but has yet to focus on the important functional sub-compartments of the neuron (*e.g.*, the synapse) or to probe for potential changes in disease states. This, in part, reflects a major limitation of protein profiling: the inherent complexity of proteomes even of subcellular compartments. We have begun to establish 2D gel maps for rat hippocampal synaptosomes and sub-fractions (synaptosol, plasma membrane and vesicle fractions) in both the resting and depolarized state. In this chapter we demonstrate how this technique can be applied to the study of kindled epilepsy.

2. METHODS

2.1 Kindling Procedure and Tissue Preparation

As described previously,[2] to induce kindled epilepsy, teflon-coated, stainless steel, bipolar twisted electrodes were implanted in the right perforant path. The minimal stimulus intensity required to produce after-discharge was determined, after which threshold stimuli (1 sec train of biphasic square wave pulses, 1 msec pulse duration, 60 Hz) were administered once/day, five days/week until completion of kindling as defined by two stage 5 seizures induced with 2 consecutive stimuli, using the Racine classification.[3] Recordings were made on an 8-channel Grass Model 8B EEG polygraph before and after each electrical stimulus. Following completion of kindling, male SD rats

* John T. Slevin, VA Medical Center, Lexington, KY 40502. Sidney W. Whiteheart and Thomas C. Vanaman, University of Kentucky Medical Center, Lexington, KY. 40536

received no further stimuli until euthanatized one month later. It was judged that differences observed between kindling and control proteomic profiles at that time point would reflect permanent changes.

Using the discontinuous Percoll gradient centrifugation method as described previously,[4] kindled and surgical control rat hippocampal synaptosomes were purified through two cycles, and resuspended in Krebs-Ringer bicarbonate buffer (KRB). Synaptosomal sub-fractions (synaptosol, particulate and vesicles) were prepared and characterized as described.[5]

For all phosphorylation studies, synaptosomes were placed in KRB with either 1.5 mM Ca^{2+} or 10 mM Mg^{2+}. Samples were depolarized by addition of K^+ (25 mM final concentration) and the process was stopped at various times (usually 30 sec) by precipitation with 10 volumes of cold acetone. Where needed, *o*-vanadate was included in buffers to inhibit phosphatase activity.[6] This procedure and the composition of all buffers have been described.[2]

2.2 Two Dimensional Gel Analysis Procedures

Samples for 2D gel analysis were precipitated with acetone and recovered by centrifugation. Precipitated samples were redissolved in a minimum volume of 6 M urea, 2 M thiourea and used to rehydrate pH 3-10 IPG strips (11 cm or 17 cm IPG). Isoelectric focusing was performed using a Bio-Rad Protean IEF cell through a series of timed voltage ramps up to a final voltage of 6,000 volts and for a fixed total of 30,000 – 50,000 volt hrs.

Second dimension separations were performed by SDS-PAGE as follows. The strip was removed from the focusing tray and equilibrated in 5 ml of buffer containing 50 mM Tris-HCl, pH 8.8, 6 M urea, 2% SDS, 30% glycerol, 1% DTT. After 10 min at room temperature, the first equilibrium buffer was removed and the strip treated again in 5 ml of the same buffer but with 1.5% iodoacetamide substituted for DTT. After a 10 min equilibration, the IPG strip was transferred to an SDS-PAGE gel, overlaid with 0.5% agarose in Laemmli buffer.[7] Separation of the proteins was performed on a 10 % SDS-PAGE gel for 60 min at 200 V. Molecular weight standards were included in a separate standards lane on every gel.

For quantification, 2D gels were stained with SYPRO Ruby (Molecular Probes) overnight, destained and scanned using an Amersham Biosciences Typhoon 9400 imager. The data acquired were analyzed using PDQuest software (BioRad). This procedure is of sufficient sensitivity to detect and quantify proteins at low nanogram quantities and is linear over a broad concentration range (up to μg). A major concern for these analyses is the reproducibility of sample application which makes absolute quantification difficult on any given gel. For this reason, comparison of intensities for a given spot from one gel to the next was done by normalizing spot intensities detected to total gel fluorescence, then comparing these relative spot intensities among different gels. Coefficient of variation for each spot was less than 2 % of spot intensity across data sets.

For most studies of protein phosphorylation, 2D gels were first stained with the phosphoprotein-specific fluorescent stain Pro-Q Diamond (Molecular Probes),[8] scanned, destained then restained with SYPRO Ruby and scanned to quantify total protein. Pro-Q Diamond detects all phosphate containing species present in the gel including phospholipids and is at least as sensitive as SYPRO Ruby. The ratio of total Pro-Q Diamond to SYPRO Ruby staining reflects extent of phosphorylation and these values can be compared across gels directly to assess changes in phosphorylation state in response to various conditions of treatment.

2.3 Mass Spectrometric Analyses of Proteins and Peptides

Protein spots were excised from the gel, destained, digested with trypsin or other agents if needed, and the resulting peptides extracted. The peptides from each spot were divided into two aliquots. One aliquot was desalted by Zip-tip and spotted to MALDI plates and mass peptide mapping performed on an ABI QStar XL by MALDI-TOF analysis. The other aliquot of each sample was subjected to LC-MS/MS analysis on the QStar XL in an automated mode. Each sample was desalted then separated on a C18 reverse phase column and peptides were eluted with a gradient of acetonitrile. QStar XL MS was programmed to first measure the masses of the eluted peptides (peptide mass map) then to automatically select the three most abundant peptide ions to perform MS/MS to obtain the sequence information.

3. RESULTS

3.1 Quantitative 2D Gel Analysis to Examine Protein Levels:

The 2D gels in Figure 1 show protein patterns obtained with cytosolic, particulate and vesicle containing fractions from purified synaptosomes prepare as described by Dunkley[4] from two right hippocampi from control rat brains. The particulate fraction contains a mixture of primarily mitochondria and synaptic plasma membrane. As can be seen, the patterns obtained for each fraction are largely unique. As expected, the cytosolic and particulate fractions are relatively complex while the vesicle fraction has a relatively limited repertoire of proteins. Although not shown, identical patterns were obtained for the same fractions prepared from the corresponding left hippocampi and these patterns were highly reproducible in similar samples from other animals run at other times. Thus, the method is sufficiently reproducible to permit comparisons between

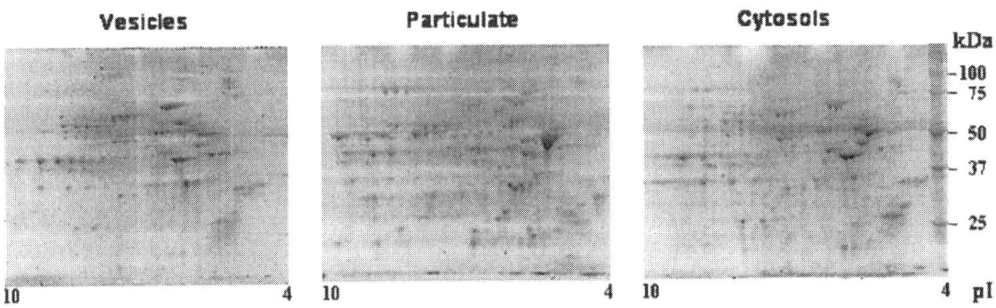

Figure 1. Two dimensional gel analyses of purified synaptosomal subfractions. Fractions were prepared and gels were run and stained with SYPRO Ruby as set forth in Methods. Samples were obtained from 2 pooled right hippocampi from naïve control animals. Superimposable patterns were obtained for each fraction purified from the left hippocampi of the same animals (data not shown).

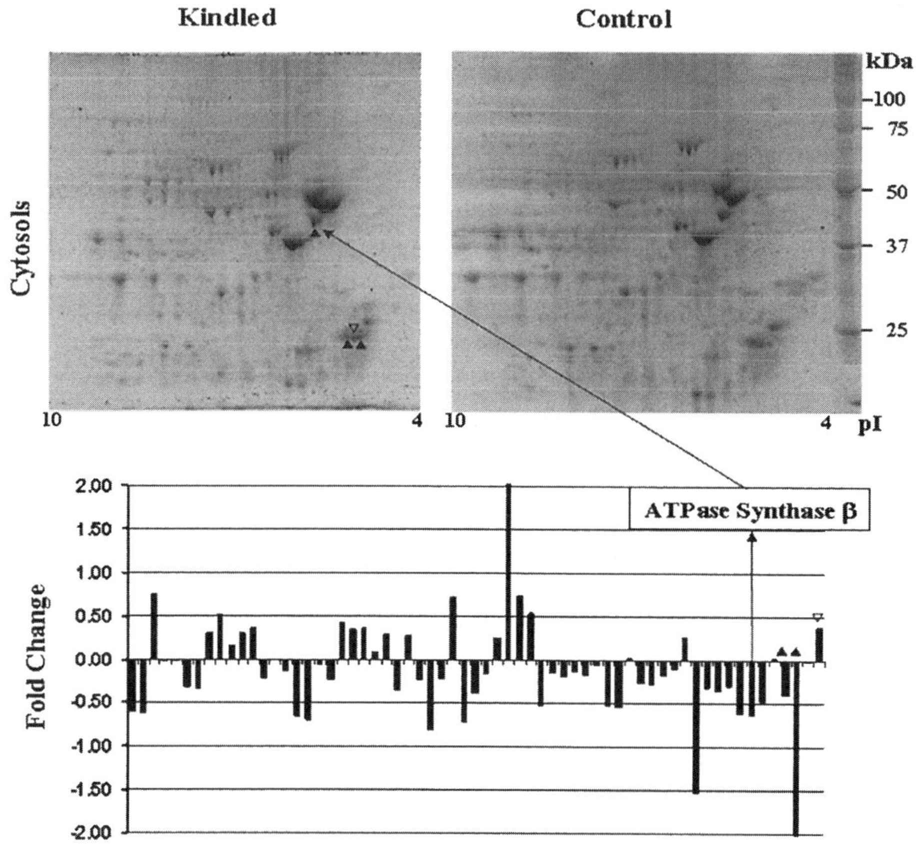

Figure 2. Two dimensional gels of cytosolic fractions prepared from kindled *vs.* control hippocampal synaptosomes. The patterns shown were obtained from the material from two pooled right (ipsilateral) hippocampi. Identical patterns and relative spot intensities were obtained from the pooled left hippocampal subfractions from the same animals. The bar graph shows the fold change for the control *vs.* kindled samples 0 = a ratio of 1 for C *vs.* K, negative values represent the fold increase in the kindled sample while a positive value shows the fold decrease. The arrow denotes the β-subunit of the mitochondrial ATP synthase identified by mass spectrometry. ▲ denotes spots whose intensities increase in kindling, Δ spots that decrease.

cohorts of control and experimental animals prepared weeks to months apart. Quantitation of individual spot intensities was also found to be extremely reproducible as discussed in the following sections.

3.2 Synaptosol Protein Complement Changes on Kindling

Our initial studies have examined the alterations in protein profiles in fully kindled (1 month after final stimulation) *vs.* surgerized control animals. A total of four control and four experimental sets of animals were analyzed in two different experiments (two right and two left hippocampi were pooled in each set). Figure 2 shows one set of 2 D gels for control *vs.* kindled cytosols from purified synaptosomes. The spot intensities for each

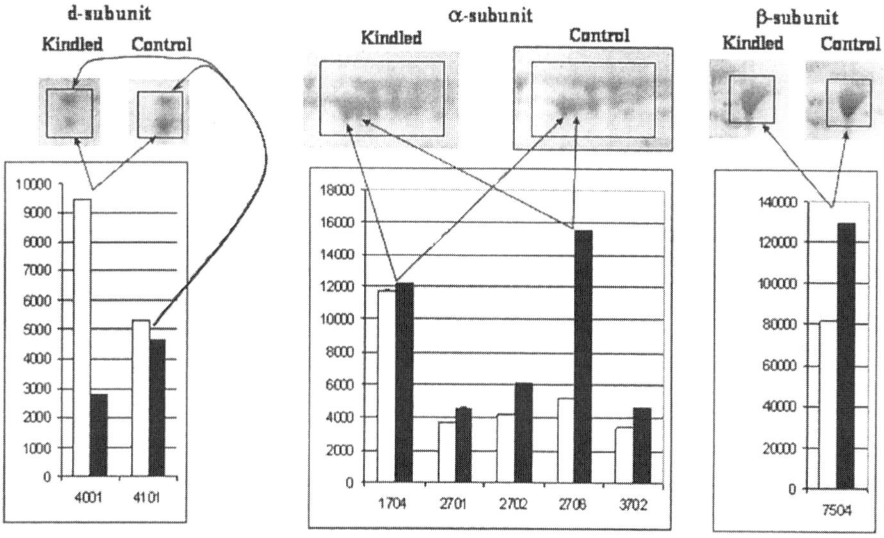

Figure 3. Kindling dependent alterations in proteins in the synaptosomal particulate fraction. Three regions of 2D gels of the particulate fraction from kindled and control hippocampi are shown containing subunits of the mitochondrial ATP synthase complex (d-subunit = # 4001, α-subunit = # 2706 and β-subunit = # 7504).

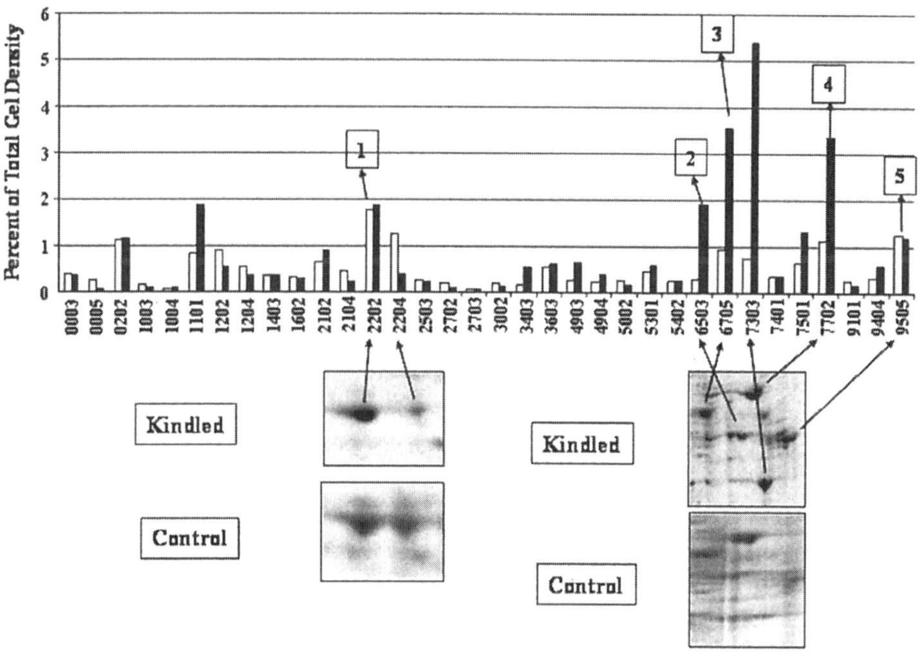

Figure 4. Vesicle H^{+}-ATPase subunits are increased in kindling. Two selected regions of 2D gel separations for kindled and control synaptic vesicle fractions are shown. The spots labeled 1-5 were identified by mass spectrometry after in gel trypsin digestion as follows: 1 = PI3 Kinase Adaptor Protein; 2 = Vesicle H^{+} ATPase B2 Subunit; 3 = NSP60; 4 = Vesicle H^{+} ATPase A Subunit; 5 = Tubulin β Chain.

corresponding set of analyses were averaged. The graph shown is the ratio of averaged relative spot intensities for the 63 most intense spots detected. As can be seen, while many spots were essentially unaltered by kindling, a larger number of spots appeared to increase in the kindled sample (downward bars) than decreased (upward bars). The spot marked with the arrow was shown by mass spectrometry to be the β-subunit of the mitochondrial ATP synthase complex present as contamination in this fraction. The increase detected in the kindled sample was also seen for the particulate fraction where the bulk of the synaptosomal mitochondria are isolated as shown below. We have also detected a similar increase in purified kindled rat brain mitochondrial fractions (data not shown). The cluster of 3 spots in the lower right of the gel (~ 25 kDa, pI ~4.5) was also of interest as this cluster shows the highest level of depolarization dependent phosphorylation detected with Pro-Q diamond stain (data not shown). As can be seen, the two spots at either end of the cluster appear to be increased on kindling (▲) while the middle spot is decreased (Δ). The former were heavily phosphorylated in depolarized synaptosomes from both kindled and control hippocampi but the latter was not. The identity of these proteins and the nature of their phosphorylation is currently under investigation.

3.3 Mitochondrial ATP Synthase Subunit Composition Is Altered On Kindling

Figure 3 shows 3 discrete regions of 2D gels obtained with purified synaptosomal membrane fractions from kindled and control hippocampi. These regions contain three subunits of the mitochondrial ATP Synthase complex, the catalytic α- (Mr ~ 58 kDa, pI ~9) and β- (Mr~56 kDa, pI ~ 5.2) subunits and the regulatory d-subunit (Mr~25 kDa, pI~7) identified by mass spectrometry analyses of in-gel trypsin digests as described in Methods. As can be seen, both the catalytic subunits appear to be substantially increased on kindling while the d-subunit is substantially decreased. Conversely, spot 1704 immediately to the left of the α-subunit spot is unchanged as is spot 4101 immediately above the d-subunit. The d-subunit has been reported to be in equilibrium in a complex-bound *vs.* free state.[9] However, our analyses did not show a corresponding increase in d-subunit content in the kindled synaptosol fraction described in section 3.2. Therefore, it would appear that the observed changes in ATP synthase subunits reflect altered protein synthesis and/or degradation.

It should be noted that Krapfenbauer et al.[10] have reported that β-subunit levels in the particulate fraction from whole rat brain are decreased by 40 % on systemic administration of kainic acid. They concluded that this is likely due to kainate induced neuronal cell death. Conversely, Yu and workers[11] showed a 2-3 fold increase in both β- and d-subunit content in rat pancreatic acinar cells undergoing cerulein-induced cell death.

3.4 Synaptic Vesicle Protein Complement Changes on Kindling

Figure 4 shows sections of the 2D gels obtained with synaptic vesicles prepared from the right (ipsilateral) hippocampi of kindled *vs.* control rats The bar graphs show a number of spots that are either increased or decreased in kindling compared to controls. A number of the spots in the selected areas shown have been identified as indicated in the legend. Of particular interest is the finding that the vesicle H^+ ATPase A and B2 subunits appear to be increased substantially as was the dihydropyramidase related protein 4

(NSP60). The fact that β-tubulin and PI3 kinase adaptor protein appear to be unchanged suggests that the observed changes are not an artifact of the analysis and may represent kindling specific adaptations.

4. DISCUSSION

We have begun to establish 2D gel maps for rat hippocampal synaptosomes and sub-fractions. While major cytoskeletal proteins such as actin and tubulin are common constituents of synaptosomes and whole cell homogenates, over 100 other protein spots appear to be enriched in the synaptosome, compared to whole hippocampus. Furthermore, the 2D gel profiles for the synaptosomal sub-compartments are largely unique from one another. As would be expected, the pattern obtained for the synaptosol most closely resembles that of the intact synaptosome while the patterns for the vesicle and particulate fractions are distinct.

A number of protein spots in the 2D gel of the hippocampal synaptic vesicle fraction are either increased or decreased in intensity relative to total gel fluorescence in the kindled sample compared to the control. The identities of a number of these proteins were determined by MALDI-TOF and/or LC-ESI-MS/MS following excision of the spot from the gel and trypsin digestion. While proteins such as β-tubulin and the PI3 Kinase adaptor protein appear to be unchanged on kindling, a major group of proteins around pI = 6 appear to be doubled in concentration in the kindled vesicle sample. These include subunits of the synaptic vesicle H^+-ATPase required for NT uptake.

Similarly, we have now quantified and identified a number of proteins isolated in the pellet fraction from disrupted kindled and control synaptosomes which includes elements of the synaptic plasma membrane as well as mitochondria and cytoskeletal remnants. To date we have detected over 20 proteins whose levels are altered by at least 2 fold in kindled samples *vs.* controls. Of particular interest, kindling appears to be associated with changes in components of energy metabolism at the synapse. Specifically, there is a significant increase in both α- and β-subunits of the mitochondrial ATP synthase complex with a corresponding loss of the regulatory d-subunit. Similar changes were observed in these components detected in the synaptosol fraction, presumably representing residual mitochondrial contamination.

These preliminary studies demonstrate that alterations in protein complements in kindled *vs.* control hippocampi can be detected on reasonable amounts of material using the combined 2D gel – mass spectrometry approach. A major limitation of protein profiling studies is the inherent complexity of proteomes even of sub-cellular compartments. However, proteomic approaches have requisite sensitivity and resolution to characterize molecular events at the synapse and alterations accompanying epileptogenesis. They allow for the screening and organization of potential mechanistic events that may not otherwise have been examined which can then be studied by traditional biochemical methodology.

5. REFERENCES

1. G. Lubec, K. Krapfenbauer, and M. Fountoulakis, Proteomics in brain research: potentials and limitations, *Prog. Neurobiol.* **69**(3), 193-211 (2003).

2. P. A. Jarvie, T. C. Logan, C. Geula, and J. T. Slevin, Entorhinal kindling permanently enhances Ca2(+)-dependent L-glutamate release in regio inferior of rat hippocampus, *Brain Res.* **508**(2), 188-193 (1990).
3. R. J. Racine, Modification of seizure activity by electrical stimulation. II. Motor seizure, *Electroencephalogr. Clin. Neurophysiol.* **32**(3), 281-294 (1972).
4. P. R. Dunkley, J. W. Heath, S. M. Harrison, P. E. Jarvie, P. J. Glenfield, and J. A. Rostas, A rapid Percoll gradient procedure for isolation of synaptosomes directly from an S1 fraction: homogeneity and morphology of subcellular fractions, *Brain Res.* **441**(1-2), 59-71 (1988).
5. S. Naito and T. Ueda, Characterization of glutamate uptake into synaptic vesicles, *J Neurochem.* **44**(1), 99-109 (1985).
6. J. A. Gordon, Use of vanadate as protein-phosphotyrosine phosphatase inhibitor, *Methods Enzymol.* **201**, 477-482 (1991).
7. U. K. Laemmli, Cleavage of structural proteins during the assembly of the head of bacteriophage T4, *Nature* **227**(259), 680-685 (1970).
8. B. Schulenberg, R. Aggeler, J. M. Beechem, R. A. Capaldi, and W. F. Patton, Analysis of steady-state protein phosphorylation in mitochondria using a novel fluorescent phosphosensor dye, *J. Biol. Chem.* **278**(29), 27251-27255 (2003).
9. S. Y. Wang, Y. Moriyama, M. Mandel, J. D. Hulmes, Y. C. Pan, W. Danho, H. Nelson, and N. Nelson, Cloning of cDNA encoding a 32-kDa protein. An accessory polypeptide of the H+-ATPase from chromaffin granules, *J. Biol. Chem.* **263**(33), 17638-17642 (1988).
10. K. Krapfenbauer, M. Berger, G. Lubec, and M. Fountoulakis, Changes in the brain protein levels following administration of kainic acid, *Electrophoresis* **22**(10), 2086-2091 (2001).
11. J. H. Yu, S. Y. Yun, J. W. Lim, H. Kim, and K. H. Kim, Proteome analysis of rat pancreatic acinar cells: implication for cerulein-induced acute pancreatitis, *Proteomics* **3**(12), 2446-2453 (2003).

Discussion

Gutierrez: Is there anything as to whether specific vesicles have a specific protein associated with them depending on the phenotype of the vesicles? In other words, can you relate a protein to a GABAergic vesicle, to a glutamatergic vesicle, or whatever?
Whiteheart: I really don't know the answer to that, whether you can actually segregate synaptic vesicles. If we had tools to segregate them, then yes we could answer that question.
Gutierrez: Do you know whether there are proteins associate with neurotransmitters?
Whiteheart: Honestly, I don't know if anybody knows. The VAMP2 molecule is the dominant SNARE associated with synaptic vesicles in the hippocampus, and it is generally used for these kinds of membrane trafficking events, in neurons. I don't think that there is a specific SNARE for specific classes of neurotransmitter containing vesicles.
Simonato: The synaptosomes are rather a unclean preparations. How did you find out how clean your preparation is and what contaminanted the final preparation?
Whiteheart: Typically what we do is run the synaptosomes through two Percol gradients to clean them up. We use EM analysis to evaluate what is in the preparations. Usually myelin sheaths or myelin structures are one of the biggest contaminants. Other than that, they are reasonably clean by EM analysis.
Teskey: Excellent talk, I very much enjoyed it. How much of the protein in a postsynaptic density is the SNARE protein, and is there a relationship with the perforated synapse?
Whiteheart: I don't know how to answer the second question. Regarding the first question, SNARE molecules in synapses have a higher specific activity than in other tissues. The SNAREs are used for a lot of different secretion events in other systems. There is a specific set in neurons, and they are at a higher specific activity. In terms of

amounts, they are some of the minor spots on the 2D gels. We can see them, and we know which ones are which, but they are some of the minor spots. So they represent maybe 1% or ½% or something like that.

Sutula: I really enjoyed your presentation. If you look at microarrays for transcripts, there are a lot of issues about quantification. You always have to move on to some other method to determine whether an increase or decrease was real. With your small error bars and pretty dramatic on and off changes in some proteins, it looks as though this is potentially much more realible as a first pass quantification. Is that a correct impression? And thinking about what it takes to get a change in a protein recognized as significant in terms of fold changes, if this is as sensitive as it appears to be, is this going to be a better way to understand protein changes than the kind of disassociated preparations, where you have to get many fold changes for anybody to believe it?

Whiteheart: To answer your first question, it does appear to be this good and this reproducible, which frankly surprised us. The data have come as good as I have shown, and it has really been a surprise. Fluorescent methods of detection are far superior to a lot of the methods that people are commonly using. I think that the linearity of the methodology and the sensitivity of methodology are really giving us this advantage. Time will tell, as we work through the data and get some specific examples for further detailed analysis, as to how good a quantitative first pass this is going to be. But so far, it has been pretty good.

Leung: Looking at the ratio of the contralateral versus the ipsilateral side is a concern, because the AD probably did spread to the other side after a while, even if not in the beginning of kindling. I guess you have used fully kindled animals.

Whiteheart: When we were doing this analysis, we tried to figure out a way to normalize for animal to animal variation. We consistently saw that there was an accumulation of SNARE complexes. That was an absolute measure, but we were getting a lot of animal to animal variation. So the interhemispheric ratio was the only way to normalize that we could come up with. We realize the limitations of the calculation in regard to whether AD spreads into the other hemisphere, but this was the best that we could come up with to allow us to normalize for animal to animal. We used at least five animals per data point.

AMYGDALA METABOTROPIC GLUTAMATE RECEPTORS AND KINDLING

Patricia Shinnick-Gallagher*

1. INTRODUCTION

Kindling is the progressive development of intense temporal lobe and generalized seizures induced by repeated subconvulsive electrical stimuli applied to certain brain structures.[1] Kindling is considered an animal model of human complex partial (temporal lobe) seizures[2, 3] and substantial evidence suggests that the amygdala is involved in temporal lobe epilepsy.[4, 5] The amygdala is one of the areas most readily kindled[6-8] and the basolateral nucleus of the amygdala is a brain area commonly selected to induce kindling.[8-10]

Kindling causes dramatic changes in the electrophysiology of amygdala neurons. Spontaneous and evoked epileptiform bursting are recorded in vitro in amygdala neurons after kindling in vivo.[11-14] The kindling-induced epileptiform activity in the basolateral amygdala is mediated in part by enhanced ionotropic glutamatergic transmission.[13-15]

Responses and second messenger pathways activated by metabotropic glutamate receptor (mGluR) agonists are also altered in kindling. Eight of these G-protein coupled receptors have been cloned and are divided in to three groups based on their pharmacology and second messenger coupling.[16, 17] Excitatory amino acid (EAA)-stimulated phosphotidylinositide (PI) hydrolysis is increased in the amygdala after kindling indicating that mGluRs may be involved.[18, 19] This effect is long-lasting and could be reproduced with the prototypical mGluR agonist, (1S,3R)-1-aminocyclopentane-1,3-dicarboxylic acid (ACPD).[20] Kindling also greatly influences mGluR actions on membrane currents and synaptic transmission in the amygdala.

* Patricia Shinnick-Gallagher, University of Texas Medical Branch, Department of Pharmacology and Toxicology, Galveston, Texas 77550.

2. MGLUR-INDUCED MEMBRANE HYPERPOLARIZATIONS RECORDED IN AMYGDALA NEURONS FROM CONTROL AND KINDLED ANIMALS

Activation of mGluRs in the amygdala induces membrane hyperpolarizations and depolarizations mediated through different mechanisms and receptors. The mGluR agonist, ACPD, hyperpolarizes BLA neurons and decreases membrane resistance by increasing K conductance.[21] The ACPD hyperpolarizations are recorded in 89% of BLA neurons showing spike firing that accommodates in response to a continued depolarizing current injection (Figure 1). The hyperpolarizations or outward currents induced by ACPD are concentration-dependent with an ED50 of 34μM and order of potency of (2S, 3S, 4S)-alpha-(carboxycyclopropyl)glycine (LCCG-I) > ACPD > (s)-4-carboxyphenyl-glycine = (RS)-4-carboxy-3-hydroxyphenylglycine (4C3HPG) > L-amino-phosphono-butyric acid (LAP4) > (1S,3S)-ACPD.[22] The non-selective antagonist, (+/-)-alpha-methyl-4-carboxyphenylglycine (MCPG) also reduced mGluR agonist-induced hyperpolarization. The mGluR outward current is mediated through a large-conductance calcium-dependent potassium (BK) conductance since the small-conductance calcium-activated potassium channel blocker, apamin, has no effect whereas BK channel blockers, iberiotoxin and charybdotoxin, as well potassium channel blockers, tetraethylammonium and 4-aminopyridine, block the response. The mGluR hyperpolarization is also blocked by pertussis toxin PTX suggesting mediation by G-proteins.[12] The mGluR mediating this response is still not clear. The pharmacological data suggests a group II mGluR[12] but the agonists are non-selective, and the calcium dependence of the response suggests mediation by group I that release intracellular calcium rather than group II mGluRs.

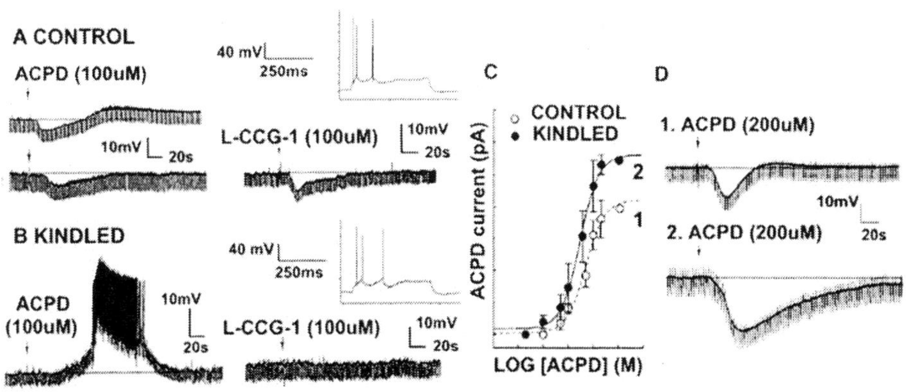

Figure 1. Kindling blocked ACPD (100μM)- and LCCG-1 (100μM)-induced hyperpolarizations in accommodating amygdala neurons.(A, B) and enhanced concentration response relationships for ACPD K^+ currents having a decreased conductance. Agonists applied by microdrop to control slices (A) and to kindled slices (B). Insets show accommodating response to 400 ms (0.5nA, depolarizing current injection). Downward lines are electrotonic potentials elicited by 200ms, 0.1nA current injection. C. Graph of inward currents vs. concentration in control (open circles) and kindled (closed circles) slices. D. Amygdala responses in control (1, upper) and kindled (2, lower) slices. Downward deflections are currents elicited by a 300ms, -5mV voltage step. Tetrodotoxin (TTX, 1μM) was present throughout. Vhold=-60mV. Reprinted with permission.[12]

Analyses of the hyperpolarizing response in kindling showed dramatic differences (Figure 1B). LCCG-I hyperpolarized 89% of accommodating neurons in slices from implanted, non-stimulated control rats but only 8% of accommodating neurons in slices from amygdala-kindled animals.[12] Similarly, the ACPD-induced hyperpolarization / depolarization in control accommodating neurons but in kindled BLA neurons evoked only membrane depolarization.[12] These lasting changes in mGluR responses after kindling reflect a down-regulation a group I mGluR subtype mediating the hyperpolarizing current response which would contribute to more excitable amygdala neurons after kindling.

3. MGLUR-INDUCED MEMBRANE DEPOLARIZATIONS AND INWARD CURRENTS IN SLICES FROM CONTROL AND KINDLED ANIMALS

The ACPD induced hyperpolarization/depolarization in control accommodating neurons or a depolarization/inward current in non-accommodating control BLA neurons is similar to that mediated by group I mGluR1 or 5 subtype in other CNS neurons. After kindling the EC50 of the ACPD-induced inward current is shifted to the left and the maximum amplitude is increased (Figure 1C, D).[12] The hyperpolarizing response reduced by kindling and the increase in depolarizing response may reflect an alteration in the balance between inhibition and excitation and may contribute to the transition to epileptiform bursting in kindled neurons.

The ACPD-induced inward current was associated with a decrease in potassium conductance; however, in Ba^{2+} and Cs^{+} solution, the inward current was enhanced and not accompanied by a change in conductance suggesting that two currents contributed to the inward current activated by ACPD.[23] A group I agonist, quisqualate (QUIS), and the group I specific agonist, (S)-3,5-dihydroxy-phenylglycine (DHPG), indicating this response was mediated through a group I mGluR. The latter inward current was dependent on Na^{+} and intracellular Ca^{2+} and blocked by Li^{+} and intracellularly applied Na^{+}-Ca^{2+} exchange inhibitory peptide suggesting the inward current is due to group I mGluR activation of Na^{+}-Ca^{2+} exchange.[23]

The postsynaptic mGluR-induced inward current mediated by Na^{+}-Ca^{2+} exchange was analyzed in slices of control (naïve and sham-operated) and amygdala-kindled rats.[24] In control neurons, the endogenous neurotransmitter, glutamate, ACPD, QUIS, and DHPG all induced activation of the Na^{+}-Ca^{2+} exchange current (Figure 2).[24] QUIS was more potent than either DHPG or glutamate (apparent EC50 = 19 μM, 57 μM, and 0.6 mM, respectively) in activating the Na^{+}-Ca^{2+} exchange current with the maximum response of DHPG equal to half of the other agonists suggesting partial agonist action. The selective mGlu5 agonist, (R,S)-2-chloro-5-hydroxyphenyl-glycine (CHPG; EC50=2.6 mM) also induced the exchange current. In slices from kindled animals, the maximum value of the DHPG (but not full agonists, QUIS or CHPG) inward current was shifted upward suggesting enhanced efficacy of a partial not a full agonist.[24] Alternatively, the increased DHPG response may be caused by insertion of new mGluR1s. These data suggest an upregulation of the Na^{+}-Ca^{2+} exchange current mediated through group I mGluRs.

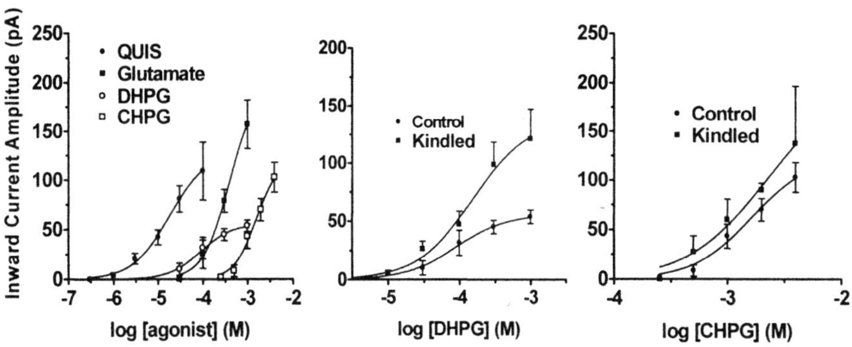

Figure 2. Concentration-response relationships for mGluR agonist-induced inward currents in control slices (left) are altered in kindled slices (middle, right). Values obtained from 3 to 18 BLA neurons. Currents were recorded in APV (50μM), CNQX (30μM/100μ in QUIS), and TTX (1μM). Vhold=-60mV. Reprinted with permission.[24]

4. KINDLING ENHANCES THE SENSITIVITY OF PRESYNAPTIC MGLUR RECEPTORS

In the amygdala two pharmacologically distinct groups of presynaptic mGluRs mediate depression of synaptic transmission. These presynaptic inhibitory mGluRs were examined three to five days after the last kindled seizure in brain slices from control and amygdala-kindled animals using whole-cell voltage-clamp and current-clamp recordings. The non-selective mGluR agonist LCCG-1 dose-dependently depressed excitatory glutamatergic transmission evoked by stimulating the lateral amygdala (LA, Figure 3).[25] The EC50 shifted leftward from 36 nM in control to 1.2nM in kindled neurons. The group III selective mGluR agonist, LAP4, was less potent with EC50 values of 297nM in control and 10.8nM in kindled neurons. These data indicated a 28- to 30-fold increase in potency in kindling. Neither LCCG-I (0.1nM-10μM) nor LAP4 (1nM-50μM) altered membrane conductance. The mGluR antagonists (2S,3S,4S)-2-methyl-2-(carboxycyclo-propyl)-glycine (MCCG, 100μM) and (S)-2-methyl-2-amino-4-phosphonobutyrate (MAP4, 100μM) selectively reversed the inhibitory effects of LCCG-1 and LAP4, suggesting mediation by group II and group III mGluRs, respectively.[25] Their exquisite sensitivity in kindling suggests that presynaptic group II- and group III-like mGluRs agonists may be useful drugs for suppression of excessive synaptic activation in temporal lobe epilepsy.

G-protein coupled metabotropic glutamate receptors (mGluRs) are implicated in various forms of neuroplasticity. The sensitivities of mGluR agonists were analyzed in different models of synaptic plasticity, kindling and chronic cocaine treatment, at another amygdala synapse, the BLA to central amygdala (CeA), to determine if changes in mGluRs were common mechanisms in neuroplastic events.[26] The non-selective mGluR agonist LCCG1 was less potent at the BLA to CeA synapse (EC50=66nM) than at the LA-BLA synapse (EC50=36nM) whereas the group III agonist LAP4 depressed neurotransmission more potently at the BLA-CeA (EC50=55nM) than at the LA-BLA

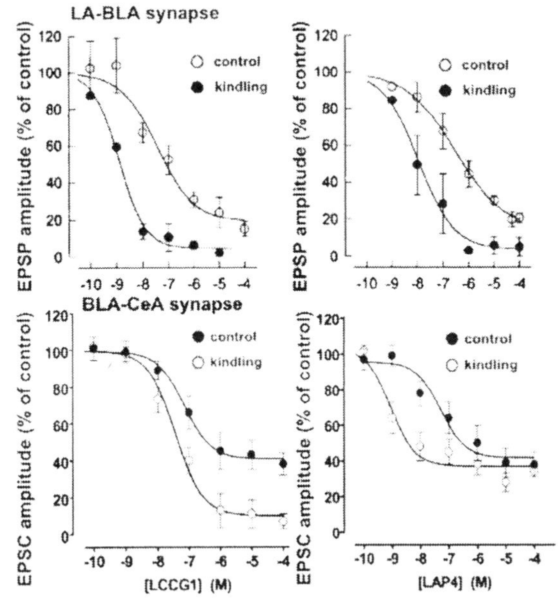

Figure 3. Kindling shifts concentration-response relationships for LCCG-I and LAP4 at two amygdala synapses. Reprinted with permission.[25, 26]

(EC50=300nM) synapse. After chemical treatment by chronic cocaine exposure or after electrical stimulation in kindling, the receptor sensitivities for mGluRs agonists were altered in opposite ways. Groups II and III agonists were ineffective 24 hours after the last cocaine treatment while three to seven days after the last kindled seizure LCCG1 and LAP4 concentration response relationships were shifted to the left suggesting an enhanced sensitivity. These data indicated that mGluRs in different models of synaptic plasticity have distinct mechanisms.[26] The enhanced sensitivity to group II and III mGluR agonists at various central synapses in kindling is similar to that recorded at the LA to BLA synapse in the amygdala but the magnitudes and mechanisms are different as illustrated in Figure 3. Specifically, the magnitude of inhibition of the mGluR agonists is greater in the BLA-CeA synapse and kindling shifts the EC 50s of LAP4 to the left at both synapses while in the BLA-CeA pathway the primary effect is an increase in the maximal inhibition suggesting different mechanisms mediate effects at these synapses. These differences may be due to activation of different subtypes of mGluRs which lack specific probes. These data further support the use of mGluRs in the treatment of epilepsy to fine-tune neuronal excitability and synaptic transmission.

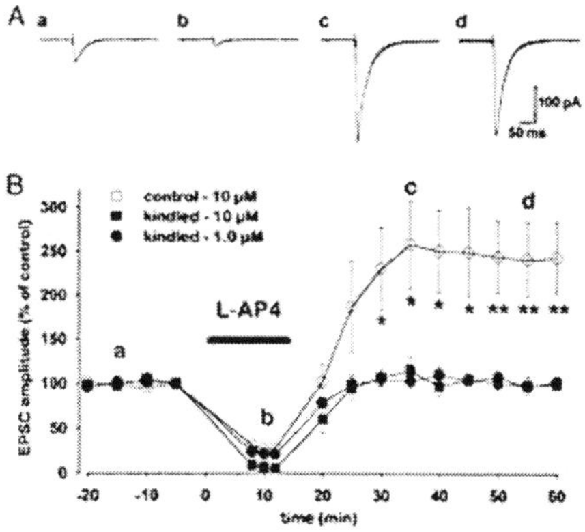

Figure 4. Kindling blocks LAP4-LTP. Reprinted with permission.[27]

5. LOSS OF GROUP III MGLUR - MEDIATED POTENTIATION WITH KINDLING

MGluRs are known to induce long-term potentiation (LTP) or long-term depression. Our laboratory reported for the first time that group III mGluR agonists elicit LTP (Figure 4).[27] LAP4 (10μM) initially depressed and then facilitated excitatory glutamatergic transmission in the LA-BLA synapse by 260%. LTP was measured 15 min after washout of LAP4, lasted for >45 min, and was not accompanied by postsynaptic membrane changes. LAP4-LTP was prevented by the group III mGluR antagonist MAP4 (100μM) but not the group II mGluR antagonist, MCCG, (100μM), indicating mediation by a group III mGluR. The LAP4-LTP was abolished after kindling. LTP was also not observed after superfusing the group II mGluR agonist LCCG-1 (1, 10μM) in either control or kindled neurons.[27] If the underlying mechanisms and functional significance of LAP4-LTP are similar to other forms of plasticity, the loss of LAP4-LTP in kindled neurons may be a neurobiological marker of learning and memory deficits in kindled animals and in epilepsy patients.

6. MGLUR AGONISTS AND ANTAGONISTS BLOCK KINDLED BURSTING

Metabotropic glutamate receptors (mGluRs) are implicated in both the activation and inhibition of epileptiform bursting activity in seizure models. Kindling enhances synaptic strength at the LA to BLA synapse, resulting in synaptically-driven bursts at low stimulus intensity.[14] MGluR agonists, ACPD and DHPG , also evoked bursting in BLA neurons from amygdala-kindled rats (Figure 5 A-D) but not in control neurons.[28] Neither the group II agonist, LCCG-I, nor the group III agonist, LAP4, induced bursting in either control or kindled neurons. The agonist-induced bursting was inhibited by the non-selective antagonists, MCPG, and (S)-4-carboxy-3-hydroxyphenylglycine (4C3HPG), which also has a partial agonist action. In contrast, 4C3HPG but not MCPG abolished afferent-induced bursting (Figure 5 E-H). Stimulating afferents with increasing intensity elicited greater epileptiform bursting in kindled neurons but not in the presence of

4C3HPG, suggesting that 4C3HPG had reversed kindled bursting to control action potential firing.[28] These data illustrate the potential for group I antagonists and group II and III mGluR agonists as antiepileptic drug therapy.

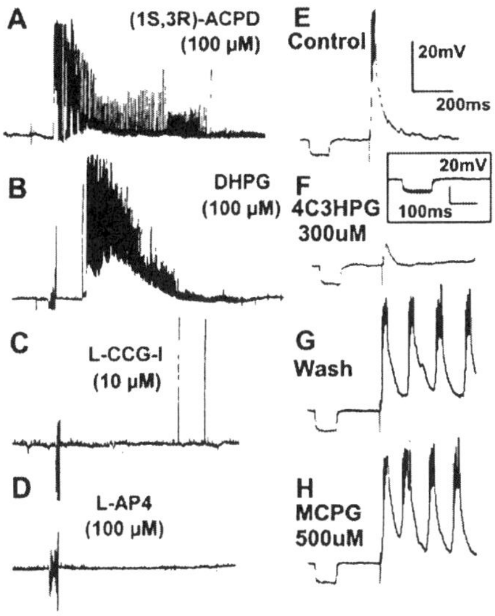

Figure 5. Group I but not II and II mGluR agonists induce bursting in kindled neurons (A-D) and mGluR antagonists have no effect on afferent-induced kindled bursting (G) unless antagonist has group II agonist (E) activity. Reprinted with permission.[28]

In summary, this laboratory has recorded profound changes in mGluR responses in the amygdala of kindled animals, specifically:

(1) a loss of mGluR-induced membrane hyperpolarization/outward current which is coupled via G-proteins to a large Ca^{2+}-dependent K conductance (BK);[12]

(2) an increase in a mGluR-induced membrane depolarization/inward current associated with a decrease in conductance;[12, 24]

(3) an increase in efficacy of group I-induced membrane depolarization/inward current coupled to the Na^{+}/Ca^{2+} exchanger;[24]

(4) a leftward shift in the ED_{50} of the dose-response relationship for inhibitory presynaptic mGluRs indicating a 30-50 fold increase in sensitivity for group II and III mGluRs with the maximal inhibitory response to group II agonists increased.[25, 26]

(5) a kindling-induced block of a group III mGluR-LTP;[26]

(6) kindling results in group I mGluR- and afferent induced bursting;[24]

(7) a block of afferent-induced bursting with group I antagonist / partial agonist group II mGluR agonists, indicating mGluRs have significant therapeutic potential as anticonvulsant drugs.[24]

7. ACKNOWLEDGEMENTS

This work was supported by NS24643 and MH58327.

8. REFERENCES

1. G.V.Goddard,Development of epileptic seizures through brain stimulation at low intensity, *Nature* **214**, 1020-1021 (1967).
2. P.S.Albright and W.M.Burnham, Development of a new pharmacological seizure model: effects of anticonvulsants on cortical- and amygdala-kindled seizures in the rat, *Epilepsia* **21**, 681-689 (1980).
3. M.Sato, R.J.Racine, and D.C.McIntyre, Kindling: basic mechanisms and clinical validity, *Electroencephalogr. Clin. Neurophysiol.* **76**, 459-472 (1990).
4. H.K.Wolf, A.F.Aliashkevich, I.Blumcke, O.D.Wiestler, and J.Zentner, Neuronal loss and gliosis of the amygdaloid nucleus in temporal lobe epilepsy. A quantitative analysis of 70 surgical specimens, *Acta Neuropathol. (Berl)* **93**, 606-610 (1997).
5. R.Kalviainen, T.Salmenpera, K.Partanen, P.Vainio, P.Riekkinen, Sr., and A.Pitkanen, MRI volumetry and T2 relaxometry of the amygdala in newly diagnosed and chronic temporal lobe epilepsy, *Epilepsy Res.* **28**, 39-50 (1997).
6. D.P.Cain, K.A.Desborough, and D.J.McKitrick, Retardation of amygdala kindling by antagonism of NMD-aspartate and muscarinic cholinergic receptors: evidence for the summation of excitatory mechanisms in kindling, *Exp. Neurol.* **100**, 179-187 (1988).
7. D.Y.Zhao and S.L.Moshe, Deep prepiriform cortex kindling and amygdala interactions, *Epilepsy Res.* **1**, 94-101 (1987).
8. W.Loscher, U.Ebert, U.Wahnschaffe, and C.Rundfeldt, Susceptibility of different cell layers of the anterior and posterior part of the piriform cortex to electrical stimulation and kindling: comparison with the basolateral amygdala and "area tempestas", *Neuroscience* **66**, 265-276 (1995).
9. D.A.Hosford, M.Simonato, Z.Cao, N.Garcia-Cairasco, J.M.Silver, L.Butler, C.Shin, and J.O.McNamara, Differences in the anatomic distribution of immediate-early gene expression in amygdala and angular bundle kindling development, *J. Neurosci.* **15**, 2513-2523 (1995).
10. G. LeGal LaSalle, Amygdaloid kindling in the rat: regional differences and general properties, in: *Kindling 2,* edited by J. Wada (Raven Press, New York, 1981), pp. 31-44.
11. E.K.Asprodini, D.G.Rainnie, A.C.Anderson, and P.Shinnick-Gallagher, In vivo kindling does not alter after hyperpolarizations (AHPs) following action potential firing in vitro in basolateral amygdala neurons, *Brain Res.* **588**, 329-334 (1992).
12. K.H.Holmes, N.B.Keele, and P.Shinnick-Gallagher, Loss of mGluR-mediated hyperpolarizations and increase in mGluR depolarizations in basolateral amygdala neurons in kindling-induced epilepsy., *J Neurophysiol.* **76**, 2808-2812 (1996).
13. P.-W.Gean, P.Shinnick-Gallagher, and A.Anderson, Spontaneous epileptiform activity and alteration of GABA- and NMDA- mediated neurotransmission in amygdala neurons kindled in vivo, *Brain Res.* **494**, 171-181 (1989).
14. D.G.Rainnie, E.K.Asprodini, and P.Shinnick-Gallagher, Kindling-induced long-lasting changes in synaptic transmission in the basolateral amygdala, *J. Neurophysiol.* **67**, 443-454 (1992).
15. Y.Shoji, E.Tanaka, S.Yamamoto, H.Maeda, and H.Higashi, Mechanisms underlying the enhancement of excitatory synaptic transmission in basolateral amygdala neurons of the kindling rat, *J. Neurophysiol.* **80**, 638-646 (1998).
16. D.D.Schoepp, D.E.Jane, and J.A.Monn, Pharmacological agents acting at subtypes of metabotropic glutamate receptors, *Neuropharmacology* **38**, 1431-1476 (1999).
17. P.J.Conn and J.P.Pin, Pharmacology and functions of metabotropic glutamate receptors., *Ann. Rev. Pharmacol.* **37**, 205-237 (1997).
18. K.Akiyama, N.Yamada, and M.Sato, Increase in ibotenate-stimulated phosphatidylinositol hydrolysis in slices of the amygdala/pyriform cortex and hippocampus ofrat by amygdala kindling, *Exp. Neurol.* **98**, 499-508 (1987).
19. K.Akiyama, N.Yamada, and S.Otsuki, Lasting increase in excitatiry amino acid receptor-mediated polyphosphoinositide hydrolysis in the amygdala/pyriform cortex of amygdala-kindled rats, *Brain Res.* **485**, 95-101 (1989).

20. K.Akiyama, A.Daigen, N.Yamada, T.Itoh, I.Kohira, H.Ujike, and S.Otsuki, Long-lasting enhancement of metabotropic excitatory amino acid receptor-mediated polyphosphoinositide hydrolysis in the amygdala/pyriform cortex of deep prepiriform cortical kindled rats, *Brain Res.* **569**, 71-77 (1992).
21. D.G.Rainnie, K.H.Holmes, and P.Shinnick-Gallagher, Activation of postsynaptic metabotropic glutamate receptors by trans-ACPD hyperpolarizes neurons of the basolateral amygdala, *J. Neurosci.* 1-13 (1994).
22. K.H.Holmes, N.B.Keele, V.L.Arvanov, and P.Shinnick-Gallagher, Metabotropic glutamate receptor agonist-induced hyperpolarizations in rat basolateral amygdala neurons: receptor characterization and ion channels, *J. Neurophysiol.* **76**, 3059-3069 (1996).
23. N.B.Keele, V.L.Arvanov, and P.Shinnick-Gallagher, Quisqualate-preferring metabotropic glutamate receptor activates Na(+)-Ca2+ exchange in rat basolateral amygdala neurones, *J. Physiol.* **499 (Pt 1)**, 87-104 (1997).
24. N.B.Keele, F.Zinebi, V.Neugebauer, and P.Shinnick-Gallagher, Epileptogenesis up-regulates metabotropic glutamate receptor activation of sodium-calcium exchange current in the amygdala, *J. Neurophysiol.*. **83**, 2458-2462 (2000).
25. V.Neugebauer, N.B.Keele, and P.Shinnick-Gallagher, Epileptogenesis in vivo enhances the sensitivity of inhibitory presynaptic metabotropic glutamate receptors in basolateral amygdala neurons in vitro, *J. Neurosci.* **17**, 983-995 (1997).
26. V.Neugebauer, F.Zinebi, R.Russell, J.P.Gallagher, and P.Shinnick-Gallagher, Cocaine and kindling alter the sensitivity of group II and III metabotropic glutamate receptors in the central amygdala, *J. Neurophysiol.* **84**, 759-770 (2000).
27. V.Neugebauer, N.B.Keele, and P.Shinnick-Gallagher, Loss of long-lasting potentiation mediated by group III mGluRs in amygdala neurons in kindling-induced epileptogenesis, *J. Neurophysiol.* **78**, 3475-3478 (1997).
28. N.B.Keele, V.Neugebauer, and P.Shinnick-Gallagher, Differential effects of metabotropic glutamate receptor antagonists on bursting activity in the amygdala, *J. Neurophysiol.* **81**, 2056-2065 (1999).

Discussion

Gale: Yesterday there was talk about the different paradigms that will lead to better kindling, or less better kindling, and maybe no kindling at all if the stimulations are not on the right kind of schedule. I wonder if you have looked at any other source of stimulation paradigms where perhaps you end up not getting as good kindling, and whether the anticonvuslant components might be increased or even the other extreme. If you give maximal electroshock seizures repeatedly you end up getting tolerance, not kindling. Maybe you could actually see sort of more overwhelming effect of the anticonvulsant mechanisms versus the proconvulsant mechanisms.
Gallagher: We are actually looking at some of these effects at different frequencies in in vitro and not in vivo. They maybe somewhat frequency dependent.
McNamara: I think that is very interesting line of work that you have been at for a while. The lack of an LAP4 induced LTP... LAP4 is acting through what group of GLU receptors – I, II, III?
Gallagher: It is presynaptic group III.
McNamara: One way of thinking about that is it as if occluded so that it has already occurred in the kindled condition.
Leung: Very nice work. I just wonder what happened to synaptic transmission in either amygdala with all those changes, and perhaps what the physiological conditions are in which the MGluR1 receptors are activated. Do you also use some synaptic activation other than direct activation of those receptors?
Gallagher: Oh yes, we were measuring the effects on synaptic transmission.
Leung: But you still applied the drug, right?

Gallagher: Yes, we still applied the drug. You are asking whether we can release glutamate endogenously?
Leung: Yes, do you have to release a lot of glutamate to activate any of those groups of receptors?
Gallagher: We haven't as yet been able to do that. Theoretically, if there is more glutamate released in kindling they would be activated, just because of the excess of glutamate, but we have not been able to test that.

INDUCTION OF B_1 BRADYKININ RECEPTORS IN THE KINDLED BRAIN

Michele Simonato, Manuela Mazzuferi, Donata Rodi, Anna Binaschi, Simona Mantovani, Brice Ongali, and Rejean Couture*

1. INTRODUCTION

Knowledge of the cellular and molecular mechanisms underlying the various forms of epilepsy could drive the search for new drugs and lead to improved therapies. The proposal of J.H. Jackson, put forward more than a century ago, that seizures are caused by "occasional, sudden, excessive, rapid and local discharges of gray matter" still provides a useful framework for research.

Within this framework, the question then becomes what makes the epileptic brain hyperexcitable. One hypothesis may be that nervous cells within the epileptic tissue are (or become) more sensitive to excitatory chemical messages within their environment. Accordingly, a mechanism of epileptic hyperexcitability may be the increased expression and/or biological activity of receptors mediating excitatory effects.

In this study, we tested whether a bradykinin (BK) receptor subtype named B_1 may be involved in kindling hyperexcitability. BK and its related peptides are autacoids. Their functions in cardiovascular homeostasis, in contraction and relaxation of smooth muscles, in inflammation and nociception are well established.[1-3] The BK system is also involved in disease states like asthma, allergies, rheumatoid arthritis, cancer, endotoxin and pancreatic shock.[1,5] Furthermore, the kinins are neuromediators of central neuronal pathways associated with the autonomic control of blood pressure and nociception.[3,4] However, their involvement in neurological diseases has not been intensively investigated thus far.

The rational for undertaking this study lays in the following characteristics of B_1 BK receptors:[5-9] (1) under normal conditions, they are scarcely represented in the central nervous system; (2) their concentration in peripheral tissues is also low, but it is highly increased under pathological conditions, such as inflammation and pain; (3) these receptors mediate excitatory effects in the peripheral nervous system.

* Michele Simonato, Manuela Mazzuferi, Donata Rodi, Anna Binaschi, Simona Mantovani, Department of Clinical and Experimental Medicine (Section of Pharmacology) and Neuroscience Center, University of Ferrara, 44100 Ferrara, Italy. Brice Ongali, Rejean Couture, Department of Physiology, Université de Montréal, C.P. 6128, Succursale Centre-ville, Montréal, Québec, Canada H3C 3J7.

In order to pursue our aim, we first compared B_1 BK receptor density in the normal and in the kindled brain. Second, we examined the effect of B_1 receptor activation on glutamate release in the normal and kindled hippocampus, using a microdialysis approach.

2. METHODS

2.1 Animals

Male Sprague-Dawley rats (300-350 g) were kept in controlled room temperature (22-24°C) and humidity (55-65%), with free access to food and water and maintained under a 12:12 hours light-dark cycle. For kindling, a twisted bipolar electrode was implanted into the right amygdala (co-ordinates: 4.8 mm lateral and 0.8 mm posterior to bregma; 8.3 mm deep from dura) under ketamine (130 mg/kg i.p.) anesthesia. Animals were allowed to recover for 7 days after surgery and then stimulated once a day with a single 1 second train of bipolar pulses (0.5+0.5 ms, 60 Hz, 25% above after-discharge threshold). Kindling criteria, i.e. 3 consecutive class 4 or 5 seizures, were reached after 15±1 stimulations (staging according to Racine, 1972[10]). Rats were sacrificed 7 days after the last stimulation. Age-matched control rats have been submitted to the same surgical procedure and to daily handling, but they were not electrically stimulated (sham-stimulated group).

2.2 Quantitative autoradiography

Rats were killed by decapitation under light diethyl ether anesthesia. Their brains and spinal cords were immediately removed, frozen in 2-methyl-butane cooled in dry ice and stored at -80°C until use. Matched whole brains or spinal cords were mounted together (2 brains or 4 spinal cords per gelatine bloc) and serially cut into 20 μm thick sections on a cryostat. Sections were thaw-mounted on gelatine/chromium potassium sulfate-coated slides and stored at –80°C. Sets of three slides were used for total binding and sets of two slides for non-specific binding.

For B_1 receptor autoradiography, sections were thawed at room temperature and pre-incubated for 30 seconds in 25 mM PIPES buffer (pH 7.4; 4°C). Thereafter, slides were incubated for 90 minutes at room temperature in 25 mM PIPES buffer containing: 1 mM 1,10-phenanthroline, 1 mM dithiothreitol, 0.014% bacitracin, 0.1 mM captopril, 0.2% bovine serum albumin and 7.5 mM magnesium chloride in the presence of 150 pM [^{125}I]HPP-des-Arg^{10} Hoe-140.[11] The non-specific binding was determined in the presence of 1 μM unlabelled ligand. The concentration of radioligands was chosen based on previous studies[12,13] and on pilot experiments. At the end of incubation, slides were transferred sequentially through four rinses of 4 minutes each in 25 mM PIPES (pH 7.4; 4°C), dipped for 15 seconds in distilled water (4°C) to remove the excess of salts, and air-dried. Films were juxtaposed to the slides in the presence of [^{125}I]-microscales and exposed at room temperature for 3 days. Densitometric readings were performed with an image analysis system (MCIDTM, Imaging Research, Ontario, Canada). A standard curve from [^{125}I]-microscales was used to convert density levels into fentomoles per milligram of tissue (fmol/mg tissue). Specific binding was determined by subtracting superimposed digitalized images of non specific labeling from total binding.

2.3 Microdialysis

Rats were anaesthetized with halothane and a vertical concentric probe, endowed with a 3 mm cuprophan dialyzing membrane (molecular weight cut-off 10000), was inserted in a previously positioned guide cannula (coordinates: 4.9 mm lateral and 3.4 mm posterior to bregma) and fixed with dental cement. Furthermore, an electrode was coupled with the guide cannula, the tip of which was positioned along the dialyzing membrane; therefore, the recording site was at the middle of the probe length, 1 mm away from the membrane surface.[14] Twenty-four h after implantation, the microdialysis probe was perfused at a flow rate of 2 μl/min with a modified Ringer solution (composition in mM: $CaCl_2$ 1.2; KCl 2.7, NaCl 148 and $MgCl_2$ 0.85, pH 6). Samples were collected every 10 min, starting 6 h after the onset of perfusion. After collection of 4 basal samples, B_1 receptor agonists and antagonists were added to the perfusion buffer. Stimulation was applied as 10 min perfusion with Ringer solution containing 100 mM K^+. The correct placement of the guide cannula, electrode, and probe in the hippocampus and of the bipolar electrode in the amygdala were verified by histological examination.

The endogenous glutamate concentration in the microdialysis samples was measured analyzing an aliquot of the samples by HPLC coupled to fluorimetric detection after precolumn derivatization with o-phthalaldehyde. The system consisted of a Beckman 125 pump, a Triathlon autosampler (Spark Holland), a Chromsep analytical column (Chrompack) and a fluorescence spectrophotometer RF-551 (Shimatzu). The aquisition and analysis of the chromatograms was performed by a computer controlled system (Beckman Gold System).

3. RESULTS

3.1 Receptor autoradiography

B_1 BK receptors have been found at very low density throughout the brain of sham-stimulated control rats (approximately 0.2 fmol/mg tissue). However, B_1 receptor binding sites were dramatically increased in kindled compared to sham-stimulated animals. The most striking increases were observed in the structures associated with amygdaloid stimulation: amygdala, piriform cortex and hippocampus (2- to 3-fold increases; Figs. 1 and 2). Significant increases were also observed in other forebrain regions, such as the hypothalamus and the neocortex, but not in other structures, such as the cerebellum and the spinal cord (data not shown).

3.2 Glutamate release *in vivo*

Basal glutamate release in kindled animals was found to be significantly higher that in sham-stimulated controls (3,82±0,58 vs. 2,27±0,44 pmol/10 min). High K^+-evoked glutamate release was also greater in kindled animals (probe perfusion of 100 mM K^+ for 10 minutes prompted an approximately 2-fold increase of glutamate release in sham-stimulated rats, vs. an approximately 3-fold increase in kindled rats). It seems plausible that these changes contribute to the latent hyperexcitability of the kindled brain.

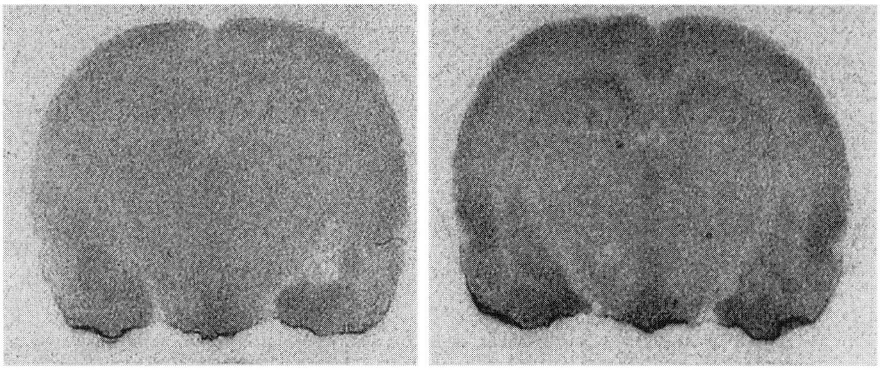

Figure 1. B_1 BK receptor density is highly increased in multiple forebrain areas of the kindled brain. Autoradiographic localization of brain B_1 BK receptors at the level of the dorsal hippocampus. Shown are autoradiograms representing total binding of [^{125}I]HPP-des-Arg^{10}-Hoe 140 (150 pM) in control, sham-stimulated (left) and kindled (right) rats.

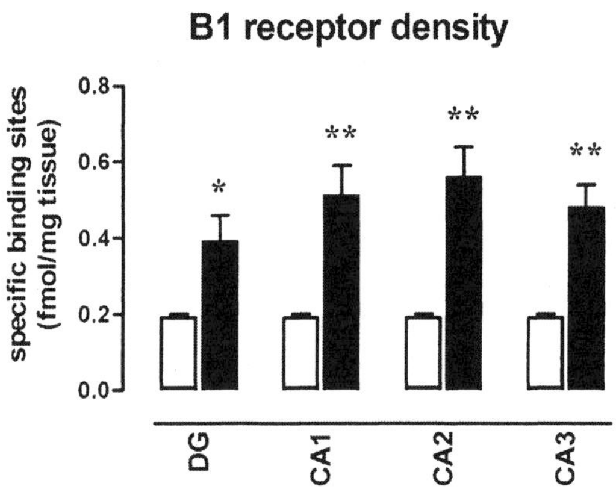

Figure 2. B_1 BK receptor density is highly increased in the hippocampus of kindled rats. Specific binding sites for B_1 BK receptors in hippocampal subareas of control (open bars) and kindled (solid bars) rats. The columns represent the means±s.e. of the values in fmol/mg of tissue. Data are the means±s.e. of 4 rats. $^{**}P<0.01$, $^{*}P<0.05$; Student's *t*-test for unpaired data. DG: dentate gyrus.

The B_1 receptor agonist Lys-des-Arg^9-BK, perfused in the probe at a concentration of 1 μM, induced a significant increase of basal glutamate release in kindled but not in control rats (Fig. 3A). This effect is B_1 receptor-mediated, because the selective B_1 antagonist AcLys[D-βNal^7,Ile^8]des-Arg^9-bradykinin[15] (R-715, 10 μM in the perfusion buffer), which did not produce any effect when applied alone, completely prevented the effect of Lys-des-Arg^9-BK (Fig. 3A). Lys-des-Arg^9-BK, however, could not increase high K^+-evoked glutamate release, while the B_1 antagonist R-715 significantly reduced it (Fig. 3B).

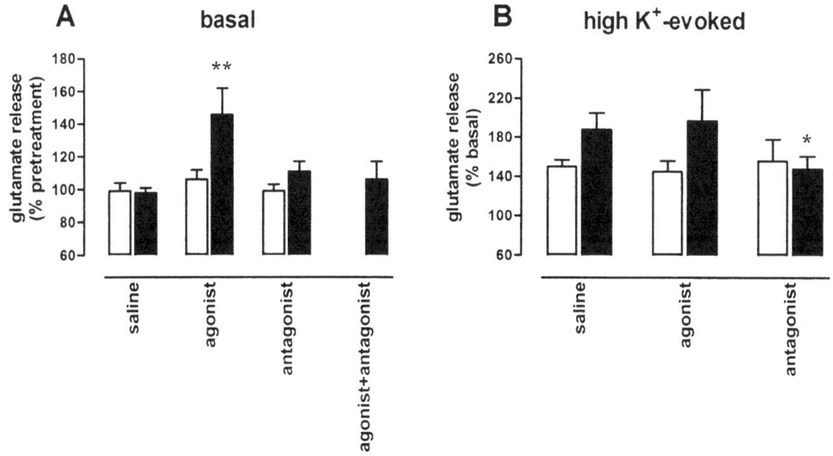

Figure 3. The B_1 receptor agonist Lys-des-Arg^9-BK increases basal glutamate release in the kindled, but not in the control, hippocampus, while the B_1 receptor antagonist R-715 decreases high K^+-evoked glutamate release in the kindled, but not in the control, hippocampus. Effect of Lys-des-Arg^9-BK (agonist, 1 μM in the perfusion buffer) and of R-715 (antagonist; 10 μM in the perfusion buffer) on basal (A) and high K^+-evoked (B) glutamate release from the hippocampus of control (open bars) and kindled (solid bars) rats Data are the means±s.e. of 5-7 experiments. **$P<0.01$, *$P<0.05$; Mann-Whitney U test.

4. DISCUSSION

The present study demonstrates marked increases of B_1 receptor specific binding sites in multiple kindled brain regions, notably those related to the electrical amygdaloid stimulation. These data are in agreement with similar results obtained in a different model (pilocarpine) using a different experimental approach (immunohistochemistry).[16] B_1 BK receptor densities were not significantly altered in the cerebellum and spinal cord of kindled, compared with control, rats.

We have also begun to evaluate the functional consequences of the changes in B_1 BK receptor binding in kindled rats. We chose to study glutamate release, because BK increases glutamate release from astrocytes and calcium levels in neurons.[17] The data we have obtained, in agreement with the autoradiography data and with previous *in vitro* observation,[18] show that the B_1 receptor agonist Lys-des-Arg^9-BK increases basal glutamate release from the hippocampus of kindled, but not of control, rats. This effect is B_1 BK receptor-mediated, being fully prevented by a selective antagonist (R-715). BK produced endogenously during depolarization seems to contribute, via B_1 receptors, to the robust increase in glutamate release evoked by high K^+ in the kindled hippocampus, because the B_1 receptor antagonist R-715 significantly reduces it. Whether these effects of B_1 receptor activation are exerted directly on glutamatergic neurons, indirectly via interconnected neurons, or are glia-mediated remains to be established. However, the observation of B_1 receptors in neuronal cells[16] supports the notion that they are, at least partially, direct.

As stated in the Introduction, while the involvement of BK in peripheral disease states (such as inflammation) is well established, its implication in central nervous system diseases is still debated. However, the present data, together with other indirect evidence, suggest that BK may play a proepileptic role: (1) BK exerts depolarizing effects on neurons;[5] (2) it has been found in brain areas thought to be critical for the development of temporal lobe epilepsy;[4] (3) this study and others[18,19] show that B_1 BK receptors, that are essentially absent in the brain under normal conditions, are induced in multiple critical brain areas in epilepsy models; (4) in contrast, B_2 BK receptors have been reported to be down-regulated in one model (kindling) and up-regulated in another (pilocarpine);[16,19] (5) *in vitro*[18] and *in vivo* (this study) evidence support the notion that BK causes, via activation of these newly-produced B_1 receptors, an increase in the release of glutamate from the hippocampus of kindled, but not of control, rats.

The mechanism of B_1 receptor induction in epilepsy models is uncertain. B_1 receptor expression is mainly associated with inflammatory conditions. This process has been well characterized and it thought to be secondary to the production of pro-inflammatory cytokines: B_1 receptors are induced by IL-1β,[1,2] via protein kinase C, protein tyrosine kinases, MAP kinase and NF-kB.[1,20-22] Changes in cytokine expression and increases in TNF-α and NF-kB expression have been reported to occur in epilepsy.[23-25] In particular, a rapid up-regulation of IL-1β has been observed.[26] Furthermore, it has been demonstrated that the administration of IL-1β causes exacerbation of seizures, whereas treatment with the natural interleukin receptor antagonist (IL-RA) prevents the occurrence of convulsions.[27] These observations suggest that the cascade of molecular events primed by limbic seizures include increased IL-1β expression, IL-1 receptor activation, induction of B_1 receptors and increased glutamate release.

However, this hypothesis is challenged by the observation that B_2 BK receptors undergo an opposite regulation (i.e. they are down-regulated) in the kindling model:[19] since both B_1 and B_2 receptors appear to exert primary excitatory effects on neurons,[4,5] it is difficult to anticipate the net consequences of B_1 receptor up-regulation and B_2 receptor down-regulation in cell populations directly involved in kindling development. In another epilepsy model, however, up-regulation of both B_1 and B_2 BK receptors has been reported to occur.[16] In this event, the BK system would be anticipated to contribute substantially to the development of hyperexcitability.

In conclusion, the present data suggest that the B_1 BK receptors may play a role in the physiopathology of epilepsy, and may represent a new interesting therapeutic target. Tools are available to challenge this idea both pharmacologically (using B_1 and B_2 receptor antagonists) and genetically (using B_1 and B_2 receptor knock-outs).

5. REFERENCES

1. F. Marceau, J.F. Hess, and D.R. Bachvarov, The B1 receptors for kinins, *Pharmacol. Rev.* **50**(3), 357-386 (1998).
2. J.B. Calixto, D.A. Cabrini, J. Ferriera, and M.M. Campos, Kinins in pain and inflammation, *Pain* **87**(1), 1-5 (2000).
3. R. Couture, M. Harrisson, R.M. Vianna, and F. Cloutier, Kinin receptors in pain and inflammation, *Eur. J. Pharmacol.* **429**(1-3), 161-176 (2001).
4. R. Couture, and C.J. Lindsey, in: *Handbook of Chemical Neuroanatomy. Peptide Receptors, Part 1, vol. 16*, edited by R. Quirion, A. Bjorklund, and T. Hokfelt (Elsevier Science, Amsterdam, The Netherlands, 2000); pp. 241-300.
5. K.D. Bhoola, C.D Figueroa, and K. Worthy, Bioregulation of kinins: kallikreins, kininogens, and kininases, *Pharmacol. Rev.* **44**(1), 1-80 (1992)
6. D. Regoli, and J. Barabé, Pharmacology of bradykinin and related kinins, *Pharmacol. Rev.* **32**(1), 1-46 (1980).
7. F. Marceau, Kinin B1 receptors: a review, *Immunopharmacology* **30**(1), 1-26 (1995).
8. K. Walker, M. Perkins, and A. Dray, Kinins and kinin receptors in the nervous system. *Neurochem. Int.* **26**(1), 1-16 (1995).
9. L.F. Donaldson, M R. Hanley, and A.C. Villablanca, Inducible receptors. *Trends Pharmacol. Sci.* **18**(5), 171-181 (1997).

10. R.J Racine, Modification of seizure activity by electrical stimulation. II. Motor seizure, *Electroencephalogr. Clin. Neurophysiol.* **32**(3), 281-294 (1972).
11. R.M. Vianna, B. Ongali, D. Regoli, J.B. Calixto, and R. Couture, Up-regulation of kinin B1 receptor in the lung of streptozotocin-diabetic rat: autoradiographic and functional evidence, *Br. J. Pharmacol.* **138**(1), 13-22 (2003).
12. C. Murone, G. Paxinos, M.J. McKinley, B.J. Oldfield, W Muller-Esterl, F.A. Mendelsohn, and S.Y. Chai, Distribution of bradykinin B2 receptors in sheep brain and spinal cord visualized by in vitro autoradiography, *J. Comp. Neurol.* **381**(2), 203-218 (1997).
13. F. Cloutier, H. de Sousa Buck, B. Ongali, and R. Couture, Pharmacologic and autoradiographic evidence for an up-regulation of kinin B_2 receptors in the spinal cord of spontaneously hypertensive rats, *Br. J. Pharmacol.* **135**(7), 1641-1654 (2002).
14. M. Marti, G. Bregola, M. Morari, A. Gemignani, and M. Simonato, Somatostatin release in the hippocampus in the kindling model of epilepsy: a microdialysis study, *J. Neurochem.* **74**(6), 2497-2503 (2000).
15. F. Gobeil, W. Neugebauer, C. Filteau, D. Jukic, S. Nsa Allogho, L.H. Pheng, X.K. Nguyen-Le, D. Blouin, and D. Regoli, Structure-activity studies of B1 receptor-related peptides. Antagonists. *Hypertension* **28**(5), 833-839 (1996).
16. G.A. Arganaraz, J.A. Silva Jr, S.R. Perosa, L.G. Pessoa, F.F. Carvalho, J.L. Bascands, M. Bader, E. da Silva Trindade, D. Amado, E.A. Cavalheiro, J.B. Pesquero, and M. Naffah-Mazzacoratti, The synthesis and distribution of the kinin B1 and B2 receptors are modified in the hippocampus of rats submitted to pilocarpine model of epilepsy, *Brain Res* **1006**(1), 114-125 (2004).
17. V. Parpura, T.A. Basarsky, F. Liu, K Jeftinija, S Jeftinija, and P.G Haydon, Glutamate-mediated astrocyte-neuron signaling, *Nature* **369**(6483), 744-747 (1994).
18. G. Bregola, K. Varani, S. Gessi, L. Beani, C. Bianchi, P.A. Borea, D. Regoli, and M. Simonato, Changes in hippocampal and cortical B1 bradykinin receptor biological activity in two experimental models of epilepsy, *Neuroscience* **92**(3), 1043-1049 (1999).
19. B. Ongali, M.M. Campos, G. Bregola, D. Rodi, D. Regoli, G. Thibault, M. Simonato, and R. Couture, Autoradiographic analysis of rat brain kinin B1 and B2 receptors: normal distribution and alterations induced by epilepsy, *J. Comp. Neurol.* **461**(4), 506-519 (2003).
20. M.M. Campos, G.E.P Souza, and J.B. Calixto, In vivo B1 kinin-receptor up-regulation. Evidence for involvement of protein kinases and nuclear factor kappaB pathways, *Br. J. Pharmacol.* **127**(8), 1851-1859 (1999).
21. J.F. Larrivée, D.R. Bachvarov, F. Houle, J. Landry, J. Huot, and F. Marceau, Role of mitogen-activated protein-kinases in the expression of the B1 kinin receptors induced by tissue injury, *J. Immunol.* **160**(3), 1419-1426 (1998).
22. J.P. Schanstra, E. Bataillé, M.E.M. Castano, Y. Barascud, C. Hirtz, J.B. Pesquero, C. Pecher, F. Gauthier, J.P. Girolami, and J.L. Bascands, The B1-agonist [des-Arg10]-kallidin activates transcription factor NF-kB and induces homologous upregulation of the bradykinin B1 receptor in cultured human lung fibroblasts, *J. Clin. Invest.* **101**(10), 2080-2091 (1998).
23. B.C. Albensi, Potential roles for tumor necrosis factor and nuclear factor NF-kB in seizures activity. *J Neurosci. Res.* **66**(2), 151-154 (2001).
24. S.M. Allan, and N.J. Rothwell, Cytokines and acute neurodegeneration, *Nat. Rev. Neurosci.* **2**(10), 735-744 (2002).
25. Y. Yaari, and H. Beck, "Epileptic neurons" in temporal lobe epilepsy, *Brain Pathol.* **12**(2), 234-239 (2002).
26. A. Vezzani, M. Conti, A. De Luigi, T. Ravizza, D. Moneta, F. Marchesi, and M.G. De Simoni, Interleukin-1beta immunoreactivity and microglia are enhanced in the rat hippocampus by focal kainate application: functional evidence for enhancement of electrographic seizures, *J. Neurosci.* **19**(12), 5054-5065 (1999).
27. A. Vezzani, D. Moneta, M. Conti, C. Richichi, T. Ravizza, A. De Luigi, M.G. De Simoni, G Sperk, S Andell-Jonsson, J. Lundkvist, K. Iverfeldt, and T. Bartfai, Powerful anticonvulsant action of IL-1 receptor antagonist on intracerebral injection and astrocytic overexpression in mice, *Proc. Natl. Acad. Sci. USA* **97**(21), 11534-11539 (2000).

Discussion

Gale: I would like to compliment you on forging new territories without having to find them with gene arrays. I think that it is a testimony to how hypothesis driven research and putting pieces together from what we know is a great thing to do. What other stimuli might up regulate this system? Would just plain activity or changing the environment or stress or any number of stimuli up regulate the system?

Simonato: Peripherally, of course, these peptides are involved in inflammation, pain control, and allergy. There are some data coming out regarding the central nervous system as well. As you know, there are data, for example, regarding inflammatory events that are associated with stroke, and apparently there may be an up regulation of B1 receptors in the late phase of stroke-associated inflammation. Inflammation in stroke is bad initially and good in the late phase, and if indeed the bradykinin system is biphasic in terms of recruitment of different receptor subtypes, with B1 receptors expressed in the late phase, that would provide an alternative to regular anti-inflammatory drugs that would do good and bad at the same time.

Vezzani: You said that bradykinin is an inductor of B1 receptors and the B1 receptors are present both on neurons and astrocytes after seizure. Where are the receptors up regulating, in which population, and from which cells is bradykinin released?

Simonato: These are good questions. As for the second question, where bradykinin comes from, I don't really know. However, bradykinin has been found in brain homogenates, and in the cerebrospinal fluid. As for B1 receptors, we used an autoradiographic approach and cannot tell where they are, but there is a paper that just came out from the group of Cavalheiro and Pesquero that largely confirms our results in the pilocarpine model using a immunohistochemistry approach. They see B1 receptors on neurons in the hippocampus.

Vezzani: Do they affect voltage-dependent calcium current, which is the mechanism for increasing glutamate release?

Simonato: That is difficult to tell at the moment. The signaling pathways that are activated by these receptors, as far as it is known in peripheral tissues, involve activation of phospholipase C and phospholipase A2. That could imply a number of different 2nd and 3rd messenger systems, including NO for example.

Schwartzkroin: Let me pursue this a little more and ask whether you think the presence of reactive glia in the area that you are looking at would make the effect bigger? That is, do you think that getting glutamate release from lots of glia might play a significant role?

Simonato: You are asking if the glutamate release is glial or neuronal? It could be both as a matter of fact. What we know is that potassium-evoked glutamate release is almost totally calcium-dependent and TTX-sensitive. Basal release is calcium-dependent and TTX-sensitive by approximately 50%. Therefore, it may be partly glial.

Schwartzkroin: At the time that you look at this phenomenon in your kindled animals, what is the situation in respect to gliosis in that region?

Simonato: It is difficult to tell, because the major source of gliosis around there is the mcrodialysis probe itself.

Schwartzkroin: If one thinks just of general seizure foci, there is often gliotic phenotype, but I didn't think that there was that much of a gliotic response in the kindling model.

Simonato: No, I would say that it is far more the probe than kindling per se that may contribute to that.

Avanzini: In kindled animals do you find quite widespread induction? I wonder if during the kindling process you could dissociate somehow and see whether there is more regional expression or it depends on the seizure again.

Simonato: Yes, that is an important question. So far, we have performed two experiments to address this point. The first experiment was really encouraging: we found an up regulation in the amygdala, entorhinal cortex, and hippocampus, but not in other regions. The second experiment was not as good. Other experiments are in progress.

Wasterlain: You have done really nice work as always. I personally believe that the peptides are a really good therapeutic target and the effect is very powerful. You showed that endorphin are also involved, and we find that tachykinins and galanin are involved. The big problem with using those as chronic therapeutic targets is the internalization of the receptors. So does the B1 receptor also internalize?

Simonato: B1 receptors do not easily internalize.

GALANIN AND KINDLING

Merab Kokaia*

1. INTRODUCTION

Galanin is a 29 aminoacid residue neuropeptide (30 amionoacids in humans) widely distributed throughout central nervous system and intestines[1,2] that has been shown to be involved in regulation of numerous processes; such as feeding, nociception, nerve regeneration, memory, neuroendocrine release, intestine secretion and contractility (see [2]). In the brain, galanin is normally present in cholinergic and noradrenergic neuronal populations of the medial septum and locus coeruleus, respectively.[3-10] It is transported and subsequently released by the extensive axonal projections of these neurons into a wide range of forebrain regions, including the hippocampus. In addition, intrahippocampal colchicine administration, which blocks fast axonal transport and mitosis, produces scattered galanin-immunoreactive neurons and glial cell bodies in the hippocampus.[2,11] Galanin acts via 3, so far identified, distinct G-protein-linked receptor subtypes: GalR1, GalR2 and GalR3.[12-16] Similar to galanin, GalR1 and GalR2 (but not GalR3) mRNAs are abundantly expressed in many brain areas.[17,18] The detailed mapping of galanin mRNA, its immunoreactivity and binding sites demonstrated that for the most part mRNA expression matches with observed immunoreactivity in the cells, and that receptor binding sites correlate well with galanin-immunoreactive fiber bundles in various regions of the rat brain (for details see[2]). Activation of GalR1 and GalR3 has been shown to inhibit adenylyl cyclase, decrease cAMP concentration, and open G-protein-coupled inwardly rectifying K^+ channels. GalR2 stimulates phospholypase C and PKC action.[13,19,20]

The physiolgical effects of galanin signaling in the brain, particularly in the hippocampus, are mostly inhibitory. Galanin either decreases excitatory transmitter release by activating ATP-dependent K channels in the hippocampal neurons,[21] or hyperpolarizes CA3 pyramidal neurons by opening K channels.[22] Moreover, it has been shown that galanin inhibits slow cholinergic EPSPs in the CA1 pyramidal neurons,[23] as well as LTP in the Schaffer collateral-CA1 synapses[24] in hippocampal

* Section of Restorative Neurology, Wallenberg Neuroscience Center, BMC A-11, Lund University Hospital, SE-221 84, Lund, Sweden. Tel.: +46 46 222 05 47, Fax: +46 46 222 05 60E-mail: Merab.Kokaia@neurol.lu.se

slices. Taken together, these data on the inhibitory role of galanin raises the possibility that it could have a suppressant effect on seizure activity in the brain. Indeed, intraventricular administration[25] and hilar infusion[26] of galanin in rats was shown to exert anticonvulsant action on picrotoxin-induced seizure activity and status epilepticus (SE), respectively. This effect was attenuated by the galanin receptor antagonist M35. Already 3 h after SE, induced by either electrical stimulation of perforant path in the dentate gyrus or Li-pilocarpine administration, galanin fiber immunoreactivity was almost completely absent in the hilus. Immunoreactivity was partially restored 1 week after SE but galanin-positive fiber density was still significantly lower as compared to control animals. At 12 h after SE, galanin immunoreactive cell bodies started to appear bilaterally in the hilus, mostly in the subgranular zone, increasing in numbers at 24 h, at which time they were evenly distributed throug-out the entire hilus. No galanin-immunoreactive cells were observed in the granule cell layer or CA3 field. The number of galanin expressing cells gradually decreased by 3 and 7 days later. These data indicated that galanin might act as an endogenous antiepileptic agent, whereby the depletion of galanin in the hippocampus by SE may promote epileptic activity, while galanin up-regulation in hilar neurons could counteract seizures.

2. GALANIN AND ITS RECEPTOR GENE MANIPULATIONS

The role of galanin as an endogenous antiepileptic agent was further substantiated by data from galanin gene knock-out (KO) mice. These mice showed increased propensity to develop SE induced either by perforant path stimulation or by systemic kainate injection.[27] For example, while 30 min of perforant path stimulation was unable to induce SE in any of the wild-type (WT) mice, all galanin KO mice developed SE. Galanin KO mice also showed significantly longer afterdischarge durations and more severe pentylenetetrazol-induced seizures than WT mice.[27]

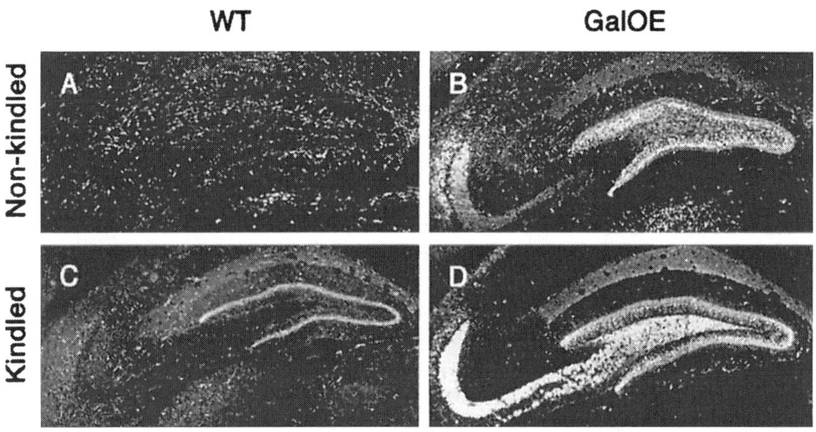

Figure. 1. Basal and seizure-evoked levels of galanin-LI in the hippocampus of GalOE mice are higher as compared with WT mice. Note that galanin-LI is detectable only in dispersed fibers in WT mice (*A*). In contrast, galanin-LI is high in granule cell layer, mossy fibers, and hilus of GalOE mice (*B*). After kindling-induced seizures, galanin-LI is increased in the supragranular layer and stratum lacunosum moleculare of CA1 in WT mice (*C*) and is much more markedly elevated in granule cell layer, hilus, and mossy fibers of GalOE animals (*D*). From Kokaia et al.[28]

To further assess for the potential antiepileptic role of endogenous galanin, a gain-of-function approach was used by generating mice over-expressing the galanin gene (GalOE mice) under the DBH promoter, which over-expresses galanin in noradrenergic cells where it is normally present. In these mice, 60 min of perforant path stimulation, which normally induced SE in WT mice, led to only brief seizure activity.[27] GalOE mice exhibited higher threshold for afterdischarge activation and less severe pentylenetetrazol seizures.

In line with these observations, GalR1 receptor gene KO mice have been shown to exhibit spontaneous tonic-clonic seizures with 25% penetrance of all animals.[29] However, this was observed only in KO mice of a mixed background strain C57BL/6J:129/Sv and not in KO mice of the 129/Sv background strain. In the sub-population of the mixed strain KO mice, which developed spontaneous seizures, a complex differential regulation of various neuropeptides in the hippocampus have been reported.[30] However, it remains unclear to what extent this distinct pattern of regulation contributes to the development of spontaneous seizure activity. Recently, sub-acute knockdown of the GalR2 receptor by continuous infusion of a peptide nucleic acid antisense oligonucleotide against GalR2 mRNA into the dentate gyrus of rats was shown to reduce GalR2 binding by 50 % and increase the severity of perforant path stimulation induced SE.[31] Interestingly, this treatment also reduced SE-induced proliferation of neuronal progenitors in the subgranular layer of the dentate gyrus, suggesting a novel role for galanin in regulation of plastic reorganization of the hippocampal formation after seizures.[31]

All these data strongly supported the role of galanin as an endogenous antiepileptic agent. However, in order to use galanin as a possible therapeutic approach, it might be necessary to increase galanin content ectopically in neuronal systems that do not normally express the protein. We asked whether ectopically over-expressed galanin could also exert an antiepileptic action similar to what has been reported for the over-expressing galanin in its native NAergic system.[27] To address this question we used transgenic mice over-expressing galanin under PDGF-B promoter. In these GalOE mice, we measured up to an 8-fold increase of galanin in the cerebral cortex and hippocampus with by radioimmunoassay (RIA).[28] The detailed immunohistochemical and in situ hybridization analyses of slices from GalOE mice revealed widespread over-expression of galanin in different brain areas. High levels of galanin immunoreactivity were observed in the dentate granule cell layer, hilus, and particularly in the mossy fibers of the hippocampal formation and the outer molecular layer of the piriform cortex. Galanin immunoreactivity was also seen in cortical cell bodies of layers 4-5, and in mitral cells of the olfactory bulb. Recurrent daily seizure activity induced by electrical kindling further increased galanin immunoreactivity in mossy fibers, dentate granule cells and cortical neurons (Fig. 1).

This massive over-expression of galanin, both in basal conditions and after seizures, had a dramatic effect on kindling by increasing the number of stimulations required to reach generalized seizure activity (Fig. 2A, C). This delay in kindling rate started to occur already at the early stages of epileptogenesis (grade 2-3), and was associated with shorter afterdischarge durations in GalOE mice (Fig. 2B), suggesting that local hippocampal events could be a contributing factor. The maintenance of kindled state was not affected by galanin over-expression, since both strains exhibited a similarly high mean seizure grades in response to the test stimulations 4-6 weeks after the last kindling session. However, the threshold to induce seizures during these test stimulations was significantly higher in GalOE as compared to WT mice (Fig. 2 D), again suggesting alterations in local hippocampal mechanisms.

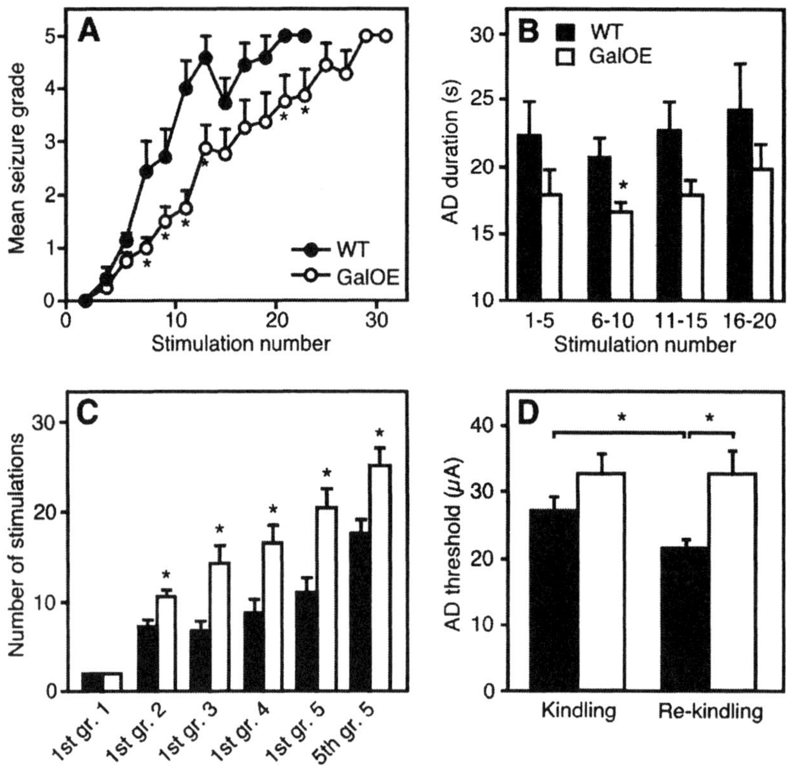

Figure 2. Kindling epileptogenesis is suppressed in GalOE mice. (*A*) GalOE mice show lower mean seizure grades in response to consecutive kindling stimulations as compared with WT animals. (*B*) Mean afterdischarge (AD) duration in response to five consecutive stimulations during kindling epileptogenesis is shorter in GalOE mice (*C*) GalOE mice require more kindling stimulations to reach different seizure grades. (*D*) AD threshold when kindled animals are subjected to new kindling stimulations after 4-6 weeks (rekindling) is decreased in WT but not in GalOE mice. WT, $n = 7$; GalOE, $n = 8$. *, Significant difference between WT and GalOE mice (unpaired t test, $P < 0.05$) Data are mean ± SEM. From Kokaia et al.[28]

We hypothesized that galanin over-expression induced alterations in synaptic transmission locally in the hippocampal circuitry, thus affecting kindling development in GalOE mice. Since mossy fibers have long been implicated in kindling,[32,33] and the most dramatic over-expression of galanin was observed inmossy fibers of GalOE mice, we explored whether glutamate release from these fibers could have been affected by galanin. To address this question we used a prominent property of mossy fibers, which is to increase glutamate release in response to high frequency stimulation, so-called frequency facilitation.[34] In this paradigm, in hippocampal slices, field excitatory postsynaptic responses in the stratum lacunosum of the CA3 substantially increase in amplitude when low frequency mossy fiber stimulation is switched to high frequency stimulation. We found that frequency facilitation was less pronounced in mossy fibers of GalOE mice as compared WT mice (Fig. 3). This was readily reversed by the galanin receptor antagonist M35, suggesting that during high frequency stimulation over-expressed galanin could be released from the mossy fibers and subsequently suppress further release of glutamate by feedback activation of galanin receptors on these fibers. These data provided the first evidence that ectopically over-expressed galanin in the excitatory glutamatergic fibers can be released in an activity-dependent manner and interact with the presynaptic galanin receptors, thus limiting further glutamate release. This property of galanin could be particularly useful in controlling epileptic seizures, because one could hypothesize that ectopically expressed galanin would be released in sufficient amounts only during high frequency seizure activity suppressing glutamate release (and seizures), while low frequency synaptic transmission during normal brain activity would not be disturbed.

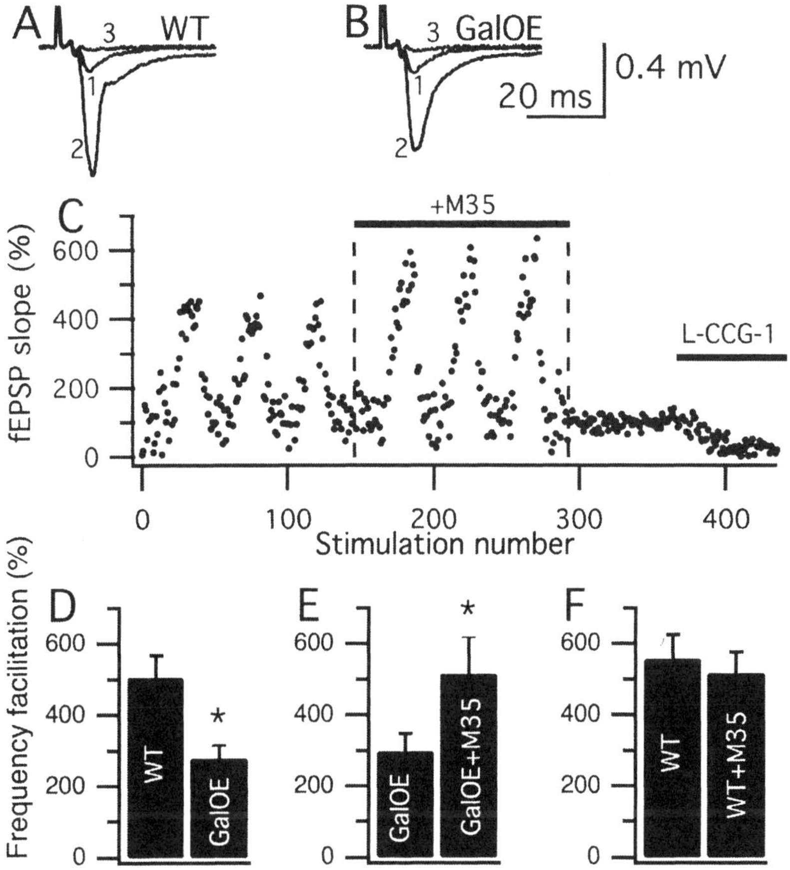

Figure 3. Frequency facilitation is decreased in mossy fiber synapses of GalOE mice. (*A*) Traces of fEPSPs (average of 10) recorded in stratum lucidum of CA3 during low-frequency (0.1 Hz; trace1) and high-frequency (1 Hz; trace 2) stimulation of mossy fibers, and after bath application of L-CCG-I (10 μM; trace 3) in WT animals. Note the increased fEPSPs at high frequency stimulation (trace 2). fEPSPs are blocked by L-CCG-I (trace 3) indicating that they are predominantly generated by mossy fibers. (*B*) Same recordings as in *A* in GalOE mice. Note smaller fEPSP during high-frequency stimulation (trace 2) in GalOE mice as compared with WT mice in *A*. Picrotoxin (100 μM) was present in the bath solution. (*C*) Frequency facilitation of mossy fiber-CA3 synapse fEPSPs (measured as fEPSP initial slope) expressed as percentage of baseline before and during bath application of the putative galanin receptor antagonist M35, and of L-CCG-I. Dashed lines indicate time lapse of 10 min between the experiments. During L-CCG-I application, stimulation frequency was 0.5 Hz. Note increase and blockade of fEPSPs during M35 and L-CCG-I application, respectively.(*D*) fEPSP frequency facilitation of mossy fiber-CA3 cell synapses in WT ($n = 30$ experiments, eight mice) and GalOE ($n = 22$ experiments, eight mice). (*E* and *F*) fEPSP frequency facilitation of mossy fiber-CA3 cell synapses in GalOE (*E*) and WT (*F*) mice before and during bath application of M35 ($n = 16$ experiments, five mice; and 18 experiments, five mice, respectively). Data are means ± SEM. Before fEPSP initial slope measurements were made, traces recorded after bath application of L-CCG-I (3) were always subtracted from those recorded during low (1) or high (2) frequency stimulation of mossy fibers. From Kokaia et al.[28]

It still remains to be explored how ectopic galanin is stored in the vesicles, how it is released from the cells, and which gene promoters would be optimal for its regulated expression. However, these findings suggest possible novel therapeutic strategies to dampen epileptic activity by ectopic overexpression of modulatory neuropeptides in excitatory neuronal systems.

3. VIRAL VECTOR TRANSFECTION OF THE GALANIN GENE

How can the galanin gene be expressed in certain brain areas, particularly in excitatory systems, to affect seizures in patients? The answer came in a recent publication by Habermann et al.[35] The authors used adeno-associated virus (AAV) to transfer a construct of the galanin gene and the constitutive secretory signal sequence for the laminar protein fibronectin to the rat brain. This approach enabled constitutive release of galanin from the transfected cells that significantly increased the threshold for the induction of focal seizures in the inferior colliculus. This seizure suppressant effect was reversible and dissipated when galanin gene expression was suppressed by administration of doxycycline in the drinking water, in a tetracycline-off gene regulation system.[36] Intrahippocampal infusion of the same AAV viral vector was unable to suppress kainate-induced seizures but protected hilar neurons from excitotoxic death, which is characteristic feature of this model. A similar approach using the AAV-galanin vector, without the constitutive secretory signal sequence, was used by another group[37] that demonstrated that intrahippocampal expression of the gene construct resulted in decreased total time spent in kainate-induced seizure activity. However, no protection from excitotoxic cell death in the CA3 area of the hippocampus was observed in this study. In summary, these two studies provide a proof-of-principle that AAV vector-based gene transfer could be a possible strategy for delivering galanin to distinct brain areas, such as the epileptic focus, to suppress intractable epileptic activity in patients.

4. CONCLUDING REMARKS

Epilepsy is a devastating neurological disease affecting about 50 million people in the world. Pharmacological treatment of epilepsy is the most common practice, directed towards increasing GABAergic inhibition, or interfering with excitatory processes such as glutamatergic transmission and sodium channel activation. However, 30 % of patients remain refractory to the treatment. Therefore, developing new strategies using gene transfer for galanin, which would modulate excitatory transmission in an activity-dependent manner exclusively during high frequency epileptic discharges, is well warranted. Many questions, however, still remain to be answered. For example, patients with epilepsy already have established focal or recurrent seizures, and it is not clear to what extent the viral vector gene transfer would suppress these seizures. Would such treatment also affect crucial brain functions, such as learning and memory?[24] While further studies are needed to address these issues, recent advances in the field already now suggest that therapeutic approaches based on ectopic viral vector gene transfer for neuropeptides, such as galanin, could be a promising tool for future epilepsy treatment.

5. AKNOWLEDGEMENTS

This work was supported by the Swedish Research Council. Special thanks to Dr. Paul Mohapel and Katherine Jakubs for critically reading the manuiscript.

6. REFERENCES

1 Kaplan L. M., Spindel E. R., Isselbacher K. J., and Chin W. W., Tissue-specific expression of the rat galanin gene, *Proc. Natl. Acad. Sci. U S A.* **85,** 1065-1069 (1988).
2. Jacobowitz D. M., Kresse A., and Skofitsch G., Galanin in the brain: chemoarchitectonics and brain cartography-a historical review, *Peptides* **25,** 433-464 (2004).
3. Senut M. C., Menetrey D., and Lamour Y., Cholinergic and peptidergic projections from the medial septum and the nucleus of the diagonal band of Broca to dorsal hippocampus, cingulate cortex and olfactory bulb: a combined wheatgerm agglutinin-apohorseradish peroxidase-gold immunohistochemical study, *Neuroscience* **30,** 385-403 (1989).
4 Bartfai T., Bertorelli R., Consolo S., Diaz-Arnesto L., Fisone G., Hokfelt T., Iverfeldt K., Palazzi E., and Ogren S. O., Acute and chronic studies on functional aspects of coexistence, *J. Physiol. (Paris).* **83,** 126-132 (1988).
5. Fisone G , Wu C. F., Consolo S., Nordstrom O., Brynne N., Bartfai T., Melander T., and Hokfelt T., Galanin inhibits acetylcholine release in the ventral hippocampus of the rat: histochemical, autoradiographic, in vivo, and in vitro studies, *Proc. Natl. Acad. Sci. U S A.* **84,** 7339-7343 (1987).
6. Skofitsch G., and Jacobowitz D. M., Immunohistochemical mapping of galanin-like neurons in the rat central nervous system, *Peptides* **6,** 509-546 (1985).
7. Skofitsch G., and Jacobowitz D. M., Quantitative distribution of galanin-like immunoreactivity in the rat central nervous system, *Peptides* **7,** 609-613 (1986).
8. Melander T., Staines W. A., Hokfelt T., Rokaeus A., Eckenstein F., Salvaterra P. M., and Wainer B. H., Galanin-like immunoreactivity in cholinergic neurons of the septum-basal forebrain complex projecting to the hippocampus of the rat, *Brain Res.* **360,** 130-138 (1985).
9. Melander T., Hokfelt T., and Rokaeus A., Distribution of galaninlike immunoreactivity in the rat central nervous system, *J. Comp. Neurol.* **248,** 475-517 (1986).
10. Melander T., Staines W. A., and Rokaeus A., Galanin-like immunoreactivity in hippocampal afferents in the rat, with special reference to cholinergic and noradrenergic inputs, *Neuroscience* **19,** 223-240 (1986).
11. Ubink R., Calza L., and Hokfelt T., 'Neuro'-peptides in glia: focus on NPY and galanin, *Trends Neurosci.* **26,** 604-609 (2003).
12. Burgevin M. C., Loquet I., Quarteronet D., and Habert-Ortoli E., Cloning, pharmacological characterization, and anatomical distribution of a rat cDNA encoding for a galanin receptor, *J. Mol. Neurosci.* **6,** 33-41 (1995).
13. Habert-Ortoli E., Amiranoff B., Loquet I., Laburthe M., and Mayaux J. F., Molecular cloning of a functional human galanin receptor, *Proc. Natl. Acad. Sci. U S A.* **91,** 9780-9783 (1994).
14. Howard A. D., Tan C., Shiao L. L., Palyha O. C., McKee K. K., Weinberg D. H., Feighner S. D., Cascieri M. A., Smith R. G., Van Der Ploeg L. H., and Sullivan K. A., Molecular cloning and characterization of a new receptor for galanin, *FEBS Lett.* **405,** 285-290 (1997).
15. Smith K. E., Forray C., Walker M. W., Jones K. A., Tamm J. A., Bard J., Branchek T. A., Linemeyer D. L., and Gerald C., Expression cloning of a rat hypothalamic galanin receptor coupled to phosphoinositide turnover, *J. Biol. Chem.* **272,** 24612-24616 (1997).
16. Wang S , Hashemi T., He C., Strader C., and Bayne M., Molecular cloning and pharmacological characterization of a new galanin receptor subtype, *Mol. Pharmacol.* **52,** 337-343 (1997).
17. O'Donnell D., Ahmad S., Wahlestedt C , and Walker P., Expression of the novel galanin receptor subtype GALR2 in the adult rat CNS: distinct distribution from GALR1, *J. Comp. Neurol.* **409,** 469-481 (1999).
18. Mennicken F., Hoffert C., Pelletier M., Ahmad S., and O'Donnell D., Restricted distribution of galanin receptor 3 (GalR3) mRNA in the adult rat central nervous system, *J. Chem. Neuroanat.* **24,** 257-268 (2002).
19. Wang S., Hashemi T., Fried S., Clemmons A. L., and Hawes B. E., Differential intracellular signaling of the GalR1 and GalR2 galanin receptor subtypes, *Biochemistry* **37,** 6711-6717 (1998).
20. Branchek T. A., Smith K. E., Gerald C., and Walker M. W., Galanin receptor subtypes, *Trends Pharmacol. Sci.* **21,** 109-117 (2000).
21. Zini S., Roisin M. P., Langel U., Bartfai T., and Ben-Ari Y., Galanin reduces release of endogenous excitatory amino acids in the rat hippocampus, *Eur. J. Pharmacol.* **245,** 1-7 (1993).
22. Xu Z. Q., Ma X., Soomets U., Langel U., and Hokfelt T., Electrophysiological evidence for a hyperpolarizing, galanin (1-15)-selective receptor on hippocampal CA3 pyramidal neurons, *Proc. Natl. Acad. Sci. U S A.* **96,** 14583-14587 (1999).
23. Dutar P., Lamour Y., and Nicoll R. A., Galanin blocks the slow cholinergic EPSP in CA1 pyramidal neurons from ventral hippocampus, *Eur. J. Pharmacol.* **164,** 355-360 (1989).
24. Sakurai E., Maeda T., Kaneko S., Akaike A., and Satoh M., Galanin inhibits long-term potentiation at Schaffer collateral-CA1 synapses in guinea-pig hippocampal slices, *Neurosci. Lett.* **212,** 21-24 (1996).

25. Mazarati A. M., Halaszi E., and Telegdy G., Anticonvulsive effects of galanin administered into the central nervous system upon the picrotoxin-kindled seizure syndrome in rats, *Brain Res.* **589**, 164-166 (1992).
26. Mazarati A. M., Liu H., Soomets U., Sankar R., Shin D., Katsumori H., Langel U., and Wasterlain C. G., Galanin modulation of seizures and seizure modulation of hippocampal galanin in animal models of status epilepticus, *J. Neurosci.* **18**, 10070-10077 (1998).
27. Mazarati A. M., Hohmann J. G., Bacon A., Liu H., Sankar R., Steiner R. A., Wynick D., and Wasterlain C. G., Modulation of hippocampal excitability and seizures by galanin, *J. Neurosci.* **20**, 6276-6281 (2000).
28. Kokaia M., Holmberg K., Nanobashvili A., Xu Z. Q., Kokaia Z., Lendahl U., Hilke S., Theodorsson E., Kahl U., Bartfai T., Lindvall O., and Hokfelt T., Suppressed kindling epileptogenesis in mice with ectopic overexpression of galanin, *Proc. Natl. Acad. Sci. U S A.* **98**, 14006-14011 (2001).
29. Jacoby A. S., Hort Y. J., Constantinescu G., Shine J., and Iismaa T. P., Critical role for GALR1 galanin receptor in galanin regulation of neuroendocrine function and seizure activity, *Mol. Brain Res.* **107**, 195-200 (2002).
30. Fetissov S. O., Jacoby A. S., Brumovsky P. R., Shine J., Iismaa T. P., and Hokfelt T., Altered hippocampal expression of neuropeptides in seizure-prone GALR1 knockout mice, *Epilepsia* **44**, 1022-1033 (2003).
31. Mazarati A., Lu X., Shinmei S., Badie-Mahdavi H., and Bartfai T., Patterns of seizures, hippocampal injury and neurogenesis in three models of status epilepticus in galanin receptor type 1 (GalR1) knockout mice, *Neuroscience* **128**, 431-441 (2004).
32. Lynch M., and Sutula T., Recurrent excitatory connectivity in the dentate gyrus of kindled and kainic acid-treated rats, *J. Neurophysiol.* **83**, 693-704 (2000).
33. Ben-Ari Y., and Represa A., Brief seizure episodes induce long-term potentiation and mossy fibre sprouting in the hippocampus, *Trends Neurosci.* **13**, 312-318 (1990).
34. Cremer H., Chazal G., Carleton A., Goridis C., Vincent J. D., and Lledo P. M., Long-term but not short-term plasticity at mossy fiber synapses is impaired in neural cell adhesion molecule-deficient mice, *Proc. Natl. Acad. Sci. U S A.* **95**, 13242-13247 (1998).
35. Haberman R. P., Samulski R. J., and McCown T. J., Attenuation of seizures and neuronal death by adeno-associated virus vector galanin expression and secretion, *Nature Med.* **9**, 1076-1080 (2003).
36. Baron U., and Bujard H., Tet repressor-based system for regulated gene expression in eukaryotic cells principles and advances, *Methods Enzymol.* **327**, 401-421 (2000).
37. Lin E. J., Richichi C., Young D., Baer K., Vezzani A., and During M. J., Recombinant AAV-mediated expression of galanin in rat hippocampus suppresses seizure development, *J. Neurosci.* **18**, 2087-2092 (2003).

Discussion

Shwartzkroin: In your over-expressors, what did the amygdala look like in terms of galanin expressions and have you tried amygdala kindling?
Kokaia: Yes, there was an overexpression of immunoreactivity in amygdala as well, but we didn't try kindling. I think that would be interesting to look at.
Corcoran: These are really interesting data, beautiful study. Do you know what the status of noradrenaline is in the over-expressor mice? Do they differ in the release in noradrenaline, or in the characteristics of the noradrenergic response to seizures?
Kokaia; No, we do not know. That's an interesting question of course, because in the overexpressing line where galanin was over-expressed under DBH promoter, there was an over-expression in the noradrenergic fibers, and perhaps the release of galanin would affect the noradrenaline release from these fibers by a feedback mechanism. In our case there was no change in galanin expression in the noradrenergic fibers; it was mostly ectopic expression that was increased. So from that point of view, we didn't expect that there would be some change in noradrenaline release. But it would be interesting to look at it.
Moshé: After there is a seizure, do you expect to see a rise in galanin? And how do you think that we can approach that from a clinical perspective? How do we increase the galanin ectopically in the brain so that one can protect from the seizures.
Kokaia: That allows me to show additional slides that I omitted due to lack of time. These are not our data, but recently there was a paper showing that actually the

AAV-galanin virus injected into the inferior colliculus could increase the threshold for induction of the wild-running seizures in rats. This effect was reversible because they had a doxicycline-regulated system, and when they gave doxicycline to the animals, the effect on seizures was reversed. Another example is also from a recent paper, where they injected AAV-galanin virus into the hippocampus, so that there was an over-expression of galanin in the mossy fiber system. On this slide you can also see some cells over-expressing galanin in the hilus, and kainate-induced seizures were less severe in these animals. So this could be one way of doing it.

KINDLING, NEUROTROPHINS AND AXON-GUIDANCE FACTORS

Ronald J. Racine, Margaret Fahnestock, and Bin Xu*

1. INTRODUCTION

Graham Goddard[1] initially introduced kindling as a model for memory. Although most frequently utilized as an epilepsy model, particularly for temporal lobe epilepsy, kindling is occasionally used as a model for drug-induced plasticity, slowly developing neuropathologies other than epilepsy, and a variety of structural reorganizations of brain circuitry.[2,3] Since our interest in the kindling phenomenon (and other models of epilepsy) is largely driven by our interest in neural plasticity, we have chosen to focus on kindling-induced neuronal reorganization. These kindling-induced effects may not have much relevance for epilepsy, but they are themselves interesting and potentially important phenomena and provide an excellent model for the study of mechanisms of activity-dependent neural growth.

2. KINDLING AS A MODEL FOR ACTIVATION-INDUCED NEURAL GROWTH

2.1. Seizures and Axonal Sprouting

One of the sequelae of seizure activity is axonal sprouting and synaptic reorganization. These neural growth effects have been investigated most thoroughly in the mossy fiber pathway that originates in dentate gyrus granule neurons. Many researchers have used Timm stains to visualize the axonal sprouting in this pathway following kindling or status epilepticus seizures. The normal target of the mossy fiber pathway is the apical dendritic layer of the area CA3 pyramidal neurons, with a light targeting of the basal dendrites. Following kindling or status epilepticus, there is an increase in the axonal collaterals and synaptic terminals in the basal dendritic layer of CA3 and in the apical dendritic layer of the dentate gyrus granule cells themselves. Although the process no doubt begins almost immediately, the changes are large enough

*Ron J Racine and Bin Xu, Department of Psychology, McMaster University, Hamilton, Ontario, L8S 4K1. Margaret Fahnestock, Department of Psychiatry and Behavioural Neuroscience, McMaster University, Hamilton, Ontario L8N 3Z5

to measure starting about 4-5 days after seizure activity terminates.[4] They probably reach asymptotic levels about 4 weeks after the inducing treatment, unless spontaneous seizures develop (which likely trigger more sprouting and prolong the growth period).

We have chosen a multi-pronged strategy for investigating the mechanisms that drive neural growth in the adult brain. The usual starting point is the observation of the growth phenomena themselves. Like others, we have focused much of our efforts on the study of mossy fiber sprouting, but sprouting effects have also been observed in other hippocampal pathways and in other structures. Li et al.[5] and Wodbye et al.,[6] for example, have shown kindling-induced synaptic reorganization in layer II/III of the piriform cortex. A reasonable next step is to look for the sufficient and necessary triggers for these effects. Our work with anatomical effects induced by nonepileptiform stimulation[7] falls into this category. We do not yet know the minimal conditions required for induction of axonal sprouting, but we know that it can be induced by stimulation that triggers LTP.

Since our interest is primarily in activation-induced neural growth in the adult brain, neural damage must be considered as a trigger for axonal sprouting and synaptogenesis.

2.2. Neural Damage and Axonal Sprouting

It is well established that excess activation can lead to neural damage. Prolonged bouts of status epilepticus can lead to damage in a number of sites, including the hippocampus.[8] Large numbers of repeated evoked seizures, if sufficiently strong, can also produce cell death.[9]

Furthermore, neural damage can trigger axonal sprouting and synapse formation, particularly into sites left vacant by degenerating axons.[10] Thus, it is quite possible that all of the axonal sprouting seen in seizure models is triggered by neural damage. This has been the implicit assumption of many researchers in this area.[11] From research done in our own labs, we can rule out gross damage (reflected by cell loss) as a critical factor for triggering axonal sprouting. The Long-Evans hooded rat is highly resistant to seizure-induced damage. We have never seen cell loss with standard kindling protocols or with extensive kindling beyond the standard protocol in this strain. Short bouts of status epilepticus, which can lead to hilar cell loss in the Wistar strain, also produce no detectable cell loss in the Long-Evans rats.[12] However, prolonged status epilepticus can lead to heavy damage in any of the tested rat strains.

Most experiments have utilized measures of gross damage (typically cell loss reflected in Nissl stains). It is possible that the axonal sprouting in the Long-Evans strain is triggered by more subtle forms of damage. Some axon terminals, for example, could degenerate leaving the neurons intact. This could be a sufficient trigger for sprouting of nearby axon collaterals. Arguing against this as a sole explanation for sprouting, in the absence of gross damage, is the demonstration of sprouting effects accompanying nonepileptogenic stimulation and the induction of long-term potentiation (LTP).[7, 13] Of course, it could still be argued that LTP is itself a pathological phenomenon that could trigger subtle damage. Perhaps the most convincing evidence that synaptogenesis can be triggered in the absence of damage is that it can also be induced by experience (e.g., behavioral training).[14] Consequently, we believe that much of the sprouting triggered by kindling is activation-induced, rather than damage induced.

2.3. Exploring Mechanisms

When investigating mechanisms, things get complicated rather quickly beyond the stimulation stage. The electrical effects of the stimulation must be translated into molecular signals within the activated cells. There are a variety of immediate and early effects of the stimulation, any or all of which could contribute to the initiation of neural growth. Because of their known effects on axonal growth, we have chosen to focus on the role of neurotrophic factors, which include a number of possible contenders. Those that have been shown to play a role in activation-induced sprouting include the neurotrophins (NGF, BDNF, and NT-3) and TGF-β family members such as GDNF.

3. NEUROTROPHIC FACTORS

The neurotrophins are secreted proteins with their own sets of receptors, and they can affect a variety of growth-related processes, including neuronal differentiation, neurite outgrowth, and neuronal survival. They also have effects on neurotransmitter synthesis, synaptic plasticity, and excitability. The primary receptors for the neurotrophins are the low-affinity receptor, $p75^{NTR}$, to which they all bind, and a subset of high-affinity transmembrane receptor tyrosine kinases called Trks. Although there is some crosstalk in binding specificities, each neurotrophin has a primary Trk receptor (TrkA for NGF, TrkB for BDNF, and TrkC for NT-3)[15]. GDNF binds to a different set of receptors, including two high-affinity receptors (Ret and $p140^{NCAM}$) and two low-affinity co-receptors GFRα1 and GFRα2.[16]

3.1. NGF

If any of the neurotrophic factors are involved in activation-induced neural growth, then one or more of them should be expressed following growth-inducing seizure activity. NGF mRNA[17] and protein[18] are upregulated in several brain regions, including the hippocampus, following kindling. To determine a causal relationship, the next step is to attempt to manipulate NGF function and monitor the effects on a measure of activation-induced neural growth.

The administration of antibodies to NGF has been shown to retard amygdala kindling.[19] It also blocks mossy fiber sprouting.[20] Infusing NGF directly into the ventricles facilitates both amygdala[21] and hippocampal[22] kindling and increases mossy fiber sprouting[21]. There is a clear interaction between neural activation and neurotrophin activity. NGF infusion in the absence of seizures actually decreases Timm staining density.[21]

There are many downstream pathways by which neurotrophins can affect cell function and neural growth. To verify that the neurotrophin is actually acting via its (known) receptors, Rashid et al.[23] infused a peptide that inhibits NGF function by destabilizing the NGF dimer, thus interfering with receptor binding. It had the same effect as the NGF antibody; both kindling and sprouting were suppressed, suggesting that the effects are receptor mediated. More recently, we have tested more specific peptide inhibitors of TrkA and $p75^{NTR}$ in an attempt to determine the relative roles of these receptors in the effects of NGF on kindling and mossy fiber sprouting. We found that

TrkA activation but not p75 ^{NTR} activation is involved in kindling, while both TrkA and p75 ^{NTR} activation are involved in mossy fiber sprouting.[24]

Moving from the receptor to one of the downstream effector pathways of TrkA, we found that an inhibitor of Ras inhibited kindling, kindling-induced mossy fiber sprouting, and kindling-induced hilar expansion[25], further confirming the role of Trk activation in these effects.

It is somewhat surprising that NGF is involved in mossy fiber sprouting, since its receptors are found primarily on cholinergic neurons and not on dentate granule neurons.[26] However, it is known that cholinergic agonists and antagonists exert a modulatory effect on kindling similar to those of the NGF manipulations reported above.[27, 28] Furthermore, Adams et al.[28] found that cholinergic agonists and antagonists facilitate and depress, respectively, kindling-induced mossy fiber sprouting. NGF infused into the ventricles increases both choline acetyltransferase (ChAT) activity and choline transport.[29] Infusion of anti-NGF antibodies, on the other hand, reduces ChAT immunostaining in basal forebrain.[20] So, the NFG effects on the mossy fiber system may very well be mediated via cholinergic pathways.

3.2. BDNF

BDNF supports growth and survival of a variety of neurons.[30, 31] Kindling was found to increase BDNF[17] and TrkB[32] mRNA expression, as well as TrkB protein[32] in the dentate gyrus. However, TrkB mRNA and protein[32] were only increased following rapid kindling; no increase in TrkB was found following standard chronic kindling protocols.[33]

BDNF enhances synaptic transmission and neuronal excitability. For example, BDNF enhances synaptic responses at Schaffer collateral-CA1 synapses in hippocampal slices,[34] and increases the frequency of spontaneously initiated action potentials in hippocampal neurons in dissociated culture.[35] Elevating BDNF levels by overexpression of BDNF[36] or by infusion of BDNF into the hippocampus over a 2-week period[37] can induce or enhance seizure activity.

Similarly, manipulations that reduce BDNF signaling can suppress epileptogenesis. Amygdala kindling was retarded in mice heterozygous for a deletion of the BDNF gene.[38] Transgenic mice overexpressing truncated trkB, a dominant negative receptor of BDNF, showed reduced epileptogenesis in response to kainite.[39]

On the other hand, kindling is also suppressed if BDNF is infused directly into the hippocampus,[40, 41] and Reibel et al.[42] found a facilitation of kindling when antisense oligodeoxynucleotides were injected into the hippocampus to block BDNF effects.

One of the reasons for these apparently contradictory results is that TrkB protein can be downregulated by chronic high levels of BDNF. This downregulation of the TrkBreceptor would be expected to reduce BDNF signaling despite the excess of BDNF, thus suppressing (rather than facilitating) kindling. We tested this by comparing the effects of continuous (via osmotic pump) and intermittent injections of BDNF. As expected, the chronic infusions retarded kindling. They also reduced TrkB and phosphorylated Trk immunoreactivity. Intermittent injections, in comparison, facilitated kindling and had no effect on receptor expression.[12]

BDNF has little or no effect on activation-induced sprouting in the mossy fiber pathway. Transgenic mice that over express BDNF do not show altered Timm staining density in the hippocampus.[43] Neither chronic infusions nor bolus injections of BDNF affect mossy fiber sprouting in our hands.[12] However, chronic BDNF infusions can induce seizure activity and increase mossy fiber sprouting.[37] The increased activation could lead to increased sprouting, which can make interpretation of sprouting effects difficult. Transgenic mice that overexpress truncated trkB to *reduce* BDNF signaling show reduced status-epilepticus-induced mossy fiber sprouting, and these results are also consistent with the weaker epileptic activity in these animals[39] (Lahteinen et al., 2002).

3.3. NT-3

NT-3 is widely distributed in the CNS.[44] It has received less attention, partly because NT-3 mRNA is actually decreased following kindling.[45] Mice that are heterozygous for a deletion of the NT-3 gene show a 30% reduction of basal NT-3 mRNA levels and also show a slowing in kindling rate. However, these mice also showed an enhancement of kindling-induced BDNF and TrkB mRNA expression.[46] Chronic kindling appears to have no effect on on TrkC protein in the hippocampus.

Because the decrease in NT-3 mRNA expression in kindled animals suggests different mechanisms for NT-3's influences on kindling, we tested the effects of direct infusion of NT-3 into the ventricles. Both perforant path kindling and mossy fiber sprouting were inhibited.[33] However, chronic NT-3 infusion led to a decrease in kindling-induced Trk phosphorylation and downregulation of TrkA and TrkC protein, so TrkC as well as TrkA receptors may play an excitatory role in kindling and mossy fiber sprouting. It is also interesting that NT-3 by itself (no seizure activation) increased Timm staining density in the hippocampus, while NGF by itself led to a decrease in Timm staining density.

3.4. GDNF

GDNF is another neurotrophic factor with effects on a variety of cells.[47] Kindling increases the expression of GDNF mRNA, and the GDNF receptors RET and GFRα1, in the hippocampus, amygdala and parietal cortex.[48] GFRα2, in contrast, is increased in the piriform cortex but decreased in the hippocampus and amygdala by kindling stimulations. Mice lacking GFRα2 show a suppression of hippocampal kindling.[49] However, Martin et al.[50] found that GDNF administration into the lateral ventricle suppresses kainate-induced seizures and associated cell loss. Furthermore, direct infusion of GDNF into the ventricles retards perforant path kindling, and blocks mossy fiber sprouting and hilar expansion without affecting AD duration.[51] Whether GDNF infusion downregulates its receptors has not yet been tested.

Although GDNF uses a different receptor and downstream effector system than the neurotrophins, its effects on our measures are somewhat similar to those of the neurotrophins BDNF (which suppresses kindling) and NT-3 (which suppresses both kindling and mossy fiber sprouting).

3.5. FGF

Neither acidic nor basic fibroblast growth factor mRNA was affected by amygdala kindling according to Sato, et al.,[52] but Bregola et al.[53] found an interesting expression pattern relative to BDNF. Following the initial kindling stimulation to the hippocampus, FGF-2 mRNA levels were increased in the hippocampus, cortex and hypothalamus. BDNF mRNA was not yet affected. Fully kindled seizures, on the other hand, produced a strong increase in BDNF, but *not* FGF-2, mRNA levels in the hippocampus and contralateral cortex. Both were increased in the ipsilateral cortex. The authors suggest that FGF-2 may be more critically involved at the early stages of kindling. The functional effects of increasing or decreasing FGF-2 levels has not been tested.

4. AXON GUIDANCE FACTORS

Neurotrophins may play a role in promoting axonal sprouting and growth, but it is the *pattern* of that growth and the termination targets that determine whether the sprouting will contribute to, or compensate for, functional pathology. There are a variety of axon guidance factors that have been shown to determine synaptic termination patterns in developing brain. Many of these appear to continue this role, although to a lesser extent, in the adult brain.

4.1 Synaptic Termination Gradients

Mossy fibers are not distributed uniformly across their termination fields. Along both longitudinal and transverse axes within the hippocampus, there are gradients of Timm staining. The differences are seen along the blades of the dentate gyrus, along the CA3-CA1 component of the hippocampal gyrus, and along the dorsal/ventral axis of the hippocampus.[21, 33, 37, 54] Kindling and status epilepticus induce sprouting in all of these areas while maintaining or emphasizing the normal graded patterns. At least one set of axon guidance factors show distribution gradients along these same axes.

4.2 EphA/ephrinA

The Eph family of guidance factors and their ligands, the ephrins, continue to be expressed in the adult brain, and they play a role in the brain's response to injury.[55] Both receptors and ligands are membrane bound, and both appear to have signaling functions in their respective cells. There are two subclasses of Eph/ephrins, EphA/ephrinA and EphB/ephrinB. We have focused on the EphA subclass, which in mammals has 8 known receptors and 5 known ligands (EphA/ephrinA1-5). There is only partial specificity between ligand and receptor; many of the ligands bind to several of the receptors.

Of particular importance to the determination of synaptic termination patterns are the graded distributions of the receptors and ligands. Since the Eph/ephrin interaction is often one of repulsion, the gradients are often in opposing directions, insuring that axons with high levels of receptors will migrate toward fields of neurons containing low levels of ligand. However, the interactions can also be based on attraction.[56]

If an immunoadhesin antagonist of EphA receptors or of ephrinA ligands is infused into the ventricles, kindling is retarded or accelerated, respectively, and the pattern of mossy fiber sprouting is altered.[57] The ephrinA5 immunoadhesin, containing the extracellular domain of ephrinA5, binds EphA5 receptors, activating them, but blocks binding of other ephrinAs, inhibiting their signaling. This prevents the normal repulsion of the sprouting mossy fibers, allowing them to sprout into regions where they normally do not (ventral hippocampus and area CA3a) and decreasing the usual sprouting into dorsal hippocampus and area CA3c. This tends to flatten the gradients of mossy fiber sprouting.[57] Activation of ephrinAs by infusion of a soluble EphA5 immunoadhesin increases the repellent signal, resulting in reduced sprouting. Figures 1 shows the graded longitudinal distributions of Timm-stained pathways and EphA5/ephrinA3 expression in the hippocampus, and Figure 2 shows the transverse gradients.

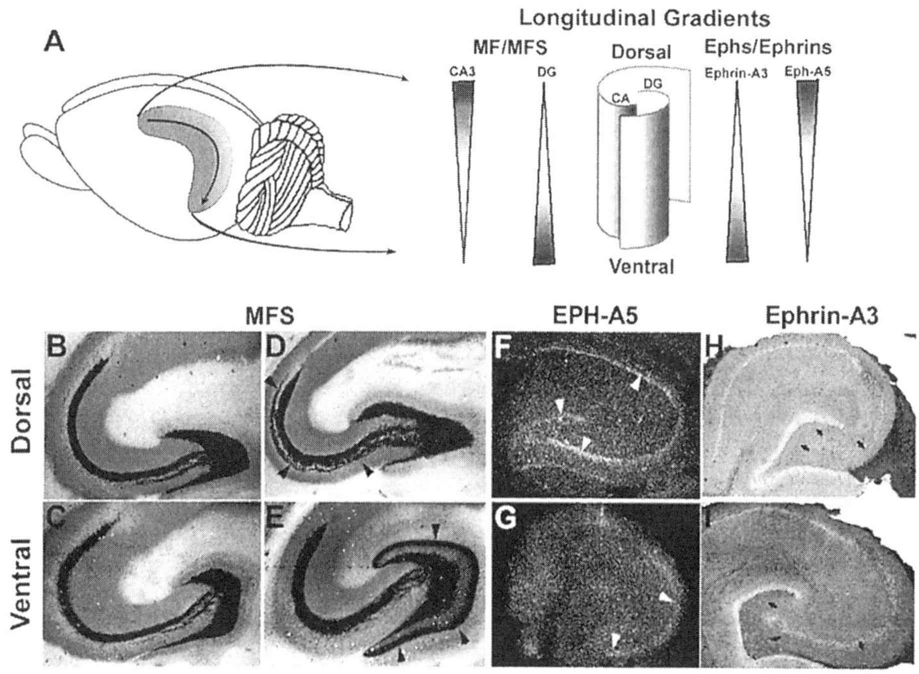

Figure 1. The graded longitudinal distributions for Timm granule density in the MF pathway and EphA5/ephrinA3 in hpc. **A:** The longitudinal gradients. **B-E:** Timm staining in dorsal and ventral sections for control animal and following status epilepticus. **F-G:** EphA5 in dorsal and ventral sections. **H-I:** Ephrin A in dorsal and ventral sections.

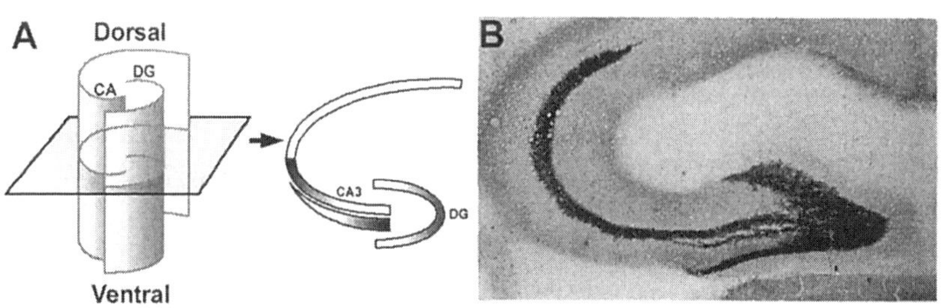

Figure 2. The graded transverse distribution of Timm granule density in the MF pathway.

4.3. Semaphorins

Neuropilin-2 is a semaphorin receptor that also has a repellant interaction with its ligand. Kindling and kainate-induced status have both been found to increase neuropilin-2 mRNA levels and neuropilin-2 immunoreactivity in the whole brain at 24 hours but not at 2 weeks after the last stimulation or injection.[58] Barnes et al.[59] monitored several semaphorins with riboprobes and *in situ* hybridization and found that kainate-induced status reduced mRNA expression for both semaphorin 3C and 3F in the hippocampus. The reductions lasted for at least 1 month. Less enduring effects were found for semaphorin 3A, 3C and 4C. The influence of the semaphorins on kindling-induced mossy fiber sprouting has not been investigated.

5. IMPLICATIONS FOR EPILEPSY

5.1. Excitation/Inhibition Balance

Whatever the underlying mechanisms, epileptic seizures presumably develop as a result of an imbalance between excitation and inhibition. Although changes in both probably contribute, deficiencies in GABA systems may be the more critical.[60] However, it is not clear how activation- (or damage-) induced synaptogenesis contributes to the balance. Most attention has been paid to the sprouting seen in the mossy fiber system, particularly sprouting back into the dentate gyrus. There is some evidence that this sprouting can lead to increased excitation in granule cells,[61] but not much evidence that it actually contributes to the epileptic state. Mossy fiber sprouting does not appear to be required for the development or progression of seizures,[62] or for the development of spontaneous seizures following status epilepticus.[63] However, the sprouting effect occurs within other hippocampal pathways as well as other brain structures,[5] so the question of how the overall balance between excitation and inhibition is affected is still unanswered.

5.2. The Development of Spontaneous Seizures

A major advantage of the status epilepticus models for the study of epilepsy is that they can produce an epileptic state (in which spontaneous seizures are generated) with much less effort that is required with the kindling model. The advantage for the study of sprouting is that it produces massive sprouting with little effort on the part of the researcher. The disadvantage is that it can also produce damage. The causal relationships can get lost in the noise, given the number of effects produced by excess activation. Longer bouts of status produce more neural damage,[64] more sprouting, and more spontaneous seizures,[65] but both the damage and the sprouting may follow from the excess activation, rather than contributing to the development of spontaneous seizures. Similarly, there is a good correlation between damage and synaptogenesis,[66] but, as we have seen, synaptogenesis can be triggered in the absence of damage.

6. IMPLICATIONS FOR RECOVERY OF FUNCTION FOLLOWING NEURAL DAMAGE

It is far too early to speculate about the relevance of the activation-induced axonal sprouting research to recovery of function following brain damage. However, the research thus far indicates that it might be useful to determine how much control can be exerted over the extent and pattern of axonal sprouting (and neurogenesis) via controlled interactions between activation, neurotrophic factors, and axon guidance factors. Some of the research currently underway in our laboratories is focused directly on this problem.

7. REFERENCES

1. G. V. Goddard, Development of epileptic seizures through brain stimulation at low intensity, *Nature* **214**, 1020-1021 (1967).
2. K. Morimoto, M. Fahnestock, and R.J. Racine, Kindling and status epilepticus models of epilepsy: rewiring the brain, *Pro. Neurobiol.*, in press. (2004).
3. R. E. Adamec, Partial kindling and behavioral pathologies, *Int. Rev. Neurobiol.* **45**, 409-434 (2001).
4. T. Sutula, H.Xiao-Xian, J. Cavazos, and G. Scott, Synaptic reorganization in the hippocampus induced by abnormal functional activity, *Science* **239**, 1147- 1150 (1988).
5. S. Li, I. Reinprecht, M. Fahnestock, and R. J. Racine, Activity-dependent changes in synaptophysin immunoreactivity in hippocampus, piriform cortex, and entorhinal cortex of the rat, *Neuroscience* **115**, 1221-1229 (2002a).
6. D. P. Woldbye, T. G. Bolwig, J. Kragh, and O. S. Jorgensen, Synaptic degeneration and remodelling after fast kindling of the olfactory bulb, *Neurochem. Res.* **21**, 585-593 (1996).
7. B. Adams, M. Lee, M. Fahnestock, and R. J. Racine, Long-term potentiation trains induce mossy fiber sprouting, *Brain Res.* **775**, 193-197 (1997a).
8. J.E. Schwob T. Fuller, J.L. Price, J.W. Olney, Widespread patterns of neuronal damage following systemic or intra-cerebral injections of kainic acid – a histological study, *Neuroscience* **5**, 991-1014 (1980).
9. J.E. Cavazos, I. Das, T.P. Sutula, Neuronal loss induced in limbic pathways by kindling: evidence for induction of hippocampal sclerosis by repeated brief seizures. *J. Neurosci.* **14**, 3106-3121 (1994)
10. J.J. Ramirez, S.P. Findlestein, J. Keller, W. Abrams, M. N. George, and T. Parackh, Basic fibroblast growth factor enhances axonal sprouting after cortical injury in rats, *NeuroReport* **10**, 1201-1204 (1999).
11. M. Mikkonen, H. Soininen, R. Kalvianen, T. Tapiola, A. Ylinen, M. Vapalahti, L. Paljarvi, and A. Pitkanen, Remodeling of neuronal circuitries in human temporal lobe epilepsy: increased expression of highly polysialylated neural cell adhesion molecule in the hippocampus and the entorhinal cortex, *Ann. Neurol.* **44**, 923-934 (1998).
12. B. Xu, B. Michalski, R. J. Racine, and M. Fahnestock, The effects of brain-derived neurotrophic factor (BDNF) administration on kindling induction, Trk expression and seizure-related morphological changes, *Neuroscience,* in press (2004).
13. M.L. Escobar, E.J. Barea-Rodriguesz, B.E. Derrick, J.A. Reyes, and J.L. Martinez Jr., Opioid receptor modulation of mossy fiber synaptogenesis: independence from long-term potentiation, *Brain Res.* **751**, 330-335 (1997).
14. B. Kolb, and I.Q. Whishaw, Brain plasticity and behavior, *Annu. Rev. Psychol.* **49**, 43-64 (1998).
15. E. J. Huang, and , L. F. Reichardt (2001), Neurotrophins: roles in neuronal development and function, *Annu. Rev. Neurosci.* **24,** 677-736 (2001).
16. Airaksinen, M. S., and Saarma, M., The GDNF family: signalling, biological functions and therapeutic value, *Nature Rev. Neurosci.* **3,** 383-394 (2002).
17. P. Ernfors, J. Bengzon, Z. Kokaia, H. Persson, and O. Lindvall, Increased levels of messenger RNAs for neurotrophic factors in the brain during kindling epileptogenesis, *Neuron* **7**, 165-176 (1991).
18. J. Bengzon, S. Soderstrom, Z. Kokaia, M. Kokaia, P. Ernfors, H. Persson, T. Ebendal, and O. Lindvall, Widespread increase of nerve growth factor protein in the rat forebrain after kindling-induced seizures,. *Brain Res.* **587**, 338-342 (1992).
19. T. Funabashi, H. Sasaki, and F. Kimura, Intraventricular injection of antiserum to nerve growth factor delays the development of amygdaloid kindling, *Brain Res.* **458**, 132-136 (1988).
20 C. E. Van der Zee, K. Rashid, K. Le, K. A. Moore, J. Stanisz, J. Diamond, R. J. Racine, and M. Fahnestock, Intraventricular administration of antibodies to nerve growth factor retards kindling and blocks mossy fiber sprouting in adult rats, *J. Neurosci.* **15**, 5316-5323 (1995).
21. B. Adams, M. Sazgar, P. Osehobo, C. E. Van der Zee, J. Diamond, M. Fahnestock, and R. J. Racine, Nerve growth factor accelerates seizure development, enhances mossy fiber sprouting, and attenuates seizure-induced decreases in neuronal density in the kindling model of epilepsy, *J. Neurosci.* **17**, 5288-5296 (1997b).
22. S. Reibel, Y. Larmet, B. T. Le, J. Carnahan, C. Marescaux, and A. Depaulis, Brain-derived neurotrophic factor delays hippocampal kindling in the rat, *Neuroscience* **100**, 777-788 (2000a).

23. K. Rashid, C. E. Van der Zee, G. M. Ross, C. A. Chapman, J. Stanisz, R. J. Riopelle, R. J. Racine, and M. Fahnestock, A nerve growth factor peptide retards seizure development and inhibits neuronal sprouting in a rat model of epilepsy, *Proc Natl Acad Sci U S A* **92**, 9495-9499 (1995).
24. S. Li, H. U. Saragovi, H. Nedev, R. J. Racine, and M. Fahnestock, Differential actions of nerve growth factor receptors TrkA and $p75^{NTR}$ in a rat model of epilepsy, Submitted.
25. S. Li, H. U. Saragovi, R. J. Racine, and M. Fahnestock, A ligand of the p65/p95 receptor suppresses perforant path kindling, kindling-induced mossy fiber sprouting, and hilar area changes in adult rats, *Neuroscience* **119**, 1147-1156 (2003).
26. P.A. Lapchak, D.M. Arujo, S. Carswell, and F. Hefti, Distribution of [125I]nerve growth factor in the rat brain following a single intraventricular injection: correlation with the topographical distribution of trkA messenger RNA-expressing cells, *Neuroscience* **54**, 445-460 (1993).
27. V. Westerberg, and M. E. Corcoran, Antagonism of central but not peripheral cholinergic receptors retards amygdala kindling in rats, *Exp. Neurol.* **95**, 194-206 (1987).
28. B. Adams, L. Vaccarella, M. Fahnestock, and R. J. Racine, The cholinergic system modulates kindling and kindling-induced mossy fiber sprouting, *Synapse* **44**, 132-138 (2002).
29. F. Hefti, A. Dravid, and J. Hartikka, Chronic intraventricular injections of nerve growth factor elevate hippocampal choline acetyltransferase activity in adult rats with partial septo-hippocampal lesions, *Brain Res.* **293**, 305-309 (1984).
30. B. Knusel, J.W. Winslow, A. Rosenthal, L.E. Burton, D.P.,Seid, K. Nikolics, and F. Hefti, Promotion of central cholineregic and dopaminergic neuron differentiation by brain-derived neurotrophic factor but not neurotrophin-3, *Proc. Natl. Acad. Sci. U. S. A.* **88,** 961-965 (1996).
31. F. Barnabe-Heider, and F. D. Miller, Endogenously produced neurotrophins regulate survival and differentiation of cortical progenitors via distinct signaling pathways, *J. Neurosci.* **23,** 5149-5160 (2003).
32. J. P. Merlio, P. Ernfors, Z. Kokaia, D. S. Middlemas, J. Bengzon, M. Kokaia, M. L. Smith, B. K. Siesjo, T. Hunter, O. Lindvall, and et al. Increased production of the TrkB protein tyrosine kinase receptor after brain insults, *Neuron* **10**, 151-164 (1993).
33. B. Xu, B. Michalski, R. J. Racine, and M. Fahnestock, Continuous infusion of neurotrophin-3 triggers sprouting, decreases the levels of TrkA and TrkC, and inhibits epileptogenesis and activity-dependent axonal growth in adult rats, *Neuroscience* **115**, 1295-1308 (2002).
34. H. Kang, and E. M. Schuman, Long-lasting neurotrophin-induced enhancement of synaptic transmission in the adult hippocampus, *Science* **267**, 1658-1662 (1995).
35. M. McLean Bolton, A. J. Pittman, and D. C. Lo, Brain-derived neurotrophic factor differentially regulates excitatory and inhibitory synaptic transmission in hippocampal cultures, *J. Neurosci.* **20**, 3221-3232 (2000).
36. S. D. Croll, C. Suri, D. L. Compton, M. V. Simmons, G. D. Yancopoulos, R. M. Lindsay, S. J. Wiegand, J. S. Rudge, and H. E. Scharfman, Brain-derived neurotrophic factor transgenic mice exhibit passive avoidance deficits, increased seizure severity and in vitro hyperexcitability in the hippocampus and entorhinal cortex, *Neuroscience* **93,** 1491-1506 (1999).
37. H. E. Scharfman, J. H. Goodman, A. L. Sollas, and S. D. Croll, Spontaneous limbic seizures after intrahippocampal infusion of brain- derived neurotrophic factor, *Exp. Neurol.* **174**, 201-214 (2002).
38. M. Kokaia, P. Ernfors, Z. Kokaia, E. Elmer, R. Jaenisch, and O. Lindvall, Suppressed epileptogenesis in BDNF mutant mice, *Exp. Neurol.* **133**, 215-224 (1995).
39. S. Lahteinen, A. Pitkanen, T. Saarelainen, J. Nissinen, E. Koponen, and E. Castren, Decreased BDNF signalling in transgenic mice reduces epileptogenesis, *Eur. J. Neurosci.* **15**, 721-734 (2002).
40. P. Osehobo, B. Adams, M. Sazgar, Y. Xu, R. J. Racine, and M. Fahnestock, Brain-derived neurotrophic factor infusion delays amygdala and perforant path kindling without affecting paired-pulse measures of neuronal inhibition in adult rats, *Neuroscience* **92**, 1367-1375 (1999).
41. Y. Larmet, S. Reibel, J. Carnahan, H. Nawa, C. Marescaux, and A. Depaulis, Protective effects of brain-derived neurotrophic factor on the development of hippocampal kindling in the rat, *Neuroreport* **6**, 1937-1941 (1995).
42. S. Reibel, Y. Larmet, J. Carnahan, C. Marescaux, and A. Depaulis, Endogenous control of hippocampal epileptogenesis: a molecular cascade involving brain-derived neurotrophic factor and neuropeptide Y, *Epilepsia* **41** Suppl 6, S127-133 (2000b).
43. X. Qiao, C. Suri, B. Knusel, and J. L. Noebels, Absence of hippocampal mossy fiber sprouting in transgenic mice overexpressing brain-derived neurotrophic factor, *J. Neurosci. Res.* **64,** 268-276 (2001).
44. R. Katoh-Semba, Y. Kaisho, A. Shintani, M. Nagahama, and K. Kato, Tissue distribution and immunocytochemical localization of neurotrophin-3 in the brain and peripheral tissues of rats, *J. Neurochem.* **66,** 330-337 (1996).

45. J. Bengzon, Z. Kokaia, P. Ernfors, M. Kokaia, G. Leanza, O. G. Nilsson, H. Persson, and O. Lindvall, Regulation of neurotrophin and trkA, trkB and trkC tyrosine kinase receptor messenger RNA expression in kindling, *Neuroscience* **53**, 433-446. (1993).
46. E. Elmer, M. Kokaia, P. Ernfors, I. Ferencz, Z. Kokaia, and O. Lindvall, Suppressed kindling epileptogenesis and perturbed BDNF and TrkB gene regulation in NT-3 mutant mice, *Exp. Neurol.* **145**, 93-103 (1997a).
47. P. A. Lapchak, P. J. Miller, and S. Jiao, Glial cell line-derived neurotrophic factor induces the dopaminergic and cholinergic phenotype and increases locomotor activity in aged Fischer 344 rats, *Neuroscience* **77**, 745–752 (1997).
48. Z. Kokaia, M. S. Airaksinen, A. Nanobashvili, E. Larsson, E. Kujamaki, O. Lindvall, and M. Saarma, GDNF family ligands and receptors are differentially regulated after brain insults in the rat, *Eur. J. Neurosci.* **11**, 1202-1216 (1999).
49. A. Nanobashvili, M. S. Airaksinen, M. Kokaia, J. Rossi, F. Asztely, K. Olofsdotter, P. Mohapel, M. Saarma, O. Lindvall, and Z. Kokaia, Development and persistence of kindling epilepsy are impaired in mice lacking glial cell line-derived neurotrophic factor family receptor alpha 2, *Proc Natl Acad Sci U S A* **97**, 12312-12317 (2000).
50. D. Martin, G. Miller, M. Rosendahl, and D. A. Russell, Potent inhibitory effects of glial derived neurotrophic factor against kainic acid mediated seizures in the rat, *Brain Res.* **683**, 172-178 (1995).
51. S. Li, B. Xu, D. Martin, R. J. Racine, and M. Fahnestock, Glial cell line-derived neurotrophic factor modulates kindling and activation-induced sprouting in hippocampus of adult rats, *Exp. Neurol.* **178**, 49-58 (2002b).
52. K. Sato, K. Kashihara, K. Morimoto, and T. Hayabara, Regional increases in brain-derived neurotrophic factor and nerve growth factor mRNAs during amygdaloid kindling, but not in acidic and basic growth factor mRNAs, *Epilepsia* **37**, 6-14 (1996).
53. G. Bregola, L. Frigati, S. Zucchini, and M. Simonato, Different patterns of induction of fibroblast growth factor-2 and brain-derived neurotrophic factor messenger RNAs during kindling epileptogenesis, and development of a herpes simplex vector for fibroblast growth factor-2 gene transfer in vivo, *Epilepsia* **41** Suppl 6, S122-126 (2000).
54. B. Xu, D. C. McIntyre, M. Fahnestock, and R. J. Racine, Strain differences affect the induction of status epilepticus and seizure-induced morphological changes, *Eur. J. Neurosci.* in press (2004).
55. J. G. Flanagan, and P. Vanderhaeghen, The ephrins and Eph receptors in neural development, *Annu. Rev. Neurosci.* **21**, 309-345. (1998).
56. K. Kullander, and R. Klein, Mechanisms and functions of Eph and ephrin signalling, *Nature Rev. Mol. Cell. Biol.* **3**, 475-486 (2002).
57. B. Xu, S. Li, A. Brown, R. Gerlai, M. Fahnestock, and R. J. Racine, EphA axon guidance molecules regulate neuronal plasticity and activity-dependent axonal growth in adult CNS, *Mol. Cell. Neurosci.* **24**, 984-999 (2003).
58. S. Shimakawa, S. Suzuki, R. Miyamoto, K. Takitani, K. Tanaka, T. Tanabe, E. Wakamiya, F. Nakamura, M. Kuno, S. Matsuura, Y. Watanabe, and H. Tamai, Neuropilin-2 is overexpressed in the rat brain after limbic seizures, *Brain Res.* **956**, 67-73 (2002).
59. G. Barnes, R. S. Puranam, Y. Luo, and J. O. McNamara, Temproal specific patterns of semaphorin gene expression in rat brain after kainic acid-induced status epilepticus, *Hippocampus* **13**, 1-20 (2003).
60. D. A. Coulter, Epilepsy-associated plasticity in gamma-aminobutyric acid receptor expression, function, and inhibitory synaptic properties, *Int. Rev. Neurobiol.* **45**, 237-52 (2001).
61. G. Golarai, and T.P. Sutula, Functional alterations in the dentate gyrus after induction of long-term potentiation, kindling, and mossy fiber sprouting, *J. Neurophysiol..* **75**, 343-353 (1996).
62. B. M. Longo, and L. E. Mello, Blockade of pilocarpine- or kainate-induced mossy fiber sprouting by cycloheximide does not prevent subsequent epileptogenesis in rats, *Neurosci. Lett.* **226**, 163-166 (1997).
63. P. A. Williams, J. P. Wuarin, P. Dou, D. J. Ferraro, and F. E. Dudek, Reassessment of the effects of cycloheximide on mossy fiber sprouting and epileptogenesis in the pilocarpine model of temporal lobe epilepsy, *J. Neurophysiol.* **88**, 2075-2087 (2002).
64. K. Inoue, K. Morimoto, K. Sato, N. Yamada, and S. Otsuki, Mechanisms in the development of limbic status epilepticus and hippocampal neuron loss: an experimental study in a model of status epilepticus induced by kindling-like electrical stimulation of the deep prepyriform cortex in rats, *Acta Med. Okayama* **46**, 129-139 (1992).
65. H. Klitgaard, A. Matagne, J. Vanneste-Goemaere, and D. G. Margineanu, Pilocarpine-induced epileptogenesis in the rat: impact of initial duration of status epilepticus on electrophysiological and neuropathological alterations, *Epilepsy Res.* **51**, 93-107 (2002).

66. L. S. Chen, J. G. Wong, P. K. Banerjee, and O. C. Snead, 3rd. Kainic acid-induced focal cortical seizure is associated with an increase of synaptophysin immunoreactivity in the cortex, *Exp. Neurol.* **141**, 25-31 (1996).

Discussion

Kokaia: I have a questioning concerning GDNF. We have observed that GDNF receptor knockout mice are slower in kindling. So the question is, do you think that GDNF is also pro-epenogenic but if you use it chronically it down regulates the receptors?
Racine: That would be my guess, yes.
Simonato: I have a technical question. You have mentioned the focal administration of BDNF. How is it done exactly?
Racine: With a cannula into the hippocampus, using either continuous infusions or microinjections. In the case of microinjections, you can't give very many without causing some damage; so we kept it to three. And they were spread out over the course of the kindling.
Wasterlain: Ron, does the sprouting induced by all these different agents and by seizures go to the same target? Or does it vary by category?
Racine: It can vary by activation pattern. For example, if you stimulate and kindle in the piriform cortex, you get more sprouting ventrally, at least in our hands, whereas kindling in other areas will shift the balance in favour of dorsal sprouting. The infusion of an immunoadhesin that affects the interaction of endogenous ephrins and their EphA receptors flattens out these gradients; so it's possible to actually control the pattern of sprouting. We are currently applying microstimulation directly to the mossy fiber system dorsally and ventrally and looking at local manipulations of the axon guidance factors, to see how precisely we can control the pattern of the sprouting.
Wasterlain: And would you draw the bottom line as to what you think it tells us about sprouting and kindling.
Racine: I don't think it tells us much about kindling, but it does tell us something about sprouting. The axon guidance factors appear to be playing a role in determining the pattern of activation-induced axon sprouting. The story with neurotrophins is also complex. We have performed some work, that I didn't talk about, involving manipulation of the individual NGF receptors (TrkA and P75). So we know that NGF acts via the receptors, and we have some idea which ones are involved in kindling and which ones are involved in sprouting. We've also looked downstream from the receptors, and we are generally attempting to tease apart some of those downstream mechanisms. However, the bottom line is that I still don't know much.
Gutierrez: The laboratory rat is a good model of the rat, meaning that the wild rats in the country would normally have sprouted mossy fibers. So lab rats have to be trained or stimulated to display sprouting.
Racine: I am not entirely sure what you mean that they will have sprouted fibers. I consider this particular sprouting to be a pathological phenomenon. Do you mean they have higher background Timm density? Yes, well, as I said, with LTP, with experience, with training, you can produce these effects. That doesn't surprise me. Wild animals are raised in an enriched environment compared to lab animals.

CONDITIONAL DELETION OF TRKB PREVENTS EPILEPTOGENESIS IN THE KINDLING MODEL

James O. McNamara, Xiao-Ping He, and Robert Kotloski*

1. INTRODUCTION

Epilepsy is a common and frequently devastating neurological disorder, affecting approximately 1% of the population. Among the diverse forms, limbic epilepsy (synonyms include complex partial epilepsy, temporal lobe epilepsy, psychomotor epilepsy) in particular is the most devastating in adults for three reasons: (1) it is common, accounting for approximately 40% of all cases of adult epilepsy; (2) limbic seizures are often quite resistant to available anticonvulsant drugs;[1] and (3) the attacks induce impairment of consciousness, thereby limiting driving, maintaining employment, etc. Therapy is only symptomatic in that available drugs inhibit seizures in some individuals but do not modify the disease itself.

Epileptogenesis is the process by which a normal brain becomes epileptic. Understanding the mechanisms of limbic epileptogenesis may lead to effective pharmacologic prevention of this disease. The British neurologist, William Gowers, observed worsening of the epileptic condition in some patients and suggested that seizures themselves may beget seizures.[2] Discovery of the kindling model by Graham Goddard and his colleagues[3] provided experimental evidence in support of Gowers' idea; in this model, repeated induction of brief, focal seizures by application of initially subconvulsive electrical stimuli eventually results in prolonged, intense focal and tonic-clonic seizures. Once established, this enhanced sensitivity to electrical stimulation persists for the life of the animal.

The question arises as to how pathological neuronal activity in the form of focal seizures might be transduced into the cellular and molecular mechanisms underlying epileptogenesis. It seems plausible that extracellular signaling molecules might be pivotal in the transduction process. We suspect that the neurotrophin, BDNF, promotes limbic epileptogenesis by activation of its cognate receptor, TrkB. The rationale for this

* James O. McNamara, M.D., Carl R. Deane Professor and Chair, Department of Neurobiology, Professor of Medicine (Neurology), 401 Bryan Research Building, Research Drive, Box 3676, Duke University Medical Centre, Durham, NC 27710

idea includes the following. BDNF promotes dendritic outgrowth of cortical neurons and also promotes long-term potentiation of excitatory synaptic transmission,[4-10] plasticities which have been implicated in epileptogenesis.[11-14] Limbic seizures induce marked increases of BDNF expression.[15] Importantly, increased activation of TrkB has been identified in the mossy fiber pathway of hippocampus,[16, 17] the activation occurring at the same site and time as the increased BDNF expression. Intraventricular infusion of TrkB receptor bodies that scavenge ligands and presumably limit activation of TrkB partially inhibited kindling development.[18] Although mice lacking both *BDNF* alleles in the germline die shortly after birth, elimination of one *BDNF* allele is not lethal but is sufficient to partially inhibit kindling development.[19]

2. DELETION OF BDNF AND KINDLING

The evidence above led us to hypothesize that BDNF mediates activation of TrkB in the hippocampus during limbic epileptogenesis and that BDNF-mediated activation of TrkB is necessary for epileptogenesis in the kindling model. By crossing mice with floxed alleles of *BDNF* or *TrkB* to mice in which Cre recombinase is driven by a *Synapsin1* promoter (*Syn-Cre), BDNF* or *TrkB* alleles were selectively eliminated from a subset of central nervous system neurons, and the mutant mice are viable in adulthood. We tested these hypotheses by quantifying epileptogenesis and hippocampal TrkB activation in wild type and *Syn-Cre* conditional *BDNF* -/- and *TrkB* -/- mutant mice.[20]

Analysis of *β-galactosidase* expression in *Rosa-26* reporter mice as well as *BDNF* expression in *BDNF* -/- mice demonstrated widespread expression of the Cre recombinase in pyriform cortex, nuclei within amygdala, thalamus, hypothalamus, and brainstem.[20] Within the hippocampus in particular, high levels of Cre activity were evident in the dentate granule and CA3 pyramidal cells with lower levels of activity in the CA1 pyramidal cells as revealed by in situ hybridization for *BDNF*. The elimination of BDNF was confirmed by ELISA as well as immunohistochemistry.[20]

Contrary to our hypothesis, *BDNF* -/- mice exhibited only a modest impairment of limbic epileptogenesis. That is, approximately 50% more stimulations were required to induce the 3rd consecutive Class 4 or 5 seizure in *BDNF* -/- compared to *BDNF* +/+ mice (14.8±0.9 and 22.5±3.8 for +/+ and -/- mice respectively, Figure 1 top $p<0.05$). Although the number of stimulations required to induce behavioral seizures in the -/- mice was modestly increased, no differences were evident in the increase of electrographic seizure duration in that increases of 300-400% were detected in both +/+ and -/- mice.[20] No significant differences in current required to induce the initial electrographic seizure were evident.[20]

3. DELETION OF trkB AND KINDLING

We previously demonstrated enhanced activation of the BDNF receptor, TrkB, in the mossy fiber pathway of hippocampus during limbic epileptogenesis as evident in increased tyrosine phosphorylation at residue 515.[16, 17] Because BDNF can activate TrkB and because the increased activation of TrkB coincides both temporally and spatially with

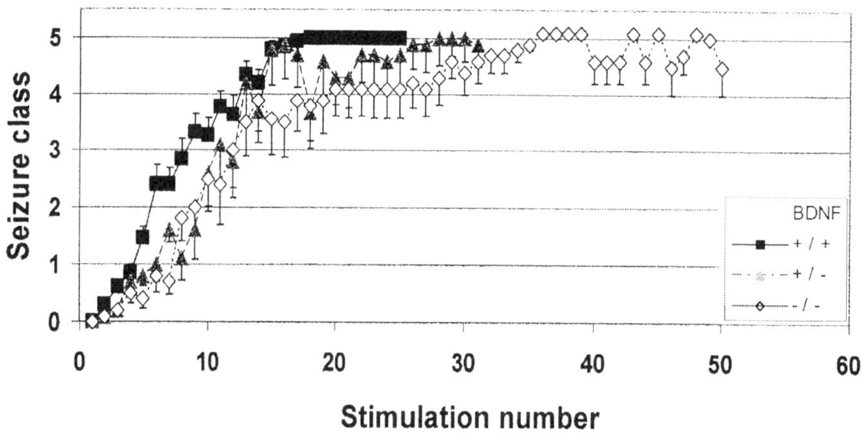

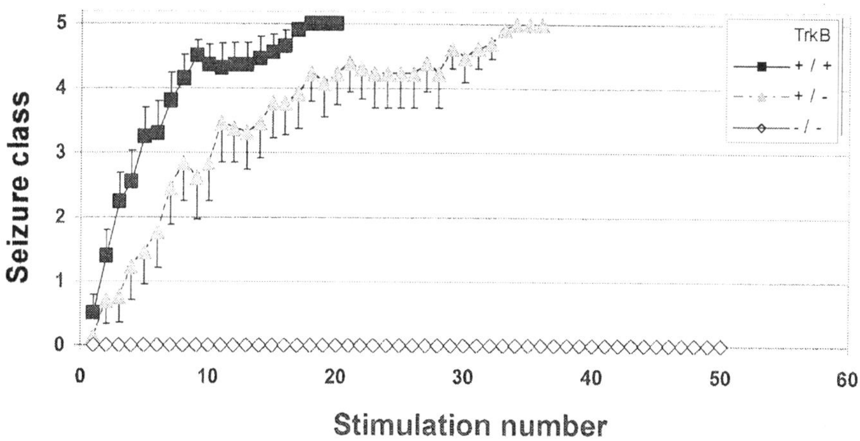

Figure 1 **Top:** Partial inhibition of kindling development in *BDNF* -/- mice. Kindling development is presented as behavioral seizure class (y-axis) in +/+ (n=21), +/- (n=10) and -/- (n=10) mice.. Stimulation number (x-axis) refers to the number of stimulations that evoked an electrographic seizure with duration of at least 5 sec. **Bottom:** Striking inhibition of kindling development in *TrkB* -/- mice. Kindling development is presented as behavioral seizure class (y-axis) in +/+ (n=16), +/- (n=16) and -/- (n=4) mice. All data are presented as mean±SEM. *: $p<0.05$, **: $p<0.01$, one-way ANOVA with post hoc Bonferroni's test.

the increased BDNF expression during limbic epileptogenesis,[15, 16, 21] we hypothesized that BDNF mediated the enhanced activation of TrkB.[16, 17] Contrary to our hypothesis, a seizure evoked in *BDNF* -/- mice evoked increased p-Trk immunoreactivity in the mossy fiber pathway similar to that of +/+ mice as revealed by quantitative immunohistochemical studies.[20] The fact that TrkB in particular exhibited increased phosphorylation was verified by immunoprecipitation and Western blot experiments.[20]

The enhanced activation of TrkB despite striking reductions of BDNF raised the possibility of compensatory responses of other neurotrophins that might activate TrkB in the *BDNF* -/- mice. Measurements of NT-4 and NT-3 content in hippocampal homogenates revealed a significant increase (39%) of constitutive expression of NT-3 protein in hippocampus in -/- compared to +/+ mice (529±21 and 379±19 ng/g for -/- and +/+, respectively, $p<0.01$). No differences in NT-4 content were detected (not shown). This raises the interesting possibility that the compensatory increase of NT-3 expression contributes to TrkB activation and limbic epileptogenesis in the *BDNF* -/- mice.

The persistent enhanced phosphorylation of TrkB associated with the persistence of limbic epileptogenesis in the *BDNF* -/- mice provided additional circumstantial evidence that TrkB is necessary for limbic epileptogenesis. We suspected that compensatory responses of ligands may be less efficient or absent when targeting the neurotrophin receptor itself. We therefore crossed the *Syn-Cre* mice to mice in which the *TrkB* allele was floxed and verified that TrkB was eliminated in a pattern similar to that of BDNF when crossed to the same line of *Syn-Cre* mice. In contrast to the *BDNF* -/- mice, it was not possible to induce limbic epileptogenesis in *TrkB* -/- mice as evident in the absence of evoked behavioral seizures even after 48-50 stimulations (Figure 1 bottom). Measures of evoked electrographic seizure duration also revealed inhibition of epileptogenesis. For example, in contrast to the 200% increase of electrographic seizure duration evident by the tenth stimulation of the *TrkB* +/+ mice, only a 100% increase of electrographic seizure duration was evident by the 21st stimulation of the *TrkB* -/- mice and no further increase was observed even after 48-50 stimulations.[20] Importantly, the development of kindling in the +/+ mice in this experiment was similar to that in the +/+ mice in the BDNF experiment as evident in progressive increase of behavioral seizure intensity and electrographic seizure duration induced by similar numbers of repeated stimulations (Figure 1 top and bottom). Significant increases were identified in the current required to trigger the initial electrographic seizure in the *TrkB* -/- mice (610±18, 675±16, and 960±22 μA, in +/+, +/-, and -/- mice respectively, $p<0.05$). Importantly, differences in current intensity required to evoke the initial electrographic seizure do not correlate with differences in epileptogenesis as measured in the kindling paradigm.[22] In the present study, subsets of +/+ mice exhibited elevated current intensities similar to those of the -/- mice, yet epileptogenesis proceeded in a pattern both quantitatively and qualitatively indistinguishable from the subset of +/+ and +/- mice with lower current intensities. Importantly, the duration of the initial evoked electrographic seizure was not significantly different between the TrkB +/+ and -/- mice (12±2 and 9±2 sec respectively). Together these findings indicate that elevated current thresholds evident in the *TrkB* -/- mice are not sufficient to explain the impairment of epileptogenesis.

Although no overt histological abnormality or defective motor function was detected in the *TrkB* -/- mice, we were concerned that the absence of *TrkB* impacted neuronal structure and/or function such that the *TrkB* -/- mice were incapable of exhibiting the behavioral endpoint of kindling, namely a tonic-clonic seizure. To address this possibility, electroconvulsive shocks were administered by transcorneal stimulation, and seizure occurrence and duration were recorded by an investigator blinded to genotype. No differences in duration of tonic flexion or tonic hindlimb extension were detected in *TrkB* -/- mice compared to wild type control mice. Likewise no differences were detected in the threshold of current required to evoke a predominantly clonic seizure. Thus the *TrkB* -/- mice can exhibit the behavioral endpoint of kindling, a tonic-clonic seizure as well as a predominantly clonic seizure with features similar to Class 1 and 2 kindled seizures.[20]

4. CONCLUSIONS

These findings demonstrate that elimination of the TrkB receptor itself prevents the plastic response of epileptogenesis in the kindling model while preserving the ability to express both electrographic and behavioral seizures. Although many pharmacological and genetic perturbations partially inhibit epileptogenesis in the kindling model, to our knowledge this study is the first to demonstrate that TrkB, or any other signaling pathway, is necessary for epileptogenesis in this model. These findings render TrkB an attractive molecular target for developing small molecules that might be used to prevent epileptogenesis in humans at high risk. Importantly, several questions must first be addressed. Knowing whether elimination of TrkB selectively in the mature brain is sufficient to prevent epileptogenesis is critical to the development of TrkB as a target for anti-epileptogenic therapies. Pinpointing the population(s) of neurons implicated in this process would provide an anatomic locale for elucidating the structural and functional consequences of TrkB activation required for epileptogenesis. Elucidating these consequences may also inform efforts aimed at prevention. That is, if TrkB activation in anatomically restricted subsets of neurons is required for limbic epileptogenesis, targeting therapeutic intervention to these sites may limit unwanted effects of global inhibition of TrkB activation. Answering these questions and determining whether limiting TrkB activation prevents epileptogenesis in additional models will be required to fully understand the status of TrkB as a target for anti-epileptogenic therapies.

5. ACKNOWLEDGEMENTS

This work was supported by NS grant 17771 (J.O.M.).

6. REFERENCES

1. S. Arroyo, M.J. Brodie, G. Avanzini, C. Baumgartner, C. Chiron, O. Dulac, J.A. French, and J.M. Serratosa, Is refractory epilepsy preventable?, *Epilepsia* **43**, 437-44 (2002).
2. W.R. Gowers, Epilepsy and other chronic convulsive diseases (London: Churchill, 1881).
3. G.V. Goddard, D.C. McIntyre, and C.K. Leech, A permanent change in brain function resulting from daily electrical stimulation, *Exp Neurol* **25**, 295-330 (1969).
4. A.K. McAllister, L.C. Katz, and D.C. Lo, Neurotrophin regulation of cortical dendritic growth requires activity, *Neuron* **17**, 1057-1064 (1996).
5. S.L. Patterson, T. Abel, T.A. Deuel, K.C. Martin, J.C. Rose, and E.R. Kandel, Recombinant BDNF rescues deficits in basal synaptic transmission and hippocampal LTP in BDNF knockout mice, *Neuron* **16**, 1137-114 (1996).
6. M. Korte, H. Kang, T. Bonhoeffer, and E. Schuman, A role for BDNF in the late-phase of hippocampal long-term potentiation, *Neuropharmacology* **37**, 553-559 (1998).
7. L. Minichiello, M. Korte, D. Wolfer, R. Kuhn, K. Unsicker, V. Cestari, C. Rossi-Arnaud, H.P. Lipp, T. Bonhoeffer, and R. Klein, Essential role for TrkB receptors in hippocampus-mediated learning, *Neuron* **24**, 401-414 (1999).
8. D. Muller, Z. Djebbara-Hannas, P. Jourdain, L. Vutskits, P. Durbec, G. Rougon, and J.Z. Kiss, Brain-derived neurotrophic factor restores long-term potentiation in polysialic acid-neural cell adhesion molecule-deficient hippocampus, *Proc Natl Acad Sci U S A* **97**, 4315-4320 (2000).

9. B. Xu, W. Gottschalk, A. Chow, R.I. Wilson, E. Schnell, K. Zang, D. Wang, R.A. Nicoll, B. Lu, and L.F. Reichardt, The role of brain-derived neurotrophic factor receptors in the mature hippocampus: modulation of long-term potentiation through a presynaptic mechanism involving TrkB, *J Neurosci* **20**, 6888-6897 (2000).
10. L. Minichiello, A.M. Calella, D.L. Medina, T. Bonhoeffer, R. Klein, and M. Korte, Mechanism of TrkB-mediated hippocampal long-term potentiation, *Neuron* **36**, 121-137 (2002).
11. J.V. Nadler, B.W. Perry, and C.W. Cotman, Selective reinnervation of hippocampal area CA1 and the fascia dentata after destruction of CA3-CA4 afferents with kainic acid, *Brain Res* **182**, 1-9 (1980).
12. T. Sutula and O. Steward, Facilitation of kindling by prior induction of long-term potentiation in the perforant path, *Brain Res* **420**, 109-117 (1987).
13. G. Golarai and T.P. Sutula, Functional alterations in the dentate gyrus after induction of long-term potentiation, kindling, and mossy fiber sprouting, *J Neurophysiol* **75**, 343-353 (1996).
14. I. Spigelman, X.X. Yan, A. Obenaus, E.Y. Lee, C.G. Wasterlain, and C.E. Ribak, Dentate granule cells form novel basal dendrites in a rat model of temporal lobe epilepsy, *Neuroscience* **86**, 109-120 (1998).
15. P.J. Isackson, M.M. Huntsman, K.D. Murray, and C.M. Gall, BDNF mRNA expression is increased in adult rat forebrain after limbic seizures: temporal patterns of induction distinct from NGF, *Neuron* **6**, 937-948 (1991).
16. D. K. Binder,, M.J. Routbort, and J.O. McNamara, Immunohistochemical evidence of seizure-induced activation of trk receptors in the mossy fiber pathway of adult rat hippocampus, *J Neurosci* **19**, 4616-4626 (1999).
17. X.P. He, L. Minichiello, R. Klein, and J.O. McNamara, Immunohistochemical evidence of seizure-induced activation of trkB receptors in the mossy fiber pathway of adult mouse hippocampus. *J Neurosci* **22**, 7502-7508 (2002).
18. D.K. Binder, M.J. Routbort, T.E. Ryan, G.D. Yancopoulos, and J.O. McNamara, Selective inhibition of kindling development by intraventricular administration of TrkB receptor body. *J Neurosci* ***19***, 1424-1436 (1999).
19. M. Kokaia, P. Ernfors, Z. Kokaia, E. Elmer, R. Jaenisch, and O. Lindvall, Suppressed epileptogenesis in BDNF mutant mice, *Exp Neurol* **133**, 215-224 (1995).
20. X.P. He, R. Kotloski, S. Nef, B.W. Luikart, L.F. Parada, and J.O. McNamara, Conditional deletion of TrkB but not BDNF prevents epileptogenesis in the kindling model, *Neuron* 43, 31-42 (2004).
21. C.M. Gall, Seizure-induced changes in neurotrophin expression: implications for epilepsy, *Exp Neurol* **124**, 150-166 (1993).
22. R.J. Racine, Modification of seizure activity by electrical stimulation. II. motor seizure, *Electroencephalogr Clin Neurophysiol* **32**, 281-294 (1972).

Discussion

Post: Let's put your data together with Ron Racine's and the early work showing that antidepressants, which increased BDNF chronically, can have diverse effects on kindling, with some of the antidepressants actually suppressing kindling. Would that be a way for you to get your molecules cooked up to chronically up-regulate that pathway so that it actually has less target in related epileptogenesis?

McNamara: There are a lot of ways of regulating BDNF expression, and I think they are all tied into, as you point out, with antidepressant effects. The three major antidepressants, electroshock therapy, SSRIs, and in particular voluntary physical activity, all increase BDNF expression in the granule cells. However, I think what we need is some drugs that are highly specific. We need a magic bullet.

Teskey: I just want to take my hat off to you for sticking to those very frustrating and negative early experiments. My question is, when you did get seizures from out of the trkB null mice, what did the seizures look like? I know they didn't grow, but did they have the same amplitude and frequency?

McNamara: Good question. In terms of response to electrical stimulation of the amygdala of the trkB nulls, their afterdischarge threshold is higher than in the wild types. However, the duration of the first afterdischarge is indistinguishable from the wild types. It did grow, but obviously to a much lesser extent – it was truncated. That is really important, because they were fully capable of generating the electrographic seizure.
Buzsaki: I would be very grateful is someone could tell what aspect of an afterdischarge is responsible for the long lasting effect of kindling. In an ill-fated study which we never completed, this is what we did. We triggered an afterdischarge and before what we call the seizure termination oscillation occurs, we terminated the seizure with a very brief septal stimulation. So you've got the same duration of the afterdischarge, the spiking part, but where the seizure terminates, the postical discharge and secondary afterdischarge are gone. And in that group of animals, we didn't see kindling effect. In other words, somehow the second part of the afterdischarge was more critical than the previous part.
McNamara: Were you first stimulating the electrode in the hippocampus?
Buzsaki: First we stimulated the perforant path or the hippocampus, and you can stop the seizure any time with a brief septal stimulation. So the question I have is, what is the factor in the afterdischarge that has the long lasting effect? And second, do you swear to God that the afterdischarge in your knockout animals was exactly the same as in the wild-type animal?
McNamara: The answer to your first question is that I used to think that the consequence of that afterdischarge was increased expression of BDNF. Now I don't think that, I think that there are all these compensatory responses, but my guess is that you have to activate trkB. That is a hypothesis, that you must somehow activate trkB as a consequence of that afterdischarge. In answer to your second question, what is our read out of an afterdischarge, it's crude, it's called the EEG. But I can tell you if you just look at the EEG these initial afterdischarges are no different, they are identical. That is the level of resolution.
Wasterlain: Jim, beautiful work. If one destroys the hippocampus and kindles from the amygdala, you eventually can kindle – you would find a way around it. I would have expected that you would have slowed kindling, but not completely prevented it, in the trkB nulls. But here you seem to have had a couple of early stages and then totally stopped the spread of discharge. So how do we reconcile this?
McNamara: That's a good question. If I could just restate it, what is the site within the forebrain from which trkB must be eliminated? The answer is that I don't know, but it is something we're working on.
Adamec: Jim, lovely work. With regard to the afterdischarge, you could be more quantitative by just doing power-spectral analysis. That is a way at getting at this more quantitatively. My question is whether you have tried kindling from a focus other than the amygdala.
McNamara: No.
Adamec: I think that's worth doing before you generalize and say that you have found something that blocks kindling and epileptogenesis.

NEUROPEPTIDE Y AND ITS RECEPTORS IN KINDLING EPILEPTOGENESIS

Cristina Richichi, Ramla Benmaamar, Marco Gobbi, En-Ju D. Lin, Matthew J. During, Gunther Sperk and Annamaria Vezzani*

1. INTRODUCTION

Neuropeptide Y (NPY)-mediated neurotransmission undergoes plastic changes during hippocampal or amygdaloid kindling in rats as shown by changes in peptide cellular expression and distribution, its release and receptor subtype plasticity (for review see [37, 51]). Electrophysiological and pharmacological evidence in *in vitro* and *in vivo* models of seizures indicate that NPY has predominant inhibitory actions on excitatory glutamate-mediated neurotransmission and it displays anticonvulsant properties in acute models of seizures.[37, 51] Information about its functional role in epileptogenesis is more limited although recent evidence points to an antiepileptogenic action of this peptide.[4, 38] In this chapter, we will review the available evidence concerning the functional role of endogenous NPY in kindling and we will address the contribution of its receptor subtypes in mediating its antiepileptogenic effects providing novel findings on the involvement of Y2 and Y5 receptor subtypes.

2. EXPRESSION OF NPY AND NPY RECEPTOR SUBTYPES IN KINDLING

2.1 NPY

In the hippocampus, NPY is constitutively expressed in GABA interneurons, a subset of which also contains somatostatin.[28] Brief convulsive stimuli, such as those used in electroconvulsive shock and kindling, markedly upregulate NPY expression in hippocampal GABA- and somatostatin-containing interneurons.[22, 44, 52] Thus, increased NPY expression occurs in neuronal populations that establish symmetric, inhibitory synaptic contacts with granule and pyramidal cell dendrites. This

* Cristina Richichi, Marco Gobbi, Annamaria Vezzani; Dept of Neuroscience, Mario Negri Inst for Pharmacol Res, Milano 20157, Italy; Ramla Benmaamar, NYU School of Medicine, Dept of Pathology, 10016 New York, USA; En-Ju Lin, Matthew J. During, Dept of Molecular Med, Univ of Auckland, Auckland 1020, New Zealand; Gunther Sperk, Dept of Pharmacol, Med Univ of Innsbruck, Innsbruck 6020, Austria

phenomenon may result in enhanced inhibition on principal hippocampal cells, likely representing an endogenous mechanism aimed at counteracting neuronal hyperexcitability. In models of status epilepticus, NPY containing neurons in the polymorphic cell layer of the dentate gyrus are particularly sensitive to seizure-induced cell loss.[47] The loss of these inhibitory neurons has been suggested to contribute to enhanced excitability of granule cells and to the onset of spontaneous seizures. A transient increase in NPY and its mRNA is observed in granule cells and their mossy fibers after hippocampal kindling stimulation leading to stage 2 or stage 5 seizures, while a prolonged and lasting increase of NPY is found in these neurons and their axonal projections in spontaneously epileptic rats.[8, 44, 47] Expression of NPY in granule cells is suppressed by anticonvulsant treatment,[31, 50] suggesting that *de novo* NPY synthesis likely reflects the level of activation of these neurons during seizures. When sprouting occurs, NPY accumulates in the aberrant mossy fibres projections into the inner molecular layer[47, 50] or in the outer molecular layer and hilus in human tissue from TLE patients.[12, 17, 32] Ionotropic glutamate receptors appear to be involved in triggering NPY neosynthesis in interneurons,[22, 43] while type I agonists of metabotropic glutamate receptors stimulate NPY mRNA expression in granule cells.[43] BDNF may act as a transcriptional factor promoting NPY neosynthesis in hippocampal principal neurons.[50] The increase in hippocampal NPY in kindled tissue is associated with enhanced ability of overexpressing neurons to release this peptide as shown by *ex vivo* evidence in slices taken from fully kindled rats and exposed to depolarizing concentrations of KCl.[40]

2.2 NPY receptors

Changes in peptide expression in kindling are accompanied by receptor subtypes modifications (Fig. 1). Thus, binding to, and mRNA expression of, postsynaptic Y1 receptors were significantly decreased in the molecular layer of the dentate gyrus, representing receptor dowregulation on dendrites of granule neurons. On the contrary, presynaptic Y2 receptor binding and mRNA expression were upregulated on mossy fibre terminals and in granule cell bodies respectively,[20] reflecting *de novo* synthesis of receptor protein. Y1 and Y2 receptor changes occurred at stages 2 and 5 of kindling and persisted for at least 30 days after kindling acquisition. Similar changes in NPY receptors have been found in spontaneously seizing rats[27, 41, 42] and in humans affected by TLE.[17]

Y5 receptor distribution in the hippocampus is still unclear; thus, in spite of abundant expression of mRNA,[33] very low specific binding has been measured in the hippocampus.[14] Y5 receptors are decreased after kainate-induced seizures in pyramidal and granule cell layers,[6] while they are up-regulated in hippocampal kindling[29, 49] (see also Fig. 4). Differently from Y1 and Y2 receptors, modifications in Y5 receptor sites are transient and restricted to the stimulated hippocampus (Fig. 4), thus likely representing an immediate consequence of local seizure activity rather than plastic alterations occurring in epileptogenic tissue.

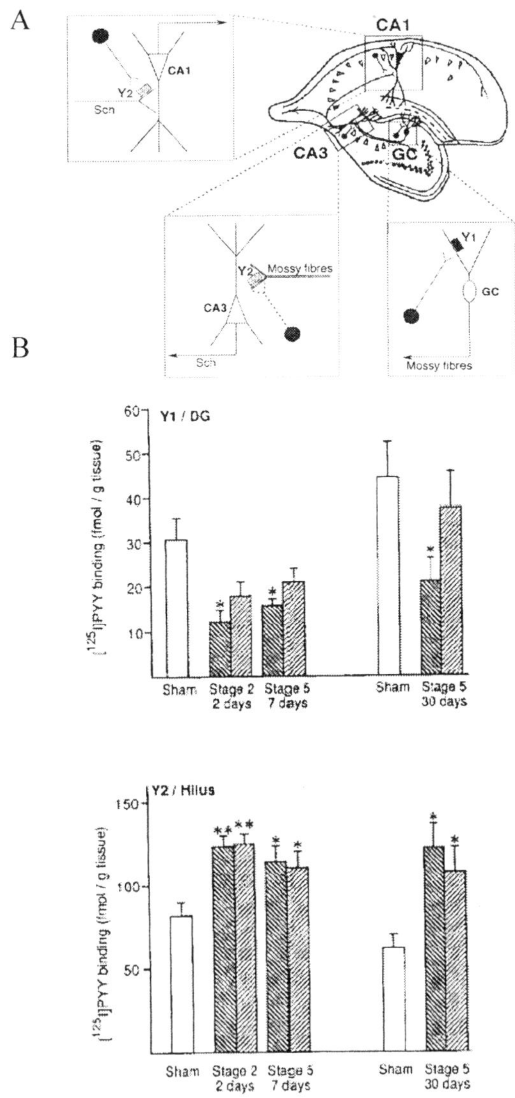

Figure 1. Schematic representation of Y1 and Y2 receptor localization in the hippocampus (panel A; from ref. 51). Y2 receptors are predominantly located presynaptically on glutamatergic nerve terminals (Schaffer collaterals and mossy fibres); Y1 receptors are present on granule cell dendrites (see also ref. 33). In B, quantitative analysis of ^{125}I-PYY specific binding to Y1 and Y2 receptors in the dorsal hippocampus (from ref. 20). Note the upregulation of Y2 receptor binding in the hilus while Y1 receptor binding is decreased in the molecular layer of the dentate gyrus.

3. ANTIEPILEPTOGENIC EFFECTS OF NEURON SPECIFIC ENDOGENOUS OVEREXPRESSION OF NPY

Transgenic rats expressing 5 extra copies of the human NPY gene showed a specific brain overexpression of NPY in CA1 pyramidal neurons and in the subiculum.[48] A similar ectopic expression pattern is transiently induced by seizures in naïve rats.[47] Transgenic rats were less susceptible to kindling epileptogenesis than wild-type littermates as demonstrated by a 65% increase in the number of electrical stimuli required to induce stage 5 seizures in NPY overexpressing rats (Table 1). Duration of the first afterdischarge in the stimulated hippocampus did not change, while secondary afterdischarge was reduced in transgenic rats only until stage 2 was reached. Fluoro-Jade analysis of brain sections did not reveal any degenerating neurons in fully-kindled wild-type or transgenic rats. NPY overexpressing rats also displayed decreased sensitivity to kainate-induced EEG and behavioral seizures. A major portion of CA1 pyramidal neurons innervates the subicular complex,[1] where NPY may reduce excitation through Y2 receptor[14] as well as by Y5 receptor activation.[23] Thus, the action of NPY in target areas of CA1 pyramidal neurons could significantly contribute to delay kindling rate by inhibiting seizure generalization in transgenic rats.

Table 1. Kindling epileptogenesis in transgenic rats overexpressing NPY in CA1 pyramidal cells and in the subiculum

	Wild-type		Transgenic
		AD threshold (μA)	
	149 ± 20		120 ± 11
		Kindling rate	
Stage 1-2	4 ± 1		5 ± 1
Stage 3	10 ± 2		15 ±3
Stage 4-5	23 ± 3		38 ±4**
		AD duration (sec)	
ADI at st 2	23 ± 3		22 ± 1
ADII	11 ± 2		3 ± 1*
ADI at St 5	29 ± 3		27 ± 1
ADII	11±2		8±2

Data are the mean±SE (n= 9). AD threshold represents the average threshold current inducing primary afterdischarge (AD) in the stimulated hippocampus. Kindling rate represents the number of electrical stimuli delivered to the hippocampus which aew required to induce each of the observed behavioral stages (Racine, 1972). ADI and ADII represent average duration of primary and secondary AD respectively. *$p<0.05$; **$p<0.01$ by Student's t-test vs wild-type. In this experiment a classical kindling protocol was adopted (from ref. 48).

In a subsequent study, the effect of long-lasting NPY overexpression in the rat hippocampus was addressed using adeno-associated viral (AAV) vectors carrying the human NPY gene (Fig. 2) (from ref. 39). Rats overexpressing NPY in hilar interneurons, pyramidal and granule cells and their mossy fibres were less susceptible to kainate-induced EEG seizures and they did not show either EEG or behavioral status epilepticus as compared to empty-vector injected control rats.

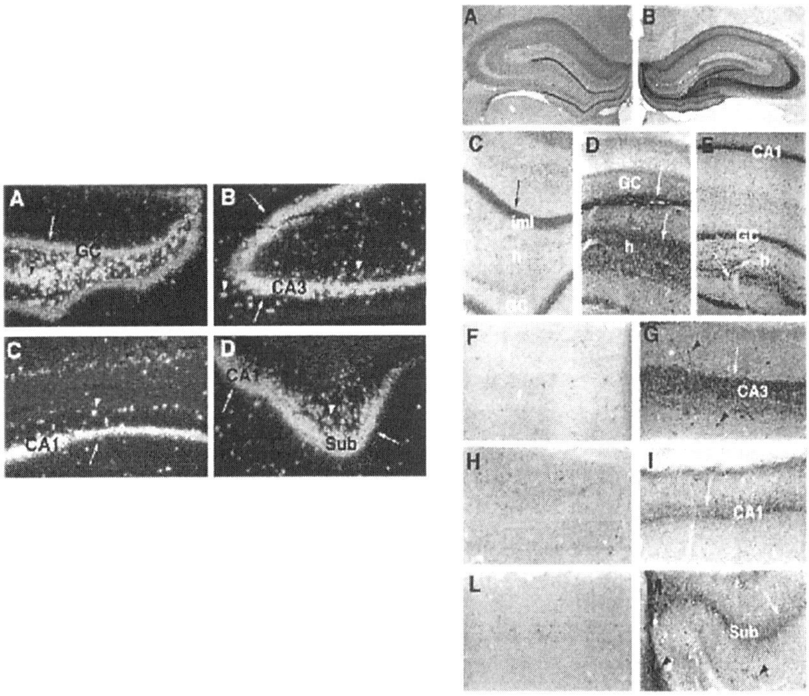

Figure 2 . Expression of NPY in the injected hippocampus of a representative rat 8 weeks after infusion of AAV-NPY vector (from ref. 39). Left panel represents *in situ* hybridization analysis of NPY mRNA in granule neurons (A), pyramidal cells (B,C) and subiculum (D) of AAV-NPY injected rat. These neurons do not express the NPY transcript in normal rat brain (not shown). Left panels represent immunohistochemistry of NPY in the AAV-NPY injected hippocampus (B, G, I, M) vs contralateral side (A,C,F,H,L). Panel D is a high magnification of the dentate gyrus (see panel B); panel E represents Nissl staining of the injected hippocampus.

Using a rapid kindling protocol, rats overexpressing NPY showed a slower kindling rate and reduced cumulative afterdischarge in the stimulated hippocampus than age-matched control rats. In addition, the threshold current needed to induce the first afterdischarge in the stimulated hippocampus was increased by 30% on average (Table 2). These findings indicate that endogenous NPY overexpression in neuronal populations that are activated by seizures has a significant inhibitory effect on local hippocampal excitability and on the spread of seizures from their site of onset.

Table 2. Kindling epileptogenesis in AAV-NPY injected rats overexpressing NPY in pyramidal cells, granule cells and hilar interneruons

	Control		AAV-NPY
		AD threshold (μA)	
	127±4		180±21*
		Kindling rate	
Stage 1	1.1 ± 0.1		1.3 ± 0.1
Stage 2	1.4 ± 0.3		2.4 ± 0.7
Stage 3	6.2± 1.8		12.8 ± 2.1*
Stages 4-5	11.4 ± 2.5		25.5 ± 4.2**
		Cumulative AD duration (min)	
AD	14.9 ±2.0		9.9 ± 1.0*

Data are the mean±SE (n= 9). AD threshold represents the average threshold current inducing primary afterdischarge (AD) in the stimulated hippocampus. Kindling rate represents the number of electrical stimuli delivered to the hippocampus which are required to induce each of the observed behavioral stages[36]). Cumulative AD duration was reckoned in each rat by adding together the durations of every primary AD in the stimulated hippocampus during the rapid kindling protocol (40 stimuli). *p<0.05; **p<0.01 by Student's t-test vs control (AAV-empty) (from ref. 39).

4. ROLE OF NPY RECEPTORS IN KINDLING EPILEPTOGENESIS

4.1 Y1 receptors

Y1 receptors are specifically located on granule cell dendrites in the hippocampus, and their pharmacological blockade with selective antagonists provides anticonvulsant effects on kainate-induced EEG seizures.[19, 49] Benmaamar et al.[4] showed that hippocampal kindling was delayed during the infusion in the rat hippocampus of an antisense oligonucleotide directed against Y1 receptor mRNA. When infusion was stopped (after attaining stage 2), kindling progression was faster. These authors measured NPY Y1 receptors by Western blot and showed that delayed kindling progression during antisense delivery was associated with a marked loss of Y1 receptor levels, whereas kindling acceleration upon antisense withdrawal was associated with enhanced Y1 receptor levels likely representing a rebound adaptive effect. These data suggest that Y1 receptors mediate a mild facilitatory effect of NPY on epileptogenesis and highlight the presence of an excitatory component of NPY actions. Mice with a genetic deletion of the Y1 receptor gene did not show any change in kindling parameters.[4]

4.2 Y2 receptors

Y2 receptors are presynaptically located on glutamatergic terminals of Schaffer collaterals and mossy fibres. We studied kindling in C57/SVJ mice with a deletion of the Y2 receptor gene. *In situ* hybridization analysis of mRNA showed that Y1 or Y5 receptor expression was not significantly modified in Y2 KO mice (Fig. 3). We did not detect any significant change in the rate of kindling development (Fig. 3) or in the afterdischarge duration (not shown) except for a trend towards kindling facilitation during the first 11 electrical stimulations. On the other hand, the same mice were more susceptible to kainate-induced EEG seizures (not shown), thus supporting an important role of these receptors in mediating inhibitory effects of NPY.

The lack of changes in kindling in Y1 or Y2 receptor knock-out mice may be due to functional recruitment of remaining NPY receptors (i.e. Y5) or to compensatory upregulation of other inhibitory systems during kindling which may substitute for the missing function of the ablated receptor subtypes.

Intrahippocampal application of [ahx5-24] NPY, a centrally truncated analog of NPY with selective agonist activity at Y2 receptors,[9] afforded significant dose-dependent protection from kainate-induced seizures in wild-type mice by delaying the onset of seizures and by reducing the total number and duration of convulsions.[15] Y2 receptors mediate inhibitory effects of NPY on glutamate release in the hippocampus.[10, 21, 26] This action likely represents the pivotal mechanism whereby NPY inhibits seizure activity.

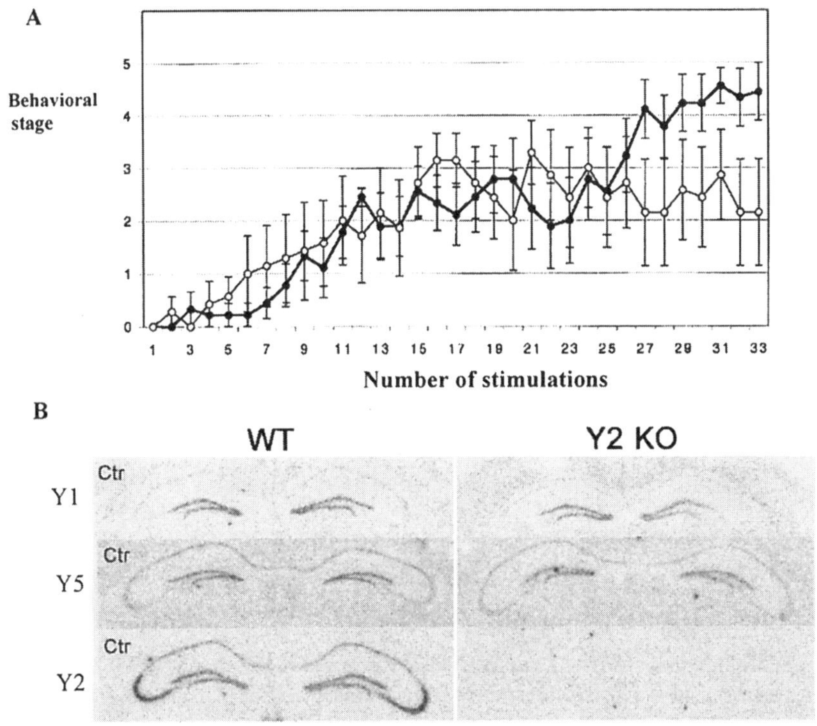

Figure 3. A. Kindling progression in Y2 C57BL6/SVJ129 knock-out (KO) mice versus wild-type controls (n=15). The achievement of the various behavioral stages during hippocampal stilulations is reported in the graph. **B.** Micrographs depict *in situ* hybridization analysis of NPY receptors mRNA in wild-type and Y2 KO mice.

4.3 Y5 receptors

We have studied the role of Y5 receptors in kindling using the selective antagonist GW438014A (IC50 = 210 nM; ref. 11), a small heterocycle molecule that crosses the blood-brain barrier, and the selective peptide agonist $Ala^{31}Aib^{34}$ NPY (IC50 = 6.0 nM; ref. 7). Intraperitoneal injection of GW438014A (10 mg/kg), 30 min before the beginning of a rapid-kindling protocol, significantly accelerated the rate of kindling as compared to vehicle-injected rats. Thus, the numbers of electrical stimuli required to reach stages 3 and 5 of kindling were reduced by 50% and 25% respectively. The average afterdischarge duration in the stimulated hippocampus was prolonged by 2-fold. Conversely, kindling rate was delayed by intracerebroventricular administration of 24 nmol $Ala^{31}Aib^{32}$ NPY. Thus, the number of stimuli necessary to reach stages 2 and 3 of kindling was increased by 3- and 4-fold respectively (Fig. 4). During the stimulation protocol (40 stimuli) none of the rats treated with the Y_5 agonist showed stages 4-5 seizures. Twenty-four h after the last kindling stimulation, thus during the re-test session, Y_5 agonist- or antagonist-treated rats had stages 4-5 seizures, like their controls. In rats treated with both the antagonist and the agonist, kindling rate was similar to vehicle—injected rats.[3]

These data indicate that Y_5 receptors mediate inhibitory actions of NPY in kindling, but they display anticonvulsant rather then antiepileptogenic effects upon agonist stimulation. The lack of changes in local afterdischarge duration during agonist receptor stimulation *vis-à-vis* the remarkable delay in the occurrence of generalized clonic behavioral convulsions suggests that Y_5 receptors affect seizure generalization rather than local hippocampal excitability. This finding is in accordance with earlier studies showing that Y_5 receptors do not mediate inhibition of epileptiform activity in rat hippocampal slices.[23] Suppression of behavioral kindled seizures in the absence of changes in local afterdischarge duration has been previously described using classical anticonvulsant drugs.[30] Since inhibitory effects of Y_5 receptors on excitatory neurotransmission have been reported in the rat subiculum,[23] we suggest Y5 receptors in this region may play a role in retarding kindling by dampening the excitatory output drive from the hippocampus. However, we cannot exclude the possibility that extrahippocampal Y_5 receptors (i.e. in amygdala and temporal cortex, where these receptors are abundant) may play a role in the observed effects.

5. CONCLUDING REMARKS

Distinct changes in the expression of various neuropeptides have been reported in experimental models of seizures. In most instances, the activation of excitatory pathways and glutamate receptor subtypes precedes the activation of neuropeptide systems in brain. This evidence suggests that neuropeptide expression reflects the activity of individual neuronal populations during seizures (Gall et al, 1990). Most of these modifications are associated with changes in their releasable pool. In this respect, neuropeptides are released from neurons at frequencies of stimulation higher than those required by classical neurotransmitters, implying that their enhanced efflux preferentially occurs during conditions of elevated neuronal activity such as those occurring during seizures.[24] Since peptides profoundly affect neuronal responses to glutamate and in some instances to GABA, extracellular changes in neuropeptides may profoundly affect neuronal excitability and synaptic transmission.

The specific changes in the NPY system are likely to represent endogenous neuroprotective mechanisms as suggested by the following considerations: the net

A.

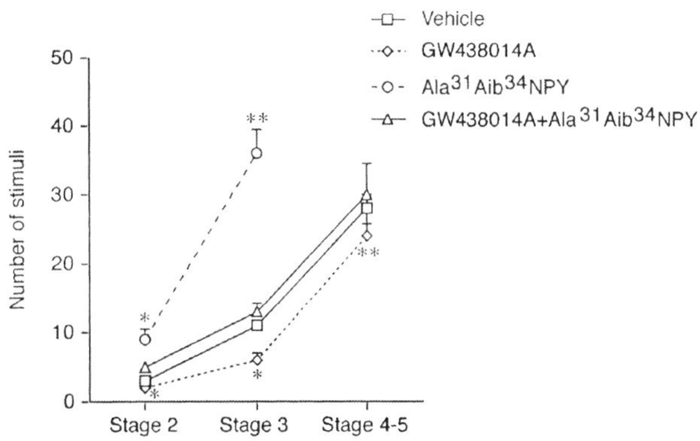

B.

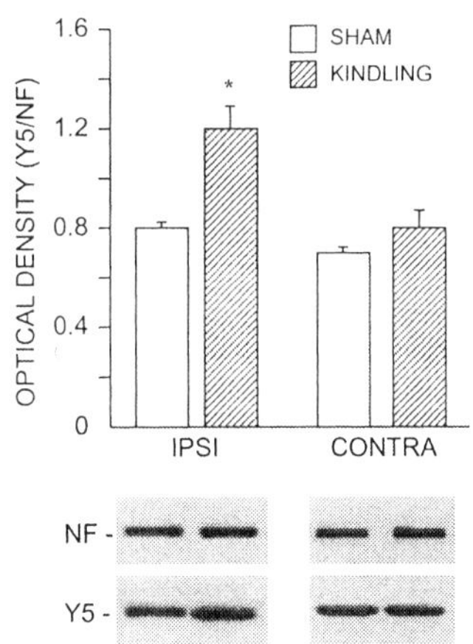

Figure 4. **A.** Left panel depicts kindling progression in rats receiving 10 mg/kg, i.p. GW438014A, a selective Y5 receptor antagonist, or 24 nmol $Ala^{31}Aib^{34}NPY$ icv, a selective Y5 receptor agonist, or their combination vs control rats treated with the corresponding vehicles. GW was given 30 min before the beginning of a rapid kindling protocol; AlaAibNPY was given 10 min before and 75 min after the beginning of kindling. Kindling was retarded by the Y5 agonist while it was accelerated by the Y5 antagonist. Rats treated with both drugs were similar to controls (from ref. 3). **B.** Right panel depicts representative Western blot analysis of Y5 receptors (see ref. 49) and Neurofilament M (NF) which was used as an internal standard in each sample. Data in bargraphs represent quantification of optical density (means±SE; n=3). Rats and their sham control were killed 24 h after rapid kindling acquisition. Note the increase in the level of Y5 receptor protein which is restricted to the stimulated hippocampus while no changes were found in the contralateral hippocampus or one week after kindling completion (not shown). *$p<0.05$ vs sham (rats implanted with electrodes but not stimulated) by Fisher 's test.

effect of NPY when bath applied to rodent[5, 15, 26, 34] or human[9, 35] brain slices or intracerebrally injected to rodents[46, 53] is inhibitory on excitatory synaptic transmission and seizures; agonist-mediated stimulation of Y2 receptors mediates anaticonvulsant effects of NPY,[10, 15, 26, 34] while Y1 receptor subtypes mediate a minor excitatory component,[4, 19] Y2 and Y1 receptors are up- and down-regulated respectively in epileptic tissue from experimental models[20, 27, 42] and humans.[17] Finally, mice with NPY gene deletion[2, 16] or inactivation[13] show enhanced susceptibility to seizures, while endogenous overexpression of NPY[39, 48] protects from seizures and delays kindling epileptogenesis. Recent evidence also showed a neuroprotective role of NPY on excitotoxic neuronal cell death induced by glutamate analogs in organotypic slice cultures[45] and enhanced neurogenesis in the hippocampus mediated by Y1 receptor stimulation.[25]

These findings underline the importance of the NPY system in controlling pathological neuronal hyperactivity in the hippocampus and suggest that NPY receptors may represent good targets for pharmacological intervention aimed at inhibiting seizures and, more importantly, to delay the epileptogenic process and the long-term sequelae associated with the epileptic state.

6. ACKNOWLEDGEMENTS

The authors are grateful to Dr Annette Beck-Sickinger and Dr Herber Herzog for providing us with the Y2 receptor agonists and Y2 knock-out mice respectively. Part of the work reviewed in this manuscript was supported by Fondazione Cariplo (AV), HFSP RG0045/2000-B (AV, GS) and the New Zealand Health Research Council (MJD).

7. REFERENCES

1. D. G. Amaral, and M. P. Witter, The three-dimensional organization of the hippocampal formation: a review of anatomical data, *Neuroscience* **31,** 571-591 (1989).
2. S. C. Baraban, G.. Hollopeter, J. C. Erickson, P. A. Schwartzkroin, and R. D. Palmiter, Knock-out mice reveal a critical antiepileptic role for neuropeptide Y, *J Neurosci* **17,** 8927-8936 (1997).
3. R. Benmaamar, C. Richichi, M. Gobbi, A. J. Daniels, A. G. Beck-Sickinger, and A. Vezzani, Neuropeptide Y Y_5 receptors inhibit kindling acquisition in rats, *Reg Pept*, in press (2004).
4. R. Benmaamar, B. T. Pham-Le, C. Marescaux, T. Pedrazzini, and A. Depaulis, Induced down-regulation of neuropeptide Y-Y1 receptors delays initiation of kindling, *Eur J Neurosci.* **18,** 768-774. (2003)
5. M. Bijak, Neuropeptide Y reduces epileptiform discharges and excitatory synaptic transmission in rat frontal cortex in vitro, *Neuroscience* **96,** 487-494 (2000).
6. G. Bregola, Y. Dumont, A. Fournier, S. Zucchini, R Quirion, and M. Simonato, Decreased levels of neuropeptide Y5 receptor binding sites in two experimental models of epilepsy, *Neuroscience* **98,** 697-703 (2000).
7. C. Cabrele, H.A. Wieland, N. Koglin, C. Stidsen, and A.G. Beck-Sickinger, , Ala31-Aib32: identification of the key motif for high affinity and selectivity of neuropeptide Y at the Y5-receptor, *Biochemistry* **41,** 8043-8049 (2000).
8. R. S. Chafetz, W. K. Nahm, and J. L. Noebels, Aberrant expression of neuropeptide Y in hippocampal mossy fibers in the absence of local cell injury following the onset of spike-wave synchronization, *Mol Brain Res.* **31,** 111-121 (1995).
9. W. F. Colmers, N. Pronchuk, C. Torok-Both, M. Ho, K Aronyk, J. McKean, T. Snyder, D. B Sinclair, M. Javidan, and A. G. Beck-Sickinger, Neuropeptide Y2 and other receptors inhibit synaptic excitation in epileptic human brain, *Epilepsia* **38** [Suppl 8], 12 (1997).
10. W. F. Colmers, K. Lukowiak, and Q. J. Pittman, Presynaptic action of neuropeptide Y in area CA1 of the rat hippocampal slice, *J Physiol* **383,** 285-299 (1987).
11. A. J. Daniels, M. K. Grizzle, R. P. Wiard, J. E. Matthews, and D. Heyer, Food intake inhibition and reduction in body weight gain in lean and obese rodents treated with GW438014A, a potent and selective NPY-Y5 receptor antagonist, *Regul Pept.* **106,** 47-54 (2002).
12. N. C. de Lanerolle, J. H. Kim, R. J. Robbins, and D. D. Spencer, Hippocampal interneuron loss and plasticity in human temporal lobe epilepsy, *Brain Res.* **495,** 387-395 (1989).

13. S. DePrato Primeaux, P. V. Holmes, R. J. Martin, R. G. Dean, and G. L. Edwards, Experimentally induced attenuation of neuropeptide-Y gene expression in transgenic mice increases mortality rate following seizures, *Neurosci. Lett.* **287,** 61-64 (2000).
14. Y. Dumont, A Fournier, and R. Quirion, Expression and characterization of the neuropeptide Y Y5 receptor subtype in the rat brain, *J. Neurosci.* **18,** 5565-5574 (1998).
15. B El Bahh, S. Balosso, T. Hamilton, H. Herzog, A. Beck-Sickinger, G. Sperk, A. Vezzani, and W. F. Colmers, Anticonvulsant response to NPY is mediated by the Y2 receptor, submitted (2004).
16. J C Erickson, K. E. Clegg, and R. D Palmiter, Sensitivity to leptin and susceptibility to seizures of mice lacking neuropeptide Y, *Nature* **381,** 415-421 (1996).
17 S. Furtinger, S. Pirker, T. Czech, C. Baumgartner, C., Ransmayr, and G. Sperk, Plasticity of Y1 and Y2 receptors and neuropeptide Y fibers in patients with temporal lobe epilepsy, *J. Neurosci,* **21,** 5804-5812 (2001).
18. C. Gall, J. Lauterborn, P. Isackson, and J. White, Seizures, neuropeptide regulation, and mRNA expression in the hippocampus, *Prog. Brain Res.* **83**, 371-390 (1990).
19. M. Gariboldi, M. Conti, D Cavaleri, R. Samanin, and A. Vezzani, Anticonvulsant properties of BIBP3226, a non-peptide selective antagonist at neuropeptide Y Y1 receptors, *Eur. J. Neurosci.* **10**, 757-759 (1998).
20. M. Gobbi, M. Gariboldi, C. Piwko, D. Hoyer, G. Sperk, and A. Vezzani, Distinct changes in peptide YY binding to, and mRNA levels of, Y1 and Y2 receptors in the rat hippocampus associated with kindling epileptogenesis, *J. Neurochem.* **70**, 1615-1622 (1998).
21. S. Greber, C. Schwarzer, and G. Sperk, Neuropeptide Y inhibits potassium-stimulated glutamate release through Y2 receptors in rat hippocampal slices in vitro, *Br. J. Pharmacol.* **113**, 737-740 (1994).
22. B. Gruber, S. Greber, E. Rupp, and G. Sperk, Differential NPY mRNA expression in granule cells and interneurons of the rat dentate gyrus after kainic acid injection, *Hippocampus* **4**, 474-482 (1994).
23. M. W. Ho, A. G. Beck-Sickinger, and W. F. Colmers, Neuropeptide Y5 receptors reduce synaptic excitation in proximal subiculum, but not epileptiform activity in rat hippocampal slices, *J. Neurophysiol.* **83**, 723-734 (2000).
24. T. Hokfelt, O. Johansson, A. Ljungdahl, J. M. Lundberg, and M. Schultzberg, Peptidergic neurones, *Nature* **284**, 515-521 (1980).
25. O. W. Howell, H. E. Scharfman, H. Herzog, L. E. Sundstrom, A. Beck-Sickinger, and W. P. Gray, Neuropeptide Y is neuroproliferative for post-natal hippocampal precursor cells, *J Neurochem.* **86**, 646-659 (2003).
26. G J Klapstein, and W. F. Colmers, Neuropeptide Y suppresses epileptiform activity in rat hippocampus in vitro, *J. Neurophysiol.* **78**, 1651-1661 (1997).
27. N. Kofler, E. Kirchmair, C. Schwarzer, and G Sperk, Altered expression of NPY-Y1 receptors in kainic acid induced epilepsy in rats, *Neurosci. Lett.* **230**, 129-132 (1997).
28 C. Kohler, L. G. Eriksson, S. Davis, and V. Chan-Palay, Co-localization of neuropeptide tyrosine and somatostatin immunoreactivity in neurons of individual subfields of the rat hippocampal region, *Neurosci. Lett.* **78,** 1-6 (1986).
29 J. Kopp, A. Nanobashvili, Z. Kokaia, O. Lindvall, and T. Hokfelt, Differential regulation of mRNAs for neuropeptide Y and its receptor subtypes in widespread areas of the rat limbic system during kindling epileptogenesis, *Mol. Brain Res.* **72**, 17-29 (1999).
30. E. W. Lothman, R. A. Salerno, J. B. Perlin, and D. L. Kaiser, Screening and characterization of antiepileptic drugs with rapidly recurring hippocampal seizures in rats, *Epilepsy Res.* **2**, 367-379 (1988).
31. J. Marksteiner, R. Prommegger, and G. Sperk, Effect of anticonvulsant treatment on kainic acid-induced increases in peptide levels, *Eur J Pharmacol* **181**, 241-246 (1990).
32 G. W. Mathern, T. L. Babb, J. K. Pretorius, and J. P. Leite, Reactive synaptogenesis and neuron densities for neuropeptide Y, somatostatin, and glutamate decarboxylase immunoreactivity in the epileptogenic human fascia dentate, *J. Neurosci.* **15,** 3990-4004 (1995).
33. R. M. Parker and H Herzog, Regional distribution of Y-receptor subtype mRNAs in rat brain, *Eur. J. Neurosci.* **11**, 1431-1448 (1999).
34. J. Qian, W. F. Colmers, and P. Saggau, Inhibition of synaptic transmission by neuropeptide Y in rat hippocampal area CA1: modulation of presynaptic Ca2+ entry, *J. Neurosci.* **17**, 8169-8177 (1997).
35. R. P. Patrylo, A. N. van den Pol, D. D. Spencer, and A. Williamson, A., NPY inhibits glutamatergic excitation in the epileptic human dentate gyrus, *J. Neurophysiol.* **82**, 478-483 (1999).
36. R. J. Racine, Modification of seizure activity by electrical stimulation. II. Motor seizure *Electroenceph. Clin. Neurophysiol.* **32**, 281-294 (1972)
37. J. P. Redrobe, Y. Dumont, J. A. St-Pierre, and R, Quirion, Multiple receptors for neuropeptide Y in the hippocampus: putative roles in seizures and cognition, *Brain Res.* **848**, 153-166 (1999).
38. S Reibel, R. Benmaamar, Y. Larmet, S. P. Kalra, S. C. Marescaux, and A. Depaulis, Neuropeptide Y delays hippocampal kindling in the rat, *Hippocampus* **13**, 557-560 (2003).

39. C. Richichi, E. J. Lin, D. Stefanin, D. Colella, T Ravizza, G. Grignaschi, P. Veglianese, G. Sperk, M. J. During, M. and A. Vezzani, Anticonvulsant and antiepileptogenic effects mediated by adeno-associated virus vector neuropeptide y expression in the rat hippocampus, *J. Neurosci.* **24**, 3051-3059 (2004).
40. M. Rizzi, A. Monno, R. Samanin, G. Sperk, and A. Vezzani, Electrical kindling of the hippocampus is associated with functional activation of neuropeptide Y-containing neurons, *Eur. J. Neurosci.* **5**, 1534-1538 (1993).
41 C. Roder, C. Schwarzer, A. Vezzani, M. Gobbi, T. Mennini, and G. Sperk, Autoradiographic analysis of neuropeptide Y receptor binding sites in the rat hippocampus after kainic acid-induced limbic seizures. *Neuroscience* **70**:47-55 (1996).
42 C. Schwarzer, N. Kofler, and G. Sperk, Up-regulation of neuropeptide Y-Y2 receptors in an animal model of temporal lobe epilepsy, *Mol. Pharmacol.* **53**, 6-13 (1998).
43. C. Schwarzer, and G. Sperk, Glutamate-stimulated neuropeptide Y mRNA expression in the rat dentate gyrus: a prominent role of metabotropic glutamate receptors, *Hippocampus* **8**, 274-288 (1998).
44. C. Schwarzer, J. M. Williamson, E. W. Lothman, A. Vezzani, and G. Sperk, Somatostatin, neuropeptide Y, neurokinin B and cholecystokinin immunoreactivity in two chronic models of temporal lobe epilepsy, *Neuroscience* **69**, 831-845 (1995).
45. A. P. Silva, P. S. Pinheiro, A. P. Carvalho, C. M. Carvalho, B. Jakobsen, J. Zimmer, and J. O Malva, Activation of neuropeptide Y receptors is neuroprotective against excitotoxicity in organotypic hippocampal slice cultures, *FASEB J.* **17**, 1118-1820 (2003).
46. M. Smialowska, M. Bijak, M. Sopala, and K. Tokarski, Inhibitory effect of NPY on the picrotoxin-induced activity in the hippocampus: a behavioural and electrophysiological study, *Neuropeptides* **30**, 7-12 (1996).
47. G. Sperk, J. Marksteiner, B. Gruber, R. Bellmann, M. Mahata, and M. Ortler, Functional changes in neuropeptide Y- and somatostatin-containing neurons induced by limbic seizures in the rat, *Neuroscience* **50**, 831-846 (1992).
48. A Vezzani, M. Michalkiewicz, T. Michalkiewicz, D. Moneta, T. Ravizza, C Richichi, M Aliprandi, F. Mule, L. Pirona, M. Gobbi, C. Schwarzer, and G. Sperk, Seizure susceptibility and epileptogenesis are decreased in transgenic rats overexpressing neuropeptide Y, *Neuroscience* **110**, 237-243 (2002).
49. A. Vezzani, D. Moneta, F. Mule, T. Ravizza, M. Gobbi, and J. French-Mullen, Plastic changes in neuropeptide Y receptor subtypes in experimental models of limbic seizures, *Epilepsia* **41** (Suppl 6), S115-121 (2000).
50. A. Vezzani, T. Ravizza, D. Moneta, M. Conti, A. Borroni, M. Rizzi, R. Samanin, and R. Maj, Brain-derived neurotrophic factor immunoreactivity in the limbic system of rats after acute seizures and during spontaneous convulsions: temporal evolution of changes as compared to neuropeptide Y, *Neuroscience* **90,** 1445-1461 (1999a).
51. A. Vezzani, G. Sperk, and W. F. Colmers, Neuropeptide Y: emerging evidence for a functional role in seizure modulation, *Trends Neurosci.* **22**, 25-30 (1999b).
52. O. Zachrisson, A. A. Mathe, C. Stenfors, and N. Lindefors, Limbic effects of repeated electroconvulsive stimulation on neuropeptide Y and somatostatin mRNA expression in the rat brain, *Mol. Brain Res.* **31**, 71-85 (1995).
53. D. P. Woldbye, P. J. Larsen, J. D. Mikkelsen, K. Klemp, T. M. Madsen, and T. G. Bolwig, Powerful inhibition of kainic acid seizures by neuropeptide Y via Y5- like receptors, *Nature Med.* **3**, 761-764 (1997)

Discussion

Simonato: Regarding your virus data, with a single injection in the dorsal hippocampus, were you kindling from the hippocampus in the same area where you have over-expression of NPY or far away from there?
Vezzani: What we did was inject the viral factor bilaterally in the dorsal and the ventral hippocampus of the rat. We waited 2-8 weeks, when there is maximal of the expression of the peptide, and then we kindled in one dorsal hippocampus. The four hippocampi, dorsally and ventrally, were over-expressing the peptide.
Simonato: The other question is regarding theY5 receptor. Most of the data supporting the involvement of the Y5 receptor in seizure control are related to a acute phenomena. Have you tested your drugs in a classical kindling paradigm?
Vezzani: No, we have not.

Schwartzkroin: I am curious about the NPY expression in the hilar interneurons after kindling. In other models of TLE, those cells tend to be very vulnerable and die. Do they disappear at all in the kindling model, and is there any relationship between their disappearance and the development of kindling?
Vezzani: In the protocol used in this experiment, we provoked only a few generalized clonic seizures in the animals. In this condition, you do not see any loss of these neurons. However, we and others have reported that if you increase the number of generalized clonic seizures, you can induce a loss of hilar interneurons and these animals develop spontaneous seizures.
Sutula: You commented that manipulations of NPY seem to affect generalization, and the earlier phases of kindling didn't seem to differ with the manipulations. Do you have any ideas about what that may represent? If it were an anticonvulsant effect, you would expect it to be seen right away, rather than later on or at a more advanced stage of the process.
Vezzani: What we have observed is that actually stage 2 is not very much affected in the over-expressing animals. The difference of course is when stage 3 appears. I really don't know why, but possibly you need a certain level of hyperexcitability in order to recruit the neuropeptide system in controlling epileptogenesis.

CONTRIBUTION OF PRE KINDLING AFFECTIVE STATE TO HEMISPHERIC DIFFERENCES IN THE EFFECTS ON ANXIETY OF BASOLATERAL AMYGDALA KINDLING

Robert Adamec, Tanya Shallow, Jacqueline Blundell, Paul Burton*

1. INTRODUCTION

The most commonly agreed upon forms of affective psychopathology accompanying seizure disorders in humans are depression and anxiety.[1,2] A variety of factors impact on affect in human epileptics, complicating determination of the contribution of epileptic pathophysiology to psychopathology. Factors include family history, medication, psychosocial factors and premorbid psychiatric condition[1]. Limbic kindling has been proposed as a model of complex partial seizure disorder with secondary generalization.[1,3-5] Kindling of limbic foci produces lasting changes in animal behavior, including anxiety like behavior (ALB)[1,6-9] and cognition.[10] The lasting changes in ALB produced by kindling are consistent with the idea that seizures change affective behavior. Nevertheless, the nature of behavioral change produced by kindling is also complicated by a number of clinically relevant factors.

Kindling may increase or decrease ALB in rodents[7,11] and in felines.[12] Factors contributing to the anxiogenic or anxiolytic effects of kindling include the hemisphere of the focus, AP plane location of the focus in a given amygdala nucleus, and pre kindling anxiety level (premorbid affective state).[7] Previous work suggests that kindling of left basolateral amygdala (BLA) is anxiolytic, while kindling of right BLA is anxiogenic.[7,11] New findings described in this chapter suggest this conclusion must be modified, when premorbid affective state is taken into account.

*R. Adamec et al., Memorial University, St. John's, NL, A1B3X9, Canada

Consideration of premorbid affective state is of obvious clinical significance. Moreover, its contribution may help to explain some of the discrepancies in findings of effects of kindling on rodent affect. In the same animal, Adamec and Shallow[13] were able to assess ALB in the elevated plus maze (EPM) prior to and following kindling of the medial amygdala. Normally retesting in the EPM increases ALB on second test.[14] Adamec and Shallow showed that testing rats in different novel rooms in different EPM's with an inter test interval of three weeks prevented the increase in ALB normally seen with repeated testing.[14] Pre morbid and post kindling ALB was measured using ratio time, an index of open arm exploration.[15] This is the ratio of time spent in the open arms to time spent in all arms. The smaller this ratio, the more anxious the rat. Adamec and Shallow divided their rats into those with greater or lesser ratio times on the first of two tests prior to kindling, but after electrode implantation in the right anterior medial amygdala. Foci locations were carefully chosen to match those where kindling was shown to be anxiogenic in the EPM.[11] It was found that kindling was anxiogenic (reduced ratio time) only in initially less anxious rats. Moreover, there was a non significant trend for ratio times of initially more anxious rats to increase following kindling (an anxiolytic like effect). One wonders if the baseline ratio times had been lower if the effects of kindling might have been anxiolytic. These data suggest that premorbid anxiety state interacts with kindling to determine behavioral outcome, and that the outcome may be either anxiogenic, anxiolytic or no effect depending on the level of anxiety at the time of kindling.

Given these findings, it is important to test this idea in other amygdala foci. Therefore, recent work examined the role of prekindling baseline ALB on the behavioral effects of kindling of right and left BLA. Behavior in the EPM was studied with an eye to replicating the methods of Adamec and Shallow.[13] Nevertheless, kindling alters other measures of defensive behavior including risk assessment,[11] and defensive response to handling.[16] Therefore, we also examined risk assessment and defensive response to handling. The work described below is also presented in more detail elsewhere.[17]

2. METHODS RIGHT/LEFT KINDLING

Male Wistar rats (n=108) were randomly assigned to one of three experimental groups: an unimplanted handled control (n=36); sham kindled implanted control (n=36); and kindled group (n=36). Half were assigned to the right and half to the left BLA kindling experiments. Unimplanted controls were handled as other groups. Handled and sham kindled controls did no differ so they were separately combined for right and left BLA groupings. Pre kindling behavior testing (Test 1) occurred 1 week after electrode implantation. The hole board and EPM were used to test rodent activity, exploration and anxiety. Post kindling behavior testing (Test 2) took place seven days after completion of kindling to 4 stage 5 seizures. As observed in the past, kindling had little effect on activity or exploration in the hole board. In addition to ratio time, another measure taken in the EPM was risk assessment (as described elsewhere[17]). Frequency of risk assessment was divided by time spent in the closed arms of the EPM to produce relative frequency risk assessment. In addition to measures in the EPM, a common measure of resistance to capture was taken in the hole board. As the rat was picked up with a gloved hand in the hole board and transferred to the EPM, response to pick up was scored on a scale of 0-6, from no reaction to a violent defensive attack. Some studies of extensive amygdala kindling in hooded rats have reported kindling increases this defensive resistance to capture measure.[18,19]

Location of electrodes in all three stereotaxic planes of the Paxinos and Watson atlas[20] were quantified (see methods[11,17]). There were no statistically significant differences in focus location in the right and left BLA of the present study (Table 1).

Table 1. Electrode locations (means ± SEM) in the Right and Left Basolateral Amygdala (BLA)

Hemisphere	Anterior-Posterior [1] Plane	Lateral [2] Plane	Vertical Plane
Present Study			
Right BLA	2.15 ± .07	4.92 ± .07	8.25 ± .09
Left BLA	2.36 ± .08	4.88 ± .08	8.33 ± .10
Adamec and Morgan			
Left BLA	2.37 ± .08	4.68 ± .10	8.90 ± .11

[1] Data are in mm posterior to bregma

[2] Lateral position is in mm lateral to the midline, vertical is mm below the dura.

To assess effects of pre kindling behavioral state on response to right BLA kindling, rats were divided into high and low anxiety pre kindling groupings. This was done by finding a mean ratio time (.439) of the normally distributed scores on the first test over all animals. Rats were classified as being below the mean (more anxious) or equal to or above the mean (less anxious) on test 1. This classification created the mean split factor (M) in subsequent analyses of effects of right BLA kindling. The ratio time split used in the left BLA groups was .57. Derivation of this number arose from consideration of the findings in right BLA kindling and is described below.

3. PREMORBID AFFECTIVE STATE AND RIGHT BLA KINDLING

Kindling increased open arm exploration (ratio time) in rats which were more anxious prior to kindling, an anxiolytic like effect. In contrast kindling decreased open arm exploration in rats which were less anxious prior to kindling, an anxiogenic like effect (Figure 1 Right BLA; Test [T, Pre-Post Kindling] x Mean Split [M]; $F\{1,15\} \geq 39.96$, $p<.0001$; mean contrasts, Fisher's LSD, $p<.05$). In contrast kindling increased risk assessment equally in rats more and less anxious before kindling (Test [Pre-Post Kindling] Effect only, $F\{1,15\}=14.07$, $p<.002$). This finding is consistent with a body of data indicating changes in risk assessment and open arm exploration are controlled by separate neural systems.[21,22]

Kindling also interacted with premorbid anxiety in its effects on defensive resistance to capture (T x M; $F\{1,15\}=10.44$, $p<.006$; Figure 1 Right BLA, Capture Score). Kindling reduced defensive resistance to capture only in rats less anxious before kindling. This suggests the kindling induced increase in EPM anxiety in this group is accompanied by more passivity in response to being picked up.

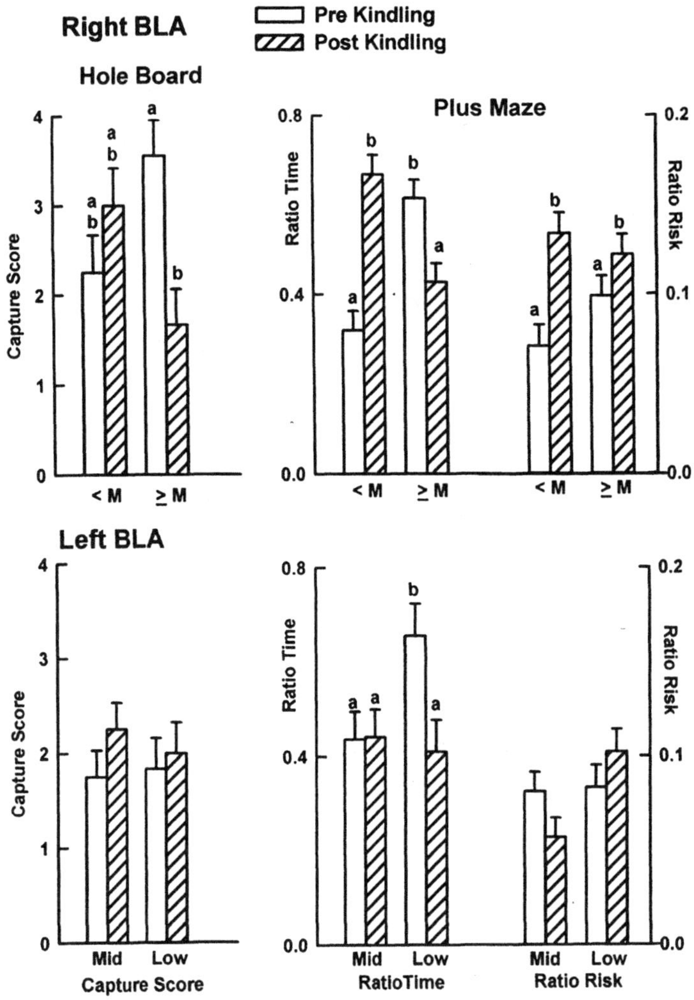

Figure 1. Mean + SEM for defensive response in hole board and EPM are plotted separately for rats kindled in the right and left BLA. Values before (pre) and 7 days after (post) kindling are plotted. "M" refers to mean split for right BLA (see text) and Mid and Low in left BLA refer to Mid Anxiety and Low Anxiety groupings based on pre kindling ratio time criteria in the text. Within a plot of a given measure, unmarked means, or means marked with the same letter do not differ. Means marked differently differ, and means marked with two letters fall between means bearing those letters.

4. PREMORBID AFFECTIVE STATE AND LEFT BLA KINDLING

Previous findings suggest that right BLA kindling is anxiogenic, and left BLA kindling is anxiolytic in the EPM.[11] However, present findings indicate that effects of right BLA kindling on EPM anxiety depend on level of pre kindling ratio time. It is possible that a similar rule applies to left BLA kindling. From the right BLA kindling study above, one might expect that BLA kindling of rats with pre kindling ratio times at or above .57 would be anxiogenic (reduce ratio time). The value of .57 is the average of pre kindling Test 1 ratio times from less anxious to be kindled rats and Test 1 and Test 2 ratio times of less anxious controls, all of which do not differ. Moreover, in the original study of Adamec and Morgan showing anxiolytic effects of left BLA kindling.[11] kindled controls had low ratio times (mean of .21) and left BLA kindling increased ratio times to a mean of .34. The value of .34 does not differ from post kindling ratio times of rats made more anxious by right BLA kindling seen above. These data suggest anxiolytic effects of left BLA kindling might depend on low open arm exploration tendencies, and that kindling drives ratio time to some mid range value. Therefore, we attempted to split left BLA rats into three pre kindling baseline groupings: High Anxiety with Test 1 ratio times of .21 or below; Low Anxiety with Test 1 ratio times of .57 or above and Mid Anxiety with Test 1 ratio times in between. We predicted that kindling of left BLA would reduce ratio time in Low Anxiety baseline rats (an anxiogenic effect); increase ratio time in High Anxiety baseline rats (an anxiolytic effect replicating Adamec and Morgan); and have no effect in Mid Anxiety baseline rats. There were no high anxious rats in the present study. So rats were divided into pre kindling Low and Mid Anxiety groups.

As predicted left BLA kindling was anxiogenic in Low Anxiety groups, decreasing ratio time (Test [T, pre, post kindling] x Split [Low, Mid Anxiety Baseline] interaction, $F\{1,12=4.80$, $p<.01$; Figure 1 Left BLA; mean contrasts, Fisher's LSD, $p<.05$). In contrast, left BLA kindling did not affect ratio time in Mid Anxiety Baseline rats (Figure 1 Left BLA). It is of interest that the pre and post kindling ratio times of Mid Anxiety Baseline rats do not differ from the ratio times of left BLA kindled rats in Adamec and Morgan[11](mean $\pm$ SEM in Adamec and Morgan = .34 $\pm$ .05 vs .43 $\pm$ .05 present data). Therefore, the lack of change following kindling in these animals may be because they were already at a level of open arm exploration to which left BLA kindling would drive them, as predicted.

Left BLA kindling had no effect on risk assessment. This differs from the increase in risk assessment following right BLA kindling. Also unlike right BLA kindling, left BLA kindling had no effect on resistance to capture (Figure 1 Left BLA). Resistance to capture scores were quite low (between 1 and 2) and within the range of capture scores of right BLA kindled rats (compare Right, Left BLA Figure 1). It is possible that the lack of effect of left BLA kindling on capture score is because they were already as low as kindling might have driven them.

Different effects of kindling in the right and left BLA groupings cannot be attributed to location of the focus. Comparison of focus locations in three planes in above and below mean split right BLA kindled groups did not differ. The same holds true for Low and Mid Anxiety left BLA kindled groups. Furthermore, left and right BLA groups did not differ in kindled focus locations (Table 1). Finally none of the sub groups kindled in right and left BLA, or comparing right and left BLA, differed in number of seizures to first stage 5 seizure or duration of the fourth stage 5 seizure (see Adamec et al. for details[17]).

In both right and left BLA kindling studies, rats served as their own controls. This was justified by the fact that controls and to be kindled rats did not differ on Test 1 in both right and left BLA studies. Nor was repeated testing of controls in the EPM anxiogenic. Moreover, the effects of kindling cannot be attributed to changes in response on repeated testing (see Adamec et al. for details.[17]

5. DISCUSSION

5.1 The Interaction of Pre Kindling Baseline and Kindling on ALB

Using a comparable paradigm, Adamec and Shallow[13] reported that right medial amygdala kindling was anxiogenic (reduced ratio time) in rats with above pre kindling median baseline, consistent with present findings. In contrast kindling was without effect in rats with below median pre kindling baseline.[13] Baseline ratio times of medial amygdala kindled rats were all within the range where Adamec and Morgan[11] found anxiolytic effects of left BLA kindling (below .21). Given present findings, one might have expected increased ratio times in both medial amygdala kindled groups. It appears that the nature of the interaction between baseline ratio time and kindling effects on behavior may vary with the amygdala nucleus kindled.

5.2 Left BLA Kindled Rats in Present and Past Studies

The anxiogenic effects of kindling left BLA in initially Low Anxiety rats differs from the Adamec and Morgan report of anxiolytic effects of left BLA kindling.[11] Nevertheless, anxiogenic effects were predicted on behavioral grounds under the assumption that pre kindling baseline anxiety level, not hemisphere, was the critical determinant of outcome of basolateral amygdala kindling on behavior in the EPM. The importance of baseline anxiety level was strengthened by the finding of no anatomical or kindling parameter differences between kindled foci in right and left BLA in the present study.

Moreover, there were no kindling parameter differences between the present study and Adamec and Morgan. Present study focus locations did not differ from Adamec and Morgan in AP or lateral plane positions. There was a difference in vertical plane location, however. Adamec and Morgan foci were located deeper in and near the left ventral basolateral amygdala (BLV) (bonferroni protected t tests, all $p < .05$, $t\{11\} = 3.93$; $t\{13\} = 3.10$, Table 1). Therefore it cannot be entirely ruled out that the difference in response to left BLA kindling in the two studies might be related to the difference in depth of the kindled foci in the BLA.

5.3 Effects of Kindling on Resistance to Capture and Risk Assessment

The kindling induced decrease in resistance to capture following right BLA kindling differs from the increases in this measure reported following far more extensive kindling in left BLA of hooded rats.[16,23] Strain, hemisphere and kindling procedural difference may account for this discrepancy.

In the present study, kindling of the right BLA increased risk assessment in all groups but kindling of similar foci in left BLA was without effect on risk assessment. Therefore effects of kindling on this measure are independent of pre kindling baselines of open arm exploration in EPM. These findings are consistent with studies showing open arm exploration and risk assessment can be changed independently of each other.[13, 22, 24] Taken together, present and past findings suggest that the nature of the effects of kindling on risk assessment are focus location specific. Increases, no change and decreases are possible, depending on the hemisphere, amygdala nucleus and AP plane of the focus within a nucleus.[11, 13, 24] It is also clear from present and past studies that the interaction of premorbid baseline behavior and effects of kindling does not apply to risk assessment, and may be specific to open arm exploration in the EPM.

6. CONCLUSIONS

The results of this study replicate and extend previous findings of lasting and selective changes in ALB following amygdala kindling. The data confirm the importance of baseline anxiety level on effects of kindling on anxiety measured by open arm exploration in the EPM. The work shows for the first time that kindling of the same sites in the right BLA produces opposite behavioral effects, depending on pre kindling anxiety level. High anxiety rats become less so and low anxiety rats become more so following kindling. The same may be true for left BLA kindling, when data from the present and previous studies are taken into account. Therefore, the previously reported anxiolytic effects of left BLA kindling may not reflect a functional hemisphere difference. These conclusions must be tempered by some differences in anatomical location of kindled foci between the present and past studies. More effort is required in BLA to determine the relative contribution of electrode location and baseline ALB on response to kindling.

Therefore, a complex set of factors appear to contribute to the effects of kindling on behavior. Nevertheless, one of them appears clearly to be pre kindling behavioral disposition. Perhaps dispositional neural correlates exists in rodents which change amygdala function differently as a consequence of amygdala kindling. In fact, high and low anxiety in EPM has been bred in Wistar rats which show elevated ACTH and corticosterone responses to stress.[25] Moreover, differences in CCK function have been related to dispositional differences in EPM,[26] and CCK mRNA is altered in amygdala by kindling.[27] Further work testing this disposition hypothesis seems warranted.

7. ACKNOWLEDGMENTS

This work was supported by Canadian Institute of Health Research (CIHR) Grant (MOP 49490) to R. Adamec.

8. REFERENCES

1. R. E. Adamec, Does kindling model anything clinically relevant?, *Biol. Psychiatry* **27**, 249-279 (1990).
2. B. P. Hermann and S. Whitman, Behavioral and personality correlates of epilepsy: a review, methodological critique, and conceptual model, *Psychol. Bull.* **95**, 451-497 (1984).
3. M. E. Corcoran, Contributions of Graham Goddard to kindling, *Neurosci. Biobehav. Rev.* **13**, 251-252 (1989).
4. J. O. McNamara, The neurobiological basis of epilepsy, *Trends in Neurosciences* **15**, 357-359 (1992).
5. R. M. Post, S. R. Weiss and A. Pert, Implications of behavioral sensitization and kindling for stress- induced behavioral change, *Adv. Exp. Med. Biol. [2LU].* **245**, 441-463 (1988).
6. R. E. Adamec, Amygdala kindling and anxiety in the rat, *NeuroReport* **1**, 255-258 (1990).
7. R. Adamec, ***Amygdala kindling and rodent anxiety,*** in: *Kindling 5*, edited by M. E. Corcoran and S. L. Moshe (Plenum Press, New York, 1998), pp.327-348.
8. V. Helfer, C. Deransart, C. Marescaux and A. Depaulis, Amygdala kindling in the rat: Anxiogenic-like consequences, *Neuroscience* **73**, 971-978 (1996).

9. L. E. Kalynchuk, J. P. Pinel, D. Treit and T. E. Kippin, Changes in emotional behavior produced by long-term amygdala kindling in rats, *Biol. Psychiat.* **41**, 438-451 (1997).
10. D. K. Hannesson and M. E. Corcoran, The mnemonic effects of kindling, *Neurosci. Biobehav. Rev.* **24**, 725-751 (2000).
11. R. E. Adamec and H. D. Morgan, The effect of kindling of different nuclei in the left and right amygdala on anxiety in the rat, *Physiol. Behav.* **55**, 1-12 (1994).
12. R. Adamec, Behavioral and epileptic determinants of predatory attack behavior in the cat, *Can. J. Neurol. Sci.* **2**, 457-466 (1975).
13 R. Adamec and T. Shallow, Effects of baseline anxiety on response to kindling of the right medial amygdala, *Physiol. Behav.* **70**, 67-80 (2000).
14. S. E. File, The interplay of learning and anxiety in the elevated plus- maze, *Behav. Brain Res.* **58**, 199-202 (1993).
15. S. E. File, The contribution of behavioural studies to the neuropharmacology of anxiety, *Neuropharmacology* **26**, 877-886 (1987).
16. L. E. Kalynchuk, J. P. J. Pinel and D. Treit, Characterization of the defensive nature of kindling-induced emotionality, *Behav. Neurosci.* **113**, 766-775 (1999).
17. R. Adamec, T. Shallow and P. Burton, Anxiolytic and Anxiogenic Effects of Kindling - Role of Baseline Anxiety and Anatomical Location of the Kindled Focus on Response to Kindling of the Right and Left Basolateral Amygdala, *Behav. Brain Res.*(2004, under revision).
18. J. P. Pinel, D. Treit and L. I. Rovner, Temporal lobe aggression in rats, *Science* **197**, 1088-1089 (1977).
19. L. E. Kalynchuk, J. P. J. Pinel and D. Treit, Long-term kindling and interictal emotionality in rats: effect of stimulation site, *Brain Res.* **779**, 149-157 (1998).
20. Paxinos, G. and Watson, C. *The Rat Brain in Stereotaxic Coordinates* (Sydney, Academic Press 1982)..
21. R. Adamec, Transmitter systems involved in neural plasticity underlying increased anxiety and defense - Implications for understanding anxiety following traumatic stress, *Neurosci. Biobehav. Rev.* **21**, 755-765 (1997).
22. R. E. Adamec, P. Burton, T. Shallow and J. Budgell, NMDA receptors mediate lasting increases in anxiety-like behavior produced by the stress of predator exposure - Implications for anxiety associated with posttraumatic stress disorder, *Physiol. Behav.* **65**, 723-737 (1999).
23. L. E. Kalynchuk, Long-term amygdala kindling in rats as a model for the study of interictal emotionality in temporal lobe epilepsy, *Neurosci. Biobehav. Rev.* **24**, 691-704 (2000).
24. R. Adamec and T. Shallow, Rodent anxiety and kindling of the central amygdala and nucleus basalis*Physiol. Behav.* **70**, 177-187 (2000).
25. R. Landgraf, A. Wigger, F. Holsboer and I. D. Neumann, Hyper-reactive hypothalamo-pituitary adrenocortical axis in rats bred for high anxiety-related behaviour, *J. Neuroendocrinol.* **11**, 405-407 (1999).
26. Wunderlich, G. R., Raymond, R., DeSousa, N. J., Nobrega, J. N., and Vaccarino, F. J., Decreased CCKB receptor binding in rat amygdala in animals demonstrating greater anxiety-like behavior, *Psychopharmacology* **164**, 193-199 (2002).
27. L X. Zhang, M. A. Smith, S. Y. Kim, J. B. Rosen, S. R. Weiss and R. M. Post, Changes in cholecystokinin mRNA expression after amygdala kindled seizures: an in situ hybridization study, *Mol. Brain Res.* **35**, 278-284 (1996).

Discussion

Wasterlain: Interesting to see that resistance to capture seems to be decreased by kindling. How do you relate that to the findings on aggression, which is increased by seizures?
Adamec: It may be useful to compare the different models. I remember talking to Pierre Gloor about this, and he thought epileptic "so called" aggressivity was actually extreme fearfulness. It is possible what Lisa Kalynchuk sees is more relevant to that clinical manifestation where these animals are so defensive that they actually attack in defence of themselves, whereas the strategy these rats are adopting is a passive one. They are passively freezing, and it appears as though they are allowing themselves to be picked up. I think it really may be a manifestation of a less intense form of defensiveness.

Gale: You sorted the animals by groups with cutoffs. Did you do any correlations within groups to see whether there was really an inverse correlation across all the animals based on where they started and where they finished?
Adamec: No, I didn't, but that is a good idea.
Schwartzkroin: How generalizable are your behavioral results to the point of stimulation, which anatomical site you kindled, and to the generation of seizure activity as opposed to just stimulation in a particular site?
Adamec: I'll address the second question first. We didn't do the necessary controls, but it has been shown that partial kindling to stage 3 in rodents can produce very long increases in anxiety-like behavior, but in a more posterior basolateral amygdala placement. In the cat, we initially showed that it is not necessary to kindle an actual motor seizure; all we did was produce repeated seizure discharges. But we did show that long-lasting changes in affective disposition in the cat required several seizures - you really had to partially kindle. It was not due to a single seizure during ADT determination or frequency of stimulation. I can't say that for sure here, but based on that history it at least requires a number of ADs for this to occur. I am not certain that it requires actual kindling to stage 5, and my guess is that partial kindling might do the same. With regard to kindling location, one of the issues that I brought up at the last kindling 5 conference was the importance of the location of the focus. We have found that the AP plane, for example, appears to be important depending on the nucleus: More anterior locations can be anxiogenic, more posterior locations can be anxiolytic, and in the central nucleus it is reversed. The actual location of the focus may be important. We have a new paper that looked slightly more posterior to this site in the basolateral amygdala, which produces strong broadband anxiolytic effects measured in a variety of test. I am not sure whether it is really anatomical locus or baseline. The other thing is that I am not entirely sure that we have explained Adamec and Morgan's data on the left basolateral amygdala, because there is a focus difference from that study to this one, in that our electrodes were slightly higher and are in the body of the basolateral nucleus, whereas in that study they were infringing upon the basolateral ventral nucleus. We have data in the right amygdala saying it is anxiolytic; so it is a complex question.

EFFECTS OF KINDLING ON SPATIAL MEMORY
Characteristics and mechanisms

Darren K. Hannesson, Ken Wolfe, and Michael E. Corcoran*

1. INTRODUCTION

Epilepsy is a chronic disorder defined by the recurrence of spontaneous seizures affecting ~2% of the population in North America.[1] There are a variety of distinct subtypes of the disorder with the most common, temporal lobe epilepsy (TLE), accounting for about 55% of all cases.[1, 2] While the seizures themselves can be significantly debilitating, the clinical impact of TLE is exacerbated by the fact that about half of all sufferers exhibit serious disturbances of affect and/or memory function.[3-6] At present, the mechanisms of these TLE-associated psychological impairments are unknown and, although recent studies have begun to explore the potential effectiveness of behavioral interventions,[7, 8] there is no treatment currently available. Consequently, these problems are often neglected clinically and have a significant adverse effect on patient quality of life.[9] There is a clear need, then, for research to provide a better understanding of the bases for epilepsy-associated psychological disturbances and to develop effective treatment strategies. Clinical studies, however, are challenging because of the large number of potentially relevant but difficult to control variables. Thus, progress relies heavily on work with suitable animal models. The present chapter reviews studies on one such model – kindling in rats, and describes characteristics of spatial memory changes in this model, their relation to kindling-related variables, and potential mediating mechanisms.

* Darren K. Hannesson, Canadian Centre for Behavioural Neuroscience, Department of Psychology and Neuroscience, University of Lethbridge, Lethbridge, AB, Canada, S4S OA2. Ken Wolfe, Michael E. Corcoran, Department of Anatomy and Cell Biology, University of Saskatchewan, Saskatoon, SK, Canada, S7N 5E5.

1.1. TLE and memory dysfunction.

There is strong evidence that TLE is associated with memory dysfunction, that this dysfunction can be relatively specific, and that it is clinically significant. Impairment of a variety of mnemonic functions is observed more frequently in patients with epilepsy compared to other comparable populations,[6] and is most common in a subset of patients with TLE[5, 6, 10]. Disturbances are seen in both verbal[5, 10, 11] and non-verbal memory[11-13] and are apparent at the onset of TLE.[14] The impairment is typically restricted to specific types of memory functions,[13, 15, 16] and can occur against a backdrop of relatively intact cognitive functioning.[8, 15, 17] Patients frequently complain about memory impairments,[9, 18, 19] and if anything, underestimate their severity relative to that apparent on laboratory memory tests.[18, 19] Finally, the clinical significance of memory dysfunction in epilepsy is underscored by the fact that memory problems adversely impact day to day functioning,[8] and correlate inversely with measures of quality of life.[3, 9]

The mechanisms underlying memory dysfunction in epilepsy are unknown. Some contribution is likely made by damage to structures involved in memory. However, not all TLE patients with memory problems have detectable lesions,[14, 20] and measures of structural pathology are highly variable in the extent to which they correlate with aspects of memory performance.[5, 11] Recent studies using magnetic resonance spectroscopy have shown that measures of hippocampal metabolites show a significant correlation with memory impairments,[21, 22] yet correlate weakly with conventional measures of structural changes, such as volumetry.[23] This suggests that metabolite measures reflect changes in neuronal function that are distinct from cell loss and may in fact be more strongly involved in memory impairments.[23] On the whole then, these considerations strongly suggest that factors besides cell loss make a significant contribution to TLE-associated memory dysfunction, although the nature of these factors is unknown.

1.2. Kindling as a model to study TLE-associated memory dysfunction

Kindling in rodents is an excellent preparation to use to investigate TLE-associated memory dysfunction for a number of reasons. First, kindling bears a number of similarities to TLE in that both show a propensity for involvement of medial temporal lobe structures, focal seizures capable of secondary generalization, distinct patterns of temporal lobe neuropathology, and characteristic patterns of drug sensitivity/insensitivity.[24, 25] Second, kindling is one of the most widely used preparations to study TLE and, therefore, a great deal has been discovered about its physiological, anatomical, and molecular consequences in the 30 years since its recognition [see this volume]. This knowledge provides a repository of information that can be used to assist in identifying candidate mechanisms for inter-ictal behavioral effects. Third, previous work has demonstrated that kindling is in fact associated with a range of inter-ictal behavioral effects, including memory impairments, and these resemble many of the changes also observed in TLE.[26-30] Finally, kindling enables precise experimental control over a range of potentially important variables including the focus (i.e., the brain site stimulated), severity, frequency, and number of seizures experienced as well as the interval between seizures and behavioral assessment.

2. THE EFFECTS OF KINDLING ON SPATIAL MEMORY

Several types of memory impairment have been observed following kindling. The present chapter focuses on experiments investigating kindling's effects on spatial memory.[31-43] Effects on spatial memory, defined as the ability to acquire, maintain and use information about viewer-independent spatial relations between stimuli,[44] are particularly worth studying for a number of reasons. First, spatial memory in rodents has been extensively studied and we are beginning to develop a good understanding of its anatomical, physiological, and neurochemical/pharmacological bases.[45-47] Second, there are a range of well-validated and frequently used behavioral paradigms to assess spatial memory in rodents as well as an arsenal of similarly designed control tasks that enable the assessment of the specificity of impairments to spatial memory.[45, 48] Third, spatial memory is known to critically involve the hippocampus,[44] a structure that is strongly implicated in TLE and memory dysfunction.[11] Lastly, spatial memory can be viewed as one example of a more general class of memory function known as explicit or declarative memory,[49] which represents a fundamentally important type of human memory and one that is essential for day-to-day functioning. This point is important because recent studies suggest that TLE-asssociated memory impairments may be selective to declarative memory tasks,[15, 16] including those that assess spatial memory.[13]

2.1. Behavioral characteristics

Our work and that of several other laboratories has led to considerable progress in the characterization of the nature of kindling's effects on spatial memory. Consideration of these studies suggests that the deficit is manifest in a relatively specific impairment of the earliest phase of spatial memory processing. Kindling of the dorsal hippocampus (dHPC) disrupts performance on a range of spatial memory tasks including several versions of both the radial-arm maze[37-41] and Morris water maze (MWM).[33-36, 42] The impairment is typically small and involves poorer than normal performance rather than a total failure of mnemonic function.[28] The consistency of the deficit across task variants is noteworthy and suggests that dHPC kindling reliably impairs some aspect of spatial memory. This view is corroborated by findings that kindling does not substantially impair a range of non-spatial functions that might otherwise account for spatial task deficits. Thus, dHPC kindling does not impair performance on a *non-spatial* visible platform variant of the MWM,[33, 35] a task that shares many of the same demands as the conventional MWM task, including the need for effective goal-directed swimming, sensory processing of remote visual information, and intact motivational processing. DHPC kindling protocols that impair spatial memory also do not affect key aspects of performance in the elevated plus maze or the open field,[34] demonstrating that anxiety-related behaviors are not affected. Finally, dHPC kindling does not impair either acquisition or retention of an object discrimination task in a modified water maze[35] or performance of a spontaneous object recognition task in an open field[34] demonstrating that a general cognitive or global mnemonic deficit is not produced. Results from the object discrimination task also rule out task-difficulty as an account for the pattern of tasks impaired by dHPC kindling since the object discrimination task was more difficult (i.e., it required more trials to acquire) than the spatial water maze tasks shown to be impaired.[35] Collectively the above considerations suggest that kindling produces a specific disruption of spatial memory.

Evidence also suggests that the deficit is restricted to a specific phase of memory processing. DHPC kindling disrupts within-day acquisition but not 7 or 28-day retention of a constant hidden platform location in the MWM[35] demonstrating that learning or short-term memory but not long-term memory is disrupted. Further evidence indicates that the earliest phase of spatial memory processing is primarily affected. DHPC kindling produces a delay-independent deficit in a delayed matching-to-place task in the Morris water maze across 15, 60, and 240 sec delays.[34] Since longer delays place greater demands on short-term memory, the lack of relation between delay and the impairment is best accounted for by a disruption of the earliest phase of spatial memory processing.

2.2. Relation to kindling variables

There are a number of kindling-related parameters that may impact the extent to which kindling leads to changes in spatial memory function including the site being kindled, the interval between kindling and behavioral testing, and the extent of kindling. The importance of these parameters has been rarely directly investigated but comparisons across studies provide relevant information.

Comparing the effects of kindling in different sites on spatial cognition can provide a basis for differentiating alternative accounts of the bases for these effects. If the disruptive effects of kindling on memory are mediated by characteristic changes in circuits involved in seizure propagation, then the effects of kindling on memory should be similar regardless of which site serves as the seizure focus. Alternatively, if the disruptive effects are mediated by changes local to the seizure focus, then the effects of kindling on memory depend on which site is being kindled and could reflect differences in the contributions of different structures to memory. Studies discussed thus far demonstrating impaired spatial memory after kindling have all used stimulation sites in the dHPC, predominantly in the CA1 region.[33-42] In other studies, kindling in a variety of other brain sites, including the ventral hippocampus, perforant path, perirhinal cortex, amygdala, and olfactory bulb, has been shown to spare spatial memory,[43, 50-54] except in those cases where massive numbers of kindling stimulations are administered.[31, 43] We have investigated the importance of site of kindling by comparing the effects of kindling in the dHPC, amygdala, and perirhinal cortex on identical batteries of tasks. Controlling for testing conditions is important because null findings in other studies can be difficult to interpret in light of the relatively small effect even dHPC kindling has on spatial memory. Thus, spared performance after kindling in another site could be interpreted as a lack of task sensitivity or statistical power in the absence of evidence that dHPC kindling impairs performance under identical testing circumstances. Using this approach we have found that perirhinal cortex kindling spares both acquisition and retention of a constant platform location in the standard maze[50] under conditions where dHPC kindling impairs acquisition though not retention.[35] Similarly, in another series of studies, we have found that both perirhinal cortex and amygdala kindling spare spatial working memory in a MWM delayed-match-to-place task (see **Figure 1**) that is impaired by dHPC kindling.[34] Collectively, the above observations indicate that a kindling site in the dHPC is more effective in disrupting spatial memory than kindling in other sites, which suggests that changes within this brain area may be critical in mediating such effects.

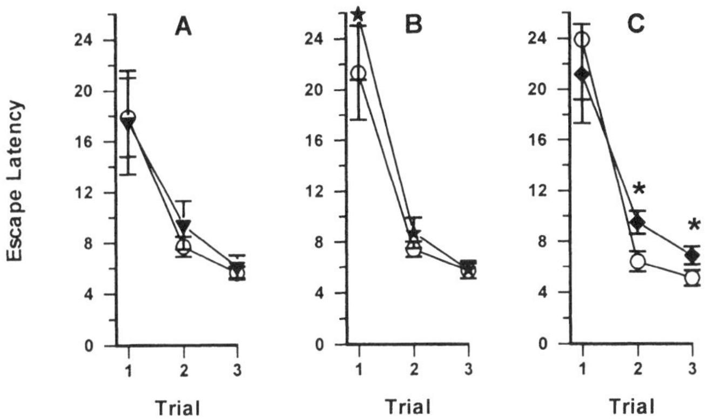

Figure 1. Latency to escape to a hidden platform in a delayed-match-to-place task after perirhinal cortex (A), amygdala (B), or dHPC (C) kindling. Data are averaged across 6 test sessions (2 each at delays of 15, 60, and 240 sec). Open circles are controls. * $p < .05$ relative to controls

The interval between kindling and behavioural testing could also be an important factor in modulating the impairment and has the potential to substantially limit prospective underlying mechanisms since many of the effects of kindling vary in their persistence following kindling. Although this variable has not been directly investigated, two important points about the interval between the last kindling stimulation and effects on memory testing can be made by comparing results across studies. First, the effects of dHPC kindling on spatial memory are not simply an acute after-effect of seizures per se because, in most studies, behavioral testing has been started 1 day to 1 week after the last seizure.[33-42] Second, these effects are very long-lasting if not permanent. Impairments have been observed during testing between 2 and 3 weeks after the last seziure[34, 40] and may persist even longer, but suitably designed studies testing performance at longer intervals following kindling and in the absence of the confounding effects of intervening learning opportunities have not been performed.

The extent of kindling achieved before behavioral testing may also be important. Kindling is an inherently progressive phenomenon and can be terminated at different stages, defined by the severity of the final seizures elicited. When kindling is stopped prior to the appearance of fully generalized seizures (i.e., stage 4 or 5 seizures)[55] it is often referred to as partial kindling whereas if it is stopped after the appearance of a few fully generalized seizures, it is often referred to as full kindling. Full dHPC kindling reliably impairs spatial acquisition in several Morris water maze tasks[33-36] whereas partial kindling spares acquisition in both the radial-arm maze and Morris water maze.[33, 38, 42] Partial dHPC kindling has been shown to disrupt spatial memory but only in those cases where some training preceded kindling thus allowing for the possibility that the disruption resulted from seizure-mediated retrograde amnesia rather than an anterograde disruption of spatial learning.[36, 38, 42] However, even in these cases, an equal or larger impairment is seen after full dHPC kindling.[33, 37] These observations suggest that *full* dHPC kindling is required to impair spatial memory or at least produces a greater impairment than partial dHPC kindling. Extended kindling (i.e., kindling in which 10 or more generalized seizures are evoked) is also effective in disrupting spatial task performance but does not require a kindling site in the dHPC[31, 43]. However, effects that require extended kindling should be viewed cautiously in terms of their specificity to spatial cognition since non-mnemonic behavioural changes are likely to be more pronounced following extended kindling,[29] and the possible contributions of such changes to spatial task performance following extended kindling have not been systematically considered. Nonetheless, these observations suggest that an interesting interaction between the site and extent of kindling may exist such that kindling in a sensitive region (e.g., the dHPC) requires only full kindling whereas kindling in less sensitive sites (e.g., the amygdala, perforant path, or olfactory bulbs) requires extended kindling to produce an anterograde spatial deficit.

An alternative way to think of extent of kindling is in terms of the number of sites from which kindled seizures have been elicited. We have recently investigated the importance of kindling extent conceived in this manner by comparing the effects of dHPC kindling, lateral entorhinal cortex kindling and bifocal kindling of both sites (i.e., entorhinal kindling followed by dHPC kindling) on acquisition and retention of spatial information in the Morris water maze (see [35] for a similar testing protocol). Our preliminary results indicate that whereas dHPC kindling disrupts acquisition but not retention, entorhinal cortex kindling disrupts retention but not acquisition (see **Figure 2**). Interestingly, bifocal kindling appears to superimpose the effects of kindling in each site by producing a clear deficit in retention and trend a towards impaired acquisition (see **Figure 2**). A milder effect of bifocal kindling relative to dHPC kindling on acquisition might be accounted for by the fact that dHPC kindling involves 40 or more dHPC stimulations whereas our bifocal kindling protocol involved between 10 and 15, a possibility which could suggest that the number of dHPC stimulations administered is a critical variable in addition to the severity of the convulsions elicited. These findings also demonstrate that kindling in different sites can selectively affect different phases of spatial memory and suggests that variability in seizure focus location could contribute to heterogeneity in memory impairments reported in epileptic patients.[11]

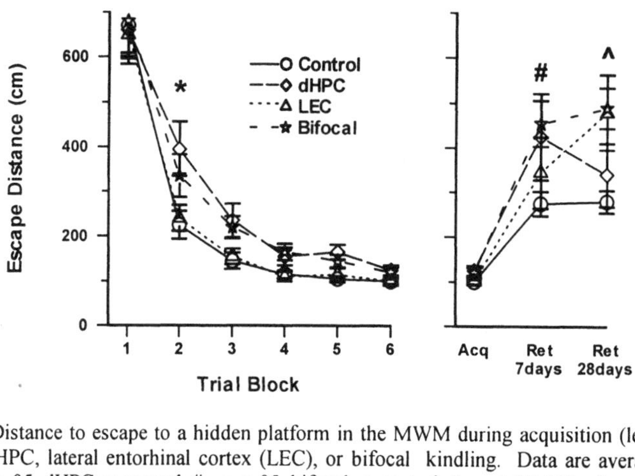

Figure 2. Distance to escape to a hidden platform in the MWM during acquisition (left) and retention (right) following dHPC, lateral entorhinal cortex (LEC), or bifocal kindling. Data are averaged across blocks of 3 trials. * - $p < .05$, dHPC vs control, # - $p < .05$, bifocal vs control, ^ - $p < .05$, bifocal vs control, LEC vs control

In review, our and others' work investigating the effects of kindling on spatial memory has lead to the following conclusions. DHPC kindling disrupts performance on a range of spatial tasks. This deficit is specific to spatial memory because the same kindling protocols that disrupt spatial task performance spare sensorimotor, motivational, emotional, and non-spatial cognitive functions as shown by unimpaired performance on a range of non-spatial tasks. The deficit in spatial memory is manifest in a disruption of the earliest phases of spatial memory processing as indicated by a deficit in within-day acquisition but not long-term retention of a platform location in the MWM and the lack of delay-dependence of the deficit in a variable-delay spatial working memory task. Finally, this deficit is preferentially induced using a dHPC kindling site, is long-lasting for at least two weeks following the completion of kindling, and generally increases with extent of kindling although this relation exhibits considerable complexity.

Considerations, then, of the nature of the effect of kindling on spatial cognition and the relation shown between this effect and kindling-related parameters can be used as a guide to identify prospective mediating mechanisms. For example, one can deduce that attractive candidate mechanisms should make a significant contribution to spatial learning/STM, should be reliably affected by kindling, and should exhibit a similar relation to the kindling-related parameters as the deficit itself.

2.3. Prospective mechanisms

Although the precise mechanisms mediating spatial memory are not definitively established, one of the most widely held views is that hippocampal synaptic plasticity plays a key role.[56, 57] According to this view, then, learning depends on mechanisms that mediate the induction and early phases of experience-dependent synaptic plasticity and memory depends on mechanisms that maintain these changes over different time intervals.[57] Given this postulated role for intra-hippocampal synaptic plasticity, we hypothesize that kindling impairs spatial learning by disrupting capacity for induction or early phase maintenance of intra-hippocampal synaptic plasticity, which results from kindling's effects on the levels or activity of one or more molecules normally involved in the signalling cascades mediating experience dependent synaptic change. In other words, we believe kindling represents an enduring form of metaplasticity,[58] which results in a shift away from the optimal "settings" for intra-hippocampal plasticity, and thereby results in alterations of functions that depend on this plasticity, such as spatial learning.

Available evidence supports the possibility that kindling may lead to subsequent changes in capacity for hippocampal plasticity as assayed by long-term potentiation (LTP) and long-term depression. In a recent study, partial dHPC kindling has been shown to impair in vitro induction of LTP using primed burst stimulation in the CA1 region of the dHPC.[59] In other studies, full amygdala kindling has been shown to impair subsequent capacity for both LTP[60] and long-term depression[61] in intra-amygdaloid pathways. Indirect evidence that dHPC kindling may disrupt subsequent capacity for hippocampal plasticity comes from observations that kindling leads to alterations in several molecules that have been implicated in LTP (e.g., [62-64]). It is also interesting to note that disturbances in hippocampal plasticity, which are associated with learning impairments, are seen in other rodent models of epilepsy[65, 66] and have in fact also been demonstrated in tissue resected from TLE patients with a hippocampal seizure focus[67]. Our current research, then, is aimed at further investigating whether kindling leads to changes in subsequent capacity for plasticity, which molecular effects mediate these changes, and how these changes impact memory and perhaps epileptogenesis itself.

3. REFERENCES

1. W.A. Hauser, J.F. Annegers, and L.T. Kurland, Prevalence of epilepsy in Rochester, Minnesota: 1940-1980, *Epilepsia* **32**, 429-45 (1991).
2. O. Devinsky, Interictal behavioral changes in epilepsy, in: *Epilepsy and Behavior*, edited by O. Devinsky and W. H. Theodore (New York; Wiley-Liss, 1991) pp. 1-21.
3. O. Devinsky, Therapy for neurobehavioral disorders in epilepsy, *Epilepsia* **45** (Suppl 2), 34-40 (1994).
4. F. Gilliam, H. Hecimovic, and Y. Sheline, Psychiatric comorbidity, health, and function in epilepsy, *Epilepsy Res*, **4** (Suppl 4), S26-30 (2003).
5. B.P. Hermann, M. Seidenberg, J. Schoenfeld, and K. Davies, Neuropsychological characteristics of the syndrome of mesial temporal lobe epilepsy, *Arch. Neurol.* **54**, 369-376 (1997).
6. B. Pedersen and M. Dam, Memory disturbances in epileptic patients, *Acta Neurol. Scand.* **74** (suppl. 109), 11-14 (1986).
7. A. Gupta and T. Naorem, Cognitive retraining in epilepsy, *Brain Inj.* **17**, 161-74 (2003).
8. P. Moore and G. Baker, The neuropsychological and emoptional consequences of living with intractable temporal lobe epilepsy: implications for clinical management, *Seizure* **11**, 224-30 (2002).
9. N. H. Engelberts, M. Klein, H. M. van der Ploeg, J. J. Heimans, H. J. Ader, M. P. van Boxtel, J. Jolles, D. G. Kasteleijn-Nolst Trenite, Cognition and health-related quality of life in a well-defined subgroup of patients with partial epilepsy, *J. Neurol.* **249**, 294-9 (2002).
10. K. Perrine and T. Kiolbasa, Cognitive deficits in epilepsy and contribution to psychopathology, *Neurol*ogy **53,** S39-S48 (1999).
11. J. J. Bortz, Neuropsychiatric and memory issues in epilepsy, *Mayo Clinic Proc.* **78**, 781-7 (2003).
12. N. Dige and G. Wik, A visual learning and memory test for preoperative evaluation of patients with temporal lobe epilepsy, *Int. J. Neurosci.* **111**, 167-74 (2001).
13. R. Schwarcz and M. P. Witter, Memory impairment in temporal lobe epilepsy: the role of entorhinal lesions, *Epilepsy Res.* **50**, 161-77 (2002).
14. M. Aikia, R. Kalviainen, and P. J. Riekkinen, Verbal learning and memory in newly diagnosed partial epilepsy, *Epilepsy Res.* **22**, 157-164 (1995).
15. R.L. Billingsley, M.P. McAndrews, and M.L. Smith, Intact perceptual and conceptual priming in temporal lobe epilepsy: neuroanatomical and methodlogical implications, *Neuropscyhologia* **16**, 92-101 (2002).
16. C. Helmstaedter, Effects of chronic epilepsy on declarative memory systems, *Prog. Brain Res.* **135**, 439-53 (2002).
17. A. Rahmann, S. Stodieck, I. W. Hustedt, and S. Evers, Pre-attentive cognitive processing in epilepsy. A pilot study on the impact of epilepsy type and anti-epileptic treatment, *Eur. Neurol.* **48**, 146-52 (2002).
18. A.R. Giovagnoli, S. Mascheroni, and G. Avanzini, Self-reporting of everyday memory in patients with epilepsy: relation to neuropsychological, clinical, pathological and treatment factors, *Epilepsy Res.* **28**, 119-128 (1997).
19. P. J. Thompson and R. Corcoran, Everyday memory failures in people with epilepsy, *Epilepsia* **33**, S18-20 (1992).
20. A. Grippo, L. Pelosi, V. Mehta, and L. D. Blumhardt, Working memory in temporal lobe epilepsy: an event-related potential study, *Eletroencephalogr. Clin. Neurophysiol.* **99**, 200-213 (1996).
21. R. C. Martin, S. Sawrie, J. Hugg, F. Gilliam, E. Faught, and R. Kuzniecky, Cognitive correlates of 1H MRSI-detected hippocampal abnormalities in temporal lobe epilepsy, *Neurology* **53**, 2052-2058 (1999).
22. S. M. Sawrie, R. C. Martin, R. E. Faught, B. Maton, J. W. Hugg, and R. I. Kuzniecky, Nonlinear trends in hippocampal metabolic function and verbal memory: Evidence of cognitive reserve in temporal lobe epilepsy?, *Epilepsy Behav.* **1**, 106-111 (2000).
23. S. M. Sawrie, R. C. Martin, R. Knowlton, E. Faught, F. Gilliam, and R. Kuzniecky, Relationships among hippocampal volumetry, proton magnetic spectroscopy, and verbal memory in temporal lobe epilepsy, *Epilepsia* **42**, 1403-1407 (2001).
24. M. E. Corcoran and S. L. Moshé, eds., *Kindling 5* (Plenum Press, New York, 1998).
25. M. Sato, R. J. Racine, and D. C. McIntyre, Kindling: Basic mechanisms and clinical validity, *Eletroencephalogr. Clin. Neurophysiol.* **76**, 459-472 (1990).
26 R. Adamec and B. Young, Neuroplasticity in specific limbic system circuits may mediate specific kindling induced changes in animal affect-implications for understanding anxiety associated with epilepsy,

Neurosci. Biobehav. Rev. **24**, 705-723 (2000).
27. H. E. Edwards, N. J. MacLusky, and W. M. Burnham, The effect of seizures and kindling on reproductive hormones in the rat, *Neurosci. Biobehav. Rev.* **24**, 753-762 (2000).
28. D.K. Hannesson and M. E. Corcoran, The mnemonic effects of kindling, *Neurosci. Biobehav. Rev.* **24**, 725-751 (2000).
29. L. E. Kalynchuk, Long-term amygdala kindling in rats as a model for the study of interictal emotionality in temporal lobe epilepsy, *Neurosci. Biobehav. Rev.* **24**, 691-704 (2000).
30. L. S. Leung, J. Ma, and R. S. McLachlan, Behaviors induced or disrupted by complex partial seizures, *Neurosci. Biobehav. Rev.* **24**, 763-775 (2000).
31. S. Cammisuli, M. P. Murphy, C. J. Ikeda-Douglas, V. Balkissoon, R. M. D. Holsinger, E. Head, M. Michael, R. J. Racine, and N. W. Milgram, Effects of extended electrical kindling on exploratory behavior and spatial learning, *Behav. Brain Res.* **89**, 179-190 (1997).
32 K. J. Feasey-Truger, L. Kargl, and G. ten Bruggencate, Differential effects of dentate kindling on working and reference spatial memory in the rat, *Neurosci. Lett.* **151**, 25-28 (1993).
33. T. H. Gilbert, R. K. McNamara, and M. E. Corcoran, Kindling of hippocampal field CA1 impairs spatial learning and retention in the Morris water maze, *Behav. Brain Res.* **82**, 57-66 (1996).
34 D. K. Hannesson, J. Howland, M. Pollock, P. Mohapel, A. E. Wallace, and M. E. Corcoran, Dorsal hippocampal kindling produces a selective and enduring disruption of hippocampally mediated behavior, *J. Neurosci.* **21**, 4443-4450 (2001).
35. D. K. Hannesson, P. Mohapel, and M. E. Corcoran, Kindling of the dorsal hippocampus selectively disrupts spatial learning/short-term memory, *Hippocampus* **11**, 275-286 (2001).
36. D. K. Hannesson, A. E. Wallace, M. Pollock, S. M. Corley, P. Mohapel, and M. E. Corcoran, The relation between extent of dorsal hippocampal kindling and DMTP performance in the Morris water maze, *Epilepsy Res.* **58**, 145-154 (2004).
37. L. S. Leung, K. A. Boon, T. Kaibara, and N. K. Innis, Radial maze performance following hippocampal kindling, *Behav. Brain Res.* **40**, 119-129 (1990).
38. L. S. Leung, D. Brzozowski, and B. Shen, Partial hippocampal kindling affects retention but not acquisition and place but not cue tasks on the radial arm maze, *Behav. Neurosci.* **110**, 1017-1024 (1996).
39. L. S. Leung and B. Shen, Hippocampal CA1 evoked response and radial 8-arm maze performance after hippocampal kindling, *Brain Res.* **555**, 353-357 (1991).
40. L. S. Leung, D. Zhao, and B. Shen, Long-lasting effects of partial hippocampal kindling on hippocampal physiology and function, *Hippocampus* **4**, 696-704 (1994).
41. F. H. Lopes da Silva, J. A. Gorter, and W. J. Wadman, Kindling of the hippocampus induces spatial memory deficits in the rat, *Neurosci. Lett.* **63**, 115-120 (1986).
42. R. J. Sutherland, L. S. Leung, M. P. Weisend, J. Schlife, and R. J. McDonald, An evaluation of the effect of partial hippocampal kindling on place navigation by rats in the Morris water task, *Psychobiology* **25**, 126-132 (1997).
43. T. Sutula, S. Lauersdorf, M. Lynch, C .Jurgella, and A. Woodard, Deficits in radial arm maze performance in kindled rats: Evidence for long-lasting memory dysfunction induced by repeated brief seizures, *J. Neurosci.* **15**, 8295-8301 (1995).
44. J. O'Keefe and L. Nadel, *The Hippocampus as a Cognitive Map* (Oxford University Press, Oxford, 1978).
45. R. D'Hooge and P. P. De Deyn, Applications of the Morris water maze in the study of learning and memory. *Brain Res. Rev.* **36**, 60-90 (2001).
46. K. Nakazawa, T. J. McHugh, M. A. Wilson, and S. Tonegawa, NMDA receptors, place cells and hippocampal spatial memory, *Nature Rev. Neurosci.* **5**, 361-372 (2004).
47. M. Shapiro, Plasticity, hippocampal place cells, and cognitive maps, *Arch. Neurol.* **58**, 874-881 (2001).
48 H. Hodges, Maze procedures: the radial-arm and water maze compared, *Cog. Brain Res.* **3**, 167-181 (1996).
49. H. Eichenbaum, The hippocampus and mechanisms of declarative memory, *Behav. Brain Res.* **103**, 123-133 (1999).
50. M. E. Corcoran, L. L. Armitage, T. H. Gilbert, D. K. Hannesson, and P. Mohapel, Kindling and spatial cognition, in: *Kindling 5*, edited by M. E. Corcoran and S. L. Moshé, (Plenum Press: New York, 1998), pp 377-393.
51. G. L. Holmes, A. Chronopoulos, C. E. Stafstrom, M. A. Mikati, S. J. Thurber, P. A. Hyde, and J. L.Thompson, Effects of kindling on subsequent learning, memory, behavior, and seizure susceptibility, *Dev. Brain Res.* **73**, 71-77 (1993).
52. R. K. McNamara, R. D. Kirkby, G. E. dePape, and M. E. Corcoran, Limbic seizures, but not kindling, reversibly impair place learning in the Morris water maze, *Behav. Brain Res.* **50**, 167-175 (1992).
53. S. A. Nieminen, J. Sirvio, K. Teittinen, A. Pitkanen, M. M. Airaksinen, and P. Riekkinen, Amygdala

kindling increased fear-response, but did not impair spatial memory in rats, *Physiol. Behav.* **51**, 845-849 (1992).
54. G. B. Robinson, H. A. McNeill, and G. D. Reed, Comparison of the short- and long-lasting effects of perforant path kindling on radial maze learning, *Behav. Neurosci.* **107**, 988-995 (1993).
55. R. J. Racine, Modification of seizure activity by electrical stimulation: II. Motor seizure, *Eletroencephalogr. Clin. Neurophysiol.* **32**, 281-294 (1972).
56. M. A. Lynch, Long-term potentiation and memory. *Physiol, Rev.* **84**, 87-136 (2004).
57. R. G. Morris, E. I. Moser, G. Riedel, S. J. Martin, J. Sandin, M. Day, and C. O'Carroll, Elements of a neurobiological theory of the hippocampus: the role of activity-dependent synaptic plasticity in memory, *Phil. Trans. Royal Soc. Lond. B Biol. Sci.* **358**, 773-786 (2003).
58. W. C. Abraham and W. P. Tate, Metaplasticity: a new vista across the field of synaptic plasticity, *Prog. Neurobiol.* **52**, 303-323 (1997).
59. L. S. Leung and C. Wu, Kindling suppresses primed-burst-induced long-term potentiation in hippocampal CA1, *NeuroReport* **14**, 211-214 (2003).
60. V. Neugebauer, N. B. Keele, and P. Shinnick-Gallagher, Loss of long-lasting potentiation mediated by group III mGluRs in amygdala neurons in kindling-induced epileptogenesis. *J. Neurophys.* **78**, 3475-3478 (1997).
61. S. J. Wang and P. W. Gean, Long-term depression of excitatory synaptic transmission in the rat amygdale, *J. Neurosci.* **19**, 10656-10663 (1999).
62. K. Akiyama, M. Ono, I. Kohira, A. Daigen, T. Ishihara, and S. Kuroda, Long-lasting increase in protein kinase C activity in the hippocampus of amygdala-kindled rat, *Brain Res.* **679**, 212-220 (1995).
63. J. M. Bronstein, D. B. Farber, and C. G. Wasterlain, Regulation of type-II calmodulin kinase: functional implications, *Brain Res. Rev.* **18**, 135-147 (1993).
64. B. S. Meldrum, M. T. Akbar, and A. G. Chapman, Glutamate receptors and transporters in genetic and acquired models of epilepsy, *Epilepsy Res.* **36**, 189-204 (1999).
65. M. Krug, M. Koch, G. Grecksch, and K. Schulzeck, Pentylenetetrazol kindling changes the ability to induce potentiation phenomena in the hippocampal CA1 region, *Physiol. Behav.* **62**, 721-727 (1997).
66. M. Lynch, U. Sayin, J. Bownds, S. Janumpalli, and T. Sutula, Long-term consequences of early postnatal seizures on hippocampal learning and plasticity, *Eur. J. Neurosci.* **12**, 2252-2264 (2000).
67. H. Beck, I. V. Goussakov, A. Lie, C. Helmstaedter, and C. E. Elger, Synaptic plasticity in the human dentate gyrus, *J. Neurosci.*, **20**, 7080-7086 (2000).

Discussion

Gale: Congratulations on a very thorough analysis of so many components that are so rarely delved into, particularly in terms of the behaviour. What I was most impressed with was your effort to sort out the learning components and the processing components versus the memory components. I was impressed that you did not get a delay-dependent effect. So I was confused by your conclusion that it was a deficit in spatial memory when, in fact, it seems to me that you ruled out spatial memory and instead identified some sort of processing deficit. In that context, I was wondering if you looked at contextual cue processing in general, space being just one example of that, and temporal processing being another example. Did you look at a simple spatial alternation task, either a rewarded spatial alternation task or spontaneous alternation task, to see if these animals have a deficit?

Hannesson: No, we haven't looked at that. Our prediction is that what we have is a general deficit in hippocampal-dependent functions, and I would expect an impairment on those sorts of things.

Leung: Very comprehensive study, very nice. I still don't understand why there can be such general spread of AD to field CA1 as well in perforant path or in your lateral entorhinal kindling, and yet there seems to be some difference if you stimulate dorsal hippocampus versus elsewhere.

Hannesson: Yes, it is a very curious finding that you can stimulate the ventral hippocampus, the perforant path, or the entorhinal cortex, which we think of as a very tightly integrated set of circuitry, yet we get these different effects. There does seem to be something special that goes on when your electrode is in the dorsal hippocampus, and it may relate to changes in AD threshold that you don't get when you kindle with an electrode elsewhere.

A POTENTIAL ROLE FOR THE HIPPOCAMPUS IN THE EXPRESSION OF KINDLING-INDUCED FEAR

Lisa E. Kalynchuk and Amanda J. Wintink*

1. INTRODUCTION

A number of interictal behavioral co-morbidities accompany temporal lobe epilepsy. These co-morbidities range from fear and anxiety to depression and memory loss. For many patients, these behavioral disturbances pose a greater problem than the seizures themselves, because they can disable patients to the point where they cannot work or sustain normal relationships. Surprisingly, relative to the seizures themselves, these behavioral problems have received little experimental attention.

The focus of this paper is to review our recent research using the kindling model to study the interictal fear and anxiety associated with temporal lobe epilepsy. Kindling provides a particularly useful way to study these behavioral problems because several important variables, such as the site of stimulation, the number of seizures, and the time between the final seizure and behavioral assessment can be carefully controlled.

2. FEARFUL BEHAVIOR PRODUCED BY KINDLING

Adamec[1] first reported the effect of kindling on fearful behavior. He found that partial amygdala or hippocampal kindling in cats increased defensive responding to rats, mice, or threat vocalizations from other cats. Soon after, Pinel and his colleagues[2] reported that long-term kindling of either the amygdala or hippocampus increased resistance to capture fear behavior in rats. Since these seminal experiments, the anxiogenic effects of kindling in rats have been well studied.[3, 4] For example, kindling has

---

* Lisa E. Kalynchuk, Department of Psychology, University of Saskatchewan, Saskatoon, SK, Canada.
Amanda J. Wintink, Department of Psychology, Dalhousie University, Halifax, NS, Canada.

been shown to decrease open-arm exploration on an elevated plus maze,[5-7] increase immobility in a social interaction test[6], increase fear-potentiated startle,[8] increase corticotrophin-releasing factor-induced defensive fighting,[9] and increase stress-induced stomach ulcers.[10] Furthermore, in our laboratory we have found that kindling increases acoustic startle responses, increases escape behavior from an elevated plus maze, and increases active defensive responses in a resident-intruder paradigm and social interaction paradigm.[11-14] We have also found that kindling produces intriguing changes in open field behavior. When placed into an unfamiliar open field, kindled rats show an initial freezing response, followed by hyper-locomotion and escape-related behaviors.[15, 16] At the end of the open field session, kindled rats are extremely resistant to capture from the open field, often engaging in fleeing, biting and defensive jump attack behaviors.[12, 13, 15]

There are several important parameters that affect the level of fear produced by kindling. First, the magnitude of fearful behavior depends on the site of stimulation. We have assessed fearful behavior in rats that have been kindled in a number of different brain regions including the basolateral amygdala, bed nucleus of the stria terminalis (BNST), perirhinal cortex, ventral hippocampus, claustrum, and caudate nucleus. In each case, the rats received 99 stimulations prior to behavioral testing--three suprathreshold stimulations per day (i.e., 800 μA peak-to-peak square wave pulses delivered at a frequency of 60 HZ over 1 second), 5 days per week, for 33 days. One day after the final stimulation, each rat was placed into an unfamiliar open field for 5 min. After the 5 min, an experimenter wearing a leather glove that was unfamiliar to the rat forcefully picked up the rat from above. The rat's resistance to being picked up was scored according to the following 7-point scale, adapted from Albert and Richmond[17]: 0 = easy to pick up, 1 = vocalizes or shies away from hand, 2 = shies away from hand and vocalizes, 3 = runs away from hand, 4 = runs away and vocalizes, 5 = bites or attempts to bite, 6 = launches a defensive jump attack. As shown in Figure 1, amygdala and BNST kindling produce the greatest increases in resistance to capture, hippocampal and perirhinal cortex kindling produce intermediate increases in resistance to capture, and caudate and claustrum kindling produce no significant increases in resistance to capture (data from [14, 18]). Based on these results, most of our subsequent studies have focused on amygdala kindling.

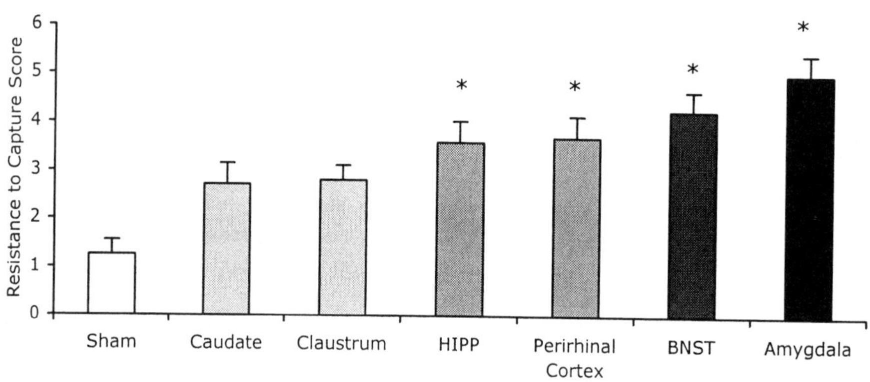

Figure 1. The mean (± S.E.M.) resistance to capture shown by rats after 99 stimulations of several different brain sites. * Significantly different from sham.

Second, the magnitude of fearful behavior produced by amygdala kindling depends on how many stimulations the rats receive. We have studied the behavioral effects of different numbers of amygdala kindling stimulations, ranging from 20 to 100 suprathreshold stimulations. As described above, the rats were tested for their resistance to capture from an unfamiliar open field one day after the final kindling stimulation. Representative data from these experiments are shown in Figure 2, and clearly show that amygdala-kindled rats display a linear increase in resistance to capture as the number of stimulations increases (data from [12, 15, 19, 20]).

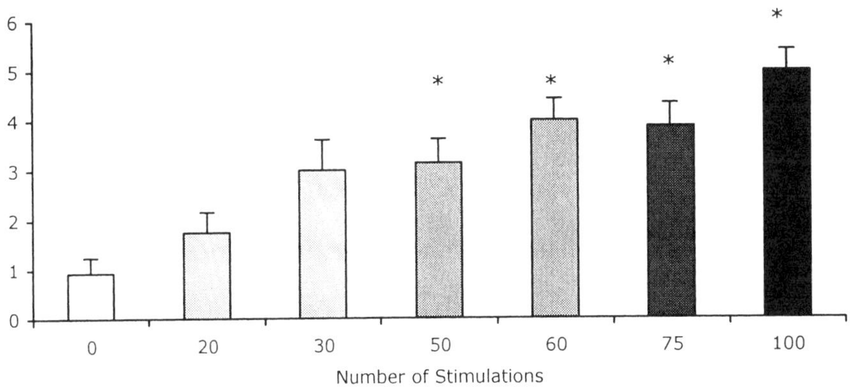

Figure 2. The mean (+ S.E.M.) resistance to capture shown by rats after different numbers of amygdala-kindling stimulations. *Significantly different from 0.

Third, the magnitude of fearful behavior produced by amygdala kindling depends on the novelty of the testing situation. We have found that rats subjected to 99 amygdala kindling stimulations are highly resistant to capture after 5 min in a novel open field, reacting on average by biting the hand of the experimenter and launching defensive jump attacks at the hand. However, these same rats are easy to pick up from their home cages or from an open field that is familiar to them (data from [13]). These data are shown in Figure 3. Similarly, we have noticed that kindled rats remain fearful following a resistance-to-capture test until they are returned to their home environment, at which time they can easily be picked up and weighed (unpublished observations). The fact that novelty is a critical factor for the expression of kindling-induced fear explains why kindled rats do not show major increases in emotional behavior during their daily kindling stimulations.[21]

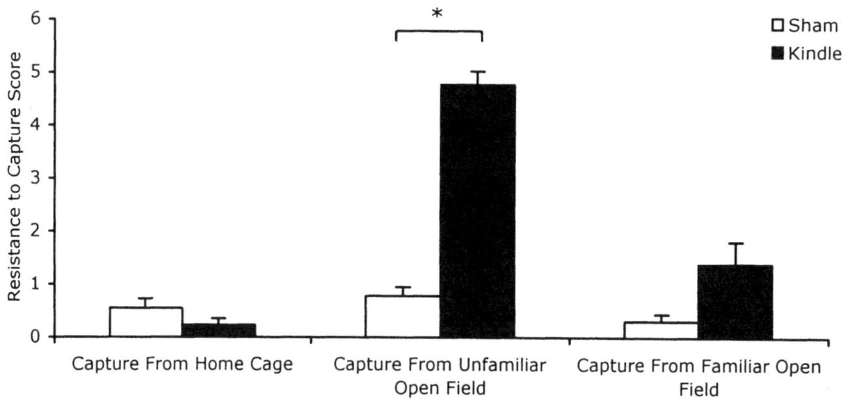

Figure 3. The mean (± S.E.M.) resistance to capture shown by rats tested under different conditions. The kindled rats showed significant increases in resistance to capture from an unfamiliar open field, but not from their home cages or from a familiar open field. *Significantly different from sham.

3. NEURAL MECHANISMS OF KINDLING-INDUCED FEAR

A central hypothesis that is driving our present work is that hippocampal dysfunction could be mediating the fearful behavior produced by kindling. Two pieces of evidence led to this idea. First, in a series of receptor binding studies, we noticed that fearful behavior in kindled rats is associated with significant alterations in a number of receptor subtypes only in the granule cell layer of the dentate gyrus.[4, 20, 22] We also noticed that the pattern of these results suggested that kindling creates an imbalance of inhibitory and excitatory receptor expression in this region. For example, 5-HT_{1A}, benzodiazepine, and GABA_A receptor binding are all significantly increased in the dentate gyrus of fearful kindled rats, whereas NMDA and AMPA receptor binding are significantly decreased.[22,23] The consequence of this receptor imbalance for neural activity is unknown, but one possibility is that it produces an inhibitory tone within hippocampal circuits. In support of this idea, Gutierrez and Heinemann[24] found that kindling increases inhibition within the dentate gyrus-CA3 pathway for at least 1 month following the last kindling stimulation. In addition, Adamec[25,26] has shown that kindling-induced defensive behavior in cats is associated with increased recurrent inhibition in the ventral hippocampus, and also that this defensive behavior is reduced by administration of a BZ receptor antagonist. Another piece of evidence implicating the hippocampus in the fearful behavior produced by kindling was that we found that FOS immunoreactivity is significantly decreased in the dentate gyrus and CA1 region of kindled rats exposed to a fear-provoking stimulus, compared to control rats exposed to the same stimulus.[16] A decrease in FOS expression is difficult to interpret, but it may indicate that exposure to an unfamiliar open field produces less hippocampal activation in kindled rats than in control rats. Taken together, these findings suggested to us that kindling might impair the capacity of the hippocampus to show adaptive plasticity, perhaps via changes in many of the receptors mentioned above.

How might decreased neuroplasticity within hippocampal circuits serve to increase fearful behavior in kindled rats? Novelty is a critical trigger for the expression of fear in long-term kindled rats: As shown in Figure 3, they are much more fearful when tested in unfamiliar situations than when tested in familiar situations.[13] The hippocampus is involved in the processing of contextual information[27] and the determination of whether a particular context requires a fearful or anxious response.[28] It also mediates the hypothalamic-pituitary-adrenal axis response to stress. That is, hippocampal glucocorticoid receptors modulate the intensity and duration of a stress response by regulating negative feedback to the hypothalamus.[29] Our data suggest that long-term amygdala kindling creates a dysfunctional hippocampus. Accordingly, when kindled rats are exposed to a novel stimulus, they may be unable to properly process contextual information about that stimulus, and unable to shut off the stress response associated with exposure to that stimulus. As a result, they respond to the stimulus in a highly emotional manner. Importantly, fearful behavior in kindled rats does appear to be related to decreased glucocorticoid receptor mRNA in the CA1 region and dentate gyrus, suggesting that stress responses may be prolonged in kindled rats.[20]

4. EFFECT OF LONG-TERM KINDLING ON A HIPPOCAMPAL-DEPENDENT TASK

One way to test our idea that long-term kindling produces hippocampal dysfunction is to assess the ability of kindled rats to perform a hippocampal-dependent spatial memory task. Past studies have revealed that amygdala-kindled rats are generally spared from deficit on spatial tasks (for a review see [30]). However, with one exception, all these studies presented data from rats tested after short-term kindling (e.g., [7, 31]). The one exception reported that amygdala-kindled rats subjected to 300 stimulations do show impaired place learning in the Morris water maze.[32] These findings suggest that long-term kindling can interfere with the acquisition of spatial information, but that this effect might depend on the number of kindling stimulations, similar to the development of kindled fear behavior.

To determine if and when amygdala-kindled rats begin to show deficits in spatial memory, we designed an experiment to repeatedly assess spatial memory during the course of kindling. We used a delayed-matching-to-place (DMTP) version of the water maze, which cannot be successfully performed by rats with hippocampal lesions.[33] The procedure was as follows. Each testing day comprised five trials. There was a 1-hour delay between trial 1 (i.e., sample trial) and trial 2 (i.e., first matching trial). The task requirements were the same from day to day, but the platform location was changed on each day. Performance was measured by the latency to find the hidden platform on trial 2. A reduction in latency to find the platform on trial 2 reflects the rats' memory of the platform location learned on trial 1. Rats were pretrained to perform the task in order to familiarize them with the testing environment and the spatial arrangement of the room. After pretraining, the rats were tested once per week throughout the 6-week kindling phase. That is, during each week of the kindling phase, the rats received kindling or sham stimulations for five consecutive days (i.e., Monday to Friday), followed by a day off (i.e., day 6), followed by water maze testing (i.e., day 7).

Our results showed that the amygdala-kindled rats were impaired on the task on weeks 3-5 of kindling (data from [34]). This is shown in Figure 4. Given that the rats received 15 stimulations per week, these findings suggest that at least 45 amygdala-kindling stimulations are required to produce spatial memory deficits. This finding supports our hypothesis that long-term kindling produces a dysfunctional hippocampus, although the fact that the kindled rats performed normally on week 6 is curious and will be further investigated.

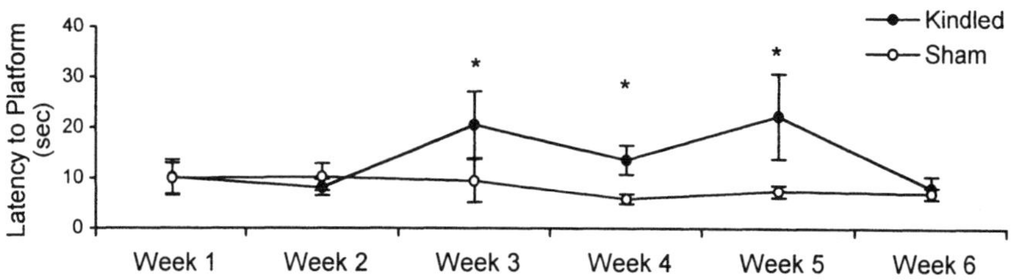

Figure 4. The mean (± S.E.M.) latency to find the hidden platform on trial 2 shown by the kindled (●) and sham-stimulated (Sham; O) rats on each week of water maze testing (Week 1 – 6). The kindled rats took significantly longer to reach the platform in Weeks 3 –5.

5. EFFECT OF ENVIRONMENTAL ENRICHMENT ON KINDLING-INDUCED FEAR

The idea that kindling-induced fear may arise as a consequence of hippocampal dysfunction has implications for the design of new treatment strategies for reducing interictal anxiety in individuals with epilepsy. For example, if the hippocampus is inhibited or less capable of adaptive plasticity as a result of kindling, then increasing hippocampal activity might reduce the fearful behaviour seen in kindled rats. We recently tested this idea by assessing whether exposure to an enriched environment during the course of kindling would reduce the magnitude of kindling-induced fear. Environmental enrichment is usually beneficial for an animal, as it facilitates performance on spatial memory tasks[35] and enhances neural and functional recovery following brain injury.[36] These benefits are thought to occur as a result of enrichment-induced changes within the hippocampus and cortex, such as increased neurotrophin levels,[35] increased neurogenesis,[37] and increased dendritic branching and spine density.[38] In our experiment, one group of rats received 99 sham stimulations and two groups of rats received 99 amygdala kindling stimulations. All the rats were housed in isolation, as per usual for our experiments, but one group of kindled rats was exposed to an enriched environment for 30-60 min per day, 6 days per week. One day after the final kindling stimulation, all the rats were tested for their resistance to capture from an unfamiliar open field. The kindling-no enrichment rats showed very high resistance to capture and the sham stimulated rats showed very little resistance to capture. However as we predicted, the kindling-enriched rats showed significantly less resistance to capture than the kindling-nonenriched rats (data from [39]). These results are shown in Figure 5. Thus, enrichment served to decrease the level of kindling-induced fear. It remains to be determined whether this protective effect is mediated by neural changes within the hippocampus, but these results do provide indirect evidence that targeting the hippocampus in a specific way may be an important new approach for treating interictal anxiety in individuals with epilepsy.

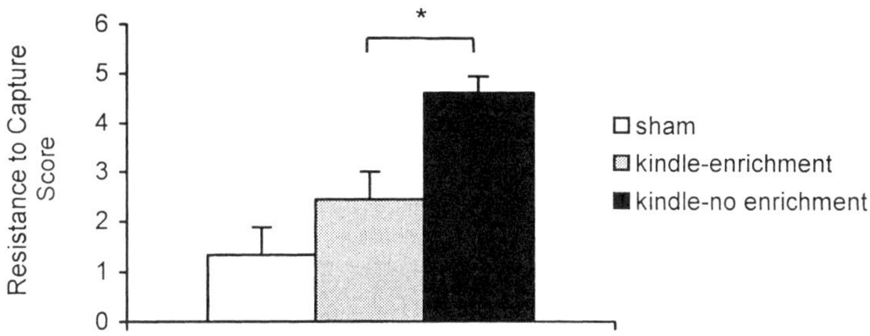

Figure 5. Effect of exposure to environmental enrichment on kindling-induced fear, expressed as the mean (± S.E.M) resistance to capture shown by kindled rats exposed to enrichment, kindled rats not exposed to enrichment, and sham-stimulated rats not exposed to enrichment during the course of kindling. The important finding is that the kindled-enrichment rats were significantly less resistant to capture than the kindled-no enrichment rats.

6. SUMMARY AND CONCLUSIONS

Kindling produces dramatic increases in fearful behavior in rats. This fear behavior depends on the site of stimulation, the number of stimulations, and the environment in which the rats are tested. Our evidence suggests that kindling-induced fear may arise as a result of neural changes that impair the normal functioning of the hippocampus. In support of this idea, kindled rats were found to be impaired on a hippocampal-dependent spatial memory task. Furthermore, exposure to environmental enrichment, which promotes hippocampal plasticity, was found to reduce the magnitude of fear behavior in kindled rats. Thus, the hippocampus may be an important target for developing new non-pharmacological interventions that decrease the behavioral co-morbidities associated with seizure occurrence.

7. ACKNOWLEDGMENTS

The Natural Sciences and Engineering Research Council of Canada (NSERC) funded this research. Lisa Kalynchuk is a recipient of a Canada Research Chair and Amanda Wintink is a recipient of a Nova Scotia Health Research Foundation postgraduate scholarship.

8. REFERENCES

1. R. E. Adamec, Behavioural and epileptic determinants of predatory attack behaviour in the cat, in *Kindling*, edited by J. A. Wada (Raven Press, New York, 1976), pp. 135-154.
2. J. P. J. Pinel, D. Treit, and L. Rovner, Temporal lobe aggression in rats, *Science* **197**, 1088-1089 (1977).
3. R. E. Adamec, Does kindling model anything clinically relevant? *Biol. Psychiatry* **27**, 249-279 (1990).
4. L. E. Kalynchuk, Long-term amygdala kindling in rats as a model for the study of interictal emotionality in temporal lobe epilepsy, *Neurosci. Biobehav. Rev.* **24**, 691-704 (2000).
5. R. E. Adamec and H. D. Morgan, The effect of kindling of different nuclei in the left and right amygdala on anxiety in the rat, *Physiol. Behav.* **55**, 1-12 (1994).

6. V. Helfer, C. Deransart, C. Marescaux, and A. Depaulis, Amygdala kindling in the rat: Anxiogenic-like consequences, *Neurosci.* **73**, 971-978 (1996).
7. S. A. Nieminen, J. Sirvio, K. Teittinen, A. Pitkanen, M. M. Airaksinen. and P. Riekkinen, Amygdala kindling increased fear-response, but did not impair spatial memory in rats, *Physiol. Behav.* **51**, 845-849 (1992).
8. J. B. Rosen, E. Hamerman, M. Sitcoske, J. R. Glowa, and J. Schulkin, Hyperexcitability: Exaggerated fear-potentiated startle produced by partial amygdala kindling, *Behav. Neurosci.* **110**, 43-50 (1996).
9. S. R. B. Weiss, R. M. Post, P. W. Gold, G. Chrousos, T. L. Sullivan, D. Walker, and A Pert, CRF-induced seizures and behavior: Interaction with amygdala kindling, *Brain Res.* **372**, 345-351 (1986).
10. P. G. Henke and R. M. Sullivan, Kindling in the amygdala and susceptibility to stress ulcers, *Brain Res. Bull.* **400**, 360-364 (1985).
11. J. Menard, M. J. Meaney, and L. E. Kalynchuk, Differential effect of long-term amygdala kindling on acoustic and fear-potentiated startle, *Soc. for Neurosci. Abs.* **28** (2002)
12. L. E. Kalynchuk, J. P. J. Pinel, D. Treit, and T. E. Kippin, Changes in emotional behavior produced by long-term amygdala kindling in rats, *Biol. Psychiatry* **41**, 438-451 (1997).
13. L. E. Kalynchuk, J. P. J. Pinel, and D. Treit, Characterization of the defensive nature of kindling-induced emotionality in rats, *Behav. Neurosci.* **113**, 766-775 (1999).
14. A. C. Davis, V. Thorne, A. Gregus, A. J. Wintink, and L. E. Kalynchuk, Differential effects of kindling different brain regions on kindling-induced emotionality, *Soc. for Neurosci. Abs.* **28** (2002).
15. A. J. Wintink, N. A. Young, A. C. Davis, A. Gregus, and L. E. Kalynchuk, Kindling-induced emotional behavior in male and female rats, *Behav. Neurosci.* **117**, 632-640 (2003).
16. L. E. Kalynchuk, A. C. Davis, A. Gregus, J. Taggart, C. C. Dodd, A. J. Wintink, and E. G. Marchant, Hippocampal involvement in the expression of kindling-induced fear in rats, *Neurosci. Biobehav. Rev.* **25**, 687-696 (2001).
17. D. J. Albert and S. E. Richmond, Septal hyperreactivity: A comparison of lesions within and adjacent to the septum, *Physiol. Behav.* **15**, 339-347 (1975).
18. L. E. Kalynchuk, J. P. J. Pinel, and D. Treit, Long-term kindling and interictal emotionality in rats: Effect of stimulation site, *Brain Res.* **779**, 149-157 (1998).
19. N. A. Young, A. J. Wintink, and L. E. Kalynchuk, Environmental enrichment facilitates amygdala kindling bu reduces kindling-induced fear in male rats, *Behav. Neurosci.* **118**, 1128-1133 (2004).
20. L. E. Kalynchuk and M. J. Meaney, Amygdala kindling increases fear responses and decreases glucocorticoid receptor mRNA expression in hippocampal regions, *Prog. NeuroPsychopharmacol. Biol. Psychiatry* **27**, 1225-1234 (2003).
21. D. P. Cain, Kindling and the amygdala, in *The Amygdala: Neurobiological Aspects of Emotion, Memory, and Mental Dysfunction*, edited by J.P. Aggleton (Wiley-Liss, New York, 1992), pp. 539-560.
22. L. E. Kalynchuk, D. M. Pearson, J. P. J. Pinel, and M. J. Meaney, Effect of amygdala kindling on emotional behavior and benzodiazepine receptor binding in rats, *ANYAS* **877**, 737-741 (1999).
23. J. C. McEachern, L. E. Kalynchuk, T. Burgmann, J. P. J. Pinel, and C. A. Shaw, Altered regulation of NMDA receptors and protein kinase C in long-term amygdala-kindled rats, *Soc. for Neurosci. Abs.* **22**, 2096 (1996).
24. R. Gutierrez and U. Heinemann, Kindling induces transient fast inhibition in the dentate gyrus-CA3 projection. *Eur. J. Neurosci.* **7**, 1371-79 (2001).
25. R. E. Adamec, Partial limbic kindling - Brain, behaviour, and the benzodiazepine receptor, *Physiol. Behav.* **54**, 531-545 (1993).
26. R. E. Adamec and C. Stark-Adamec, Partial kindling and emotional bias in the cat: lasting after-effects of partial kindling of ventral-hippocampus. I. Behavioral changes, *Behav. Neural Biol.* **38**, 223-239 (1983).
27. S. Maren, S.G. Anagnostaras, and M. S. Fanselow, The startled seahorse: is the hippocampus necessary for contextual fear conditioning? *Trends. Cog. Sci.* **2**, 39-42 (1998).
28. N. McNaughton and J. A. Gray, Anxiolytic action on the behavioral inhibition system implies multiple types of arousal contribute to anxiety, *J. Affect. Dis.* **61**, 161-176 (2000).
29. E. R. de Kloet, E. Vreugdenhil, M. S. Oitzl, and M. Joels, Brain corticosteroid receptor balance in health and disease, *Endocrinol. Rev.* **19**, 269-301 (1998).
30. D. K. Hannesson and M. E. Corcoran, The mnemonic effects of kindling, *Neurosci. Biobehav. Rev.* **24**, 725-751 (2000).
31. R. K. McNamara, R. D. Kirkby, G. E. dePape, and M. E. Corcoran, Limbic seizures, but not kindling, reversibly impair place learning in the Morris water maze, *Behav. Brain Res.* **50**, 167-175 (1992).
32. S. Cammisuli, M. P. Murphy, C. J. Ikeda-Douglas, V. Balkissoon, R. M. Damain Holsinger, E. Head, M. Michael, R. J. Racine, and N. W. Milgram. Effects of extended electrical kindling on exploratory behavior and spatial learning, *Behav. Brain Res.* **89**, 179-190 (1997).

33. R. J. Steele and R. G. Morris, Delay dependent impairment of a matching-to-place task with chronic and intrahippocampal infusion of the NMD-antagonist D-AP5, *Hippocampus*, **9**, 118-136 (1999).
34. A. J. Wintink, N. A. Young, and L. E. Kalynchuk, Watermaze performance in male and female amygdala-kindled rats, *Poster presentation at the Canadian Society for Brain, Behaviour, and Cognitive Science Annual Meeting*, Hamilton, ON (2003).
35. T. Falkenberg, A. K. Mohammed, and B. Henriksson, Increased expression of brain-derived neurotrophic factor mRNA in rat hippocampus is associated with improved spatial memory and enriched environment, *Neurosci. Lett.* **138**, 153-156 (1992).
36. M. J. Passineau, E. J. Green, and W. D. Dietrich, Therapeutic effects of environmental enrichment on cognitive function and tissue integrity following severe traumatic brain injury in rats, *Exp. Neurol.* **168**, 373-384 (2001).
37. G. Kempermann, H. G. Kuhn, and F. H. Gage, Experience-induced neurogenesis in the senescent dentate gyrus, *J. Neurosci.*, **18**, 3206-3212 (1998).
38. E. J. Green, W. T. Greenough, and B. E. Schlumpf, Effects of complex or isolated environments on cortical dendrites of middle-aged rats, *Brain Res.* **264**, 233-240 (1983).
39. L. E. Kalynchuk, A. J. Wintink, A. L. Darnbrough, S. J. Smith, and S. A. Johnson, Environmental enrichment reduces kindling-induced fear behavior in male and female rats, *Soc. Neurosci. Abs.* **30** (2004).

Discussion

Gale: This is quite exciting in terms of the interactions with the environment manipulations that you have shown. It is quite reminiscent of the effects of septal lesions, which of course make the animals pretty defensive and aggressive and quite fearful and so forth. But if you handle the animals from the time they are lesioned, they don't show these changes whatsoever. Do you think there might be some common mechanisms or some relationship to the story of septal lesions?
Kalynchuk: I had not thought of that, but it is an interesting connection. You are right in that septal lesioned animals that are handled show a lot fewer behaviors than animals that are not handled. There are extensive connections between the septum and the hippocampus; so there could be something going on there.
Gale: Did you by chance just try handling the animals daily instead of enriching them?
Kalynchuk: We only did that in the course of the first experiment, where we handled the rats instead of placing them in the spatial memory task. But it was only one day per week; so it wasn't very much.
Gale: In the septal lesion animals you really have to do it on a daily basis starting from immediately after the intervention.
Kalynchuk: I would argue that handling is a form of environment enrichment. So my guess is that that would work.
Kokaia: I have a question concerning the proliferation. First of all, what was the paradigm of the BrdU injections?
Kalynchuk: In that particular experiment we gave three injections of BrdU prior to sacrifice. One was 12 hr before the animal was sacrificed, one was 24, and the other was 36 hr before the animal was sacrificed. We used a fairly low dose of BrdU, 50 mg/kg.
Kokaia: So the injections were after all the 99 stimulations were done?
Kalynchuk: No, actually the first injection was given on morning of the last day of kindling. Then the animals had the last three stimulations. The second injection was given after those stimulations; so it was at the end of kindling. The final one was given the morning the animals were sacrificed. There were three stimulations after the first BrdU injection.

Kokaia: Did you determine whether this proliferation is related to neurogenesis and whether there was any inflammatory reaction due to so many stimulations and generalized seizures?
Kalynchuk: Thus far we have only counted the cells that were labelled with BrdU; so we can only say that it was cell proliferation and can't be sure that it was neurogenesis.

DOES INADVERTENT CONDITIONING CONTRIBUTE TO THE MAJOR FEATURES OF KINDLING?

Steven J. Barnes and John P. J. Pinel*

1. INTRODUCTION

Conventional kindling experiments appear ideal for generating inadvertent conditioned effects: A variety of stimuli (e.g., removal from the home cage, attachment to the stimulation lead, and placement in the stimulation environment) are repeatedly presented to the animal prior to each stimulation and convulsion. We recently demonstrated that the stimulation environment can have a major impact on both kindled convulsions and interictal behaviour. Rats received periodic stimulations to the basolateral amygdala (BA) in one conditional environment (CS+) and an equal number of sham stimulations (the stimulation lead was attached but no current was delivered) in a second environment (CS-) in a quasirandom sequence. As kindling progressed, the rats became more defensive in the CS+ environment than in the CS- environment; and, when they were finally stimulated in the CS- environment, their convulsions were substantially less severe than in the CS+ environment.[1] Furthermore, the kindled rats preferred the CS- environment to the CS+ environment in a subsequent conditioned place preference test.

Since this demonstration that the stimulation environment can have a significant effect on kindled convulsions and interictal behaviour, our research has focused on four questions. First, are the effects of the stimulation environment on kindled convulsions a product of Pavlovian conditioning? Second, can the kindling of brain structures other than the BA produce conditioned effects; and, if so, are such effects comparable to those of BA kindling? Third, can stimuli other than the stimulation environment function as inadvertent conditional stimuli (CSs) in conventional kindling experiments? And fourth, do inadvertent conditioned effects influence the defining features of kindling (e.g., permanence)? This paper focuses on samples of these four lines of research.

* Steven J. Barnes and John P. J. Pinel, Department of Psychology, University of British Columbia, Vancouver, British Columbia, Canada V6R 2J9.

2. GENERAL METHODS

The following is a brief description of the methods that were used in our experiments. More complete descriptions are available elsewhere.[1-3]

2.1. Kindling and Conditioning

One bipolar stimulation electrode was implanted in the site of interest in each Long-Evans rat. Following recovery from surgery, the rats underwent a regimen of training in which they received both stimulations and sham stimulations. One conditional stimulus (CS+) preceded each stimulation, and a second conditional stimulus (CS-) preceded each sham stimulation. In most of our experiments, the CS+ was the stimulation chamber, and the CS- was a similar, but distinctive, chamber. During each trial, the rats were allowed to move freely around the test chamber for a specified time interval (the preadministration interval), typically for 30 s, before the stimulation or sham stimulation was administered. Sham stimulation trials were identical to the stimulation trials except that no current was passed through the implanted electrode. Accordingly, any differences that developed in the behavior of a subject in response to the CS+ and CS- could only be attributed to the contingency. Which particular stimulus served as the CS+ or the CS- in a particular experiment was counterbalanced between subjects, and because no systematic differences ever developed between these two subgroups their data were always combined for analysis.

2.2. Measuring the Kindled Convulsions and the Interictal Behaviour

Each convulsive response was rated according to Pinel and Rovner's[4] extension of Racine's[5] limbic convulsion severity scale (class 1: facial movements only; class 2: facial movements and head nodding; class 3: facial movements, head nodding, and forelimb clonus; class 4: facial movements, head nodding, forelimb clonus, and rearing; class 5: facial movements, head nodding, forelimb clonus, rearing, and falling once; class 6: a class 5 with multiple rearing and falling episodes; class 7: a convulsion with running fits). In addition, both the latency to the onset of the convulsion and the convulsion duration were recorded; and if a class 5 convulsion or greater occurred, the number of times the rat fell during the course of the convulsion was also recorded.

The particular preadministration-interval behaviours that were quantified from videotaped recordings depended on the site of stimulation, but they always included freezing (i.e., the percent of the 15 2-s epochs in the preadministration interval during which the rat made no movements, other than those associated with breathing).

3. EXTINCTION CONFIRMS A PAVLOVIAN MECHANISM

The discrimination paradigm that we had employed in our first study[1] suggested that the observed effects were the product of Pavlovian conditioning: It is difficult to explain the differences observed in their behavior and convulsions in the two environments in any other way. To provide additional support for this Pavlovian view, we assessed the effects of extinction on the effects that come to be associated with the stimulation environment during BA kindling.

The kindling phase of the experiment was a replication of our first study[1]: Rats received 45 BA stimulations in one environment (CS+) and 45 sham stimulations in another environment (CS-), quasirandomly over 45 days. At the end of the kindling phase, the CS+ elicited more freezing than the CS- and convulsions elicited in the CS+ were more severe than those elicited in the CS-.

During the extinction phase, when the rats received another 45 stimulations and 45 sham stimulations--either as before in their original environment (the no-interchange rats) or with their original CS+ and CS- environments interchanged (the interchange rats), the no-interchange rats continued to display the effects observed during the kindling phase. In contrast, the interchange rats displayed more freezing in the former CS- than in the former CS+, and their convulsions were equally severe in the two environments.

These results confirmed that the effects of the stimulation environment on ictal and interictal behavior are the product of Pavlovian conditioning. Pavlovian mechanisms have been implicated in a variety of other progressive phenomena that result from the repeated predicted administration of physiological treatments—for example, functional drug tolerance,[6] sensitization to stimulants,[7] and development of premeal hunger.[8]

4. EFFECT OF STIMULATION SITE ON THE CONDITIONED EFFECTS OF KINDLING

4.1. Anterior Neocortex vs. Basolateral Amygdala

Does the kindling of structures other than the BA produce conditional effects; and if so, are such conditional effects different from those produced by BA kindling? Our first experiment to address this question compared the ictal and interictal effects conditioned to the stimulation environment during kindling of the BA and the anterior neocortex (AN). The AN was selected as the second kindling site because the topography of anterior neocortex kindled convulsions differs markedly from that of convulsions associated with kindling of the BA or of other limbic sites.[9-11] Our premise was that, because of this difference in topography, a comparison of BA and AN kindling would likely reveal stimulation-site-related differences in the conditioned effects of kindling--if such differences existed. Furthermore, because AN kindled convulsions are highly variable in terms of both their severity[10] and occurrence[9] from subject to subject and from stimulation to stimulation in the same subject, we presumed that an analysis of the conditioned effects of AN kindling might provide insight into the source of variability in AN kindled convulsions.

Rats received 53 stimulations to the AN or BA in one environment (CS+) and 53 sham- stimulations in another environment (CS-), quasirandomly over 54 days. Unlike BA kindling, AN kindling led to more wet dog shakes and less, rather than more, severe convulsions in the CS+. Moreover, during AN kindling, the mean number of wet dog shakes in the CS+ was negatively correlated with the mean convulsion class. This latter finding suggests that wet-dog shakes are mediated by mechanisms which inhibit the elicitation of kindled convulsions from the AN; that such effects can be conditioned; and that the previously documented variability in the severity and occurrence of convulsions during AN kindling may be due to such conditioned inhibitory effects. These results established for the first time that the pattern of the conditioned effects of the stimulation environment on kindled convulsions and interictal behavior is a function of kindling site.[2]

With extended kindling of the anterior neocortex, the topography of the convulsions becomes increasingly similar to that of "limbic" convulsions.[12] In this experiment, only 2 AN-kindled rats developed generalized convulsions that were of this "limbic" type. Interestingly, when these 2 rats were stimulated in the CS-, their convulsions were less severe than in the CS+--in this respect they were similar to BA-kindled rats, but opposite to those of all the other AN-kindled rats. These 2 rats also demonstrated strong avoidance of the CS+ in the conditioned place preference test--again, they were similar to the BA-kindled rats, but unlike the other AN-kindled rats. These observations suggest that the conditioned effects on kindled convulsions and interictal behavior may change as the topography of the convulsions changes. Moreover, they suggest that the conditioned interictal defensive behaviors are associated with kindled convulsions that are topographically "limbic" in nature.

4.2. Three Sites in the Hippocampal Complex

Next, we compared the behavioral effects--both ictal and interictal--conditioned to the stimulation environment during kindling of three different sites in the hippocampal complex: The perirhinal cortex (PRh), the ventral hippocampus (VH), and the dorsal hippocampus (DH). Does the kindling of related brain sites associated with topographically similar kindled convulsions, but vastly different rates of kindling, produce similar patterns of conditioned effects? Our working premise was that this approach would clarify the nature of the unconditional stimulus (UCS) in kindling-related conditioning: Is the UCS related more to the stimulations or to the convulsive responses?

Rats received 53 stimulations to the PRh, VH, or DH in one environment (CS+), and 53 sham stimulations in a second environment (CS-), quasirandomly over 54 days. The PRh-kindled rats displayed rapid kindling and a comparably swift emergence of interictal defensiveness conditioned to the CS+. In contrast, the VH- and especially the DH-kindled rats displayed much slower kindling and slow or no conditioning, respectively. No effects of conditioning on the convulsions, comparable to those associated with BA or AN kindling, were observed. These results confirmed the generality of kindling-related conditioned effects on interictal behaviour and the site specificity of these effects, and they suggest that the convulsive responses, rather than the stimulations, function as the UCS.

5. TEMPORAL CUES AS CONDITIONAL STIMULI FOR CONDITIONING BY AMYGDALA KINDLING

Like environmental cues, temporal cues are present in virtually all kindling experiments. The time of day at which an animal is stimulated, the order in which events occur from the time an animal is taken from its cage to the time that it receives a stimulation, and the temporal intervals between each of those events, are repeated on each trial.

In two experiments, we showed that time could serve as a CS in the conditioning of interictal behaviour by BA-kindling. In both experiments, rats received one BA stimulation and one sham stimulation each day during the light phase of their light-dark cycle for 53 days (experiment 1) or 73 days (experiment 2). The stimulations always occurred at the same time, the sham stimulations always occurred at another time, and the rats were tested in a single chamber; thus the CS+ in this experiment was the time at which the rats had previously received a stimulation, and the CS- was the time at which the rats had previously received a sham stimulation. The only difference between the two experiments was that the second employed a 150-s preadministration interval and included a peak procedure test: At the CS+ time, the rats were placed in the test chamber as usual, except they did not receive a stimulation after the 150-s interval; instead, they were allowed to move freely about the chamber for an additional 90 s (i.e., for a total of 240 s). This sort of test, known as a peak procedure, is often used to measure the degree of responding (e.g., pressing a lever) that an animal will display after it has learned that after a certain time interval it will receive a reward.[13]

In the first experiment, as kindling progressed the rats displayed more freezing at the CS+ time than at the CS- time.[3] In the second experiment, rats also began to display more freezing at the CS+ time than at the CS- time, and during the peak procedure test freezing gradually increased over the 150-s interval, and peaked at the time at which they had previously received a stimulation, at which point normal activity quickly resumed--this peak in freezing was only present at the CS+ time.[3] In this latter experiment rats had to both discriminate between the CS+ and CS- time in order to predict when it would receive a stimulation or sham- stimulation. Superimposed on this discrimination, the rats also had to utilize short-interval timing to predict the onset of the stimulation. Accordingly, the results of these two experiments indicate that rats use multiple stimuli in concert with one another to learn when and where a stimulation and subsequent convulsion will occur.

6. CONDITIONED EFFECTS CONTRIBUTE SUBSTANTIALLY TO THE PERMANENCE OF KINDLING

The neuroplastic changes that underlie kindling have been shown to be remarkably enduring—if a kindled rat is left unstimulated for several months, fully generalized convulsions are quickly elicited once the regimen of periodic stimulations recommences.[14,15] We adapted the protocol used in our previous experiments to explore the contributions of conditioned effects to the permanence of kindling.

Each rat received 25 BA stimulations in one environment (CS+) as well as 25 sham stimulations in another environment (CS-), in a quasirandom sequence over 25 days. Then, all of the rats remained in their home cages for 100 days. After this 100-day stimulation-free period, half the rats (the nonswitch rats) received another 25 stimulations and another 25 sham stimulations just as they had during the kindling phase; that is, during the retention phase they were stimulated in the CS+ and sham stimulated in the CS-. The other rats (the switch rats) also received another 25 stimulations and another 25 sham stimulations, but during the retention phase, they received both the stimulations and sham stimulations in the former CS-.

The results are presented in Figure 1. The permanence of kindling was assessed in two ways: by examining the severity of the convulsions elicited by the first stimulation following the 100-day stimulation-free period; and by calculating the percent savings in the number of stimulations required to re-attain the criterion (i.e., three fully generalized convulsions).

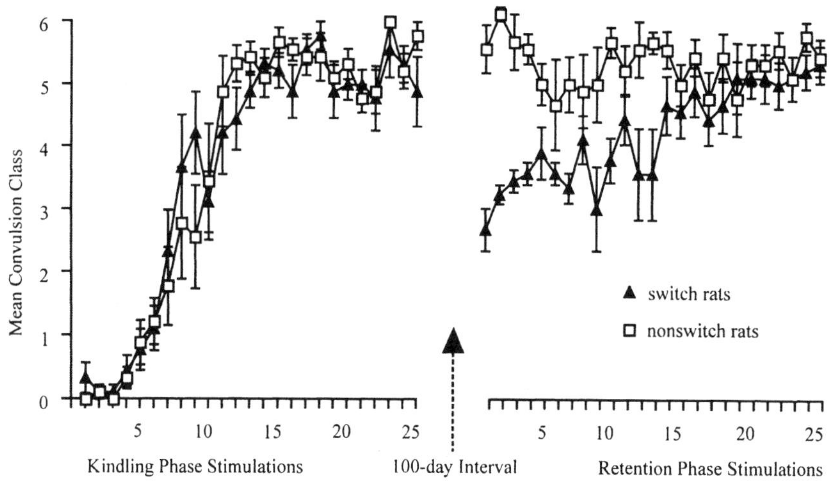

Figure 1. Mean class of the convulsions displayed by the rats in response to each of the 25 stimulations administered during the kindling phase of the experiment (left side) and to each of the 25 stimulations administered during the retention phase (right side). During the retention phase, the nonswitch rats received stimulations in the same environment in which they had been stimulated during the first phase; whereas the switch rats received stimulations in a familiar environment in which they had not been previously stimulated. The retention of kindling was substantially less in the switch condition.

The convulsions of the nonswitch subjects--the rats who were stimulated following the 100-day interval in the original CS+ environment--displayed no sign of attenuation following the interval: Their mean convulsive response following the interval was 5.6, which was not significantly different from 5.8--their mean convulsive response to the last stimulation of the kindling phase (see Figure 1). Furthermore, during the retention phase, they attained the criterion of three generalized convulsions much faster than they had during the kindling phase; their mean savings was 72%. This result confirmed that when there are no major alterations in the cues that predict the stimulations--as kindling is typically studied--kindling is indeed a permanent, or at least a very enduring, neuroplastic phenomenon.

In contrast, the kindling effect was substantially attenuated following the 100-day interval in the switch rats (see Figure 1). The first stimulation after the 100-day interval, that is, the first stimulation administered in their former CS-, elicited convulsive responses with a mean convulsion class of 2.7—significantly weaker than those observed on the same trial in the nonswitch rats. Moreover, the switch rats displayed –33% savings, which was not significantly different from no savings.

7. CONDITIONED EFFECTS CONTRIBUTE SUBSTANTIALLY TO THE TRANSFER OF KINDLING

In addition to its permanence, another significant property of kindling is that it shows transfer between stimulation sites: If an animal is first kindled from one brain site, it subsequently requires many fewer stimulations for the animal to be kindled from a second brain site, either ipsilateral or contralateral to the first site.[14] We adapted the protocol used in our previous experiments to explore the contributions of conditioned effects to the transfer of kindling.

Stimulation electrodes were surgically implanted in the left and right BA of each rat. Following recovery from surgery, each rat received 30 BA stimulations through one of their electrodes in one environment (CS+) as well as 30 sham stimulations in another environment (CS-), in a quasirandom sequence over 30 days. Then, all of the rats remained in their home cages for 10 days. After this 10-day stimulation-free period, half the rats (the nonswitch rats) received 30 BA stimulations through their second electrode and 30 sham stimulations as they had during the kindling phase; that is, during the transfer phase they were stimulated in the CS+ and sham stimulated in the CS-. The other rats (the switch rats) also received 30 stimulations to their second electrode and another 30 sham stimulations, but during the transfer phase, they received both the stimulations and sham stimulations in the former CS-.

The convulsions of the nonswitch subjects--the rats who were stimulated through their second electrode in the original CS+ environment--displayed transfer: During the retention phase, they attained the criterion of three generalized convulsions much faster than they had during the kindling phase; their mean savings was 20%. This result confirmed that when there are no major alterations in the cues that predict the stimulations--as kindling is typically studied--kindling can indeed transfer to the contralateral BA. In contrast, there was no evidence of transfer in the switch rats. The switch rats displayed 2% savings, which was not significantly different from no savings.

8. DISCUSSION

Here we have summarized sample studies from our four lines of research on the effects of Pavlovian conditioning on the ictal and interictal effects of kindling. These studies make five important points: (1) that the effects exerted by the stimulation environment on both the ictal and interictal behaviour of BA-kindled rats are a result of Pavlovian conditioning, (2) that the topography of the conditioned effects of the stimulation environment on kindled convulsions is a function of convulsion topography (i.e., cortical- vs. limbic-kindled convulsions), (3) that the topography of the conditioned effects of the stimulation environment on interictal behavior is a function of kindling site, (4) that at least two sorts of temporal cues can serve as CSs in the conditioning of interictal behaviour by BA-kindling, (5) and that two of the defining aspects of kindling--it's permanence and it's capacity to transfer between brain sites--is greatly influenced by effects conditioned to the stimulation environment.

Why is it important to consider conditioned effects in the study of kindling? There are at least three important reasons. First, our research is showing that conditioned effects are likely a major component of virtually all conventional kindling experiments, and thus a complete understanding of kindling is not likely to emerge without considering them. Our experiments on the permanence and transfer of kindling illustrate this point. Second, considering conditioned effects may resolve some long-standing kindling-related puzzles. For example, why do AN-kindled convulsions show such a remarkable degree of within- and between-subjects variability?[9,10] We believe that the answer may lie in the neurobiological mechanisms that underlie the conditioned wet-dog shakes observed during AN kindling[2]. Third, an appreciation of the fact that conditioning plays an important role in kindling supports the view that kindling may constitute a useful model of learning and memory in general, and as such may contribute to the search for its neural mechanisms--in the same way that the discovery that conditioned effects contribute to drug tolerance has focused attention on the role of the hippocampus and amygdala in the development of drug tolerance.[16]

Neuroplasticity research has largely been motivated by the potential for adapting various neuroplastic phenomena to the treatment of brain and spinal cord injury. The present experimental findings provide strong support for a caution recently issued by Döbrössy and Dunnett:[17] In the development of neuroplastic therapeutic protocols, it is a mistake to focus on molecules, neurons, and synapses at the exclusion of the patient and his or her experiences; these can have a substantial influence on the development and survival of neuroplastic change.

9. ACKNOWLEDGEMENTS

This work was supported by a grant from the Natural Science and Engineering Research Council to JPJP, and by Doctoral Research Fellowships from the Canadian Institutes for Health Research and the Michael Smith Foundation for Health Research to SJB.

10. REFERENCES

1. S. J. Barnes, J. P. Pinel, L. H. Francis, and G. S. Wig, Conditioning of ictal and interictal behaviors in rats by amygdala kindling: context as the conditional stimulus, *Behav.Neurosci.* **115**, 1065-1072 (2001).
2. S. J. Barnes, J. P. Pinel, G. S. Wig, M. C. Stuettgen, and C. H. Holzel, Stimulation site determines the conditioned effects of kindling in rats: anterior neocortex versus amygdale, *Eur.J.Neurosci.* **17**, 1671-1679 (2003).
3. S. J. Barnes, O, Magyar, J. P. Pinel, and A. Takahashi, Anticipating the attack: temporal conditioning during amygdala kindling in rats, *Behav.Neurosci.* **118**, 89-96 (2004).
4. J. P. Pinel and L. I. Rovner, L. I. Experimental epileptogenesis: kindling-induced epilepsy in rats, *Exp.Neurol.* **58**, 190-202 (1978).
5. R. J. Racine, Modification of seizure activity by electrical stimulation. II. Motor seizure, *Electroencephalogr.Clin.Neurophysiol.* **32**, 281-294 (1972).
6. S. Siegel, Morphine analgesic tolerance: its situation specificity supports a Pavlovian conditioning model, *Science* **193**, 323-325 (1976).
7. R. E. Hinson and C. X. Poulos, Sensitization to the behavioral effects of cocaine: modification by Pavlovian conditioning, *Pharmacol.Biochem.Behav* **15**, 559-562 (1981).
8. S. C. Woods and D. S. Ramsay, Pavlovian influences over food and drug intake, *Behav. Brain.Res.* **110**, 175-182 (2000).
9. W. T. Seidel and M. E. Corcoran, M. E. Relations between amygdaloid and anterior neocortical kindling, *Brain Res.* **385**, 375-378 (1986).
10. W. M. Burnham,. In: *Limbic Mechanisms: The Continuing Evolution of the Limbic System Concept,* edited by K. E. Livinston and O. Horykiwicz, (Plenum Press, New York, 1978) pp. 507-519.
11. R. J. Racine, Modification of seizure activity by electrical stimulation: Cortical areas, *Electroencaphalogr. Clin.Neurophysiol.* **39**, 1-12 (1975).
12. J. P. J. Pinel, Kindling-induced experimental epilepsy in rats: Cortical stimulation, *Exp.Neurol.* **72**, 559-569 (1981).
13. C. M. Thorpe and D. M. Wilkie, D. M., Unequal interval time-place learning, *Behav.Processes* **58**, 157-166 (2002).
14. G. V. Goddard, D. C. McIntyre, and C. K. Leech, A permanent change in brain function resulting from daily electrical stimulation, *Exp.Neurol.* **25**, 295-330 (1969).

15. Z. Dennison, G. C. Teskey, and D. P. Cain, Persistence of kindling: effect of partial kindling, retention interval, kindling site, and stimulation parameters, *Epilepsy Res.* **21**, 171-182 (1995).
16. J. M. Mitchell, A. I. Basbaum, and H. L. Fields, H. L. A locus and mechanism of action for associative morphine tolerance, *Nature Neurosci.* **3**, 47-53 (2000).
17. M. D. Dobrossy and S. B. Dunnett, The influence of environment and experience on neural grafts, *Nature Rev. Neurosci* **2**, 871-879 (2001).

Discussion

Post: That is a beautiful paper. I was wondering how different were the two environments, and whether you could have bigger effects on the permenance if you made them really different.
Barnes: I think so. In fact, in our earlier studies we were doing a place preference test, and the environments were highly similar in the sense that they were only discriminable based on the color of the chamber, because we had to have a connecting chamber and the chambers had to be placed adjacent to each other in the colony. In our later experiments, the permanence study and the transfer study, I put the chambers at opposite ends of the colony room where I was stimulating them and made them very distinctive in appearance.
Adamec: Very nice paper, Steve. These studies are done in hooded rats, correct?
Barnes: Yes.
Adamec: So, to kindle them you have to do lots of stimulations and you show numbers of 25, 45 to show your conditioning effects. What if you applied fewer stimulations – do you still get conditioning effects? Related to that, what if you tried this in Wistar rats where you need only 8 or 10 stimulations to kindle convulsions – would still expect the same Pavlovian effects?
Barnes: Likely. Perirhinal kindled animals kindle very quickly and you still see these conditioning effects. You should keep in mind that observing this conditioned freezing in the environment does not necessarily imply that you require that many stimulations to actually produce those conditions effects. This is, when you observe it is not when it actually becomes conditioned.
Adamec: It would still would be of interest to do this in Wistar rats, where a very few stimulations kindle the animal.

CONTINGENT TOLERANCE AND CROSS TOLERANCE TO ANTICONVULSANT EFFECTS IN AMYGDALA-KINDLED SEIZURES: Mechanistic and clinical implications

R.M. Post, Z-J Zhang, S.R.B. Weiss, G. Xing, and K. Obeng*

1. INTRODUCTION

1.1 Contingent Tolerance

A unique form of pharmacodynamic tolerance develops when effective anticonvulsants are given as pre-treatment before once-daily amygdala-kindled seizures in animals.[1,2] This loss of efficacy (i.e., tolerance) is *contingent* upon the drug's administration before, but not immediately after, the kindling stimulation.[3-6] This form of tolerance can be reversed after a period of seizures off anticonvulsants, or even continuing to give anticonvulsants on a once-daily basis but after a fully developed kindled seizure has already occurred.

Some of the neurochemical correlates of contingent tolerance to the anticonvulsant effects of carbamazepine have been elucidated.[5] Following normally-developed amygdala-kindled seizures, an array of changes in gene expression occur. However, in animals that have become tolerant to the anticonvulsant effects of carbamazepine, after the emergence of similarly appearing behavioral seizures (to those occurring in the anticonvulsant-free state), there is a very different profile of changes in gene expression, such that some seizure-induced changes continue to occur, but others do not.

* R.M. Post, Z-J Zhang, S.R.B. Weiss, G. Xing, and K. Obeng, Department of Health and Human Services, National Institutes of Health, National Institute of Mental Health, Biological Psychiatry Branch, Bethesda, MD

1.2 Neurochemical correlates

We have postulated that it is the loss of some of these endogenous adaptive anticonvulsant alterations that may be associated with intermittent emergence of seizures[7,8] and eventually the complete loss of carbamazepine efficacy.[5] In animals that seize in the carbamazepine-tolerant condition, there is a decreased induction of mRNA for fos-related antigens (Fras), brain-derived neurotrophic factor (BDNF), glucocorticoid receptors, thyrotropin-releasing hormone (TRH), neuropeptide Y (NPY), corticotropin-releasing factor (CRF), and the α4 subunit of gamma-aminobutyric acid subunit A ($GABA_A$) receptors. However, normal decreases in neurotrophin 3 (NT3) and dynorphin occur, along with the normal seizure-induced increases in mineralocorticoid receptors, benzodiazepine receptors, and t-butylbicyclophosphorothionate (TBPS) binding.[5] The failure of increases in TRH mRNA[9] and $GABA_A$ receptors,[10] for example, could be associated with the loss of efficacy of carbamazepine, because TRH has anticonvulsant properties when injected into the hippocampus of amygdala-kindled animals,[11] and GABA is the major inhibitory neurotransmitter in the brain.

We have previously observed that some of these seizure-induced adaptations are also important to the initial anticonvulsant effects of carbamazepine, lamotrigine, or diazepam as evidenced by the fact that these drugs become less effective if sufficient number of days from the last amygdala-kindled seizure are allowed to elapse, i.e., the "time-off" seizure effect[5] (Krupp et al., 2002, unpublished data). It appears that the transient induction of TRH peptide and $GABA_A$ receptor synthesis, and/or other putative endogenous anticonvulsant factors for the 2-4 days following an amygdala-kindled seizure, are necessary for the anticonvulsant effects of these drugs. Because some of these same adaptive changes are also lost in the carbamazepine tolerant state, they could be relevant to the relatively rapid time frames of tolerance development for carbamazepine, lamotrigine, and diazepam, as well as levetiracetam,[12] and to the much slower development of tolerance with valproate (which does not show a "time-off" effect).[5]

2. CROSS TOLERANCE

Cross tolerance from one drug to another implies at least partial overlapping mechanisms of action. It appears that the peripheral-type benzodiazepine effects of carbamazepine[13] are important to its actions because there is cross tolerance to the peripheral-type benzodiazepine ligand PK11195, as well as to carbamazepine-10,11-epoxide (which also interacts with this receptor system), but not to diazepam and clonazepam, which act at central type benzodiazepine receptors [4] (Table 1). Yet Kim et al.[14] reported cross tolerance from clonazepam to carbamazepine. Reddy & Rogawski[15] also reported unidirectional cross tolerance from the neurosteroid ganaxolone to diazepam, but not to itself. There is also no cross tolerance from carbamazepine to levetiracetam,[16] which is thought to act, in part, by inhibiting the negative modulation of the benzodiazepine-GABA-chloride ionophore complex by zinc and beta carbolines,[17] However, there is cross tolerance in the opposite direction, from levetiracetam to carbamazepine.

Interesting is the lack of cross tolerance from carbamazepine to phenytoin[5] both of which have similar efficacy in inhibiting sodium channels and sodium influx through the type-2 batrachotoxin-sensitive receptor.[18] This lack of cross tolerance would suggest that other properties of carbamazepine (compared with phenytoin) are important to its anticonvulsant effects on amygdala-kindled seizures. Carbamazepine would also not be postulated to show cross tolerance with valproate because the two show many differences in mechanisms of action; we question whether the failure to upregulate $GABA_A$ receptors at the level of the α4 subunit during carbamazepine tolerance may be associated with the loss of valproate's effects, although this remains to be directly tested.[19]

There is bidirectional cross tolerance between carbamazepine and lamotrigine, yet lamotrigine, in contrast to carbamazepine, does not convey cross tolerance

Table 1 Cross Tolerance Patterns in Contingent Anticonvulsant Effects on Amygdala-Kindled Seizures

Drug (tolerance from):	Cross tolerance to:	No-cross tolerance to:
Carbamazepine (CBZ)	PK11195 CBZ-10-11-epoxide Valproate (?via ↓ ά4 subunit of $GABA_A$	Clonazepam Diazepam Phenytoin Levetiracetam*
Lamotrigine (LTG)	Carbamazepine	Valproate MK801[a] Gabapentin[a]
Levetiracetam (LEV)	Carbamazepine*	

*Unidirection cross tolerance from LEV to CBZ; not from CBZ to LEV
[a] These drugs slow the development of tolerance to LTG

to valproate.[20] Therefore, it would be interesting to assess whether there is inhibition of seizure-induced adaptations in the α4 subunit of the $GABA_A$ receptor during lamotrigine tolerance as there is for carbamazepine and diazepam tolerance.

Lamotrigine tolerance develops relatively rapidly, and attempts to alter drug administration parameters to slow the process down have not been very successful. Instead, we have observed that drugs active at GABAergic receptor systems (such as gabapentin) and N-methyl-D-aspartate (NMDA)-glutamatergic receptor systems (such as MK-801) are capable of markedly slowing lamotrigine tolerance or reversing it transiently once it has occurred.[21]

2.1. Potential Clinical Implications of Tolerance to Anticonvulsant Effects

The development of tolerance to the anticonvulsant effects of many of these drugs in clinical epilepsy syndromes has not been well documented and is controversial. However, with respect to carbamazepine's therapeutic effects in the treatment of trigeminal neuralgia and related pain syndromes, in which the assessment of tolerance is methodologically much easier because of a greater number and consistency of episodes at baseline, there is unequivocal evidence for tolerance development in a substantial portion of patients so treated.[22, 23] Tolerance also appears to develop to a variety of drugs used in the long-term prevention of manic and depressive episodes, including carbamazepine[24, 25] and valproate,[26] as well as the non-anticonvulsant compound classically used in the prevention of episodes, lithium carbonate.[27-30]

To the extent that loss of efficacy via tolerance mechanisms plays a role in a subset of patients with seizure and other neuropsychiatric disorders who are initially responsive to these agents but then show a gradual loss of efficacy over time, the preclinical data on cross tolerance among these agents in the amygdala-kindled seizure model (Table 1) may yield testable hypotheses about which drugs will, and will not, show cross tolerance in the clinic. When tolerance has developed to a previously effective agent, one clinical strategy has been to use other drugs with different mechanisms of action and then rotate among them. Moreover, if seizures have occurred, some patients appear to re-respond to the drug to which they were previously tolerant,[25] as seen in the preclinical model of contingent tolerance reversal by a period of seizures expressed during a period of time off that medication.[3] From a hypothetical mechanistic perspective, presumably such seizures in the medication-free state would be sufficient to allow the re-induction of some of the mRNAs for endogenous anticonvulsant substances, such as TRH and $GABA_A$ receptors, that may be important to the manifestation or renewal of a drug's anticonvulsant effects.[5]

2.2. Differential Tolerance Mechanisms of Carbamazepine and Lamotrigine Despite Bidirectional Cross Tolerance

It is of considerable interest that although carbamazepine and lamotrigine show bi-directional cross tolerance, their mechanisms of tolerance development appear to differ considerably in some respects (Table 2). In particular, when tolerance has developed to the anticonvulsant effects of carbamazepine, the seizure threshold measured in the medication-free state[5] *decreases*, as one might expect. Paradoxically, during the development of lamotrigine tolerance, there is evidence of an *increased* seizure threshold measured in the medication-free state,[20] although it is possible that this increase reflects some persisting presence of lamotrigine over several days following the discontinuation of drug administration.

However, other differences are also apparent. Whereas higher doses may be more effective than lower doses in slowing the rate of tolerance development to carbamazepine,[5] higher doses of lamotrigine are, paradoxically, proconvulsant, both speeding the development of tolerance and increasing the number of stage six kindled seizures (associated with violent running attacks).[21] Moreover, whereas MK-801 slows the development of lamotrigine tolerance, it is without effect on carbamazepine tolerance[31] and, as noted previously, there is differential cross tolerance from each of these agents to valproate.

In addition, the spectrum of clinical anticonvulsant effects of lamotrigine and carbamazepine differ markedly.[32] Whereas lamotrigine is effective in *absence* seizures, carbamazepine is not and may even exacerbate them. Together, these data suggest that cross tolerance processes are sensitive to some common mechanisms of drug actions, but other important differences may not be reflected in the cross tolerance paradigm.

Table 2 Differential Effects of Carbamazepine (CBZ) and Lamotrigine (LTG) on Anticonvulsant Tolerance Development

	CBZ (15 mg/kg)	vs.	LTG (15 mg/kg)
Rapid tolerance to anticonvulsant effects (amygdala kindling)	+++		+++
Cross tolerance	+++		+++
Duration of "time-off" effect (seizures enhance efficacy)	4-5 days		4-5 days
Seizure threshold change with tolerance	Decrease		Increase (possible residual drug effect)
Using high doses	Slows tolerance		Speed tolerance and are proconvulsant
Alternating high and low doses	?		Slows tolerance
Chronic non-contingent drug dosing	Slows tolerance		?
MK-801 on tolerance development	No effect		Slows (NMDA implicated)
Cross tolerance to valproate	Yes		No
Valproate combination	Slows tolerance		?
Gabapentin augmentation (2 hrs. pretreatment)	?		Slows tolerance
(Half hr. pretreatment)	No effect		↓ Stage VI seizures
(Tolerance reversal)	?		+++

NMDA - N-methyl-D-aspartate

3. DIFFERENT ANTICONVULSANT EFFECTS AS A FUNCTION OF THE INITIAL PHASE OF KINDLING DEVELOPMENT VS. THE MID PHASE OF COMPLETED AMYGDALA-KINDLED SEIZURES

3.1. Contingent Inefficacy

Dramatic pharmacological dissociations have been demonstrated most classically by Pinel[33] in the double-dissociation of diazepam and phenytoin in their efficacy against elicited vs. spontaneous amygdala-kindled seizures. Moreover, a number of drugs such as carbamazepine, lamotrigine, and phenytoin are completely without effect on the development phase of amygdala-kindled seizures, even though they are highly effective against the completed variety.[34]

Surprisingly, pre-treatment (but not post-treatment) with these drugs prior to each amygdala-kindled seizure during the development phase prevents their effectiveness when completed kindled seizures have begun to occur, a phenomenon we have termed "contingent inefficacy"[3, 5, 35]. Moreover, there is cross-contingent inefficacy between carbamazepine and lamotrigine just as there is cross tolerance between these two compounds.

These observations are of particular interest in relation to preventive antiepileptic treatment for syndromes such as posttraumatic epilepsy, in which phenytoin and a variety of other compounds have not been shown to be effective.[36] The contingent inefficacy findings also suggest the unfortunate possibility that pretreatment with a drug that is given, but is ineffective in the initial developmental stages of some types of kindling-like epileptogenesis, may prevent its eventual effectiveness when full-blown completed seizures have occurred.[3, 5, 35] Such treatment ineffectiveness might be expected based on the observations of contingent inefficacy in the kindling paradigm to the extent that mechanisms in posttraumatic or tumor-induced epilepsy follow some of the principles of kindling-like development.

3.2. Drugs Affecting Kindling Development (Epileptogenesis)

In contrast to the drugs that are ineffective in the initial development phase of amygdala-kindled seizures, levetiracetam, diazepam, and valproate are effective in inhibiting both the development of and completed kindled seizures.[5,37-39] These drugs might offer better options for exploration in posttraumatic epilepsy prevention studies.

Interestingly, when animals are given levetiracetam during the initial phase of kindling development, and brief duration and low-seizure stage manifestations begin to occur, the animals show cross tolerance to the anticonvulsant effects of carbamazepine. That is, when the unmedicated animals are showing full-blown stage 4 or 5 kindled seizures and animals are challenged with carbamazepine, they may respond well, whereas those animals that have been pretreated with levetiracetam are completely unresponsive to the same dose of carbamazepine.[16] These data suggest that tolerance-like mechanisms that occur during amygdala-kindling development may be sufficient to show cross contingent tolerance to another agent on completed kindled seizures, as we have previously observed for valproate pretreatment during kindling development precluding carbamazepine's efficacy in the later phases of completed seizures.[5]

It is noteworthy, however, that once animals have shown complete tolerance to carbamazepine, they still show excellent responses to the anticonvulsant effects of levetiracetam,[16] indicating that carbamazepine and levetiracetam tolerance[12] share some, but not other, important mechanisms of action in order to convey this unidirectionality.

4. SUMMARY AND CONCLUSIONS

The preclinical data on tolerance development offer a considerable number of novel pathophysiological, conceptual, and treatment-related postulates that can now be directly tested in the clinic. One of these postulates is that the seizure-induced changes in gene expression may be roughly grouped into those that: (1) are part of the primary pathophysiology of the kindling process and memory trace, whereas others (2) are more transient and adaptive, representing endogenous anticonvulsant mechanisms.[7, 8] To the extent that this conceptualization is valid, it would suggest the therapeutic importance of not only inhibiting the primary pathological changes, but also enhancing those related to the endogenous anticonvulsant variety.

These same alterations may also be involved in the tolerance process because a variety of changes in gene expression appear to be suppressed in animals that are tolerant to carbamazepine or diazepam, in spite of the fact that full-blown seizures are occurring and one would expect the entire array of seizure-induced changes in mRNA to be manifest [5]. We have seen that some drug combinations (such as carbamazepine and valproate) at doses that each alone are associated with rapid tolerance development, when used in combination markedly slow the process. Similarly, when either gabapentin or MK-801 are added to lamotrigine in the suppression of amygdala-kindled seizures, lamotrigine tolerance is remarkably slowed.[21] These data suggest that in some instances in the clinic where tolerance is an issue, combination treatment may not only be initially important in re-acquiring seizure control, but also important for enhancing its duration of effectiveness (if a tolerance process had been underlying the original loss of efficacy).

Elsewhere, we have discussed other manipulations that may be important in slowing tolerance development based on the observation of contingent tolerance development to the anticonvulsant effects of carbamazepine on amygdala-kindled seizures.[5, 40] Some of these manipulations may also be usefully explored for their potential applicability in the clinic, including the potential benefit of early treatment.[41, 42]

Finally, we have mapped cross tolerances and lack thereof in the amygdala-kindling model that demonstrate interesting individual drug differences in mechanisms of action and, in some instances, surprising cross tolerance phenomena despite such differences. All of these preclinical observations are presented with the caveats that (1) they may or may not be directly applicable to a given clinical situation, and (2) the mechanisms of action and of tolerance development to a given drug may differ considerably in seizures, pain syndromes, and the recurrent affective disorders.[34, 42]

Thus, considerable caution is warranted in assessing whether some of the observations in the amygdala-kindled seizure model directly pertain to the clinic, or more likely whether some of the principles elucidated in the contingent inefficacy, contingent tolerance, and cross tolerance phenomena may be indirectly pertinent to clinical therapeutics. Such propositions can then be formulated with more precision based on the more readily available and rapidly acquired preclinical model data, and then directly tested in the clinic where appropriate.

5. REFERENCES

1. G. V. Goddard, D. C. McIntyre, and C. K. Leech, A permanent change in brain function resulting from daily electrical stimulation, *Exp. Neurol.* **25**(3), 295-330 (1969).
2. R. J. Racine, Modification of seizure activity by electrical stimulation. II. Motor seizure, *Electroencephalogr. Clin. Neurophysiol.* **32** 281-294 (1972).
3. S. R. Weiss and R. M. Post, Development and reversal of contingent inefficacy and tolerance to the anticonvulsant effects of carbamazepine, *Epilepsia* **32**(1), 140-145 (1991).
4. S. R. Weiss and R. M. Post, Contingent tolerance to carbamazepine: a peripheral-type benzodiazepine mechanism, *Eur. J. Pharmacol.* **193**(2), 159-163 (1991).
5. S. R. Weiss, M. Clark, J. B. Rosen, M. A. Smith, and R. M. Post, Contingent tolerance to the anticonvulsant effects of carbamazepine: relationship to loss of endogenous adaptive mechanisms, *Brain Res. Rev.* **20**(3), 305-325 (1995).
6. M. J. Mana, C. K. Kim, J. P. Pinel, and C. H. Jones, Contingent tolerance to the anticonvulsant effects of

carbamazepine, diazepam, and sodium valproate in kindled rats, *Pharmacol. Biochem. Behav.* **41** 121-126 (1992).

7. R. M. Post and S. R. Weiss, Ziskind-Somerfeld Research Award 1992. Endogenous biochemical abnormalities in affective illness: therapeutic versus pathogenic, *Biol. Psychiatry* **32**(6), 469-484 (1992).
8. R. M. Post and S. R. B. Weiss, A speculative model of affective illness cyclicity based on patterns of drug tolerance observed in amygdala-kindled seizures, *Mol. Neurobiol.* **13**(1), 33-60 (1996).
9. J. B. Rosen, S. R. Weiss, and R. M. Post, Contingent tolerance to carbamazepine: alterations in TRH mRNA and TRH receptor binding in limbic structures, *Brain Res.* **651** 252-260 (1994).
10. M. Clark, G. S. Massenburg, S. R. Weiss, and R. M. Post, Analysis of the hippocampal GABAA receptor system in kindled rats by autoradiographic and in situ hybridization techniques: contingent tolerance to carbamazepine, *Mol. Brain Res.* **26**(1-2), 309-319 (1994).
11. R. Q. Wan, E. C. Noguera, and S. R. Weiss, Anticonvulsant effects of intra-hippocampal injection of TRH in amygdala kindled rats, *Neuroreport* **9**(4), 677-682 (1998).
12. W. Loscher and D. Honack, Development of tolerance during chronic treatment of kindled rats with the novel antiepileptic drug levetiracetam, *Epilepsia* **41**(12), 1499-1506 (2000).
13. S. R. Weiss, R. M. Post, J. Patel, and P. J. Marangos, Differential mediation of the anticonvulsant effects of carbamazepine and diazepam, *Life Sci.* **36**(25), 2413-2419 (1985).
14. C. K. Kim, J. P. Pinel, M. M. Hudda, D. J. Wong, and A. Yung, Tolerance to the anticonvulsant effects of phenobarbital, trimethadione, and clonazepam in kindled rats: cross tolerance to carbamazepine, *Pharmacol. Biochem. Behav.* **41,** 115-120 (1992).
15. D. S. Reddy and M. A. Rogawski, Chronic treatment with the neuroactive steroid ganaxolone in the rat induces anticonvulsant tolerance to diazepam but not to itself, *J. Pharmacol. Exp. Therap.* **295**(3), 1241-1248 (2000).
16. Z. J. Zhang, G. Q. Xing, S. Russell, K. Obeng, and R. M. Post, Unidirectional cross-tolerance from levetiracetam to carbamazepine in amygdala-kindled seizures, *Epilepsia* **44**(12), 1487-1493 (2003).
17. J. M. Rigo, G. Hans, L. Nguyen, V. Rocher, S. Belachew, B. Malgrange, P. Leprince, G. Moonen, I. Selak, A. Matagne, and H. Klitgaard, The anti-epileptic drug levetiracetam reverses the inhibition by negative allosteric modulators of neuronal GABA- and glycine-gated currents, *Br. J. Pharmacol.* **136**(5), 659-672 (2002).
18. C. C. Kuo, A common anticonvulsant binding site for phenytoin, carbamazepine, and lamotrigine in neuronal Na+ channels, *Mol. Pharmacol.* **54**(4), 712-721 (1998).
19. S. R. Weiss, R. M. Post, E. Sohn, A. Berger, and R. Lewis, Cross-tolerance between carbamazepine and valproate on amygdala-kindled seizures, *Epilepsy Res.* **16**(1), 37-44 (1993).
20. E. Krupp, T. Heynen, X. L. Li, R. M. Post, and S. R. Weiss, Tolerance to the anticonvulsant effects of lamotrigine on amygdala kindled seizures: cross-tolerance to carbamazepine but not valproate or diazepam, *Exp. Neurol.* **162**(2), 278-289 (2000).
21. Z. J. Zhang, S. Russell, K. Obeng, T. Postma, G. Obrocea, S. R. Weiss, and R. M. Post, Coadministration of gabapentin or MK-801 with lamotrigine slows tolerance to its anticonvulsant effects on kindled seizures, *Pharmacol. Biochem. Behav.* **74**(3), 565-571 (2003).
22. G. H. Fromm, C. F. Terrence, and A. S. Chattha, Baclofen in the treatment of trigeminal neuralgia: double-blind study and long-term follow-up, *Ann. Neurol.* **15**(3), 240-244 (1984).
23. J. C. Taylor, S. Brauer, and M. L. Espir, Long-term treatment of trigeminal neuralgia with carbamazepine, *Postgrad. Med. J.* **57** 16-18 (1981).
24. R. M. Post, G. S. Leverich, A. S. Rosoff, and L. L. Altshuler, Carbamazepine prophylaxis in refractory affective disorders: a focus on long-term follow-up, *J. Clin. Psychopharmacol.* **10**(5), 318-327 (1990).
25. P. J. Pazzaglia and R. M. Post, Contingent tolerance and reresponse to carbamazepine: a case study in a patient with trigeminal neuralgia and bipolar disorder, *J. Neuropsychiatry. Clin. Neurosci.* **4**(1), 76-81 (1992).
26. R. M. Post, T. A. Ketter, K. Denicoff, G. S. Leverich, and K. Mikalauskas, Assessment of anticonvulsant drugs in patients with bipolar affective illness, in: *Human Psychopharmacology: Methods and Measures,* edited by I. Hindmarch and P. D. Stonier (John Wiley & Sons, Chichester, England, 1993), pp. 211-245.
27. R. M. Post, G. S. Leverich, L. Altshuler, and K. Mikalauskas, Lithium-discontinuation-induced refractoriness: preliminary observations, *Am. J. Psychiatry* **149**(12), 1727-1729 (1992).
28. R. M. Post, G. S. Leverich, P. J. Pazzaglia, K. Mikalauskas, and K. Denicoff, Lithium tolerance and discontinuation as pathways to refractoriness, in: *Lithium in Medicine and Biology,* edited by N. J.

Birch, C. Padgham, and M. S. Hughes, (Marius Press, Lancashire, UK, 1993), pp. 71-84.
29. M. Maj, R. Pirozzi, L. Magliano, and L. Bartoli, Long-term outcome of lithium prophylaxis in bipolar disorder: a 5-year prospective study of 402 patients at a lithium clinic, *Am. J. Psychiatry* **155**(1), 30-35 (1998).
30. A. Koukopoulos, D. Reginaldi, G. Minnai, G. Serra, L. Pani, and F. N. Johnson, The long term prophylaxis of affective disorders, *Adv. Biochem. Psychopharmacol.* **49,** 127-147 (1995).
31. S. R. Weiss, R. M. Post, P. Anthony, and J. Ferrer, Contingent tolerance to carbamazepine is not affected by calcium- channel or NMDA receptor blockers, *Pharmacol. Biochem. Behav.* **45**(2), 439-443 (1993).
32. H. S. White, Comparative anticonvulsant and mechanistic profile of the established and newer antiepileptic drugs, *Epilepsia* **40 Suppl 5,** S2-S10 (1999).
33. J. P. J. Pinel, Effects of diazepam and diphenylhydantoin on elicited and spontaneous seizures in kindled rats: a double dissociation, *Pharmacol. Biochem. Behav.* **18,** 61-63 (1983).
34. R. M. Post, Do the epilepsies, pain syndromes, and affective disorders share common kindling-like mechanisms?, *Epilepsy Res.* **50**(1-2), 203-219 (2002).
35. T. Postma, E. Krupp, X. L. Li, R. M. Post, and S. R. Weiss, Lamotrigine treatment during amygdala-kindled seizure development fails to inhibit seizures and diminishes subsequent anticonvulsant efficacy, *Epilepsia* **41**(12), 1514-1521 (2000).
36. A. Iudice and L. Murri, Pharmacological prophylaxis of post-traumatic epilepsy, *Drugs* **59**(5), 1091-1099 (2000).
37. J. M. Silver, C. Shin, and J. O. McNamara, Antiepileptogenic effects of conventional anticonvulsants in the kindling model of epilespy, *Ann. Neurol.* **29,** 356-363 (1991).
38. W. Loscher, D. Honack, and C. Rundfeldt, Antiepileptogenic effects of the novel anticonvulsant levetiracetam (ucb L059) in the kindling model of temporal lobe epilepsy, *J. Pharmacol. Exp. Therap.* **284**(2), 474-479 (1998).
39. S. C. Stratton, C. H. Large, B. Cox, G. Davies, and R. M. Hagan, Effects of lamotrigine and levetiracetam on seizure development in a rat amygdala kindling model, *Epilepsy Res.* **53**(1-2), 95-106 (2003).
40. R. M. Post and S. R. B. Weiss, Convergences in course of illness and treatments of the epilepsies and recurrent affective disorders, *Clin. Electroencephalogr.* **35**(1), 14-24 (2004).
41. E. H. Reynolds, Early treatment and prognosis of epilepsy, *Epilepsia* **28**(2), 97-106 (1987).
42. R. M. Post, T. A. Ketter, A. M. Speer, G. S. Leverich, and S. R. Weiss, Predictive validity of the sensitization and kindling hypotheses, in: *Bipolar Disorders: Basic Mechanisms and Therapeutic Implications,* edited by J. C. Soares and S. Gershon, (Marcel Dekker, Inc., New York, 2000), pp 387-432.

Discussion

Teskey: These are fascinating data. My understanding is that one of the mechanisms for medical refractoriness to anticonvulsants is decreased permeability of blood brain barrier to the anticonvulsants. So I am wondering if you thought about this as the mechanism for your tolerance, that the blood brain barrier becomes less permeable.
Post: Yes, we have thought about that, and we have also thought about the induction of the resistance genes and some of these genes that transport drugs out of cells. It is this kind of tolerance that is associated with the induction of those genes. Then you would have to have seizures not inducing those genes, but it is only seizures in the presence of the drug inducing them to be relevant to this paradigm. It is a very specific prediction, but it would be a very good one to look at.
Wasterlain: I think our very next speaker will have something to say about that. But to my knowledge there is very little known about the blood barrier to anticonvulsants,
Teskey: What I am suggesting is that seizures can change blood-brain barrier permeability, and in fact seizures can increase angiogenesis, and that's either making the drugs more available to the blood or less available to the cells.
Post: Right, but they would have to do it at just the right temporal relationship, and that

effect would have to wear off the next day, because we have given drug and seizures uncorrelated and that is not associated with the process.
Gale: Actually, haven't you given the drugs and seizures correlated, but in the reverse temporal order?
Post: Yes, that is what I have shown. If you do that you will reverse the tolerance process because the animals are, in effect, having a med-free seizure. Because carbamazepine has a very short half-life in the rat, they are having the seizure, and then the drug is given afterwards.
Gale: The timing you are talking about, where the drug comes first, is reminiscent of some of the things we heard this morning about conditioning. Have you thought about the possibility that these drugs may be serving as discriminative stimuli? We heard this morning that animals can become conditioned to actually augment their response with a conditioned stimulus. If these drugs are serving as a conditioned stimulus...
Post: I think that is a great idea. We saw that with context-dependent cocaine sensitization, where the whole thing was sensitized, and we naively thought that this was more hard wired. I did not know until today how potent conditioned and context elements may be for this paradigm. Immediately it brings in the possibility that this could be like a context-dependent cueing, like you can see in some memory paradigms.
Burnham: The phenomena you describe are very much paralleled in alcohol tolerance by what we call the intoxicated practice phenomenon. If you don't know Harold Kalant's work, you might want to investigate it.
Post: Thank you.

MECHANISMS OF PHARMACORESISTANCE IN THE PHENYTOIN-RESISTANT KINDLED WISTAR RAT

Heidrun Potschka and Wolfgang Löscher*

1. SUBGROUPS OF KINDLED RATS AS A MODEL OF PHARMACO RESISTANT EPILEPSY

Although several new antiepileptic drugs (AEDs) have been launched during the last two decades, pharmacoresistance of epilepsy is still a major problem in clinical neurology. About 30% of all epileptic patients do not adequately respond to drug treatment.[1] In adults, temporal lobe epilepsy (TLE) with complex-partial seizures has the poorest prognosis with up to 70% of the patients being resistant to treatment with available AEDs.[2] Although several risk factors for intractability have been identified, the mechanisms underlying intractable epilepsy have not been fully enlightened. Recent data gave evidence that pharmacodynamic mechanisms, i.e., changes in drug targets (target hypothesis),[3] as well as pharmacokinetic mechanisms, i.e., a local decrease in brain access of AEDs mediated by overexpression of multidrug transporters (multidrug transporter hypothesis),[4] may be involved. Investigations in animal models of pharmacoresistant epilepsy are helpful to increase the understanding of the underlying basis of intractability, i.e., to further substantiate the existing hypotheses, to define further mechanisms, and to examine how different mechanisms coact. Unfortunately, there are only few animal models of pharmacoresistant epilepsy. One of the most extensively characterized animal models in this respect is a pharmacoresistant subgroup of amygdala-kindled Wistar rats.

Amygdala-kindling was first proposed by us as a model to study pharmacoresistant TLE in 1986.[5] This proposal was based on the observation that amygdala-kindled seizures are more difficult to suppress than seizures induced in the maximal electroshock seizure (MES) test. In follow-up studies, using female Wistar rats, it was found that the individual response to phenytoin differs. A subgroup of fully kindled rats consistently responded to phenytoin (PHT responders), whereas another subgroup of rats never responded to phenytoin (PHT nonresponders), although the plasma concentrations were in the therapeutic range in all rats[6]. Average data from several studies show that about 20% of the animals are PHT responders,

* Heidrun Potschka and Wolfgang Löscher, Dept. of Pharmacology, Toxicology, and Pharmacy, University of Veterinary Medicine, Foundation, Buenteweg 17, 30559 Hannover, Germany

about 20% of the animals are PHT nonresponders, and the remaining 60% show a variable response[7]. Interestingly the subgroups do not differ with regard to the kindling acquisition or the severity and duration of their fully kindled seizures.

Following the identification of subgroups with different pharmacosensitivity to PHT, most clinically available AEDs were tested in these animals. Except for the novel AED levetiracetam, all AEDs were significantly less efficacious or not efficacious in PHT nonresponders compared to responders, demonstrating that the resistance extends to various other old and new AEDs (Table 1).

Several factors which could influence the pharmacosensitivity in kindled rats have been investigated since the first description of the model[8]. It was demonstrated that pharmacoresistance in this model is not due to differences in the electrode location, drug pharmacokinetics, seasonal variations in drug response, or the gender of the animals. Breeding studies in which PHT responders and nonresponders were selectively mated gave evidence that pharmacosensitivity in individual kindled Wistar rats is genetically determined.[9] The involvement of genetics was substantiated by studies, in which phenytoin responders and nonresponders were selected in 7 outbred and inbred strains[10, 11] (and unpublished data). Besides genetics, the kindling process itself may have an impact on the pharmacosensitivity. A study, in which the anticonvulsant efficacy of PHT was tested before and after kindling, indicated that pharmacosensitivity was altered by the kindling process.[12] In rats which were nonresponders after kindling, phenytoin exerted a significant anticonvulsant effect before kindling. This study indicates that kindled PHT nonresponders become pharmacoresistant through kindling-induced brain alterations. Taken together, the results indicate that the genetic background of an individual rat determines whether it becomes a responder or nonresponder by limbic epileptogenesis.

The data raised the question, which genetic differences and which kindling-induced brain alterations have an impact on the pharmacosensitivity of kindled rats.

2. CHANGES IN DRUG TARGETS

Drugs of primary choice for treatment of TLE such as phenytoin and carbamazepine are thought to act via modulation of voltage-activated sodium and calcium channels. It has been demonstrated that the properties of these channels change in the hippocampus of patients with pharmacoresistant TLE.[13-15]

To directly address the possibility that neuronal sodium currents in the hippocampus play a crucial role in the pharmacoresistance of TLE, the effect of phenyoin on voltage-activated sodium currents was compared in CA1 neurons of PHT-responding and PHT-nonresponding kindled rats.[16] In view of the potential role of calcium current modulation in the anticonvulsant action of phenytoin, we also studied the effect of phenytoin on high-voltage-activated calcium currents in CA1 neurons. Electrode-implanted non-kindled rats were used as sham controls for comparison with kindled rats. In kindled rats with in vivo resistance to the anticonvulsant effect of phenytoin (phenytoin nonresponders), in vitro modulation of sodium and calcium currents by phenytoin in hippocampal CA1 neurons did not differ to any extent from respective data obtained in phenytoin responders, i.e. phenytoin resistance was not associated with a changed modulation of the sodium and calcium currents by this drug. Compared to sham controls, phenytoin's inhibitory effect on sodium currents was significantly reduced by kindling without any difference between the responder and nonresponder subgroups. These findings indicate that pharmacoresistance in individual kindled rats is not related to a reduced pharmacological sensitivity of sodium or calcium currents. However, recent studies in tissue of pharmacoresistant patients indicated that there is a complete loss of carbamazepine's effect on voltage-activated sodium channels in granule cells of the dentate gyrus.[3] In contrast, a distinct inhibition of granule cell sodium currents by carbamazepine was detectable in tissue from pharmacosensitive patients. Based on these results, additional future experiments appear to be reasonable, i.e. experiments testing for a difference in the sensitivity of dentate granule cell sodium currents to PHT in kindled responder and nonresponder subgroups of rats.

3. MULTIDRUG TRANSPORTER OVEREXPRESSION

In view of the fact that almost 90% of patients resistant to one AED are also resistant to other AEDs, involving drugs with different mechanisms, it seems unlikely that functional changes at target sites for AEDs are the major or only factor underlying pharmaco-resistance. In PHT nonresponding kindled rats the decreased pharmacosensitivity to several AEDs reflects the clinical situation. A likely explanation for multidrug pharmacoresistance in patients and in subgroups of kindled rats is that AEDs do not reach sufficiently high brain levels despite adequate plasma levels in the "therapeutic range". Several studies indicated that the multidrug transporter P-glycoprotein (Pgp, ABCB1) is overexpressed in epileptogenic brain tissue of patients with intractable epilepsy.[17-20] Pgp functions as an efflux transporter at the blood-brain barrier.[4] Thus, it was proposed that increased expression of Pgp in epileptic tissue may play a clinically significant role by limiting local brain access of AEDs to the brain parenchyma in the epileptic focus region.[4] Furthermore, in addition to Pgp, other multidrug transporters, such as members of the multidrug resistance-associated protein (MRP, ABCC) family, were reported to be overexpressed in capillary endothelial cells or astrocytes in such tissue.[21]

Many AEDs proved to be transported by the multidrug transporter Pgp[4] (Table 1). Of several AEDs tested, levetiracetam was the only AED that seems not to be a substrate of Pgp.[22] In view of the fact that levetiracetam was the only AED without a reduced efficacy in PHT nonresponders (Table 1), this gives evidence for a correlation between transport by Pgp and anticonvulsant efficacy in the subgroup of PHT-resistant kindled rats. Thus, these data indicate that overexpression of the multidrug transporter Pgp may underly pharmacoresistance in PHT nonresponders.

Using immunhistochemistry, we demonstrated that kindling per se results in an induction of Pgp expression in different brain regions involved in seizure generation and spread.[23] Eight hours following the last kindled seizure significant increases of endothelial Pgp immunoreactivity were seen in the hippocampus, piriform cortex, and cerebral cortex. The kindling-induced upregulation appears to be transient, because no significant increases in Pgp expression were observed 1-2 weeks after a kindled seizure.[24]

Table 1. Anticonvulsant efficacy of different AEDs in PHT nonresponders and transport of AEDs by multidrug transporters

Drug	Reduction of anticonvulsant efficacy in PHT nonresponders (in % of responders)	Transport by Pgp/ABCB1	Transport by MRP2/ABCC2
Phenytoin	100%	+	+
Carbamazepine	51%	+	?
Phenobarbital	85%	+	-
Valproate	79%	?	+
Vigabatrin	46%	?	?
Lamotrigine	60%	+	-
Felbamate	75%	+	-
Topiramate	61%	+	?
Gabapentin	52%	?	?
Levetiracetam	No reduction of efficacy	-	-

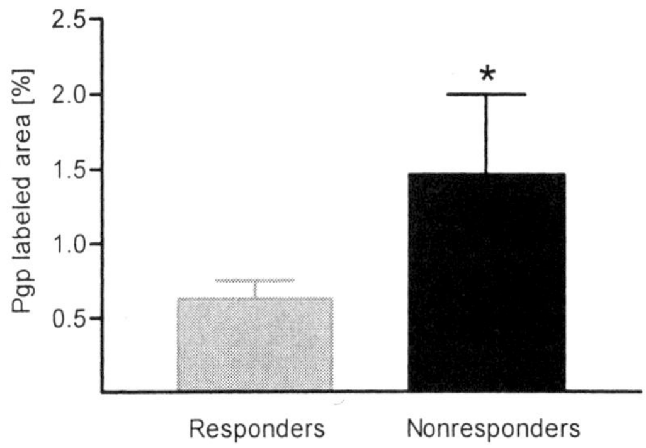

Figure 1. P-glycoprotein labeled area in the ipsilateral amygdala in % of the total area. Data are means ± SEM of PHT responders (n=7) and PHT nonresponders (n=6). Significant difference between the groups is marked by an asterisk ($p < 0.05$).

To address the question of whether there is a correlation between pharmacosensitivity to PHT and expression levels of Pgp, we analyzed Pgp expression in the amygdala of PHT responders and PHT nonresponders. Fully kindled rats were tested in four consecutive PHT drug trials in order to select PHT responders and nonresponders. Two days following the last kindled seizure, Pgp immunoreactivity was quantified using computer-assisted image analysis. In the ipsilateral amygdala the Pgp-labelled area was significantly higher (57%) in PHT nonresponders as compared to PHT responders (Fig.1). The intensity of Pgp staining was comparable in the amygdala of both groups of kindled rats. These data indicate that PHT resistance is correlated with an increased number of cells that express Pgp above detection level in the amygdala. Increased expression of Pgp at the site of stimulation may result in decreased brain access of PHT in the subgroup of PHT nonresponders and thus may contribute to pharmacoresistance in these rats. Polymorphisms in the Pgp encoding genes (in rodents two genes: *mdr1a* and *mdr1b*) may result in increased basal levels of Pgp in PHT nonresponders. Furthermore differences in the regulation of Pgp expression in response to stimuli may exist between the group of PHT nonresponders and responders, and thus PHT nonresponders may respond with a more intense Pgp upregulation to seizure activity as compared to PHT responders. In epileptic patients a correlation between polymorphisms in the Pgp encoding gene (*MDR1*) and the efficacy of AEDs has already been demonstrated[25]. To investigate if polymorphisms of Pgp do also contribute to pharmacoresistance in subgroups of kindled Wistar rats, we started to compare the *mdr1a* genotype in PHT responders and nonresponders by SSCP (single strand conformation polymorphism) analysis.

Experiments or clinical studies demonstrating that pharmacoresistance can be overcome by inhibition of multidrug transporter function would give a proof-of-principle for the multidrug transporter hypothesis. Thus, we started to test combinations of multidrug transporter inhibitors with AEDs. Because PHT proved to be transported by Pgp and MRP2 at the blood-brain barrier (Table 1), inhibitors of Pgp and of MRP2 are candidates for combination with PHT. In a preliminary experiment, the combination of phenytoin and the MRP inhibitor probenecid was tested in phenytoin-resistant kindled rats. With this combination, it was only possible to overcome phenytoin-resistance in one individual animal out of a group of phenytoin-resistant rats. This result is not surprising, because Pgp seems to play a major role in efflux transport of phenytoin at the BBB and proved to be overexpressed in PHT nonresponders. Thus, it is likely that it will be necessary to inhibit Pgp as well as MRPs in order to achieve therapeutic success. In ongoing experiments, we are testing combinations of the Pgp inhibitor XR9576, the MRP inhibitor probenecid, and phenytoin in rats that were selected to be phenytoin-resistant.

4. FURTHER MECHANISMS OF PHARMACORESISTANCE IN SUBGROUPS OF KINDLED RATS

In order to identify further mechanisms contributing to differences in pharmacosensitivity of individual kindled rats, a broad gene expression analysis was performed to determine differentially expressed genes in PHT responders and PHT nonresponders. This study was planned and performed in cooperation with UCB Pharma (Preclinical CNS Research, Braine L'Alleud, Belgium and UCB Research, Cambridge, USA). Tissue from PHT responders and PHT nonresponders was sampled in our department, gene expression analysis using gene chips was performed by Gene Logic (Gaithersburg, USA), and data analysis and validation was performed by UCB Pharma.

Two weeks following the last kindled seizure the ipsilateral temporal lobe was sampled from groups of 6 PHT responders and 6 PHT nonresponders, which were selected in four consecutive PHT drug trials. The tissue included the amygdala, the hippocampus, and parahippocampal tissue. Following RNA preparation and cDNA synthesis, fragmented cDNA from individual rats was hybridized to Affymetrix gene chips (rat genome U34 set). The data were evaluated based on data generated in a previous study, in which we determined kindling-induced changes in gene expression using gene chip analysis.[26] The expression of 49 genes differed in PHT responders and PHT nonresponders, of which the expression of 36 genes proved to be higher in PHT nonresponders, and the expression of 13 genes proved to be lower in PHT nonresponders. Interestingly, three of the differentially expressed genes are involved in nerve growth factor (NGF) receptor signaling or in modulation of NGF effects. The growth factor NGF is known to have diverse effects on the nervous system.[27]

Neurotrophic effects of NGF have been described but also antiproliferative and apoptosis enhancing effects. These effects of NGF are mediated by two receptors, i.e., the tyrosine kinase receptor (TrkA receptor) and the p75 neurotrophin receptor (p75-NTR or non-tyrosine kinase type receptor). In tissue of PHT-resistant rats, we found increased levels of mRNA coding for the proteins RAP-1 and 14-3-3-epsilon as compared to PHT-sensitive rats. Both proteins are involved in NGF/p75-NTR signalling,[27-29] and thereby modulate apoptosis. The mRNA that encodes the cytoskeletal protein 4.1N proved to be expressed at lower levels in the temporal lobe of phenytoin-resistant rats as compared to the tissue of phenytoin-sensitive rats. It has been described that 4.1N mediates the antiproliferative effect of NGF.[30] Seizure induced neurodegeneration and neuronal plasticity are considered to be relevant for the generation of a hyperexcitable network.[31] Thus, it is possible that differences in the modulation of neuronal apoptosis and plasticity by RAP-1, 14-3-3 epsilon or 4.1N have an impact on processes involved in the network generation. Resulting differences may be reflected by different levels of pharmacosensitivity.

In summary, gene expression profiling of PHT nonresponders and responders indicated that differences in neuronal degeneration and proliferation may contribute to the variance of pharmacosensitivity and to the phenomenon of pharmacoresistance in subgroups of kindled rats. In ongoing studies, we are thus addressing the question, do PHT nonresponders and responders differ with regard to the neuronal density in regions of the temporal lobe.

6. CONCLUSIONS

Investigations in subgroups of PHT sensitive responder rats and PHT resistant nonresponder rats gave no evidence that target associated mechanisms underly the variance in pharmacosensitivity. However, based on recent data from human epileptic patients, further experiments will be necessary to exclude the possibility that differences in the sensitivity of dentate granule cell sodium currents to PHT exist in subgroups of kindled rats.

Immunhistochemical expression analysis of the major multidrug transporter Pgp gave evidence for a contribution of Pgp overexpression to the phenomenon of pharmacoresistance in individual kindled rats. These data substantiate studies in pharmacoresistant epileptic patients, which showed that different multidrug transporters including Pgp seem to be upregulated in the epileptic tissue. Thus, the limitation of AED access to the epileptic focus by Pgp overexpression seems to be a mechanism of pharmacoresistance that is shared by pharmacoresistant epileptic patients and a pharmacoresistant subgroup of kindled Wistar rats. This points to the significant relevance of investigations in PHT nonresponding kindled rats, and indicates that these rats are suitable for testing new therapeutic strategies designed to overcome pharmacoresistance.

Gene expression profiling indicated that differences in neuronal degeneration and plasticity may exist that might render the kindled network more or less pharmacosensitive. Further investigations will adress the question, does the density of selected neuronal subpopulations in the temporal lobe differ between the groups of PHT nonresponders and PHT responders.

7. ACKNOWLEDGEMENTS

We thank Henrik Klitgaard, Jessie Gu, Berkley Lynch, Karl Nocka, Dina Anderson, and Feng Liang from UCB Pharma (Preclinical CNS Research, Braine L'Alleud, Belgium and UCB Research, Cambridge, USA) and Claudia Brandt from our department for their contribution to the gene expression study.

8. REFERENCES

1. P. Kwan and M.J. Brodie, Refractory epilepsy a progressive, intractable but preventable condition? *Seizure* **11**(2), 77-84 (2002).
2. I.E. Leppik, Intractable epilepsy in adults, in: *Surgical treatment of epilepsy,* edited by W.. H. Theodore, (Elsevier, Amsterdam, 1992) pp. 7-11.
3. S. Remy, S. Gabriel, B.W. Urban, D. Dietrich, T.N. Lehmann, C.E. Elger, U. Heinemann, and H. Beck, A novel mechanism underlying drug resistance in chronic epilepsy, *Ann. Neurol.* **53**(4): 469-479 (2003).
4. W. Löscher and H. Potschka, Role of multidrug transporters in pharmacoresistance to antiepileptic drugs, *J. Pharmacol. Exp. Therap.* **301**(1), 7-14 (2002).
5 W. Löscher, R. Jäckel, R., S.J. Czuczwar, Is amygdala kindling in rats a model for drug-resistant partial epilepsy? *Exp. Neurol.* **93**(1), 211-226 (1986).
6. W. Löscher and C. Rundfeldt, Kindling as a model of drug-resistant partial epilepsy: selection of phenytoin-resistant and nonresistant rats, *J. Pharmacol. Exp. Therap.* **258**(2), 483-489 (1991).
7. W. Löscher, Animal models of intractable epilepsy, *Prog. Neurobiol.* **53**(2), 239-258 (1997)
8. W. Löscher, Animal models of drug-resistant epilepsy, *NovartisFound.Symp.* **243**, 149-159 (2002)
9. U. Ebert and W. Löscher, Characterization of phenytoin-resistant kindled rats, a new model of drug-resistant partial epilepsy: influence of genetic factors, *Epilepsy Res.* **33**(2-3), 217-226 (1999).
10. S. Cramer, U. Ebert, and W. Löscher, Characterization of phenytoin-resistant kindled rats, a new model of drug-resistant partial epilepsy: comparison of inbred strains, *Epilepsia* **39**(10), 1046-1053 (1998).
11. W. Löscher, S. Cramer, and U. Ebert, Selection of phenytoin responders and nonresponders in male and female amygdala-kindled Sprague-Dawley rats. *Epilepsia* **39**(11), 1138-1147, (1998)
12. W. Löscher, E Reissmüller, and U. Ebert, Kindling alters the anticonvulsant efficacy of phenytoin in Wistar rats, *Epilepsy Res.* **39**(3), 211-220 (2002).
13. H. Beck, R. Steffens, U. Heinemann, and C E. Elger, Properties of voltage-activated Ca^{2+} currents in acutely isolated human hippocampal granule cells, *J. Neurophysiol.* **77**(3), 1526-1537 (1997).
14. H. Beck, R. Steffens, C.E. Elger, U. Heinemann, Voltage dependent Ca^{2+} currents in epilepsy, *Epilepsy Res.* **32**(1-2), 321-332 (1998).
15. G. Reckziegel, H. Beck, J. Schramm, C.E. Elger, and B.W. Urban, Electrophysiological characterization of Na^{2+} currents in acutely isolated human hippocampal dentate granule cells, *J. Physiology* **509**, 139-150 (1998).
16. M. Jeub, H. Beck, E. Siep, C. Rüschenschmidt, E.J. Speckmann, U. Ebert, H. Potschka, C. Freichel, E. Reissmüller, W. Löscher, Effect of phenytoin on sodium and calcium currents in hippocampal CA1 neurons of phenytoin-resistant kindled rats, *Neuropharmacology* **42**(1), 107-116 (2002)
17. D.M. Tishler, K.I. Weinberg, D.R. Hinton, N Barbaro, G.M. Annett, and C. Raffel, 1995, *MDR1* gene expression in brain of patients with intractable epilepsy, *Epilepsia* **36**(1), 1-6 (1995)
18. S.M. Sisodiya, J. Heffernan, and M V. Squier, Over-expression of P-glycoprotein in malformations of cortical development, *Neuroreport* **10**(16), 3437-3441 (1999)
19. S.M. Sisodiya, W.R Lin, B.N. Harding, M.V. Squier, and M. Thom, Drug resistance in epilepsy: expression of drug resistance proteins in common causes of refractory epilepsy, *Brain* **125**: 22-31 (2002)
20. E. Aronica, J.A. Gorter, G.H. Jansen, C.W.M. van Veelen, P.C. van Rijen, S. Leenstra, M. Ramkema, G.L. Scheffer, R.J. Scheper, and D. Troost, Expression and cellular distribution of multidrug transporter proteins in two major causes of medically intractable epilepsy: Focal cortical dysplasia and glioneuronal tumors, *Neuroscience* **118**(2), 417-429 (2003)
21. S M. Dombrowski, S.Y. Desai, M. Marroni, L. Cucullo, K. Goodrich, W. Bingaman, M.R. Mayberg, L. Bengez, D. Janigro, Overexpression of multidrug resistance genes in endothelial cells from patients with refractory epilepsy, *Epilepsia* **42**(1), 1501-1506 (2001)
22 H. Potschka, S. Baltes, and W. Löscher, Inhibition of multidrug transporters by verapamil and probenecid does not alter blood-brain barrier penetration of levetiracetam in rats, *Epilepsy Res.*, **58**(2-3), 85-91 (2004).

23. H.A. Volk, H. Potschka, and W. Löscher, Increased expression of the multidrug transporter p-glycoprotein in brain capillary endothelial cells and brain parenchyma of amygdala kindled rats, *Epilepsy Res.* **58**(1), 67-79 (2004)
24. U. Seegers, H. Potschka, and W. Löscher, Expression of the multidrug transporter p-glycoprotein in brain capillary endothelial cells and brain parenchyma of amygdala-kindled rats, *Epilepsia* **43**(7), 675-684 (2002)
25. A. Siddiqui, R. Kerb, M.E. Weale, U. Brinkmann, A. Smith, D.B. Goldstein, N.W. Wood, and S.M. Sisodiya, Association of multidrug resistance in epilepsy with a polymorphism in the drug transporter gene ABCB1, *New Engl. J. Medicine* **348**(15), 1442-1448 (2003).
26. J. Gu, B.A. Lynch, H. Klitgaard, S. Lu, M. Elashoff, U. Ebert, H. Potschka, and W. Löscher, The antiepileptic drug levetiracetam selectively modifies kindling-induced alterations in gene expression in the temporal lobe of rats, *Eur. J. Neurosci.* **19**(2), 334-345 (2004).
27 M T. Kimura, S. Irie, S. Shoji-Hoshino, J. Mukai, D. Nadano, M. Oshimura, and T.A. Sato, 14-3-3 is involved in p75 neurotrophin receptor-mediated signal transduction, *J. Biol. Chem.* **276**(20), 17291-17300 (2001).
28. R.D. York, H. Yao, T. Dillon, C.L. Ellig, S.P. Eckert, E.W. McCleskey, and P.J.S. Stork, Rap1 mediates sustained MAP kinase activation induced by nerve growth factor, *Nature* **392**, 622-626 (1998)
29. S. Shinoda, C.K. Schindler, J. Quan-Lan, J.A. Saugstad, W. Taki, R.P. Simon, and D.C., 2003, Interaction of 14-3-3 with Bid during seizure-induced neuronal death, *J. Neurochem.* **86**(2), 460-469 (2003)
30. K. Ye, D.A Compton, M.M. Lai, L.D.Walensky, and S.H. Snyder, 1999, Protein 4.1N binding to nuclear mitotic apparatus protein in PC12 cells mediates the antiproliferative actions of nerve growth factor, *J. Neurosci.* **19**(24), 10747-10756 (1999)
31. D.H. Lowenstein, Recent advances related to basic mechanisms of epileptogenesis. *Epilepsy Res. (Suppl.)* **11**, 45-60 (1996)

Discussion

Avanizi: Very nice data. Did you try to correlate the induction of the transport protein with the tissue concentration of the drug?
Potschka: We did that, with microdialysis in the ipsilateral amygdala. What we used in this experiment were non-selected kindled rats, and we found that when we elicit a seizure during the microdialysis procedure the concentration of phenytoin in the dialysate is lower in the animals that developed a seizure as compared to control animals. So we have evidence that there are lower concentrations of the antiepileptic drug in the epileptic focus region.
Moshé: This phenytoin resistance is universal to other agents too?
Potschka: Yes. Levetiracetam is the only antiepileptic drug that is equally effective in the group of phenytoin non-responders and responders. All the other antiepileptic drugs proved to be less efficacious in the group of phenytoin non-responders. However, there is not a complete loss of efficacy in this subgroup of kindled rats -- it is of course 100 percent for phenytoin, but for all the other antiepileptic drugs the efficacy is reduced by 50 up to 85 percent.
Moshé: So how do you explain that the PGP changes are not 100 percent?
Potschka: We believe that PGP is only one factor of pharmacoresistance. There will be other factors as well. That's why we performed the gene expression study. Furthermore the affinity of different anti-epileptic drugs for PGP is likely to differ.
Moshé: How about the genetics? You started with female rats, but you ended up with male rats.
Potschka: No, we used female rats in all our microdialysis and immunhistology experiments.
Moshé: But you said that pharmacological resistance is greater in the males.
Potschka: Yes, correct.
Moshé: So how is the genetics involved, and after the second generation what happens?

Potschka: With regard to genetics, we selectively bred with phenytoin responders or phenytoin non-responders. This is a really hard task, because kindled animals do not breed that well. When we selected the offspring animals, we observed that the offspring from responder breedings had a higher percentage of anticonvulsant responses in the repeated drug trials. So it was not that we got only responders in the offspring from responder breedings. Thus, there is evidence for an involvement of genetics. On the other hand, kindling induced changes also play a role. At the moment we are addressing the question of whether there are polymorphisms in the gene coding for P-glycoprotein or in the corresponding promotor region, because it may be that there is a difference in the regulation of expression, which may influence the effect of seizures in these groups of rats.

DEVELOPMENT OF NEW ANTICONVULSANTS USING THE KINDLING MODEL

W. McIntyre Burnham, Deborah Lonsdale, Afshin Shahzamani, Claudia Perez-Cruz, and Heather E. Edwards*

1. INTRODUCTION

In recent years, our laboratory has been using the kindling model in a search for new drugs effective against intractable epilepsy. In this chapter, I would like to discuss (1) intractable epilepsy; (2) the limbic kindling model and how it should be used in pharmacological studies related to complex partial seizures; (3) our approach to drug discovery; (4) our experiments with deoxycorticosterone and its metabolites; and (5) our experiments with progesterone and its metabolites.

2. INTRACTABLE EPILEPSY

The epilepsies are a group of neurological disorders characterized by spontaneous, recurrent seizures.[10] The causes of epilepsy are varied, and may include neoplasms, vascular anomalies, scars, stroke, genetic factors, birth trauma and brain injury.[6]

The most common therapy for epilepsy is anticonvulsant drug therapy. Many patients with epilepsy benefit from anticonvulsants, and live normal, productive lives. About 20% of the people with epilepsy, however, fail to respond to the standard medications.[21] These "drug resistant" or "intractable" patients continue to have seizures despite the best medical care. They face lives of economic hardship and social rejection.

* W. McIntyre Burnham, Deborah Lonsdale, Afshin Shahzamani, Claudia Perez-Cruz, Heather E. Edwards, University of Toronto, 1 King's College Circle, Toronto, Ontario, M5S 1A8.

2.1. Complex Partial Epilepsy

The majority of people with intractable seizures have a single type of epilepsy – complex partial epilepsy, usually of temporal lobe origin. Complex partial seizures are the most common seizures in adults,[9] and they are resistant to all of the present anticonvulsants.[14]

The control of complex partial seizures is one of the major challenges in the treatment of epilepsy today. Some complex-partial patients benefit from seizure surgery, the ketogenic diet or vagal stimulation. Many patients are not candidates for these procedures, however. Better medications are needed for these people. This has been the focus of our recent work.

2.2. West's Syndrome

We have also been interested in the seizures of West's syndrome. West's syndrome is an epileptic syndrome that begins during the first year of life.[11] Epidemiological studies suggest that over 2,000,000 cases of West's syndrome occur each year worldwide.[12]

The prognosis for West's syndrome is very poor. Up to 61 % of patients experience other types of seizures later on in life - often in the form of Lennox-Gastaut syndrome - and 85 % of patients show some degree of mental retardation.[12]

The seizures in West's syndrome are refractory to the standard antiepileptic drugs.[19] They often respond to adrenocorticotrophic hormone (ACTH), however.[2] ACTH also suppresses other forms of childhood seizure.[22] Long-term ACTH therapy, however, results in numerous and serious side effects, including hypertension, immune suppression, intracranial hemorrhage and cataracts.[3]

3. THE LIMBIC KINDLING MODEL AS A MODEL OF COMPLEX PARTIAL SEIZURES

The most widely used model for complex partial seizures (with secondary generalization) is the amygdala kindling model.[7] The pharmacological response of the kindled amygdala focus models the intractability of human complex partial seizures.[1] Used properly, the amygdala kindling model offers a preparation in which to develop new compounds effective against complex partial seizures. The hippocampal kindling model can also be used for this purpose.

Unfortunately, the amygdala kindling and hippocampal kindling models are seldom used properly. Researchers tend to report on the effect of drugs on the severity of kindled convulsions, or on the afterdischarge. Used this way, amygdala kindling does not model complex partial seizures. It models generalized convulsive seizures.[1]

Researchers also report on the effect of drugs on the rate of kindling. What this models is not clear. It may model the development of secondarily generalized seizures in patients that initially have only partial seizures. The preparation does *not* model the development of post-traumatic epilepsy, however. Anticonvulsants do retard kindling, but they do not appear to slow the development of post-traumatic epilepsy.

When Penny Albright and I were working to validate amygdala kindling as a pharmacological model for complex partial seizures, we tested drugs against two components of the seizures: (1) focal spiking in the amygdala; and (2) generalized convulsions. We found that generalized convulsions, and long afterdischarges, were suppressed by most anticonvulsant drugs – and particularly by those drugs effective against tonic-clonic seizures. Focal amygdala spiking – like complex partial seizures – was resistant to most anticonvulsants. Our criterion for focal spiking, however, was complete suppression of the afterdischarge – not just a change in duration. It is only when the model is used this way that its drug response resembles that of complex partial seizures.

4. OUR APPROACH TO DRUG DISCOVERY

Our laboratory has developed a novel approach to the discovery of new drugs for intractable seizures. We study medical treatments that are successful when the drugs fail – treatments such as the ketogenic diet and hormone therapy. We then try to discover their mechanisms of action, and to reproduce them pharmacologically.

We have discussed our studies of the ketogenic diet elsewhere.[15-17] Here I would like to discuss our experiments with hormones – and in particular the experiments that grew from our investigation of the anticonvulsant effects of adrenocorticotrophic hormone (ACTH).

5. EXPERIMENTS WITH DEOXYCORTICOSTERONE (DOC)

As mentioned above, the majority of children with West's syndrome respond to ACTH, a treatment that also suppresses other types of childhood seizures. Long-term ACTH, however, has numerous and serious side effects.

Heather Edwards, working in our laboratory as a post-doctoral fellow, was interested in trying to understand the anticonvulsant mechanism of action of ACTH. She therefore tested the anticonvulsant effects of ACTH and corticosterone – the hormone that ACTH releases – in several different seizure models. She also tested the effects of several metabolic precursors to corticosterone (see Figure 1). These precursor steroids are produced and released during the production of corticosterone. Like all steroids, they cross the blood-brain barrier and enter the brain.

In Heather's experiments, we found that neither ACTH nor corticosterone was an effective anticonvulsant.[5] Two corticosterone precursors, however, were strong, broad-spectrum anticonvulsants. These were deoxycorticosterone (DOC) and progesterone.

We had already been working on progesterone[4] - due to its involvement in catamenial epilepsy. DOC's effects, however, were new to us. Heather and, later, Claudia Perez-Cruz, therefore, began to study the anticonvulsant effects of DOC. We hoped that DOC might be as effective as ACTH as suppressing childhood seizures – without having as many side effects. We also hoped that it might be effective against complex partial seizures.

It is important to note both here and in the following discussions, that DOC and progesterone are "neuroactive steroids." They cross the blood-brain barrier and bind to receptors in the brain. It is also important to note that they have cell surface receptors, as well as the traditional nuclear steroid receptors. Since we have looked at anticonvulsant

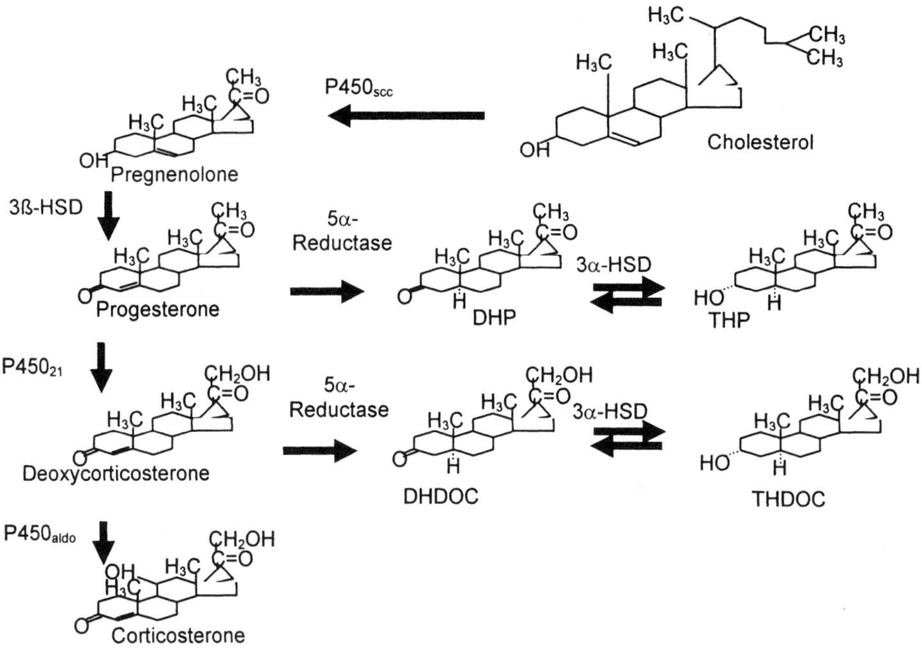

Figure 1. Steroid metabolism.

effects 15 minutes after injection, our effects relate to cell surface receptors. (Genomic effects take longer to develop.) Finally, it is important to note that both DOC and progesterone have active neurosteroid metabolites that also enter the brain and bind to cell-surface receptors.

Claudia investigated both DOC's mechanism of action and its spectrum of action. In terms of mechanism, she studied the relation of DOC's anticonvulsant effects to its primary and secondary metabolites. She worked primarily in 15-day-old rats, which had been the subjects of Heather's earlier ACTH studies. Seizure tests were done 15 minutes after injection in the hopes of maximizing the concentration of the test compound injected, although, at 15 minutes, metabolites of these compounds are present.

Claudia found that DOC's anticonvulsant effects disappeared in the presence of finasteride. Finasteride stops the conversion of DOC its metabolites. This suggested that DOC is not anticonvulsant in itself, but that it owes its anticonvulsant effects to its metabolites. Claudia therefore tested the anticonvulsant actions of DOC's first and second metabolites, dihydrodeoxycorticosterone and tetrahydrodeoxycorticosterone (DHDOC and THDOC, respectively). She found that both DHDOC and THDOC are strong anticonvulsants in 15-day old rats. It is now known that both are allosteric enhancers of GABA-ergic inhibition.

The above work was all done in the maximal pentylenetetrazol model (MMT), which is particularly responsive to steroids. Claudia also tested DOC's effects in the maximal electroshock (MES) model in the hippocampal-kindling model. DOC suppressed *generalized* convulsions in both of these models. Disappointingly, however, focal spiking in the hippocampus model was not fully suppressed, even at high doses. We concluded that DOC might be useful in the treatment of the generalized seizures of childhood. It would probably not be useful, however, in the treatment of complex partial seizures.

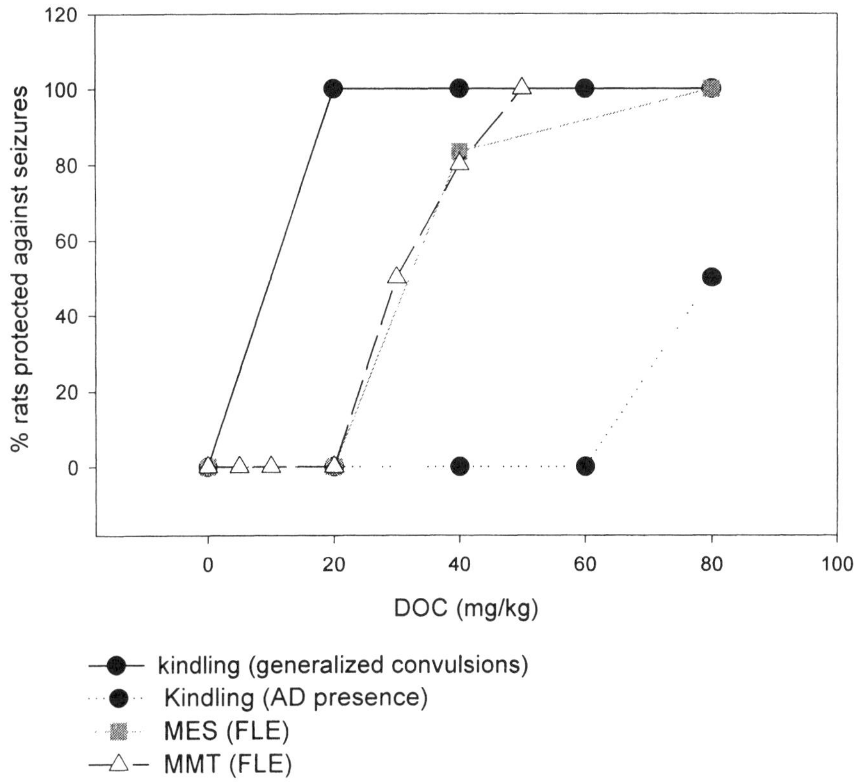

Figure 2. Dose-response data for DOC's anticonvulsant effects in 15-day-old rats on hippocampal kindled, MES and MMT seizures. Different doses of DOC were administered to pups, followed 15 min later by the seizure provoking stimulus.

6. EXPERIMENTS WITH PROGESTERONE

Investigations of the anticonvulsant effects of progesterone are ongoing in our laboratory. They have been carried out by Heather Edwards, Afshin Shahzamani and Deborah Lonsdale.

It has been known that progesterone has anticonvulsant effects ever since Selye's pioneering experiments[20] in 1941. Most experiments, however, have involved female subjects, since progesterone is thought of a female hormone. Heather, however, pointed out that both male and infant brains also have receptors for progesterone and its metabolites. We therefore went on to show that progesterone is anticonvulsant in both male and female rats, and in both adults and pups (Shahzamani et al., in preparation). It could be used for any sort of patient.

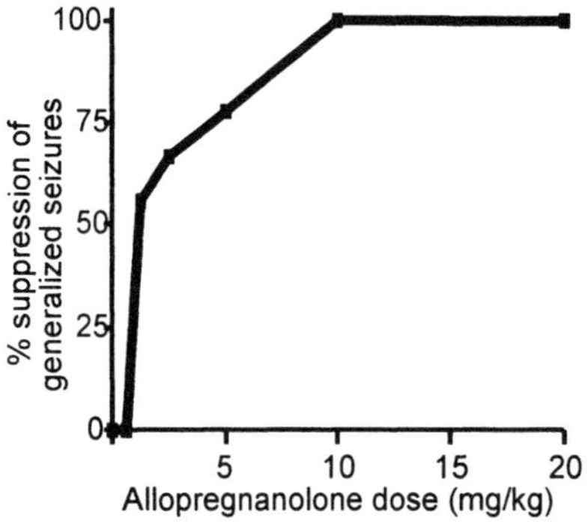

Figure 3. Allopregnanolone Dose-Response Curve at 15 min.

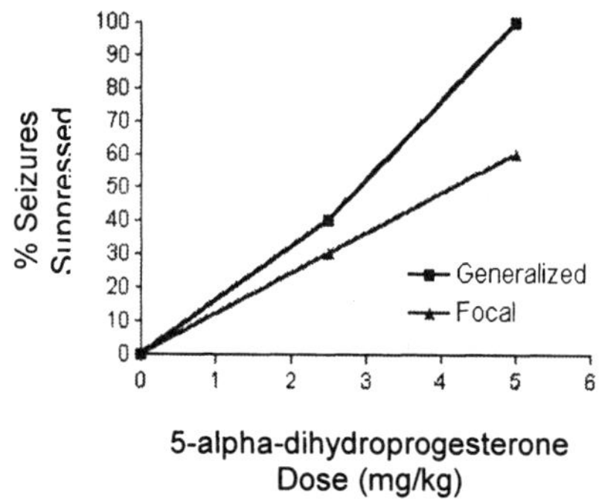

Figure 4. 5alpha-dihydroprogesterone Dose-Response Curves at 15 min (Reproduced from ref. 18, with permission).

Recent experiments from other groups have addressed progesterone's mechanism of action. Kokate et al.[13] and Frye et al.[8] have found that progesterone's anticonvulsant actions are due, at least in part, to its metabolites. Progesterone is first metabolized into 5-dihydroxyprogesterone (DHP) and then into 5-tetrahydroxyprogesterone (THP) - sometimes called "allopregnanolone." As with DOC, finasteride can block the first step in progesterone's metabolism. Inhibition of metabolism with finasteride, or a genetic deficiency in the 5alpha-reductase type I enzyme in mice, both result in a decrease in the anticonvulsant activity of progesterone.[8,13] In our hands, finasteride *completely* blocks progesterone's anticonvulsant effects (Edwards and Burnham, unpublished data), suggesting that progesterone's effects are completely due to its metabolites.

Recent investigations of the progesterone metabolites, however, have primarily focused on progesterone's second metabolite, THP. THP potentiates GABA-ergic activity, which is believed to be the mechanism of its anticonvulsant action. At higher doses, however, THP is quite sedative.

Deborah Lonsdale in our group suspected that progesterone's first metabolite, DHP, might also be anticonvulsant. She compared the effects of DHP to the effects of progesterone and THP in adult, female, amygdala-kindled rats. Tests were done 15 minutes post injection.

Progesterone was tested first and it was not very effective. It had an ED50 of 100 mg/kg against generalized kindled convulsions, but did not suppress the amygdaloid focus.[18] Subjects were not sedated at the time of seizure testing, but sedation developed later (40-60 minutes post-injection) - presumably due to the formation of metabolites. THP was stronger than progesterone. It had an ED50 of less than 5 mg/kg, against kindled generalized convulsions. Unfortunately, it did not suppress the amygdaloid focus, even at high doses associated with sedation (Lonsdale and Burnham, in preparation). DHP, however, was as effective as THP against the generalized convulsion, and it *did* suppress spiking in the amygdaloid focus. It did not produce sedation 15 minutes post injection, or at any later time interval.

Thus, DHP is the first compound we have found in almost 25 years of work that suppresses the amygdaloid focus at non-toxic doses. We are currently attempting to establish DHP's mechanism of action, and to begin its development as an anticonvulsant drug.

7. ACKNOWLEDGEMENTS

The authors would like to thank the Canadian Institutes of Health Research and the Michael Bahen Chair of Epilepsy Research for financial support, and Brian Tse and Richard Liu for help in preparation of the manuscript.

8. REFERENCES

1. P. S. Albright and W. M. Burnham, W. Development of a new pharmacological seizure model: effects of anticonvulsants on cortical- and amygdala-kindled seizures in the rat, *Epilepsia* **21,** 681-689 (1980).
2. T. Z. Baram, W. G. Mitchell, A. Tournay, O. C. Snead, R. A. Hanson, R.A., and E. J. Horton, E.J., High-dose corticotropin (ACTH) versus prednisone for infantile spasms: a prospective, randomized, blinded study, *Pediatrics* **97,** 375-379 (1996).
3. F. Dehkharghani, Myoclonic Epilepsy, in: *Handbook of Pediatric Epilepsy*, edited by J. Murphy and F. Dehkharghani, F (Marcel Dekker, USA, 1993), pp. 157-189.
4. H. E. Edwards, W. M. Burnham, A. Mendonca, D. A. Bowlby, D.A., and N. J. MacLusky, Steroid hormones affect limbic afterdischarge thresholds and kindling rates in adult female rats, *Brain Res.* **838,** 151-157 (1999).
5. H. E. Edwards, S. Vimal, S., and W. M. Burnham, The effects of ACTH and adrenocorticosteroids on seizure susceptibility in 15-day-old male rats, *Exp. Neurol.* **175,** 182-190 (2002).
6. J. Engel, Jr., Etiology as a risk factor for medically refractory epilepsy: a case for early surgical intervention, *Neurology*, **51(5),** 1243-1244 (1998).
7. R. S. Fisher, R.S., 1989, Animal models of the epilepsies, *Brain Res. Rev.* **14(3),** 245-278 (1989).
8. C. A. Frye, M. E. Rhodes, A. Walf, and J. Harney, Progesterone reduces pentylenetetrazol-induced ictal activity of wild-type mice but not those deficient in type I 5alpha-reductase, *Epilepsia*, **43(S5),** 7-14 (2002).
9. H. Gastaut, J. L. Gastaut, G. E. Goncalves e Silva, G and G. R. Fernandez Sanchez, Relative frequency of different types of epilepsy: a study employing the classification for the International League Against Epilepsy, *Epilepsia*, **16,** 457-461 (1975).
10. A, Guberman and J. Bruni, J., *Essentials of Clinical Epilepsy* (Butterworth-Heinemann, Boston, 1997).
11. R. A. Hrachovy, R.A., 2002, West's syndrome (infantile spasms), in: *Intractable seizures, Advances in Experimental Medicine and Biology Vol 497,* edited by W. M. Burnham, P. L. Carlen, and P. A. Hwang, P (Kluwer Academic/Plenum Publishers, New York, 2002), pp. 33-50.
12. D. L. Hurts, D.L., 1994, The Epidemiology of infantile spasm, in: *Infantile Spasm and West Syndrome*, edited by O. Dulac, H. Chungan, and B. Dalla Bernadina (Saunders, London, 1994), pp. .
13. T. G. Kokate, M. K. Banks, T. Magee, S. I. Yamaguchi, and M. A. Rogawski, Finasteride, a 5 alpha-reductase inhibitor, blocks the anticonvulsant activity of progesterone in mice, *J. Pharmacol. Exper. Therap.* **288,** 679-684 (1999).
14. I. E. Leppik, I.E., 2002, Intractable epilepsy in adults, in: *Intractable Seizures, Advances in Experimental Medicine and Biology Vol 497,* edited by W. M. Burnham, P. L. Carlen, and P. A. Hwang, (Kluwer Academic/Plenum Publishers, New York, 2002), pp. 1-7.
15. S. S. Likhodii, K. Musa, A. Mendonca, C. Dell, W. M. Burnham, and S. C. Cunnane, Dietary fat, ketosis and seizure protection in rats on the ketogenic diet, *Epilepsia* **41,** 1400-1410 (2000).
16. S. S. Likhodii W. M. and Burnham, On the anticonvulsant effect of acetone and the ketogenic diet, *Epilepsia* **43,** 1596 (2002).
17. S. S. Likhodii, I. Serbanescu, M. A. Cortez, P. Murphy, O. C. Snead 3rd, and W. M. Burnham, Anticonvulsant properties of acetone, a brain ketone elevated by the ketogenic diet, *Ann. Neurol.* **54,** 219-226 (2003).
18. D. Lonsdale W. M. and Burnham, The anticonvulsant effects of progesterone and 5alpha-dihydroprogesterone on amygdala-kindled seizures in rats, *Epilepsia* **44(12),** 1494-1499 (2003).

19. E. Schlumberger O. and Dulac, A simple effective and well-tolerated treatment regime for West syndrome, *Dev. Med. Child Neurol.* **36,** 863-872 (1994).
20. H. Selye, Anesthetic effects of steroid hormones, *Proc. Soc. Exp. Biol.* **46**, 116-121 (1941).
21. S. D. Shorvon, The epidemiology and treatment of chronic and refractory epilepsy, *Epilepsia* **37(S2)**, S1-S3 (1996).
22. O. C. Snead, J. W. Benton, L. C. Hosey, J. W. Swann, D. Spink, D. Martin, and R. Rej, Treatment of infantile spasm with high-dose ACTH: efficacy and plasma levels of ACTH and cortisol, *Neurology* **39,** 1027-1031 (1989).

Discussion

Teskey: Wonderful work. Mac. Does DHP convert back to progesterone or is it only unidirectional?
Burnham: As far as I know, it is unidirectional. We have actually tested that with DHDOC and THDOC. If you put in indomethacin they won't interconvert. I would guess that DHP and THP don't interconvert. Otherwise THP would have similar effects as DHP, but we'd have to do the indomethacin to be sure.
Veliskova: Do you have any evidence that the hormones are working through the membrane receptors, or are they working through the $GABA_A$ receptors? Progesterone is acting through the $GABA_A$ receptor, and the neurosteroids have their own binding site at the $GABA_A$ receptor.
Burnham: Let me answer that question in two or three ways. THDOC, DHDOC, and THP all bind to the steroid site on the $GABA_A$ complex, and they work through the GABAergic mechanism. DHP obviously does not because it has a very different spectrum of action. All the receptors that we will be dealing with here would be cell surface receptors. We always test 15 minutes after administration of the drug, and you couldn't get genomic effects that quickly.
Veliskova: Well, it has been shown that there are genomic effects as soon as 10 minutes after hormone administration.
Burnham: I was told 20 minutes. We have been reading different books! It is unlikely that we would be seeing genomic effects at this time. The DHP receptor has been described in peripheral tissue, and it is a cell surface receptor. The next step is to study it and try to find out what it does.
Wasterlain: Usually the neurosteroids are thought to bind preferentially to the delta subunits of the $GABA_A$ receptor, which is responsible for extracellular tonic currents. So does that mean that they will affect your particular type of seizures, or would it affect general tone more than the physical effects? And would that not have some drawbacks are far as sedation?
Burnham: Basically all I can say is three of the four drugs we've been working with clearly are binding to the $GABA_A$ site. They all are associated with sedation, although the therapeutic index is pretty good for DHDOC and THDOC. DHP does not seem to be binding to the GABA complex in any way, and it has totally been unassociated with sedation even at fairly high doses.

CANNABINOIDS AND KINDLING

Michael E. Corcoran, Deborah Saucier, Xia Zhang, Aaron Sheerin, and Kenneth Wolfe*

1. INTRODUCTION

For millennia there has been interest in the potential therapeutic effects of marijuana (cannabis), including its potential antiepileptic effects. A flurry of research on cannabis and seizures occurred in the 1970s, when purified tetrahydrocannabinol (THC) became available for research. The results demonstrated that a variety of seizures, including kindled seizures can be suppressed by THC, but typically at toxic doses.[1-3] It was also reported that prophylactic administration of THC can delay limbic kindling in rats[2] and cats.[3] In contrast, other studies that received less attention described proepileptic or proconvulsant effects of a variety of cannabinoids, including THC.[4-6] Because of a lack of knowledge about the mechanisms of cannabis's effects on the brain, research on cannabis and epilepsy declined to a very low level in the 1980s.

Scientific interest in the effects of cannabinoids recently revived with the discoveries that cannabinoid receptors exist in the body, including brain, and that endogenous ligands for cannabinoid receptors are present in brain. As reviewed by Pertwee,[7] 2 classes of cannabinoid receptors have been identified: the CB1 receptor, localized primarily on central neurons, although also present in the PNS; and the CB2 receptor, located mainly on immune cells. Both receptors are coupled to G proteins.[8] Recent research has also suggested that a third target site for cannabinoids exists, perhaps serving as a vanilloid receptor.[9] It is now known that several endogenous ligands, the "endocannabinoids," bind to these receptors. The best characterized ligands are anandamide and 2-arachidonoylglycerol.[7]

*Michael Corcoran and Kenneth Wolfe, Department of Anatomy and Cell Biology. Deborah Saucier and Aaron Sheerin, Department of Psychology. Xia Zhang, Neuropsychiatry Research Unit, University of Saskatchewan, Saskatoon, SK, Canada, S7N 5E5..

Work has also progressed on developing drugs that act as selective agonists and antagonists. The availability of synthetic cannabinoid agonists and antagonists has provided a pharmacological probe for the involvement of endocannabinoids and CB receptors in brain function. A wide array of natural and synthetic cannabinoids have now been tested; some act as agonists, some as partial agonists, some as antagonists, and some as inverse agonists.[7] Several recent reports have indicated that CB1 agonists suppress tonic seizures induced by pentylenetetrazol (PTZ) or electroconvulsive shock (ECS), and that PTZ and ECS-induced seizures are exacerbated by treatment with CB1 antagonists.[10-12] These results suggest that some types of seizure might benefit from treatment with cannabinoid agonists, and, further, that endocannabinoids may act to dampen seizure discharge in the CNS. Furthermore, a study by Lambert et al.[13] reported the provocative finding that tonic seizures in mice are suppressed by N-palmitoylethanolamide (PEA), a putative endocannabinoid that is implicated in inflammatory processes, at nontoxic doses.

We wished to determine whether selective cannabinoid drugs will influence kindled seizures and kindling itself. Hence we examined the effects of a CB1 agonist, HU-210, on amygdaloid seizures and kindling and on neocortical seizures in rats.

2. METHODS

2.1. Effects of a Cannabinoid Agonist on Kindled Seizures and Kindling

We report the results of two lines of experimentation, one characterizing the effects of cannabinoids on kindled seizures, and the other characterizing the effects of cannabinoids on kindling itself. All experiments were performed in male hooded rats, weighing 200 to 250 gr at the start of experimentation. For kindling, bipolar stimulating recording electrodes were implanted bilaterally into the basolateral amygdala, dorsal hippocampus, or anterior neocortex. Ten days after surgery, kindling began. Before any drug injections, we determined the threshold for afterdischarge (AD). Stimulation consisted of a 1-sec train of balanced biphasic square wave 1 msec pulses at 60 pps at an initial intensity of 10 μA (base-to-peak). Intensity was raised in gradual steps until an AD was triggered. To investigate the potential antiepileptic effects of cannabinoids, we triggered kindled seizures with stimulation at 100 μA above threshold, in order to ensure stability of seizures. To investigate the potential prophylactic effects of cannabinoids, we kindled seizures with stimulation at the threshold for AD.

Rats received ip injections of the CB1/CB2 agonist HU-210, dissolved in DMSO, at a volume of 1 ml/kg and at doses of 10, 25, 50, 100, and 500 μg/kg. Because the pharmacokinetics and half-life of HU-210 have not been described, we performed pilot studies to identify the interval at which significant drug effects were observed, and then gave the drug at this interval in the formal experiments. To characterize the antiepileptic effects of HU-210, we kindled generalized seizures with amygdaloid stimulation. After the duration and intensity of the seizures had stabilized, we gave the rats ip injections of drug or vehicle, and then triggered seizures 20 min later. Each rat received only one dose of drug, and to assess the effects of chronic administration we gave the drug once daily for 5 days. We continued to trigger kindled seizures for at least 10 days after termination of drug injections. The effects of HU-210 on seizures kindled from dorsal hippocampus and from anterior neocortex were also examined in small groups of rats.

To determine whether HU-210 can produce prophylactic effects on kindling, we gave other rats injections of the drug during amygdaloid kindling. Kindling stimulation and drug injections were administered every other day. HU-210 or vehicle was injected ip 20 min before each stimulation of the amygdala. After kindling of 3 generalized seizures, injections were terminated and seizures were triggered for 3 more days.

3. RESULTS

3.1. Effects of Cannabinoid Agonists on Kindled Seizures and Kindling

Gross behavioral observation indicated that the higher doses of HU-210 produced toxic effects on general behavior. Rats treated with either 100 or 500μg/kg of HU-210 responded with loud vocalizations when approached or handled. They were generally inactive and ataxic, and displayed flaccid muscle tone. When placed in a supine position, the rats treated with lower doses responded with strong righting and bracing reflexes, whereas the rats treated with higher doses failed to right themselves and did not display any bracing limb response. Other effects of HU-210 were observed on a variety of reflexive, cognitive, and emotional behaviors (data not shown).

When characterized as protection against clonic seizures, the acute effects of HU-210 on amygdaloid seizures displayed a steep dose-response curve. As shown in Figure 1, no protective effect was produced by 10 and 25 μg/kg, a modest protective effect was obtained with 50 μg/kg, and the maximum effect was observed at 100 μg/kg, with no further effect being seen at 500 μg/kg. Figures 2 and 3 depict the effects of HU-210 on duration of clonus, resulting in a generally similar pattern, with minimal effects evident after the first injection of 10, 25, and 50 μg/kg, and statistically significant reduction of clonus evident at 100 and 500 μg/kg.

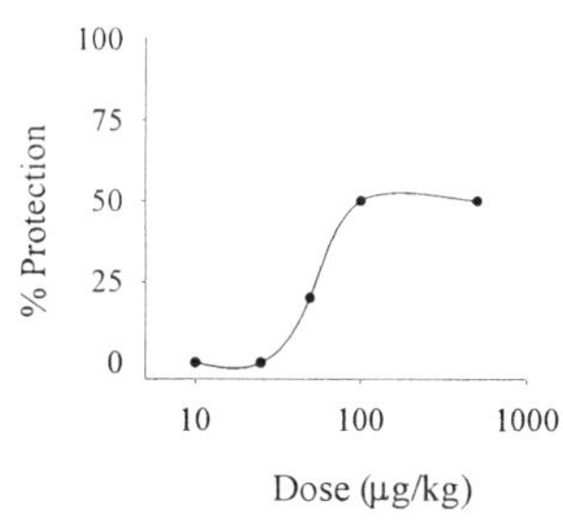

Figure 1. Log dose-response relations of the protection against clonic seizures produced by HU-210.

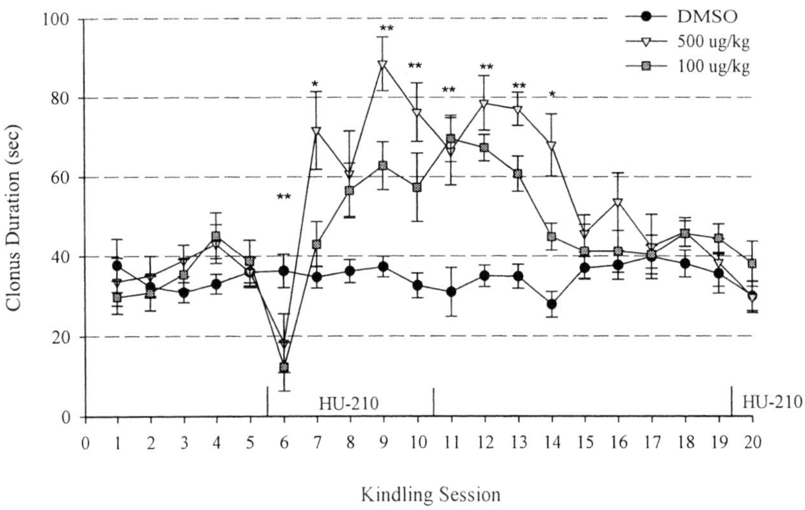

Figure 2. Effects of HU-120 on clonus duration of kindled amygdaloid seizures. * 500 μg/kg different from control, $p<0.025$; ** 500 & 100 μg/kg different from control $p<0.025$.

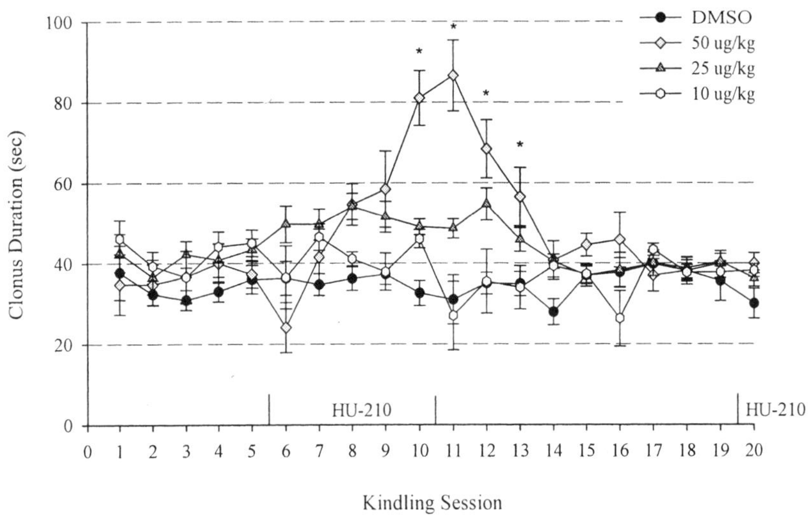

Figure 3. Effects of HU-120 on clonus duration of kindled amygdaloid seizures. * 50 μg/kg different from control, $p<0.025$.

Unexpectedly, chronic administration of HU-210 resulted in a proepileptic effect on amygdaloid kindled seizures, as indicated by the significant increase in seizure duration occurring on the second or subsequent injection of several of the doses, as shown in Figures 2 and 3. Thus tolerance to the modest acute antiepileptic effects of HU-210 developed extremely rapidly, as did apparent sensitization to the proepileptic effect of HU-210, which was evident for the higher doses by the fourth injection. Seizure durations remained elevated for 3 or 4 days after discontinuation of drug injections, but returned to baseline by the end of testing.

To examine the generality of the proepileptic effect of HU-210, we have tested the effects of HU-210 at 100 µg/kg on seizures kindled from anterior neocortex. or dorsal hippocampus in preliminary experiments No antiepileptic effect was observed at this dose, and instead an increase in duration of clonus was evident during treatment with the drug (Figures 4A and 4B).

In order to examine the potential prophylactic effects of HU-210, we administered the drug in doses of 10, 25, 50, and 100 µg/kg during amygdaloid kindling. As shown in Figure 5A and 5B, HU-210 at 10 and 25 µg/kg had no significant effect on the rate of kindling of generalized seizures, whereas doses of 50 and 100 µg/kg produced a significant acceleration of kindling. The duration of AD during kindling was also increased significantly by treatment with doses of 25, 50, and 100 µg/kg (Figure 6).

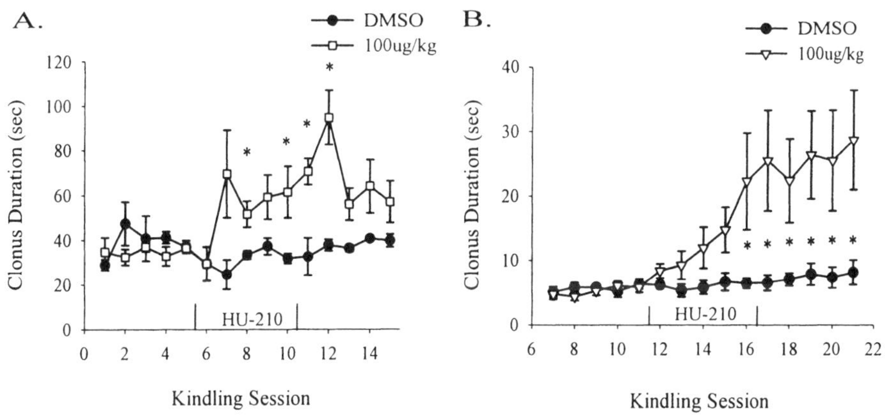

Figure 4. Effects of HU-210 on duration of clonus of seizures kindled from dorsal hippocampus (A) or anterior neocortex (B). *Significantly different from control, $p<0.025$.

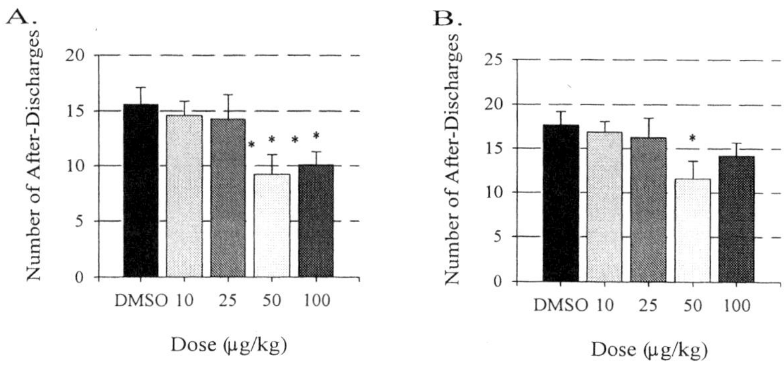

Figure 5. Effects of HU-210 on amygdaloid kindling, as measured by number of ADs to the first stage-5 seizure (**A**) or to 3 consecutive stage-5 seizures (**B**). *different from control, p 0.025.

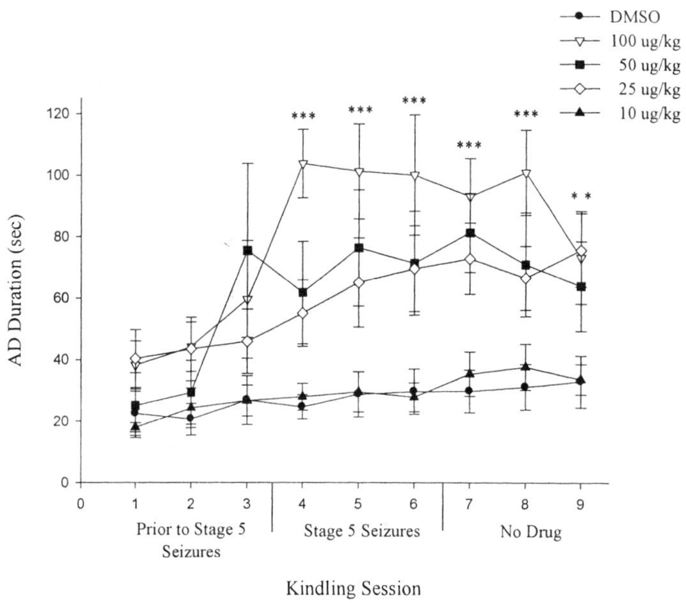

Figure 6. Effects of HU-210 on duration of AD during amygdaloid kindling. *** 100, 50, & 25 µg/kg different from control , $p<0.05$; * * 100 & 25 µg/kg different from control, $p<0.05$.

4. DISCUSSION

The present experiments demonstrate that cannabinoid agonists can produce either antiepileptic or proepileptic effects. Thus acute administration of high doses of HU-210, an agonist at CB-1 and CB2 receptors, produces modest protection against clonic amygdaloid seizures. Tolerance to the antiepileptic effect develops rapidly, and, unexpectedly, chronic administration of HU-210 results in a large increase in the duration of the clonus associated with kindled amygdaloid seizures. A similar proepileptic effect on hippocampal and neocortical seizures occurs with chronic administration of HU-210. In ongoing preliminary experiments, we are also finding (data not shown) that a proepileptic effect on amygdaloid seizures is produced by WIN 55,212-2, another CB1/CB2 agonist that reportedly suppresses tonic seizures induced by pentylenetetrazol or ECS.[10-12] Thus the effects seem not to be specific to HU-210, but extend to another selective cannabinoid agonist. Furthermore, administration of HU-210 during amygdaloid kindling results in an acceleration of the rate of kindling.

Superficially, the effects of HU-210 appear to differ from the effects of THC on kindling and kindled seizures.[2, 3] However, it is worth noting that acute antiepileptic effects of THC are seen only at toxic doses, and chronic administration results in prolongation of seizures triggered on non-drug days.[2] Although this was interpreted as an acute withdrawal syndrome, it might also reflect long-lasting proepileptic effects of THC that are unmasked in the absence of acute antiepileptic effects. It is also unclear that the retardation of kindling produced by THC is a genuine prophylactic effect, as opposed to an acute suppression of seizures.[2, 3] Thus although there are some differences in the effects of THC and HU-210, there are also similarities.

Our results also differ from other recent demonstrations of an antiepileptic effect of cannabinoid agonists against tonic seizure.[10-12] One explanation for the discrepancy is that some agonists might have additional properties, although above we cited evidence that WIN 55,212-21 affects kindled seizures similarly to HU-210. Another possibility is based on the observation that kindled limbic seizures are essentially forebrain clonic seizures, as are kindled neocortical seizures. The neural substrates of forebrain clonic seizures are thought to differ from the substrates of brainstem tonic seizures[14] and might display differential sensitivity to the effects of cannabinoids as well. A similar explanation might apply to the differential ability of the putative endocannabinoid PEA to suppress tonic but not clonic seizures, whether induced by kindling or pentylenetetrazol.[15]

It has been reported that the endocannabinoid anandamide is proconvulsant in knockout mice lacking the enzyme fatty acid amide hydrolase,[16] which catalyzes the metabolism of anandamide. Thus our observation of a proepileptic effect of a cannabinoid is not entirely unprecedented. To account for our results, we suggest the hypothesis that seizures that involve the hippocampal formation are sensitive to proconvulsant effects of cannabinoids. That is, hippocampal cannabinoids have been shown to decrease release of GABA from inhibitory interneurons, thereby disinhibiting activity in principal cells, at concentrations an order of magnitude smaller than those that decrease release of glutamate.[9] Thus, according to the hypothesis, seizures that are sensitive to hippocampal hyperactivity would be exacerbated by cannabinoid agonists. Consistent with the hypothesis, hippocampal hyperexcitability induced by kainic acid increases susceptibility to amygdaloid kindling.[17] The hypothesis is testable, suggesting, for example, that selective manipulations of hippocampal cannabinoids would affect kindled seizures and other forms of clonic seizure induced by activation of the forebrain, but not tonic seizures induced by activation of the brainstem. A corollary of the hypothesis is that kindling involves an enduring increase in the density of hippocampal CB1 receptors, and hence that seizures kindled from the forebrain may be abnormally sensitive to the effects of cannabinoids. In support of the corollary, we have obtained preliminary evidence that CB1 receptor immunoreactivity is increased in the hippocampal formation after amygdaloid kindling, measured either with immunohistochemistry[18] or Western blot (unpublished).

Finally, our data require some reconsideration of the idea that cannabinoids might be useful treatments for seizures or epilepsy. The data clearly suggest that it would be unwise to use cannabinoids clinically for treatment of seizures originating at limbic sites, whether partial or secondarily generalized. The possibility that other types of seizure could benefit from cannabinoids requires further careful investigation.

5. ACKNOWLEDGEMENTS

Supported by grants from CIHR and NSERC.

6. REFERENCES

1. J. A. McCaughran, M. E. Corcoran, and J. A. Wada, Anticonvulsant activity of delta 8- and delta 9 tetrahydrocannabinol in rats, *Pharmacol. Biochem. Behav.* **2** 227-233 (1974).
2. M. E. Corcoran, J. A. MCCaughran, and J. A. Wada, Antiepileptic and prophylactic effects of tetrahydrocannabinols in amygdaloid kindled rats, *Epilepsia*, **19** 47-55 (1978).
3. J. A. Wada, A. Wake, M. Sato, and M. E. Corcoran, Antiepileptic and prophylactic effects of tetrahydrocannabinols in amygdaloid kindled cats, *Epilepsia* **16,** 503-510 (1975).
4. P. Consroe, B. Jones, H. Laird-II, and J. Reinking, Anticonvulsant-convulsant effects of delta-9-tetrahydrocannabinol, in *The Therapeutic Potential of Marijuana*, edited by S. Cohen and R. C. Stillman (Plenum Press, New York, 1976), pp. 363-382.
5. D. Feeney, Marihuana and epilepsy: paradoxical anticonvulsant and convulsant effects, in *Marihuana: biological effects, Advances in the Biosciences, Vol. 22/23*, edited by G. G. Nahas and W. D. M. Paton (Pergamon Press, Oxford, 1979), pp. 643-657.
6. M. Segal, Central implantation of cannabinoids: induction of epileptiform discharges, *Eur. J. Pharmacol.* **27** 40-45 (1974).
7. R. G. Pertwee, Pharmacology of cannabinoid receptor ligands, *Current Med. Chem.* **6,** 635-664 (1999).
8. R. G. Pertwee and R. A. Ross, Cannabinoid receptors and their ligands, *Prostaglandins, Leukotrines and Essential Fatty Acids* **66** 101-121 (2002).
9. N. Hajos and T. F. Freund, Pharmacological separation of cannabinoid sensitive receptors on hippocampal excitatory and inhibitory fibers, *Neuropharmacology.* **43,** 503-510 (2002).
10. M. J. Wallace, R. E. Blair, K. W. Falenski, B. R. Martin, and R. J. DeLorenzo, The endogenous cannabinoid system regulates seizure frequency and duration in a model of temporal lobe epilepsy, *J Pharmacol Exp Ther.* **307,** 129-37 (2003).
11. M. J. Wallace, B. R. Martin, and R. J. DeLorenzo, Evidence for a physiological role of endocannabinoids in the modulation of seizure threshold and severity, *Eur J Pharmacol.* **452,** 295-301 (2002).
12. M. J. Wallace, J. L. Wiley, B. R. Martin, and R. J. DeLorenzo, Assessment of the role of CB1 receptors in cannabinoid anticonvulsant effects, *Eur J Pharmacol.* **428,** 51-7 (2001).
13. D. M. Lambert, S. Vandevoorde, G. Diependaele, S. J. Govaerts, and A. R. Robert, Anticonvulsant activity of n-palmitoylethanolamide, a putative endocannabinoid, in mice, *Epilepsia* **42,** 321-327 (2001).

14. R. A. Browning, Effects of lesions on seizures in experimental animals, in: *Epilepsy and the Reticular Formation: The Role of the Reticular Core in Convulsive Seizures*, edited by G. H. Fromm, C. L. Faingold, R. A. Browning, and W. M. Burnham (Alan R. Liss, Inc., New York, 1987), pp. 137-162.
15. A. H. Sheerin, X. Zhang, D. M. Saucier, and M. E. Corcoran, Selective antiepileptic effects of *N*-palmitoylethanolamide, a putative endocannabinoid, *Epilepsia* **45**, 1-5 (2004).
16. A. B. Clement, E. G. Hawkins, A. H. Lichtman, and B. F. Cravatt BF, Increased seizure susceptibility and proconvulsant activity of anandamide in mice lacking fatty acid amide hydrolase, *J Neurosci.* **23**, 3916-23, (2003).
17. S. Feldblum, S. and R. F. Ackermann, Increased susceptibility to hippocampal and amygdala kindling following intrahippocampal kainic acid, *Exp. Neurol.* **97**, 255-269 (1987).
18. X. Zhang, K. Mackie, D. K. Hannesson, P. Mohapel, A. Wallace, J. Howland, A. Boulton, and M. E Corcoran, Effects of cannabinoid drugs on kindled seizures and status epilepticus in rats, Presented at the meeting of the Society for Neuroscience, October 1999.

Discussion

Wada: You mentioned AD duration. Which part of the behavioral seizure gets prolonged during the AD?

Corcoran: The clonic seizures are longer. We don't see this until the rats are having stage 5 seizures. So when they are having generalized clonus, it is associated with the longer AD duration.

Engel: This is very disturbing, Mike. I have been telling my patients for years when they ask me if they can smoke marijuana that it is fine as long as they don't stop. But I am confused about several aspects about this. First of all, you're measuring seizure severity; so it really doesn't say anything about the effect it might have on when seizures occur. Did you look at AD threshold?

Corcoran: That's a really good question. No, we have not yet looked at AD threshold.

Engel: The first effect that you described reported is an antiepileptic effect followed by a proepileptic effect. Are you implying that it is tolerance that then causes a withdrawal proepileptic effect, or do you think that it's a direct effect? In other words, in the graphs that you show where seizure duration dips down and then goes way up, is that upswing a withdrawal effect because of the tolerance, rather than a direct effect?

Corcoran: Well, that is obviously a key question, and I do not yet know the answer.

Engel: It does not go along with the effect of the cannibinoids, which is disinhibition. So if it is a disinhibition, you wouldn't expect the initial effect to be antiepileptic. However, we know that cannabinoids produce disinhibition, and that was one of the reasons for suggesting that inhibition is a mechanism for causing hippocampal seizures.

Corcoran: Let me remind you that at 50 μg/kg there was a hint of an antiepileptic effect, but it was not significant. What was significant was the proepileptic effect that developed by about the fifth injection. So the antiepileptic effect can be uncoupled from the proepileptic effect. It could even be that the antiepileptic effect at the higher doses is not CB1 receptor mediated, and this is a testable proposition.

Engel: The alternative hypothesis is that the antiepileptic effect is a true effect and the disinhibition actually is antiepileptic in the hippocampus in this situation. But because of the rapid tolerance, the proepileptic effect is then a withdrawal effect.

Simonato: I may have missed this. Is the drug you used selective for any of the cannibinoid receptors?

Corcoran: This drug acts at the CB1 and CB2 receptors.

Simonato: Have you challenged the hypothesis using a selective antagonist?

Corcoran: Not yet.
Heinemann: Since the CB3 receptor apparently depresses glutamate release, an alternative explanation would be that, when you apply the drug, the intrinsic cannabinoids no longer bind to the CB1 receptor but instead to the CB3 receptor. Then subsequently production of anandamide or whatever cannabinoid is depressed. So did you do repeat these experiments in the presence of a CB1 antagonist?
Corcoran: No, that is an excellent idea, and, as I said, we have not done that yet.
Wasterlain: Has anyone tried this in absence seizures?
Corcoran: Not that I know of.
Burnham: Just a note that rats quickly develop tolerance to anticonvulsants, but people don't develop tolerance at all. So you can't usually go from one to the other. Anecdotally in Toronto we've got a guy who has been taking THC or marijuana and claims it has controlled his seizures for many many years.
Corcoran: There are lots of anecdotal reports of that, Mac, but the problem is that there is not much hard laboratory evidence.

LOW FREQUENCY SINE WAVE STIMULATION DECREASES THE INCIDENCE OF KINDLED SEIZURES

Jeffrey H. Goodman*

1. INTRODUCTION

Although deep brain electrical stimulation has been studied as a therapy for epilepsy since the 1930's[1] we have still not determined the optimal sites for delivery of the stimulus or the most effective stimulus parameters. Superimposed on these questions is the requirement that the ultimate therapeutic stimulation needs to be safe and not interfere with normal brain function. Given the recent success using deep brain stimulation to treat movement disorders[2] and the limited success of vagus nerve stimulation for epilepsy,[3] there has been a renewed interest in deep brain stimulation as a therapy for epilepsy. In many ways the clinical use of deep brain stimulation is moving faster than the basic research, which usually precedes and is the often the basis for the development of new clinical therapies. Nevertheless, a number of laboratories are using in vitro and in vivo animal models to develop deep brain electrical stimulation as a therapy for epilepsy.

The kindling seizure model is well suited for an investigation of the potential benefits of deep brain stimulation as a new therapy for epilepsy. One of the model's most powerful features is that the investigator has control over when the seizure will occur. This characteristic combined with distinct behavioral convulsive stages[4] and measurable electrographic activity;[5] makes kindling an attractive seizure model for an examination of the anticonvulsive properties of deep brain stimulation. Early studies by Gaito et al.[6,7] provided the initial evidence that low frequency stimulation (LFS) could interfere with kindled seizures but these studies were largely ignored or unaccepted by the kindling community. Weiss and colleagues[8] coined the phrase "quenching" to describe the decrease in kindled seizures after DC stimulation.. A report by Ullal et al.[9] suggested that

* Jeffrey H. Goodman, Center for Neural Recovery and Rehabilitation Research, Helen Hayes Hospital, West Haverstraw, New York, 10993.

LFS could raise kindled seizure threshold and Velisek et al.[10] demonstrated that LFS could interfere with kindling acquisition in immature animals. Despite these positive reports, the anticonvulsant effects of LFS on kindled seizures is still controversial. For this reason we examined the effect of preemptive LFS on kindled seizures initiated in the amygdala and the hippocampus of the rat.

2. METHODS

All animals were treated in accordance with the guidelines set by the New York State Department of Health and the National Institutes of Health for the humane treatment of animals.

2.1 Electrode Implantation

Methods for electrode implantation have been described in detail elsewhere.[11,12] Briefly, adult male Sprague-Dawley rats (275-325g) were anesthetized with a combination of ketamine and xylazine (1ml/kg, i.p.; ketamine 80mg/kg and xylazine 12mg/kg) and placed in a stereotaxic apparatus. A midline incision was made in the scalp, the calvarium was exposed and burr holes were drilled for the placement of anchor screws and electrodes. Bipolar, Teflon-coated stainless steel electrodes were stereotaxically implanted bilaterally into the basolateral amygdala or the dorsal hippocampus at the following coordinates measured from bregma: Amygdala A-P –2.3mm; M-L ± 4.5mm from the midline; and Depth: 8.5mm from skull surface; Hippocampus A-P –3.0mm; M-L ±3.7 from midline; 3.2mm from skull surface.[13] After each electrode was lowered to the appropriate coordinate, it was cemented in place with acrylic cement (Plastics1) and connected to a socket (Ginder Scientific). All animals were allowed to recover from the surgery a minimum of one week before initiation of afterdischarge (AD) threshold testing and kindling stimulation.

2.2 AD Threshold Testing

The AD threshold was determined for the left and right electrodes for each animal using a 60Hz, 1msec sine wave pulse. Amygdala animals received a 1sec stimulus and hippocampal animals received a 2sec stimulus. AD threshold testing was started with a current intensity of 25μA, and was increased in 25μA steps until an AD was observed. The electrode with the lower AD threshold and/or better recording quality was selected for control and experimental stimulation. Hippocampal animals with an AD threshold greater than 150μA and amygdala animals with an AD threshold greater than 250μA were not included in any experiments.

2.3 Effect Of LFS On Kindling Acquisition

Animals with electrodes implanted in the amygdala were randomly placed into a control or experimental group. Control animals were stimulated twice/day with a 60Hz, 400μA, 1msec sine wave pulse for 1sec. Experimental animals were stimulated in a similar manner except they received a preemptive 1Hz , 50μA sine wave stimulus for

30sec immediately before the kindling stimulus was delivered. Animals in both groups received a total of 20 stimulations. The AD duration and behavioral seizure score were measured after each stimulation and averaged. In addition, the number of stimulations required to elicit the first stage 5 seizure and for each animal to become fully kindled were also recorded. Data pertaining to the incidence of AD and stage 5 seizures were collected for both groups of animals.

2.4 Effect Of LFS On The Kindled State

Upon completion of the 20 stimulation series a crossover procedure was performed such that fully kindled animals from the control group received LFS *plus* the kindling stimulus and animals from the experimental group received only the kindling stimulus. A total of 20 additional stimulations were delivered to the animals in each group.

A separate study examined the effect of preemptive LFS in rats fully kindled in the dorsal hippocampus. Each animal was stimulated twice/day with just the kindling stimulus until fully kindled at which time a baseline of 10 stimulations was used to determine the mean incidence of stage 5 seizures. Upon completion of the 10 stimulation series, a preemptive 30sec 1Hz sine wave LFS was delivered immediately before each kindling stimulus for an additional 10 stimulations. The incidence of stage 5 seizures was measured after each stimulation.

2.5 Data Analysis

During kindling acquisition, values for AD duration and behavior seizure score were averaged after each stimulation. The data used in this analysis were generated from only those stimulations that elicited an AD. Stimulations that failed to elicit an AD were not included in these data in order to prevent a bias towards a decrease in AD duration. The statistical comparison between the control and experimental groups was made using a 2 way ANOVA with 1 repeated measure. This was followed by the Hotelling T^2 test, which allows for a comparison of individual points within the 20 stimulation series.[14] Comparisons of AD or seizure incidence between the 2 groups were made using the Fisher Exact test.

3. RESULTS

3.1 Effect Of LFS On Kindling Acquisition

Two parameters used to assess kindling acquisition, the mean number of stimulations required to elicit the first stage 5 seizure and the mean number of stimulations required for the animals to become fully kindled, were not significantly increased by the preemptive 1Hz sine wave stimulus in the experimental animals. There was a significant decrease ($p<0.01$, Hotelling T^2) in the mean AD duration towards the end of the 20 stimulation series (Figure 1A). Although there was a trend towards a decrease in behavioral seizure score in the experimental animals the change was not significant (Figure 1B).

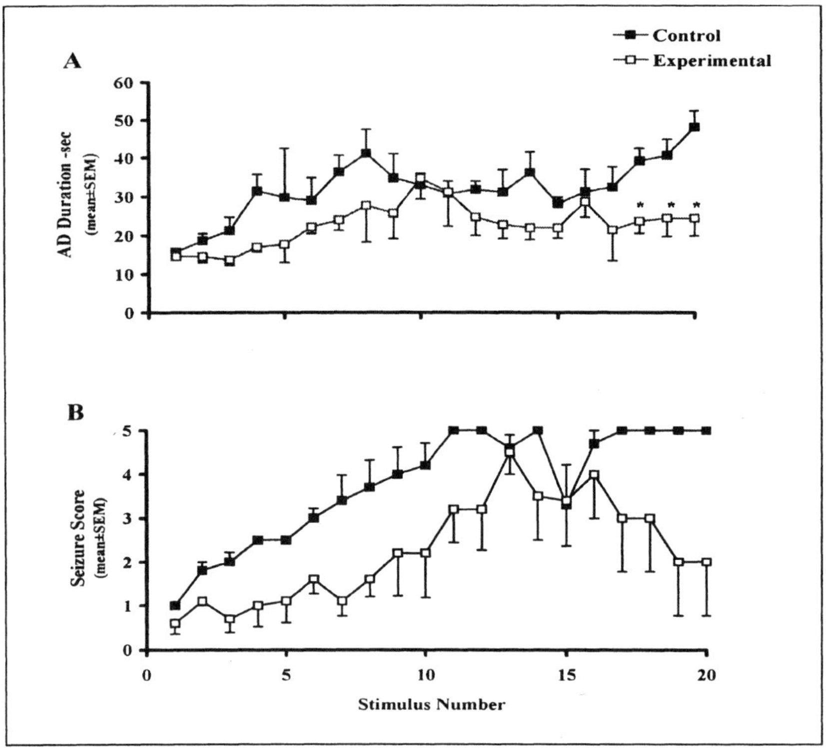

Figure 1. Effect of LFS on AD duration and behavioral seizure score during kindling acquisition. **A.** Comparison of mean AD duration after each stimulation between control (n=5) and experimental animals (n=5) stimulated with LFS. The mean AD duration for stimulations 18-20 was significantly decreased in the experimental animals (*= p<0.01, 2 way ANOVA with 1 repeated measure and Hotelling T^2 test). **B.** Comparison of mean behavioral seizure score after each stimulation between control (n=5) and experimental animals (n=5) stimulated with LFS. There was no significant difference between the groups (p>0.05, 2 way ANOVA with 1 repeated measure and Hotelling T^2).

The experimental animals exhibited a significant decrease in the incidence of AD in response to the LFS plus the kindling stimulus (Figure 2). In amygdala kindling the kindling stimulus rarely fails to elicit an AD. However, the incidence of AD decreased from 99% in the control animals to 70% in the experimental animals (p<0.0001, Fisher Exact test).

3.2 Effect Of LFS Fully Kindled Animals

Fully kindled amygdala animals underwent a crossover procedure such that the original control animals now received the LFS *plus* the kindling stimulus and the original experimental animals received only the kindling stimulus. A total of twenty stimulations were delivered in this phase of the experiment. The mean incidence of stage 5 seizures was measured for each group of amygdala kindled animals (Figure 3). The addition of the preemptive LFS to the kindling stimulus caused a significant decrease in the incidence of stage 5 seizures from 98% to 42% (p<0.0001, Fisher Exact test).

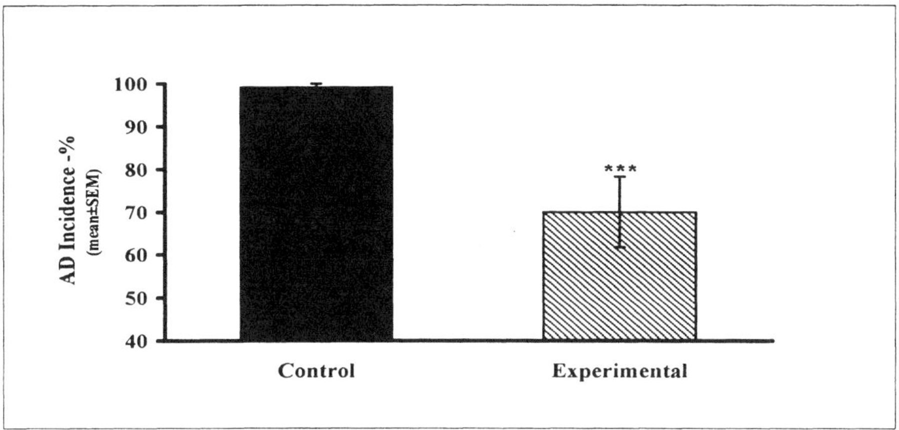

Figure 2. Effect of LFS on the incidence of stimulus-induced AD during kindling acquisition. The kindling stimulus elicited an AD 99% of the time in control animals (n=5). The addition of the preemptive LFS to the kindling stimulus decreased the incidence of AD to 70% in the experimental animals (*** = p<0.0001, Fisher Exact test).

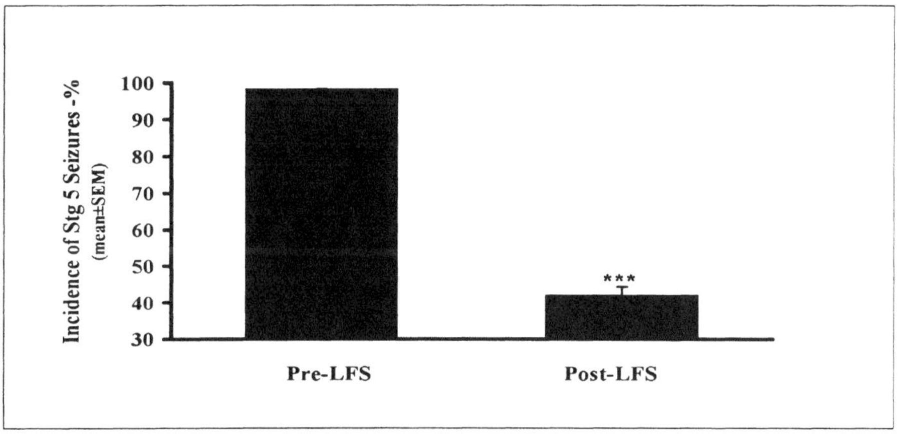

Figure 3. Effect of LFS on the incidence of stage 5 seizures in fully kindled control animals (n=5). Note the significant decrease from 98% to 42% in the incidence of stage 5 seizures when LFS was added to the kindling stimulus in fully kindled control animals (*** = p<0.0001, Fisher Exact test).

The effect of preemptive LFS was also examined in fully kindled hippocampal animals. This group of animals always exhibited a stage 5 seizure in response to the kindling stimulus. However, when the preemptive LFS was added to the kindling stimulus the mean incidence of stage 5 seizures decreased from 100% to 60% (Figure 4, p<0.0001, Fisher Exact test). The LFS decreased the number of stage 5 seizures by 40%. It should be noted that 1 animal exhibited a stage 5 seizure during the 30sec preemptive LFS. For both amygdala and hippocampal fully kindled animals when the LFS *plus* the kindling stimulus did not lead to a stage 5 seizure we seldom observed a lower stage seizure. When animals failed to exhibit a stage 5 seizure, 90% of the time the preemptive stimulus was completely successful such that there was no response to the stimulus.

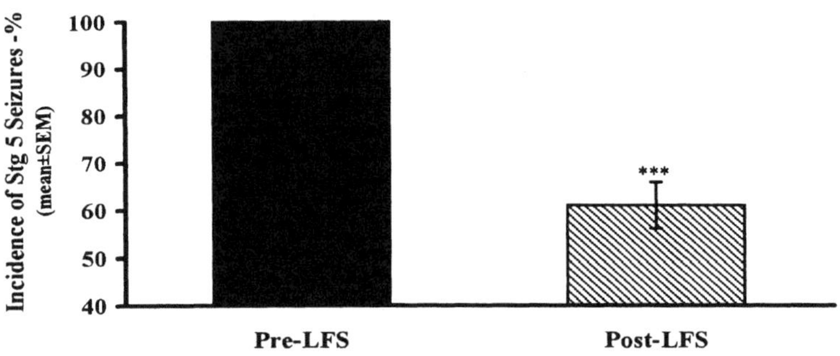

Figure 4. Preemptive LFS using sine waves (1Hz, 50μA for 30 sec) significantly decreased the incidence of stage 5 seizures by 40% in hippocampal-kindled rats (n=10, *** = p< 0.0001, Fisher Exact test). It should be noted that one animal exhibited a stage 5 seizure during the delivery of the LFS before the kindling stimulus was delivered.

4. DISCUSSION

Deep brain stimulation has great potential to become a new therapy for patients with epilepsy who are pharmacoresistant or who are not candidates for epilepsy surgery. Questions related to where to deliver the stimulus and what stimulus paradigms are likely to be effective remain unanswered. Kindling provides an excellent seizure model for studies that examine these questions.

A number of clinical and animals studies have reported decreases in seizure activity using a high frequency stimulus (HFS).[15,16] The mechanisms responsible for these effects are unknown but it has been hypothesized that the HFS inhibits the stimulated area through a depolarization block.[17] The long-term presentation of HFS has the potential to be epileptogenic as evidenced by the kindling process.[18]

An alternative way to interfere with or prevent a seizure would be to use LFS. Several in vitro hippocampal slice studies used LFS to decrease experimentally-induced seizure activity[19-21] and to prevent the transition from interictal activity to seizure-like events.[22] Albensi et al.[21] used HFS and LFS to block bicuculline-induced bursting. However, when the HFS was discontinued the bursting returned but when the LFS was discontinued the bursting did not return. Clinically, LFS has been shown to decrease interictal spiking in patients with temporal lobe epilepsy.[23,24]

The initial studies[6,7] that provided the first evidence LFS could interfere with kindled seizures were difficult to interpret and for the most part were ignored. Several years later Ullal et al.[9] suggested that LFS could raise AD threshold in amygdala and hippocampal kindled rats. More recently, Velisek et al.[10] demonstrated that low frequency **square wave** stimulation could delay kindling acquisition and shorten AD duration in immature rats. In the present study, LFS with **sine waves** induced a decrease in AD duration similar to the results reported by Velisek et al.[10] However an additional observation was the significant decrease in the incidence of AD. Amygdala stimulation rarely fails to elicit an AD and the control animals in this study exhibited an AD 99% of the time. The consistent inability of the kindling stimulus to elicit an AD can only be attributed to the preemptive presentation of the LFS.

It is interesting that the experimental rats became fully kindled in the same number of stimulations as the controls. Since the experimental rats had fewer ADs than the controls, one interpretation of the data is that the LFS accelerated epileptogenesis. There are reports of kindling with LFS but long stimulus trains with a high current intensity (1000μA) were required.[25,26] The LFS used in this study had a short duration with a low current intensity and therefore was unlikely to be a kindling stimulus. However, it is important to note that the LFS elicited one seizure in a fully kindled hippocampal rat before the kindling stimulus was delivered. When fully kindled control animals were stimulated with LFS *plus* the kindling stimulus they failed to exhibit a stage 5 seizure 40-60% of the time. Since these animals exhibited a stage 5 seizure 98% of the time when receiving only the kindling stimulus, the significant decrease in stage 5 seizures is likely due to the presentation of the LFS. The observation that LFS was effective in hippocampal kindled rats indicates that the inhibitory effect of LFS was not unique to the amygdala.

The LFS-induced decrease in kindled seizures in this study can be explained by the recent work of McIntyre et al.[27] Using the same LFS that was used in this study they observed a 200% increase in AD threshold that lasted for 2-3days after the stimulation. An elevation in seizure threshold would make it less likely a given stimulus would elicit a seizure.

The effects of LFS used in this study and others is suggestive of long-term depression (LTD) or depotentiation.[28-31] However, classic LTD consists of a minimum of 15min of 1Hz square wave stimulation.[29] In this study, seizures were successfully prevented with only 30sec of 1Hz sine wave stimulation. The contribution of the sine wave to the success of the LFS is unclear. Since more charge is delivered during sine wave stimulation it may be more potent than square wave stimulation. If kindling occurs through a potentiation of excitatory pathways, then a LFS-induced increase in seizure threshold may occur through a depotentiation-like process.

In conclusion, we have demonstrated that preemptive, LFS using sine waves can decrease the incidence of amygdala and hippocampal kindled seizures. This stimulus paradigm may be an effective therapy for some types of pharmacoresistant epilepsy.

5. REFERENCES

1. A. E. Walker, An oscillographic study of the cerebello-cerebral relationship, *J. Neurophysiol.* **1**, 16-23 (1938).
2. K. Ashkan, B. Wallace, B. Bell, and A. Benabid, Deep brain stimulation of the subthalamic nucleus in Parkinson's disease 1993-2003: where are we 10 years on?, *Br. J. Neurosurg.* **18**, 19-34 (2004).

3. S. C. Schachter and C. B. Saper, Vagus Nerve Stimulation, *Epilepsia* **39**, 677-686 (1998).
4. R. J. Racine, Modification of seizure activity by electrical stimulation. II. Motor seizure, *Electroencephalogr. Clin. Neurophysiol.* **32**, 281-294 (1972).
5. G. V. Goddard, D. C. McIntyre, and C. K. Leech, A permanent change in brain function resulting from daily electrical stimulation, *Exp Neurol.* **25**, 295-330 (1969).
6. J. Gaito, The effect of variable duration one hertz interference on kindling, *Can. J. Neurol. Sci.* **7**, 59-64 (1980).
7. J. Gaito, J. N. Nobrega, and S. T. Gaito, Interference effect of 3 Hz brain stimulation on kindling behavior induced by 60 Hz stimulation, *Epilepsia* **21**, 73-84 (1980).
8. S. R. Weiss, A. Eidsath, X. L. Li, T. Heynen, and R. M. Post, Quenching revisited: low level direct current inhibits amygdala-kindled seizures, *Exp. Neurol.* **154**, 185-192 (1998).
9. G. R. Ullal, T. Ninchoji, and K. Uemura, Low frequency stimulation induces an increase in afterdischarge thresholds in hippocampal and amygdaloid kindling, *Epilepsy Res.* **3**, 232-235 (1989).
10. L. Velisek, J. Veliskova, and P. K. Stanton, Low-frequency stimulation of the kindling focus delays basolateral amygdala kindling in immature rats, *Neurosci. Lett.* **326**, 61-63 (2002).
11. J. H. Goodman, R. W. Homan, and I. L. Crawford, Kindled seizures elevate blood pressure and induce cardiac arrhythmias, *Epilepsia* **31**, 489-495 (1990).
12. J. H. Goodman, R. W. Homan, and I. L. Crawford, Kindled seizures activate both branches of the autonomic nervous system, *Epilepsy Res.* **34**, 169-176 (1999).
13. G. Paxinos and C. Watson, *The rat brain in stereotaxic coordinates*. Fourth ed. (Academic Press, San Diego, 1998).
14. H. Hotelling, The generalization of Student's ratio, *Ann. Math. Stat.* **2**, 360-378 (1931).
15. M. Velasco, F. Velasco, A. L. Velasco, B. Boleaga, F. Jimenez, F. Brito, and I. Marquez, Subacute electrical stimulation of the hippocampus blocks intractable temporal lobe seizures and paroxysmal EEG activities, *Epilepsia* **41**, 158-169 (2000).
16. M. A. Mirski, L. A. Rossell, J. B. Terry, and R. S. Fisher, Anticonvulsant effect of anterior thalamic high frequency electrical stimulation in the rat, *Epilepsy Res.* **28**, 89-100 (1997).
17. A. Benazzouz and M. Hallet, Mechanism of action of deep brain stimulation, *Neurology* **55(12 Suppl 6)**, S13-16 (2000).
18. M. Durand and M. Bikson, Suppression and control of epileptiform activity by electrical stimulation: A Review, *Proc. IEEE* **89**, 1065-1082 (2001).
19. M. Barbarosie and M. Avoli, CA3-driven hippocampal-entorhinal loop controls rather than sustains in vitro limbic seizures, *J. Neurosci.* **17**, 9308-9314 (1997).
20. Kano, M. D'Antuono, P. d. Guzman, R. Benini, and M. Avoli, Low-frequency stimulation of the amygdala inhibits ictogenesis in the perirhinal cortex, *Epilepsia* **43**, 129 (2002).
21. B. C. Albensi, G. Ata, E. Schmidt, J. Waterman, and D. Janigro, Activation of long-term synaptic plasticity causes suppression of epileptiform in rat hippocampal slices, *Brain Res.* **998**, 56-64 (2004).
22. H. Khosravani, P. L. Carlen, and J. L. Perez Velazquez, The control of seizure-like activity in the hippocampal slice, *Biophys. J.* **84**, 687-695 (2003).
23. J. Yamamoto, A. Ikeda, T. Satow, K. Takeshita, M. Takayama, M. Matsuhashi, R. Matsumoto, S. Ohara, N. Mikuna, J. Takahashi, S. Miyamoto, W. Taki, N. Hashimoto, J. C. Rothwell, and H. Shibaski, Low-frequency electric cortical stimulation has an inhibitory effect on epileptic focus in mesial temporal lobe epilepsy, *Epilepsia* **43**, 491-495 (2002).
24. D. Tkesheleshvili, H. Zaveri, J. Krystal, S. Spencer, D. D. Spencer, and I. Cavus, Effects of 1-Hz intracranial stimulation on interictal spiking and EEG power in humans, *Epilepsia* **44**, 246 (2003).
25. M. E. Corcoran and D. P. Cain, Kindling of seizures with low-frequency electrical stimulation, *Brain Res.* **196**, 262-265 (1980).
26. D. P. Cain and M. E. Corcoran, Kindling with low-frequency stimulation: generality, transfer, and recruiting effects, *Exp. Neurol.* **73**, 219-232 (1981).
27. D. C. McIntyre, K. Gilby, and C. A. Carrington, Effect of low-frequency stimulation on amygdala-kindled afterdischarge thresholds and seizure profile in fast and slow kindling rat strains, *Epilepsia* **43**, 12 (2002).
28. S. Fujii, K. Saito, H. Miyakawa, K. Ito, and H. Kato, Reversal of long-term potentiation (depotentiation) induced by tetanus stimulation of the input to CA1 neurons of guinea pig hippocampal slices, *Brain Res.* **555**, 112-122 (1991).
29. B. R. Christie, D. S. Kerr, and W. C. Abraham, Flip side of synaptic plasticity: long-term depression mechanisms in the hippocampus, *Hippocampus* **4**, 127-135 (1994).
30. P. K. Stanton, LTP, LTD and the sliding threshold for long-term synaptic plasticity, *Hippocampus* **6**, 35-42 (1996).

31. R. Adamec, Evidence that limbic neural plasticity in the right hemisphere mediates partial kindling induced lasting increases in anxiety-like behavior: effects of low frequency stimulation (quenching?) on long term potentiation of amygdala efferents and behavior following kindling, *Brain Res.* **839**, 133-152 (1999).

Discussion

Wada: In your stimulation pre-empted group, the seizure didn't occur because the AD threshold was elevated?
Goodman: That's what I think, although I don't know that's the exact reason for it.
Wada: There i no EEG seizure?
Goodman: There was no EEG seizure. I didn't measure threshold, but Dan MacIntyre measured threshold using a similar simulation that I did. I was looking at incidence.
Wada: How long does the anticonvulsant effect last?
Goodman: The elevation in threshold can last for days. So the elevation from one stimulation seems to last for two to three days.
Wada: And it's cumulative, I take it?
Goodman: I looked with a longer duration stimulation, and it seemed to cause a larger increase in threshold and seemed to last longer.
Gale: Since you did a cross over design, you didn't actually follow up the same animals that had been exposed during the kindling to the second block.
Goodman: Yes, I did follow that up. Those animals did start to show an increase in the number of seizures that they had, but it was not significant.
Gale: If you were to continue the low frequency stimulation, would those animals, instead of having a 60% decrease in seizures, would they be more primed to show the protective effect?
Goodman: I do know that after we stop stimulating with the low frequency stimulation, the reliability of some our animals to have seizures is not ever the same – they don't consistently have stage 5 seizures after that.
Gale: So in a sense you have a longer lasting effect.
Goodman: It's altered the system, I don't know in what way.
Gale: If you were going to do this for the maximum efficacy, would you want to start during the kindling or would you want to wait until they are fulling kindled?
Goodman: I saw a 30% decrease during acquisition and about a 60% decrease in fully kindled animals; so it may work better in fully kindled animals than during acquisition. So for someone who already has epilepsy, it might work better as a way of preventing seizures.
Schwartzkroin: That's pretty neat stuff. What do you think the requirement is with respect to your yoking relationship?
Goodman: We haven't looked at that yet.
Schwartzkroin: Do you think there is anything magic about the low frequency coming just before you deliver the kindling stimulation? I'm thinking about your last goal, in seizure detection, and it seems to me unlikely that you are ever going to be able to achieve that kind of proximity.
Goodman: If you look at Dan McIntyre's data that say that threshold is elevated for days, it suggests that it should work and not have to be yoked together.

Leung: Nice work, but I have just a little concern whether you can totally eliminate the fact that the electrode polarization might be a cause of your effect. Even though you have balanced delivery of the charge, I don't think we can necessarily think that the electrodes have recovered completely at the time you deliver your high frequency train.
Goodman: Would polarization arise from such a short duration, 30 seconds? And if an electrode were polarized, how long do you think that would last?
Leung: I don't know.
Goodman: Dan McIntyre was able to stimulate these animals days later and threshold was still elevated.
Leung: Right. Perhaps using the shorter pulses would help.
Goodman: This was just our first stab at that, and I think that's a great idea
Engel: I don't think that connection is really necessary. Our closest encounter with kindling in recent years, because we use the intra-hippocampal kainate rat, happened because the seizures weren't occurring frequently enough for the studies that we wanted to do. So I suggested to Anitol Bragin that he may want to kindle the rat and maybe it would make the seizures more frequent and severe. For reasons that I don't understand, he kindled them at 3Hz, and I don't remember the parameters, except for the fact that it was 3Hz. The first time that he stimulated the animal had a full stage 5 seizure.
Goodman: So he used 3Hz to initiate the seizure?
Engel: He started kindling with 3Hz, the first stimulation produced a seizure, and every subsequent stimulation increased the ADT and the spontaneous seizures decreased. This is a chronic rat with spontaneous seizures.
Goodman: Well there are papers by Mike Corcoran's group on low frequency kindling, but in those papers they had to use very high currents for very long durations.
Wada: Does the 1 Hz stimulation have any distant effects?
Goodman: I don't have any data, but Dan McIntyre may have some data to address that.
McIntyre: We did it at the site of stimulation, and we are actually doing other experiments now where we have electrodes that are spread to span the site of kindling. If you provide that low frequency stimulation across the spanning electrode and yet check the thresholds of the kindling electrode, you will find that in many of the cases threshold will have gone up and in others it will not have changed. I haven't been able to figure out what the difference is, why some have changed and others have not.
Wada: What is the distance?
McIntyre: The distance is 4mm.
Gale: Do you think that you could do this with electrodes in the amygdala and the hippocampus, and you could kindle one and do this preemptive stimulation in another, because this could be a network phenomenon?
Goodman: I would love for that to be the case, but I have to try it.
McIntyre: If that was the case it might speak to Burchfiel's kindling antagonism, where you can have very distant negative effects. I wanted to make a comment, because Dr. Wada earlier had asked about thresholds and what kind of discharge did you have. Jeff, you were stimulating at a single value and the threshold was presumably going up and down, and in some cases you would trigger a seizure and in other cases you would not. We were actually measuring thresholds, and one of the things that we found interesting was that normally when you trigger a seizure you have a full length AD and a full convulsive seizure - you don't have little fragmented ones. But what we saw here, speaking to Juhn Wada's question, was that the low frequency stimulation would often

trigger fragmented seizures, where you would have now a short AD or even a fragmented convulsive seizure that was now half of what it would normally be. It wasn't just the threshold per se, it did seem to be impacting more on the network as well. So I think that the effect of low frequency stimulation it is probably bigger than just at the tips of the electrode.

Goodman: I should add that one of those hippocampal animals actually had a stage 5 seizure during the low frequency stimulation. There is the possibility in some situations that the low frequency stimulation could initiate a seizure, but out of 100 stimulations it happened one time.

Corcoran: Maybe I can speak to that, Jeff. One of the complications here is that, as Ron Racine has shown a long time ago and Peter Cain and I showed more recently, you can actually kindle very rapidly with low frequency square waves in the amygdala, at 3 per second, using 20 second trains, very similar to what you are using. There is a trade off with intensity. In order to trigger an AD with a low frequency train you have to increase the intensity. With stimulation at 60 per second, if the threshold for AD is 50 μA, it might be 400 to 500 μA with low frequency stimulation.

Goodman: These were during acquisition?

Corcoran: This is taking a naive rat and applying a low frequency stimulation to kindle seizures. Minabe et al. did this in cats too, by the way, and showed that the total amount of stimulation required, the total amount of charge, is less than when it is with conventional high frequency stimulation. So I think if there were to be therapeutic use for low frequency stimulation, people would want to be very careful that they adjusted the intensity to stay well below the levels that can indeed, even in the naive brain, and certainly in a kindled brain, trigger AD. I would be concerned that higher intensities in fact would reliably trigger AD.

Goodman: And I am concerned about charge density. We have started to look at whether we are causing damage. In animals that were never stimulated, just having the electrode in the tissue is causing tissue reactions. We also examined tissue 72 hrs after 6min of low frequency stimulation. We thought if we took a long presentation of our stimulation, if that did not cause damage, then chances are that the 30 sec stimulation wasn't causing damage. We found a little bit of gliosis around the tip that did not spread out far from the tip. So we are even getting changes in our controls, and we think that possibly these electrodes may actually have small bits of movement that may be irritating the tissue.

GENERAL DISCUSSION 2

Engel: I was really pleased to see all the applications of kindling and the study of behavior, and I wondered if anybody looked at lateralized effects. I don't mean left vs right, which some of you talked about. Rather, did you ever see, for example, hyperreactivity that was just unilateral. I ask this because, many years ago, we looked at the cat with intrahippocampal kainate to produce defensive rage. When the animal started having spontaneous seizures, they had defensive rage when you approached them, but only when you approached them contralateral to the epileptic hippocampus. If you approached ipsilateral they rubbed up against you and purred. I always wondered whether we should be looking for this in our patients, for unilateral behavioral changes. Has anyone ever looked at that?
Buzsaki: But if you kindle the perforant path and measure AD, you can get an enormous increase in gamma frequency, and it is strictly unilateral.
Wasterlain: We find that the animals that display continuous seizures after a bout of status epilepticus induced by pilocarpine or by perforant path stimulation are extremely aggressive. Obviously the kindled animals aren't nearly as aggressive. Do you have an explanation for that?
McIntyre: Years ago what we had done was to kindle one amygdala versus kindling both anygdalae, and we were looking at kindling as a functional lesion. When we kindled just one amygdala to a certain number of seizures, we didn't see any impairment in things that normally would be impaired after amygdala lesions, such as inhibitory avoidance. But if you did that same manipulation on the other side as well, then you had the classic impairment that looked for all the world just like an electrolytic lesion, even though it was just a kindled site. So the rat is not the best model system for that.
Engel: Sally Hazard did that in our lab years ago, and I don't even remember if she published it. But with unilateral amygdala kindling, there was no effect on response to tail pinch, or heating the tail. But if she kindled it bilaterally there was no response to heating the tail.
McIntrye: I think that my impression of kindling over the years is probably related to the fact that amygdala kindling in our hand seems to be 80% a unilateral phenomenon. Whereas if you are kindling from the hippocampus of the rat, you are kindling pretty much both sides in an nice symmetrical fashion; so if you are looking to create a behavioral change I'd go for the hippocampus because it would alter both sides of the network as opposed to just one hemisphere.
Wasterlain: I have a question for Dan McIntyre: In looking at slow kindlers and fast kindlers, does behavioral change relate to the rate of kindling?
McIntyre: The strains are derived from selecting one attribute of the animal, and it has an

emotionality aspect and a learning aspect. Ron Racine will remember, wasn't it Zaide who found that the maze bright and maze dull animals very big kindling differences that looked like our fast and slow strains? And they were appropriate in the sense of their learning capacities as well.

Wasterlain: I think people expected maze bright rats to kindle faster, but they kindled slower. That's very much like your animals that kindled slower but were actually the better learners.

Moshé: Are there any sex differences in all the paradigms? I know that Lisa Kalynchuk showed some female and male differences, but I didn't see any followup after that.

Adamec: I can't really speak for Lisa, although she has done some work looking for sex differences kindling and she doesn't find any. When I was doing some work with cats I didn't find any either.

McIntyre: We actually have done some of that with the fast and slows kindling rats. There are sex differences on a lot of these tasks. The males do better in the Morris water maze than the females. On some of the emotionality tasks, which is consistent with the human literature, the females are showing much more emotional behavioural in some of the behavioral tests then are the males, especially in the context of anxiety.

Chiba: Let me mention about my experience in regard to Pete Engel's question. In rats with midbrain lesions, during kindling, the rats were very aggressive and biting each other, and very violent. I am a psychiatrist, but not for rat. But I kindled the rats, and they are very psychotic!

Gale: Claude, just in response to your question/observation about animals that have kainate treated status epilepticus then get difficult to handle, and also about kindled animals. By the way, if you do lidecaine kindling, or cocaine kindling as Bob Post has done, in all of these cases you do see jumpiness and aggressiveness, but it turns out if you handle the animals before starting these treatments it tends not to happen. If you handle them after the treatment, it doesn't happen for sure. And we have had absolutely docile post status animals that are not tense and jumpy and hyperreactive – it is just a matter of continuing to handle them. Fifty years ago that was the observation with septal lesion animals. There is nothing worse than a septal lesioned animal, which will actually attack in the fiercest possible manor. But those animals are purely docile as long as you maintain handling and don't isolate them. So a lot of this has to do with isolation, handling and the interaction between the animals in cages, and the stress levels and so forth.

Wasterlain: It's an important factor, but I doubt it's the only one, because we simply cannot house then four per cage. Normally we house our animals four animals per cage, and they will literally kill each other.

Leung: Certainly handling helps, but if a rat is challenged by amphetamine they will show more locomotion. It's not just handling – handling may help, and maybe they are calm too, they don't get stressed out. But once they are given a challenge they cannot overcome by being calming down.

Engel: Let me just comment that years ago Sally Caldacott-Hazard looked at opiates, and if you give nalaxone to these rats before the seizure the behaviour is much worse. If you opiate addict the rats and then give them nalaxone after the seizure, they are climbing the walls.

Leung: If I could change the topic a little bit, Dr. Chiba's presentation on brainstem seizures prompted me about this question. A loss of consciousness often occur after really severe seizures, and I think it would especially happen after brainstem seizures. What has recently

interested me is that chemical inactivation of the amygdala or the hippocampus, you would infer, should lead to a slight decrease in consciousness. But the animals do not fall down, they are not sedated. But if you challenge then with an anesthetic, and it can be any type of anesthetic such as pentobarbital or halothane, then they would be sedated or anesthetized. So adding an inactivation of a structure by lidocaine, by musicmol, I infer to suppress the "consciousness." So during a seizure I wonder if what kind of conditions are those that provoke unconsciousness. Does that necessarily involve the brainstem, the arousal system?
Chiba: I don't know. After the brainstem is excited, the limbic system is subsequently active, excited, but the timing of excitation is rather important. So long-term excitation is key in the limbic involvement.
Wasterlain: We did a study that we never published using vigabatrin to inhibit GABA turnover and tried to use that method to measure GABA release of turnover in various part of the brain at various stages of kindling. We were hoping to see a change in kindling, but we did not see that. However, in every animal that was post-ictal after the seizure, we saw tremendous increase in GABA accumulation, in the presence of the GABA inhibitor. There was a several hundred fold increase in GABA accumulation. including in the substania nigra. So I think that could be part of what causes this massive hyperpolarization and transient coma.
Wada: I have two comments. One is that over the years, I always anticipated we might observe some behavioral change in the amygdaloid kindled cats. We housed at least 20 or 30 cats in a big gang cage, but we never found any aberrant behavior to humans. We did notice that when the cats were kindled bilaterally, then we began to see, within their feline society, quite aberrant behavior. We saw not only aberrant homosexual activity, but heterosexual activity, quite persistently heterosexual activity among cats particularly when they were kindled bilaterally. I wonder whether anyone of you have seem this or not. The second thing is when we kindle to the final stage convulsion, the cats became profoundly comatose, sleeping , lying, snoring, etc. But when we were examining the threshold for hypothalamic-induced affective behavior, we were profoundly impressed by the fact that immediately after the stage 5 was over and we stimulated the reticular formation or hypothalamic system, the animal would immediately stand up with coordinated act, defend, etc.. There was no coma, and their basic drive mechanisms were quite intact.
Bertram: The comment about consciousness and epilepsy reminded me a bit of a comment made by Pierre Gloor a long time ago about consciousness and epilepsy and how difficult it is actually to define. He gave some examples I have confirmed over the years as well. For example, you can have people who are in a convulsive like event, four-limb clonus and tonic-clonic activity, who remain completely aware of what is going on around them and can tell you afterwards exactly what everyone did around then when they are having those seizures, which were true clonic seizures that involved all extremities that went on for some time. You can have people who are completely responsive to the environment during or after seizure but have no recollection for a prolonged period of time of what happened then or for the period afterwards. And there are huge gradations of that, people who are otherwise unresponsive but are aware of things. So when we talk about consciousness and epilepsy, it's really more important to be more specific about whether they are aware of the environment or are just unable to respond but aware, because they are very different functions, and very different types of epilepsy.
McIntyre: You can actually get at some of that with the rats too. Years ago we did

experiments using kindling as a retrograde amnestic agent, which really is defining knocking consciousness out for the animal. One of the interesting things that we discovered is that the seizure itself actually created seeming amnesia by creating an altered state of consciousness, so that in fact you gave a convulsion and then, as Juhn Wada was describing, if you activated the animal, you could put the animal in a task, like a shuttle avoidance task, it learns that task perfectly well in that state but will only remember the task in that state. If you put the animal then in a normal state, it doesn't know anything about the task, and a normal animal then made postictal and put in that avoidance task knows nothing about it as well. I have always been of the feeling that retrograde amnesia is simply not a wiping out of memory, it's an inaccessibility of memory rendered by an altered state of consciousness.

Wasterlain: You can't forget that seizures inhibit brain protein synthesis and may disrupt consolidation.

Leung: I would just like to respond to the previous comment. I am quite aware of Pierre Gloor's influence and influential paper, about using the word consciousness, proposing that it's not really useful in that sense. But the way I like to test it is just what the general anaesthesiologist would like to do, and ask whether they have loss the righting reflex, have they lost the pain , maybe they also have amnesia. Really this a very minimal level of consciousness that we are talking about.

Engel: That results of that state dependent learning study are bizarre. Are you implying that when I have a patient on the ward who has a seizure and is postictal and I give him things to remember, when he comes to he won't, but the next time he is postictal he will? I will try it.

Gutierrez: I would like to make a comment on how the brain is receiving the information during
epileptogenesis. In the 1980s with the Dr Fernandez-Guardiola, we measured visual evoked potentials during epileptogenesis, to test whether the whole nervous system was increased in excitability during kindling. What we found is that if we recorded from the main visual pathway, the evoked potentials grew as kindling progressed, whereas if we recorded from secondary sites, for example cortical areas not directly involved in visual processing but associational areas, they were decreased. This means that excitability might be growing, but on the other hand the processing of this input was disturbed.

Wasterlain: I would like to switch to the cause of kindling. I think we have heard today a very impressive series of talks on factors associated with kindling which seem to play very important roles: trkB, BDNF, all those peptides, and Pat Gallagher's very impressive metabtropic receptor changes. Yet it seems that a few years ago we couldn't find anything wrong with the kindled brain. So now we have too many. I would like to ask Drs. Simonato or Kokaia and Racine, which of those changes are most important?

Simonato: I think that is a very key question at this point, at least with thinking of neuropeptides. In principle neuropeptides are a perfect target, because their release is limited to repetitive firing and pulse firing, and it is in principle associated with seizures. Modulation from of peptides pharmacologically could be feasible and could be associated with less side effect than other approaches. The problem is that a few years ago, the only peptide system on which we were running some data was the opiate peptide. And I think this was related to technological advances and to the fact that we were evolving some pharmacological tools to explore that system. Now this is not at the top of the list anymore. For example, only two years ago we were not making any tools to go farther into the research on bradykinin, but now in the field we have knockout animals, we have the possibility of transfecting genes, and to explore what's happening there with ectopic expression. Also I believe that pharmaceutical companies are more interested in developing new non-peptidic drugs with appropriate pharmacokinetic profiles. So I am optimistic of the possibility of finding something interesting in this field in the years to come, but right now you would put me in a very difficult position if you ask me which you would put at the top of the list.

Wasterlain: In defence of peptides, I think Pete Engel used to say we shouldn't have therapies based on glutamate or GABA because you can't shut off the mainstream. But the peptides are perfect from that view point because they are just facilitators or inhibitors.

Kokaia : You could actually induce over-expression of peptides in the excitatory glutamatergic system, and that can have an effect on glutamate release and the receptors expressed in the system. So I think that this provides new opportunities to use peptides in a way to control excitability or excitatory synaptic transmission.

Wasterlain: So we show that it does so that work you put in the beginning of your talk was done in my lab. They do regulate glutamate release very strongly of course

McIntyre: Is there any information on interactions between those glutamate and peptides? Often if instead of a single drug administration, you give a couple of lower doses of two other things, you get a big synergism. Is there any evidence of interactions for the peptides?

Vezzani: As far as I know, there is no evidence about increasing the potency of one peptide by injecting another one together with the first peptide. However, we have done some experiments with some of the somatostatin receptor agonists. Somatostatin is another anticonvulsant peptide, and we found out that if we stimulate two receptors subtypes simultaneously using selective agonists, we can improve anticonvulsant potency versus giving each receptor agonist alone. So there is a possibility of increasing the potency of one peptide by taking another one, but I don't think that anyone has done it yet.

Wasterlain: In principle what peptides give you in addition to modulation effects is regional specificity. From that viewpoint you might add to the effect by affecting different regions with different peptides. That indeed is probably one of the reasons why there are so many peptides – they do different things because they go to different places.

Vezzani: It is quite interesting that the Y2-receptors for NPY are upregulated in mossy fiber terminals and also in epiletptic tissue. For this reason I think they may represent a very good target, because there is this upregulation, The targets are there; so if we find a good drug crossing the blood- brain barrier, then we may have the chance to inhibit glutamate release in those fibers selectively. Of course, thinking about drug development and using peptide receptors as targets, there is the problem of crossing the blood-brain barrier, because the two available peptide analogues do not cross the blood-brain barrier easily. It seems that if you divide the molecules more, they lose the ability to have agonistic activity at the receptors. It is easier to develop antagonists.

Wasterlain: But isn't it possible to develop non-peptide agonists that cross the blood-brain barrier? It may not be easy, but certainly not impossible.

Simonato: I think the problem with the agonist is that most of the receptors for peptides internalize quickly and downregulate; so the risk is to have transient effects.

Bertram: Please do not misinterpret what I am about to say, because I think I agree with Claude Wasterlain that the neuromodulators are the way to go in the future. As we look at all these changes that are happening in the peptide receptors and the peptides themselves, the brain has this seemingly potent ability to reestablish equilibrium. If you try to tilt it one way, the tolerance is an example of that, how it builds up to overcome what you are trying to do to it. How much of these changes that we see really are a process of counter-balance, of trying to get things back to normal as opposed to being part of the mechanism? For example, take the $GABA_B$ receptor. Depending where you make the change or where you lose it you, can have a very different effect. Loss of postsynaptic $GABA_B$ would be considered proexcitatory, and loss of presynaptic $GABA_B$ would actually be proinhibitory. So it is difficult now to think about what the functional consequence might be until you get more of a systems perspective on it.

Wasterlain: The brain has a tremendous amount of homeostasis in it, but in the case of peptides, those reactive changes can be maladaptive, and we have found several. We have worked mostly not with kindling, but with status epilepticus. It truns out that after a half hour status, there is a profound depletion of Galanin, which has a anticonvulsant effect. We have not been able to measure the galanin turnover; so the depletion from tissue makes it likely that the release decreases, but we can't say that for sure. At the same time, if you look at tachykinins or substance P or neurokinin B, an extremely excitatory peptide, very strong proconvulsant, they are induced in novel populations including the mossy fibers and granule cells, in very large amounts and again after a relatively short time. These changes probably participate in making the status self-sustaining. I have an unshakable sense that the world makes sense, but peptides don't always make that much sense. Peptides are strange lot, and many of the changes are maladaptive, certainly the changes in tachykinins and the changes in galanin and the changes in dynorphin, Dynorphin is a presynaptic inhibitor of glutamate release in the CA3, and it is profoundly depleted by status, within a half hour. So it is true that the brain is a very homeostatic organ, but there are exceptions.

Gale: It is not just a matter of finding out what's facilitating kindling or facilitating epileptogenesis; at the same time there are all sorts of compensatory mechanisms that are at work. So it is not a unidirectional phenomenon – everything that is happening is a counter-play of opposing events. That it wins out in terms of getting a proconvulsant effect is probably maybe the tip of the iceberg, because there is these opposing influences going on underneath. We know this because if we don't set things up for the purpose of getting kindling, if we just give seizures intermittently, perhaps very much like what happens clinically, what we get is tolerance. If I give animals electroshock seizures repeatedly over a period of a week, those animals decrease their response to the electroshock, they become tolerant. So if you were just to pick randomly paradigms for stimulating animals and giving them seizures, sometimes you would see proconvulsant effects or sensitization, sometimes you would see kindling, sometimes you would see anticonvulsant effects. The key would be to understanding what the difference is between the different paradigms.

PREDISPOSED SUSCEPTIBILITY IN PRIMATE EPILEPTOGENESIS AND ANTI-EPILEPTOGENESIS

Juhn A. Wada*

1. INTRODUCTION

While kindling emphasizes a progressive process underlying partial-onset epilepsy, recognition of its variation, not only regarding ontogenesis and phylogenesis but also regarding the role of predisposition, will expand its clinical relevance. Nearly 70 years ago, Watanabe[1] showed that electrically induced generalized convulsions in dogs led to pathological changes similar to the Ammon's horn sclerosis found in the brain of patients with epilepsy. In his paper he described a remarkable individual variation in susceptibility, not only for development of the pathology but also for spontaneous seizures and status epilepticus. Variation of kindling results among five primate species, i.e., photosensitive Papio papio (Pp, Senegal) and Papio cynocephalus (Pc, Ethiopia); non-photosensitive Papio hamadryas (Ph, Kenya), Macaca fuscata (Mf, Japan) and Macaca mulatta (Mm, India), have been reviewed from a similar perspective. The technical protocol including once-a-day kindling stimulation was identical in all the species. Since the amygdala (AM) was the only brain site where all five species were kindled, this discussion will be largely, if not completely, centered on AM kindling.

2. INTERICTAL SPIKE DISCHARGE (IID)

IID is a telltale sign of epilepsy, but the clinical significance of its quantity remains unclear. In primate kindling, the transient nature of increased post-ictal IID is in good contrast to a steady state of 24-hour post-ictal IID that displays a biphasic chronological change: a progressive increase followed by an abrupt reduction. The latter reduction signals that a kindled state of susceptibility is reached and spontaneous recurrent seizures may evolve. This 24-hour post-ictal IID frequency upon completion of primary site kindling has been reviewed across species and brain sites.

*Juhn A. Wada, Neurosciences & Neurology, University of British Columbia, 2255 Wesbrook Mall, Vancouver, B.C., Canada V6T 2A1

2.1 AM (Amygdala) Site

One might suspect that a species that kindles rapidly generates more IID than those that kindle slowly. Among five species, both Pp and Pc kindled the most rapidly and at an equal rate, and experienced a similar number of generalized convulsions. Yet the IID frequency was 9.5/min in Pp while 2 of 3 and 1 of 3 Pc had no and only rare IID, respectively.[2] IID frequency of the remaining three species, i.e., Ph (2.3/min), Mf (2.4/min), and Mm (2.6/min), did not reflect their respective differences in either kindling speed or total number of Stage 4 convulsions experienced. Therefore, 24-hour post-ictal IID frequency is not related to kindling susceptibility, photosensitivity, or "severity" as represented by the number of convulsions experienced. It appears to be modulated by a species-specific predisposition, as exemplified by Pp and Pc with abundant and rare or no IID, respectively.

2.2. Cortical Site

IID in patients with an epileptogenic focus in the supplementary motor cortex (SMA) is said to be generally rare, presumably in part due to its anatomical location, although invasive study may disclose it. Among seven cortical sites kindled (orbital, prefrontal, mesial frontal, cingulate, motor, premotor, and SMA), the SMA was the only site that did not generate IID. Because spontaneous seizures were witnessed upon completion of the primary SMA site kindling, two Pp underwent two weeks of continuous video/EEG monitoring, but no IID was found while they were awake or asleep. Since Pp was the only species kindled at the SMA site, it will require confirmation whether this negative finding at the SMA is Pp-specific. However, the striking intra-species difference of IID frequency in Pp between AM and SMA sites suggests that it is modulated by predisposition, not only specific to species but also to brain anatomy. Admittedly, one cannot prove total absence of IID. However, although rare, there are patients with proven epilepsy with no or only rare IID. Therefore, Pp and Pc provide us with primate models for investigation of pathogenesis underlying abundant and rare or no spiking, respectively, in partial-onset epilepsy.

3. BILATERAL INTERICTAL DISCHARGE

Bilateral presence of IID at homotopic sites in patients with unilateral epileptogenesis is not uncommon and, indeed, bilateral IID is almost always found when intractable temporal lobe epilepsy patients are invasively studied. It may be due to either initial bilateral insult and/or progression of unilaterality to bilaterality, as some longitudinal EEG studies suggest. A recent study suggested that bilaterality is related to a larger number of seizures rather than the duration of epilepsy.[3] If, indeed, bilateral independent IID results from an unilateral epileptogenesis, it is reasonable to assume that development of an independent AD precedes that of IID at the homotopic site. During primary site AM kindling, the number of seizures experienced for development of contralateral independent AD was 96 (Pp), 110 (Pc), 126 (Ph), and 167.6 (Mf), while Mm did not develop it with 428 seizures. Yet an independent contralateral IID developed only in Pp and Ph. The findings indicate that (a) an evolution of independent AD at the homotopic site is a time-consuming process and varies according to species, and (b) development of contralateral independent IID does not depend on the number of seizures experienced alone but appears modulated by a species-specific predisposition. Viewed from this perspective, the question arises: Do those patients with bilateral independent IID with unilateral mesial sclerosis, for example, represent a subset of a population predisposed to developing independent contralateral limbic AD and IID?

4. STATUS EPILEPTICUS (SE)

Status epilepticus (SE) is a relatively infrequent but serious complication of epilepsy. It is considered as idiopathic when no inducing factor can be identified, whereas the frontal lobe has been implicated for its propensity to convulsive SE.

4.1. AM Site

In felines, non-convulsive SE occurs infrequently, but generalized convulsive SE is extremely rare unless the animals are over-kindled.[4, 5] Contrary to our anticipation, only 1/18 Pp kindled at AM developed SE. The animal was only weakly photosensitive and was a moderate kindler (27 stimulations to Stage 4 and 89 generalized convulsions experienced), while very rapidly kindled Pp (one stimulation to Stage 4 and 120 generalized convulsions) did not develop SE. Among the remaining four species, only two non-photosensitive Ph, one each of slow (121 stimulations to Stage 4 and 82 generalized convulsions) and moderate (78 stimulations to Stage 4 and 80 generalized convulsions) kindler, developed fatal SE. The fastest kindler (29 stimulations to Stage 4 and 80 generalized convulsions experienced) did not develop SE. Obviously, the sample of Ph possessed an unusually stronger propensity to SE than photosensitive Pp. The findings suggest that convulsive SE is not related to either kindling susceptibility or photosensitivity, but depends instead on an intrinsic predisposition of the baboon species concerned. Contrary to our expectation, Pp was resistant to SE under AM kindling. Could it be that Pp with genetic epileptogenic predisposition is also endowed with an efficient homeostatic synaptic mechanism to counteract the progressive build-up of epileptogenic susceptibility for the purpose of biological survival?

4.2. Cortical Site

Among seven cortical sites kindled, the premotor cortical area (PMA) was the only site where SE developed.[6] At this site, the convulsive motor effect was immediate, with very rapid Stage 4 generalization in both Pp and Mm reflecting an intrinsic property of the PMA. This is in striking contrast to the AM site, where Pp and Mm showed a remarkable difference in kindling susceptibility.[7, 8] When kindled at PMA, only Pp (3 of 5), and not Mm, developed clustering (5-6 times/day) spontaneous seizures identical to the kindled seizures. Two of them, one relatively slow-kindler (31 stimulations to stage 4 and 109 generalized convulsions experienced) and another rapid-kindler (3 stimulations to Stage 4 and 56 generalized convulsions experienced) evolved to fatal SE. However, the most rapidly kindled Pp (the first stimulation induced Stage 4 and 134 generalized convulsions experienced) developed neither clustering spontaneous seizures nor SE. It was found that their post-ictal refractory period was extremely brief (<15 min), compared to that at AM (>6 hours). This may not be surprising since PMA is within the area of an intrinsic neurophysiological abnormality in Pp. However, SE development at this site cannot be explained by this abnormality, kindling susceptibility, or the number of convulsions experienced; instead, it suggests a role played by an individual predisposition.

5. STAGE 5 KINDLED CONVULSION IN Pp AND Pc

As in partial-onset epilepsy in man, Stage 4 asymmetric generalized convulsion is the endpoint of AM kindling. The only exception is Pp/Pc that evolve to the Stage 5 bisymmetrical, bisynchronous convulsive state. The Stage 5 seizure develops at other brain sites as well and is identical to photically induced and spontaneous generalized seizures in Pp.[9, 10] It appears as though kindling "unmasks" a genetically dictated, predisposed susceptibility for this type of bisymmetrical seizure in Pp (and presumably Pc). Pp has an intrinsic bilateral abnormality in the fronto-central cortex. This cortical area is the target of AD propagation from the AM for convulsive evolution through the claustrum.[11] It seems conceivable that repeated AD bombardment from the AM enhances the state of ipsilateral fronto-central cortical excitability that in turn invigorates the latent excitability state of the homotopic area through interconnecting disinhibitory callossal (CC) mechanism. The latter is suggested by not only a remarkable amelioration of kindled seizure, but also elimination of spontaneous recurrent generalized seizure in Pp following CC bisection (CCB).

By contrast, AM kindling-resistant Mm responds to CCB with an accelerated kindling, intensified hemi-convulsion, and emission of spontaneous seizure, indicating a predominantly inhibitory role of the CC.[12] Recent clinical studies suggest a modulating role of the CC on cortical excitability state through reverberation in generalized epilepsy undergoing CCB.[13, 14] According to this model, CCB causes a reduction and increase of the fronto-central cortical excitability state in Pp and Mm, respectively. Thus, the Stage 5 bisymmetrical seizure development in Pp (and presumably in Pc) is considered to be a consequence of repeated AD propagation priming the ipsilateral and then contralateral fronto-central cortices interconnected by dis-inhibitory CC, eventually igniting and releasing the dormant pattern of predisposed bisymmetrical generalized seizure. However, the story appears more complex, since not all the Pp evolve to Stage 5. Seven of 18 Pp kindled at AM, and only 11 evolved to Stage 5. The remaining 7 reached Stage 4 earlier than the former, but despite more than double the number of stimulations during Stage 4 they did not reach Stage 5. This difference was independent of kindling susceptibility or photosensitivity.

6. SECONDARY EPILEPTOGENESIS

Focal cortical epileptogenesis causing independent IID in the homotopic site is an electrographic secondary epileptogenesis. Development of recurrent spontaneous clinical seizures originating from such a mirror focus would be unequivocal evidence of electro-clinical secondary epileptogenesis. With this in mind, all the witnessed spontaneous non-convulsive and convulsive seizures upon completion of primary site kindling but prior to secondary site AM kindling were reviewed. Those animals that had surgical procedures such as resection or bisection were excluded. Pp, Pc, Ph, and Mf (but not Mm) had spontaneous seizures. Among them, recurrent seizures were electro-clinically documented in Pp (4 of 18) and Ph (1 of 5). These seizures were all generalized convulsive in nature with no initial kindled AM participation. Seizure origin was either ipsilateral (to the kindled AM) or bilateral fronto-central cortex in PH and Pp, respectively. It is noted that an identical bisymmetrical pattern of spontaneous seizures was also documented in kindled Pp at other brain sites.

In Mm, the AD threshold is relatively low at the fronto-central area among cortical regions. This site in Pp has genetically dictated epileptogenic abnormality. Although no comparative information is available, a degree of fronto-central cortical excitability among primate species is presumed to be Pp>Ph>Mm. This assumption is based on a comparative difference in the threshold for systemic bemegride-induced generalized convulsions, i.e., PP: 4.0 mg/kg, Ph: 11.0 mg/kg, and Mm: 16.5 mg/kg. The latter roughly corresponds to susceptibility to AM kindling in the respective species: the number of stimulations required for Stage 3 convulsive evolution/ Stage 4 asymmetrical generalization are: Pp 11/22, Ph 65/87, Mm 154/248. The fronto-central cortical site ipsilateral to the stimulated AM is a target of AD propagation from the AM; it was at this fronto-central site that independent IID developed in both Pp and Ph. These observations suggest that at least one condition that promotes development of electro-clinical secondary epileptogenesis is the presence of an area of predisposed susceptibility, be it acquired or intrinsic, within a trajectory of AD propagation from the original focus.

7. LASTING SECONDARY ANTI-EPILEPTOGENESIS

7.1. AM Site

As mentioned earlier, primary site AM kindling induced independent IID in the contralateral AM only in Pp and Ph. Such a homotopic development might imply the acquisition of a higher AM excitability state. AM excitability state measured by after-discharge threshold (ADT)/ generalized seizure triggering threshold (GST) expressed in μA at the primary and secondary site is: Pp: 375/135, Ph: 470/188 and Pp 433/200, Ph 487/?, respectively. Therefore, independent IID is not associated with an increased AM excitability state as measured by ADT/GST. On the other hand, 7 of 18 Pp showed a sustained kindling inhibition despite presence of independent IID. Among them, four were continuously inhibited while three showed an isolated Stage 5 seizure within seven stimulations, only to return to a continuous kindling inhibition lasting for 100 days. The enduring nature of this failure was further confirmed by a longitudinal follow-up of one Pp for 250 days. This animal developed an isolated Stage 5 seizure on the 236th day of stimulation, only to return to an inhibited state again. Such kindling inhibition was not due to a change of the animal's condition since the Stage 5 seizure was readily reactivated at the primary site.

Among Ph with an independent IID, one animal (of 4) showed what appeared to be a partial kindling inhibition at the secondary site: the secondary site seizure began with an expected secondary site Stage 3 pattern, but gradually transformed into the primary site Stage 3 and then Stage 4 pattern. Thus, an AD induced at the secondary site AM appeared to be re-routed back to the kindled primary site hemisphere. An identical pattern was also noted in Mm (2 of 2) that did not develop an independent IID. The findings in

Ph and Mf suggest that secondary site AM kindling failed to fully access the ipsilateral fronto-central cortex. The lasting nature of this partial inhibition was suggested when an identical pattern of a spontaneous seizure originating from the secondary site AM was documented in a Mf.[15] Thus, the secondary site AM kindling inhibition extended from a partial one in Ph/Mf to a complete one in Pp. No inhibition occurred in Mm. This kindling inhibition was not necessarily related to development of independent IID. However, the presence of independent IID coupled with the lasting inhibition in Pp and Ph may be viewed as a secondary anti-epileptogenesis. Predisposition for such development at AM appears to be shared in part among Pp, Ph, and Mf. So far, no comparable secondary-site inhibition has been reported in subprimate species with conventional AM kindling.

7.2. Cortical Site

With cortical kindling at prefrontal, mesial frontal, and cingulate sites (but not at premotor, supplementary motor, and motor sites), complete kindling inhibition (without a positive transfer effect) occurred, lasting for up to 100 days.[16-18] At the mesial frontal cortex, the inhibition was found in all three primate species examined, i.e., Pp, Pc, and Mm. Among a number of brain sites examined in felines, a complete and a lasting kindling inhibition was found only at the cingulate cortex.[19] The finding suggests that the kindling inhibition occurs in both primates and felines at secondary mesial hemispheric sites where independent IID is present. It is an electrographic secondary epileptogenesis. Yet the secondary site cannot be kindled. Therefore, it must be regarded as a lasting secondary anti-epileptogenesis. This development appears to be anatomy-specific in both felines and primates at the cortical site. The underlying mechanism is not known, but a strong intrinsic inhibitory potential of the cortex is suggested by the fact that prefrontally kindled epileptogenesis erodes and spontaneous recurrent seizures dissipate within several months while Am-kindled epileptogenesis persists for three years.[20]

Realization that the mammalian brain possesses an intrinsic capacity to generate a lasting anti-epileptogenesis in response to repeated seizures has two important implications. First, patients with an active epileptogenesis in the prefrontal and mesial hemispheric area have the potential to develop bilateral dysfunction: an entrenched epileptogenesis on one side and a lasting anti-epileptogenesis in a homotopic site—both of which are attained at the expense of physiological function. The prefrontal and mesial hemispheric areas are known to participate in cognition, conation, and emotion for normal organization of behavior. This emphasizes that repeated partial-onset seizures at these sites could cause symptoms other than seizure.[21] Furthermore, the enduring nature of kindling inhibition provides a window for future exploration of an anti-epileptogenic tool or medication that is currently lacking. Characterization and identification of mechanisms underlying lasting secondary anti-epileptogenesis would be the first step towards creating new armamentaria to deploy an anti-epileptogenic capability intrinsic to the mammalian brain.

8. SUMMARY

The following points summarize the paper:

(1) Review of AM and cortical kindling in 5 different primate species suggests that an ultimate consequence of developing epileptogenesis depends on a complex interaction between repeated seizures and intrinsic species-specific, individual-specific, and anatomy-dependent predisposition.

(2) Inter-ictal spike generation and its quantity and spatial extent are independent of kindling susceptibility or "severity" as indicated by the number of generalized convulsions experienced.

(3) The state of predisposed susceptibility within a trajectory of AD propagation preferentially primes that area for development of electro-clinical secondary epileptogenesis.

(4) Anatomy-dependent resistance to interictal spike generation and propensity to status epilepticus at supplementary motor area and premotor area, respectively, are identified.

(5) The prefrontal cortex and mesial hemispheric surface have an intrinsic capacity to cause lasting homotopic secondary anti-epileptogenesis.

(6) Lasting secondary anti-epileptogenesis can cause symptoms other than seizure while providing a window for exploration of an anti-epileptogenic tool that is currently not available.

9. ACKNOWLEDGMENTS

This work was supported by grants from the Vancouver Society for Epilepsy Research and UBC Epilepsy Research Fund to J. A. Wada.

10. REFERENCES

1. E. Watanabe, Experimental study on pathogenesis of epileptic convulsive seizures, *Psychiat. et Neurol. Japonica* **40**, 1–36 (1936).
2. M. E. Corcoran, P. D. Cain, and J. A. Wada, Photically induced seizure in the yellow baboon, Papio papio, *Can. J. Neurol. Sci.*6(2), 129–131(1979).
3. J. Janszky, G. Rásonyi, Z. Clemens, R. Schulz, M. Hoppe, P. Barsi, A. Fogarasi, P. Halász, and A. Ebner, Clinical differences in patients with unilateral hippocampal sclerosis and unitemporal or bitemporal epileptioform discharges, *Seizure* **12**(8), 550–554 (2003).
4. J. A. Wada, M. Sato, and M. E. Corcoran, Persistent seizure susceptibility and recurrent spontaneous seizures in kindled cats, *Epilepsia* **15,** 465–478 (1954).
5. T. Hiyoshi, Evidence of secondary epileptogenesis in amygdaloid overkindled cats: electroclinical documentation of spontaneous seizures, *Epilepsia* **34,** 408–418 (1993).
6. H. Baba, S. Sakai, and J. A. Wada, Premotor (area 6) cortical kindling in primates: Senegalese baboon (Papio papio) and rhesus monkey, in: *Kindling 3*, edited by J. A. Wada (Plenum Press, New York, 1986), pp. 447–469.
7. J. A. Wada and T. Osawa, Spontaneous recurrent seizure state induced by daily electric amygdaloid stimulation in Senegalese baboons (Oapiuo papio), *Neurology* **26,** 273–286 (1976).
8. J. A. Wada, T. Mizoguchi, and T. Osawa, Secondarily generalized convulsive seizures induced by daily amygdaloid stimulation in rhesus monkeys, *Neurology* **28,** 1026–1036 (1978).
9. J. A. Wada, A. Terao, and H. E. Gooker, Longitudinal correlative analysis of epileptic baboon, Papio papio, *Neurology* **22,** 1271–1285 (1972).
10. H. Baba and J. A. Wada, Reciprocal inhibition between seizures induced by intermittent light stimulation and premotor cortical stimulation in Senegalese baboons, Papio papio, *Epilepsia* **28,** 645–650 (1987).

11. J. A. Wada, Forebrain convulsive mechanisms examined in the primate model of generalized epilepsy: emphasis on the claustrum, in: *Idiopathic Generalized Epilepsies: Clinical, Experimental and Genetic Aspects*, edited by A. Malafosse, P.Genton, E. Hirsch, C. Marescéaux , D. Broglin, and R. Bernasconi (John Libby, London 1994), pp. 349–374.
12. J. A. Wada, Anterior 2/3 callosal bisection: comparative observations in animals and man, in *Fundamental Mechanisms of Human Brain Function*, edited by J. Engel, Jr., G. A. Ojeman, H. O. Lueders, and P. D. Williamson (Raven Press, New York 1987), pp. 259–266.
13. T. Ono, A. Matsuo, H. Baba, and K. Ono, "Is a cortical spike discharge transferred" to the contralateral cortex via the corpus callosum? : an intraoperative observation of electrocorticogram and callosal compound action potentials. *Epilepsia* **43**(12), 1536–1542 (2002).
14. A. Matsuo, T. Ono, H. Baba, and K.Ono, Callosal role in generation of epileptiform discharge: quantitative analysis of EEGs recorded in patients undergoing corpus callosotomy. *Clinic. Neurophysiol.* **114,** 2165–2171 (2003).
15. T. Hiyoshi, J. A. Wada, T. Kudo, K. Amano, and M. Seino, Amygdaloid kindling in the Japanese monkey Macaca fuscata: experimental consideration on primate secondary epileptogenesis. *Epilepsia* **37** (Suppl. 5), 76 (1996).
16. J. A. Wada, Amygdaloid and frontal cortical kindling in subhuman primates, in: *Limbic Epilepsy and the Dyscontrol Syndrome*, edited by M. Girgis and L. G. Kiloh (Elsevier, Amsterdam, 1980). pp.133–147.
17. J. A. Wada, T. Mizoguchi, and S. Komai, Kindling epileptogenesis in orbital and mesial frontal cortical areas of subhuman primates, *Epilepsia* **26**(5), 472–479 (1985).
18. J. A. Wada and H. Tsuchimochi, Cingulate kindling in Senegalese baboons, Papio papio. *Epilepsia* **36,** 1142–1151 (1995).
19. J. A. Wada and Y. Hirayasu, Cingulate kindling induces a lasting anti-epileptogenesis. *Epilepsia*, in press.
20. J. A. Wada, Erosion of kindled epileptogenesis and kindling-induced long-term seizure-suppressive effects in primates, in: *Kindling 4*, edited by J. A. Wada (Plenum Press, New York,1990), pp. 383–396.
21. J. A. Wada: Does epilepsy cause symptoms other than seizures? *Japan J Psychiat Neurol*, **46,** 273–280 (1992).

Discussion

Gale: I am so pleased that you brought back this terribly important frame of reference, because so much of the work we do now is in rodent species and frame of reference of what goes on in the primate is extraordinarily important. Am I correct in remembering that not all the macaques could actually be kindled?

Wada: We were able to kindle them, but it took a long time, over one year, over 428 or so stimulations. I am not to sure if you continued to kindle macaques, two years or four years, that they would ever get to stage 5. I doubt it.

Gale: I think there was a big range. Some animals kindled in six month, and some took over a year. So even within that group there is tremendous differences between animals in terms of their susceptibility, and that may have some important implications.

Wada: Yes, indeed.

Wasterlain: Juhn, does the ease of kindling and the likelihood of developing spontaneous seizures corelate with how much time species spend in trees? Some of the baboons spend a lot of the time on the ground, and you would think that evolutionary pressure selection would be much stronger against epileptic organization of the brain for the species that spend a lot of time in the trees, because every seizure means death when they are in the tress, but when they are on the ground they can recover. So that might be one factor. I always thought that for many of the monkeys there is such tremendous selection pressure against epileptogenic organization of the brain. And that implication would not apply to man.

Gale: For the limbic temporal lobe partial seizures, I have watched my animals. They could be in the upper corner of the cage, and they hang on just fine. They can have a seizure that can last for actually several minutes, and it's really quite striking how they do not fall off their perches. Even when it goes bilaterally, they are still able to hang on until the point where it really become a tonic/clonic seizure. So if there were selection pressure against anything, it would be against the generalization, but certainly not the initial manifestation of the seizure.

Engel: I just want to make a comment relative to the lumper/splitter argument of the last couple of days, in support of the splitters. Juhn, you were really the first one to show that kindling had a genetic basis when you showed the difference between the Papio papio and the macaque. But then you looked at all these other species and showed that different phenomena sort out differently and are under different genetic control. All this has to argue that there are multiple mechanisms, not a single mechanism.

Wada: Exactly.

NATURAL HISTORY OF MESIAL TEMPORAL LOBE EPILEPSY WITH HIPPOCAMPAL SCLEROSIS: How does kindling compare with other commonly used animal models?

Jerome Engel, Jr.*

1. INTRODUCTION

Temporal lobe epilepsy, the most common form of human epilepsy,[1] is most often associated with hippocampal sclerosis (HS), the most common epileptogenic lesion.[2] This condition is referred to as mesial temporal lobe epilepsy (MTLE) with HS.[3] Perhaps in recognition of the importance of MTLE with HS as a major health burden worldwide, a large percentage of research on epilepsy now utilizes animal models of this disorder. Whereas kindling was the most popular model of MTLE during the 1970s and '80s, it has been largely replaced in recent years by chronic post status epilepticus models induced by excitotoxic substances such as kainic acid (KA) and pilocarpine (Pilo), or tetanic hippocampal stimulation to produce self-sustained status epilepticus (SSSE).[4] This chapter will briefly review MTLE with HS and examine the strengths and weaknesses of these various experimental models for investigating mechanisms underlying this human condition.

2. NATURAL HISTORY OF MTLE WITH HS

Classically, patients with MTLE with HS have a history of complicated febrile convulsions or other predisposing insults within the first five years of life, an increased incidence of a family history of epilepsy, onset of habitual seizures in mid to late childhood, and rare, if any, secondarily generalized seizures.[2] MTLE with HS is the most thoroughly studied human epilepsy syndrome because it is the epileptic condition most often treated surgically. Such patients presenting at an epilepsy surgery center usually

* Jerome Engel, Jr., David Geffen School of Medicine at UCLA, Los Angeles, California 90095-1769.

report a stuttering course with prolonged periods of remission or infrequent seizure before their ictal events become resistant to antiepileptic drug treatment. When seizures have continued for many years, they may have severe social and psychological impairment as a result of their disability and environmental factors; however, there is evidence that some interictal behavioral disturbances may be a direct effect of the epileptic activity in mesial temporal structures.[5]

Seizures typically begin with an aura, most commonly of epigastric rising, although other autonomic or psychic symptoms, with emotions such as fear, are also frequently encountered. Auras often occur in isolation, but when complex partial seizures follow, they commonly begin with arrest and stare, oroalimentary automatisms, and other, more complex, automatisms. Postictally, there is disorientation and amnesia for the ictal event, by definition.[6] Seizures usually begin with 5 to 7 Hz rhythmic EEG activity in one basilar temporal lead.[7] On depth recording, hypersynchronous ictal EEG onsets are characteristics in hippocampus, but low-voltage fast ictal EEG onsets also occur.[8]

The neurological exam should be normal except for material-specific memory deficits. Interictal behavioral problems, however, can be a disabling feature of the condition. The EEG shows anterior temporal interictal spikes which may be unilateral or bilateral and independent. Unilateral hippocampal atrophy, often with increased signal intensity on T2-weighted images, can be seen in most patients with high-resolution magnetic resonance imaging (MRI),[9] and the same temporal lobe is also hypometabolic with fluorodeoxyglucose positron emission tomography (FDG-PET).[10] There is a characteristic pattern of temporal lobe hyperperfusion on ictal single-photon emission computed tomography (SPECT),[11] and there is a reduction in n-acetyl-aspartate (NAA) with magnetic resonance spectroscopy (MRS).[12] All of these imaging findings reflect structural disturbances characteristic of HS, i.e., loss of principal neurons, particularly in CA1, and CA3 of the hippocampus proper, as well as loss of somatostatin and NPY-containing hilar interneurons; neuronal reorganization, most easily seen as mossy fiber sprouting; and reactive gliosis.[3] The epilepsy surgery setting has provided opportunities for invasive studies of patients with MTLE and HS, both *in vivo* during preoperative depth electrode recordings, and *in vitro* on resected tissue.[13]

A workshop was held in Istanbul, Turkey, in May, 2002, specifically to discuss whether MTLE with HS is a syndrome or a disease. Despite the fact that most clinical epileptologists can easily recognize this condition when they see it, the group of clinicians and basic scientists with particular interest and expertise in hippocampal epilepsy assembled to debate this issue were not able to identify any unique features that could justify designating MTLE with HS a disease, nor were they able to agree that it was a single syndrome; concluding that it probably represents several syndromes.[14] A number of features were discussed in detail.

2.1 Is there a genetic predisposition?

There are several genetic factors that might influence the development of MTLE with HS. There is a genetic predisposition to febrile seizures, which could in some cases result in HS and lead to MTLE.[15] There may also be a genetic predisposition to the development of HS, for instance aberrations in sodium-channel genes have been shown to predispose to hippocampal damage similar to HS in mice.[16] There is also a familial form of MTLE, and some patients in these families have refractory epilepsy with HS which appears no different from the sporadic forms of MTLE with HS.[17] The familial and sporadic forms, therefore, may in fact represent the same condition.

2.2 What are the causative factors?

Although retrospective studies identify precipitating insults, usually within the first five years of life,[18] in a high percentage of patients, such historical data are unreliable. Prospective studies are necessary, but it is extremely difficult to diagnose MTLE with HS early, and many patients with early HS do not have classical MTLE.

2.3 Is there an age specificity for seizure onset?

Patients with MTLE due to HS have seizures that begin earlier than those with MTLE due to other lesions such as neoplasms and dysplasias. Although the first habitual seizures usually begin between 4 and 16 years of age, they can begin at any age with the same pathophysiological changes and surgical outcome.

2.4 Is there a specific pathophysiology?

It is presumed, although far from proven, that an initial precipitating injury early in life results in a specific pattern of cell loss with reactive gliosis and neuronal reorganization, which cause increased inhibition,[19] and localized pathological synchronous burst firing characterized by high-frequency oscillations (150-500 Hz) termed Fast Ripples (FR).[20] These discharges might then recruit distant structures into the epileptogenic process, perhaps through a kindling mechanism, which ultimately must involve a critical mass for spontaneous seizures to occur. There may be a genetic predisposition necessary for either the specific cell loss and neuronal reorganization, or for the resultant epileptogenicity, or both. To what extent both injury and genetic predisposition are necessary is unknown. Furthermore, it is unknown which of the structural changes in mesial temporal structures are responsible for epileptogenicity, and which are nonspecific effects of epilepsy.

2.5 Is there a characteristic electroclinical ictal pattern?

Two types of electrographic ictal onsets have been identified from depth electrode recordings directly from mesial temporal structures.[8] The first consists of hypersynchronous discharges which are usually focal in hippocampus and correlate with auras or no behavioral change. The other consists of a buildup of low-voltage fast activity which may be more regional and is associated with contralateral propagation, impaired consciousness, and other clinical features of complex partial seizures. A transition from hypersynchronous ictal discharge to low-voltage fast activity is necessary before propagation can occur to produce a clinical complex partial seizure. This typically takes tens of seconds, or does not occur at all, accounting for the observation that patients with MTLE and HS commonly have auras in isolation, i.e., frequent auras that do not evolve into complex partial seizures.[2]

2.6 Are there primary and secondary forms?

MTLE with HS only is a primary form of this disorder. Dual pathology, that is MTLE with HS and another lesion, could represent: (i) a situation in which the other lesion is the primary epileptogenic disturbance and repeated discharges have resulted in a secondary form of epileptogenic HS; (ii) a situation in which the other lesion is a primary epileptogenic disturbance which has caused changes in the hippocampus resembling HS that are not epileptogenic; or (iii) two completely unrelated lesions. There is no evidence to prove or disprove any of these possibilities.

2.7 Are there characteristic interictal behavioral features?

Material-specific memory deficits, related to the side of seizure onset, are clearly defined behavioral disturbances in patients with MTLE and HS, and these appear to worsen with time and frequent seizures (see [21]). Increased incidence of other psychiatric and psychological problems, especially depression, are reported which could represent: (i) direct biological effects of repeated seizures; (ii) nonspecific effects of the initial brain injury; or (iii) consequences of environmental factors. There is evidence to suggest that all three possibilities can contribute to interictal behavioral disturbances.[5]

2.8 Are there benign and severe forms?

Only the severe forms are well characterized, because only patients with medically refractory MTLE with HS are referred to tertiary epilepsy centers as potential surgical candidates. Although one large study in a tertiary referral center found that only 11% of patients with MTLE and HS had been seizure free for the previous year,[22] another in a primary center found that 46% had been seizure free in the previous year,[23] suggesting that there are many patients with MTLE and HS who do not have medically refractory epilepsy, and who do not come to the attention of epileptologists.

2.9 Is there a latent/silent period?

The proposed pathophysiology of MTLE with HS requires time following the initial precipitating insult for the epileptogenic structural and functional changes to eventually manifest as spontaneous seizures. Such a latent period is characteristic of this disorder, but there are patients who have no latent period, and also patients who have no identifiable initial precipitating insult. Once habitual seizures begin, they often are initially responsive to antiepileptic medication and remit for many years. This silent period appears to be a feature of many patients with MTLE with HS.[24]

2.10 Is it progressive?

There is strong evidence for progressive behavioral changes, especially increasing memory deficit (see [21]). There is also evidence that contralateral interictal EEG spikes increase over time, suggesting progressive bilateral involvement.[25] The silent period, if present, also suggests progression from a drug-responsive form of epilepsy to a drug-refractory form. Hippocampal atrophy and cell loss may correlate with duration of epilepsy;[26] however, these studies are cross-sectional, and longitudinal studies are needed to clearly document progression.

2.11. Are secondarily generalized seizures characteristic?

Generalized tonic-clonic seizures are rare in patients with MTLE and HS because antiepileptic drugs used to treat this disorder are effective in preventing secondarily generalized seizures. This reflects the fact that most potential antiepileptic compounds are screened against animal models of generalized tonic-clonic seizures and absences, but not against animal models of limbic seizures.

3. ANIMAL MODELS

Experimental animal models of epilepsy are classified as: (i) models of chronic epilepsy with recurrent spontaneous seizures; (ii) models of epileptic seizures induced in normal brains; or (iii) epileptic equivalents which are component parts of epileptiform abnormalities or their underlying mechanisms.[27] The International League against Epilepsy has recently proposed definitions for epileptic seizures and epilepsy.[28] An epileptic seizure is defined as "a transient occurrence of signs and/or symptoms due to abnormal excessive or synchronous neuronal activity in the brain." It can reflect a natural response of a normal brain to insult and, although seizure models in animals with a normal brain are useful for understanding mechanisms of ictal events themselves, they provide no information about the enduring disturbances that exist interictally in the brains of individuals with epilepsy, which are responsible for the intermittent appearance of spontaneous seizures. Epilepsy is defined as "a chronic disorder of the brain characterized by an enduring propensity to generate epileptic seizures, and by the neurobiological, cognitive, psychological, and social consequences of this condition. The definition of epilepsy requires the occurrence of at least one epileptic seizure." Ideally, therefore, models of MTLE with HS should provide an opportunity to investigate the fundamental mechanisms of the enduring propensity to generate epileptic seizures. Epileptic equivalents, however, permit specific component parts of the epileptic condition to be modeled as a more cost-effective way of investigating different aspects of the underlying pathophysiologic disturbances.

3.1. Chronic epilepsy models

A variety of techniques have been used to produce recurrent seizures in animals, including freeze lesions, partially isolated cortical slabs, metals such as alumina, cobalt, tungstic acid, ferric chloride, tetanus toxin, anti-GM1 ganglioside antibodies, and genetic manipulations,[27] but those most commonly used to model MTLE with HS involve the induction of status epilepticus with excitotoxic agents such as KA and Pilo, or electrically induced SSSE. All of these techniques result in hippocampal damage resembling HS and spontaneous seizures with electroclinical features similar to those of human limbic epilepsy. Systemic KA and Pilo, as well as SSSE models, have widespread bilateral limbic system damage and frequent seizures, whereas intrahippocampal KA produces more restricted unilateral hippocampal damage and less frequent seizures, more closely mimicking the human condition. Other differences exist among these models which are compounded by peculiar characteristics of specific animals or strains. All these models, however, have been useful for investigating fundamental neuronal disturbances responsible for spontaneous seizure generation in the hippocampus.

Focal dysplasia is also a common cause of human epilepsy which can involve hippocampus, or be coexistent with HS.[29] A variety of dysplastic animal models are now available, caused by neonatal freeze lesions, prenatal radiation, and MAM.[27]

3.2. Kindling

Although kindling is a process that results in permanent epileptogenic changes in the brain, and has been used as a chronic model of MTLE, it is not truly a model of epilepsy because, as the technique usually is applied, it does not lead to spontaneous seizures. It is, rather, an approach to bring the progressive epileptogenic mechanisms that ultimately lead to spontaneous seizures under laboratory control, which is of great value for certain experimental designs. Although it provides no insight into the pathological changes that occur following an initial precipitating insult that result in synchronous burst firing, because the synchronizing mechanism is artificially applied, and it also does not provide an opportunity to study how spontaneous seizures are generated, because they are generated by artificial stimulation, it does permit careful investigation of how the very localized synchronous firing, over time, recruits downstream structures into the epileptogenic process and is useful to elucidate fundamental mechanisms of the progressive nature of epilepsy. Kindling can also be continued for prolonged periods of time to produce a chronic epileptic condition in which spontaneous seizures occur without stimulation.[30] In this situation, seizures are not usually initiated at the site of stimulation, but some distance away, attesting to the necessity of transsynaptic changes for the epileptogenic process to mature into a chronic epileptic condition. Kindling is a particularly valuable technique because the site of stimulation is known, and no other disturbances are introduced, in contrast to the status epilepticus models which are associated with widespread unilateral or bilateral limbic damage. Consequently, whatever transsynaptic changes occur with kindling that lead to spontaneous seizures must utilize mechanisms available in the normal brain.

3.3. Acute seizure models and epileptic equivalents

Although chronic models are preferred for investigations into fundamental mechanisms of MTLE with HS, they are labor-intensive and not sufficiently cost-effective for some applications, for instance screening of potential antiepileptic compounds. There is a great need to supplement the subcutaneous metrazol and maximal electroshock mouse models of absence seizures and generalized tonic-clonic seizures, with equally cost-effective models of limbic seizures, and kindling has been used for this purpose. Indeed, one of the recently introduced antiepileptic drugs, levetiracetam, proceeded to clinical trial because it was effective against kindled seizures even though it failed to block subcutaneous metrazol-induced absences and maximal electroshock-induced generalized tonic-clonic seizures in mice.[31] There are no simple seizure models better than kindling at the present time for screening potential antiepileptic compounds; however, epileptic equivalents could ultimately serve this purpose. One such equivalent, or surrogate marker of epileptogenicity, is FR. These oscillations uniquely occur in areas of hippocampus and parahippocampal structures capable of generating spontaneous seizures in patients with MTLE and HS, and also in the KA, Pilo, and SSSE chronic rat models of this condition.[20] FR are believed to reflect small populations of synchronously bursting neurons which represent the fundamental mechanisms of epileptogenesis and epileptogenicity.[32] It is possible, therefore, that FR could be used to screen for potential anti-limbic seizure compounds, to localize the epileptogenic region for surgery, and if these high frequency oscillations could be measured noninvasively, they might also be useful to predict which patients are likely to develop MTLE with HS after an initial precipitating injury, and to determine the effectiveness of antiepileptic drugs without the need to wait for another seizure to occur. Interestingly, FR cannot be recorded from limbic structures of kindled rats, presumably because the tissue is not sufficiently epileptogenic to generate spontaneous seizures.

3.4. Comparison of features of MTLE with the post-status chronic models, and limbic kindling

Table 1 lists a number of pathophysiological phenomena of the epileptogenic process in MTLE with HS for comparison with the commonly used post-status (KA, Pilo, and SSSE) chronic models, and kindling, particularly with respect to the value of these models for studying the underlying mechanisms of these independent phenomena. Although all of these phenomena can, and most do, exist in all patients with MTLE and HS, under ordinary circumstances it is only possible to study the endpoint of their manifestations, because only patients with severe epilepsy are surgical candidates and subjects for invasive research. Animal models are necessary to investigate the evolution of these phenomena during progressive epileptogenic processes. Each phenomenon conceivably represents at least one, and perhaps even multiple different, fundamental neuronal processes, which are variably also influenced by maturational, genetic and acquired predisposing factors. It is helpful to organize them into three broad groups. The first group consists of the initial effects of insult, including cell loss, neuronal reorganization, and synchronous neuronal bursting with FR. These can be best studied with the chronic models, and most are not relevant to kindling, although sub-threshold kindling, which produces afterdischarge only with repeated stimulation, must induce local changes at the tip of the stimulating electrode. Neuronal reorganization, and in some cases cell loss, occur in kindling. Overkindling, which results in spontaneous seizures, might also produce synchronous bursting and FR, although this has not yet been demonstrated. The second group consists of transsynaptic progressive effects of these changes which recruit downstream structures, resulting in an evolution of clinical seizure types, and in some cases secondary epileptogenesis, pharmacoresistance, a breakdown in protective inhibitory mechanisms causing status epilepticus, and interictal behavioral disturbances. All of these are most easily investigated with kindling, because exquisite experimental control of individual phenomena can be achieved. The third group that needs to be separated from the others consists of phenomena relevant to spontaneous seizure generation, which include not only mechanisms which cause seizures to occur, but those natural homeostatic mechanisms which maintain the interictal state, and those which limit seizure propagation and determine the clinical manifestations of seizures. Kindling again is not an appropriate model to study these mechanisms as usually practiced because spontaneous seizures do not occur.

Some interesting differences between MTLE with HS and these models are apparent in this table. For instance, the effective insult to produce hippocampal sclerosis in the human occurs during a critical period early in life, usually under the age of five, whereas status-induced hippocampal damage is difficult to obtain in the immature rat and is much more readily produced in the adult. Extensive studies relevant to the second group of phenomena in Table 1 have been carried out with kindling and are well documented in other chapters in this book, but very little has been done to examine these phenomena in the status-induced chronic models. Contralateral secondary epileptogenesis could be investigated in the intrahippocampal KA rat, but hippocampal damage and seizures are bilateral to begin with in the other models. Interictal behavior has been studied in the intrahippocampal KA cat.[5] With respect to the third group, fully kindled limbic seizures invariably progress through alls stages, whereas seizures in human MTLE with HS, and in the chronic models, often consist only of simple partial events, and in the human and intrahippocampal KA rat, secondarily generalized seizures are rare. Mechanisms which limit ictal propagation, therefore, appear to be less effective in kindling.

Table 1. Discrete pathophysiological epileptogenic phenomena and opportunities for investigation in MTLE, post-status chronic animal models, and kindling.

	MTLE with HS	Post-status chronic models	Kindling
Insult	Y <5 years	Y Adult	N
Cell Loss	Y+	Y+++	?
Synaptic Reorg.	Y+	Y+++	Y+
Synchr. Bursting (FR)	Y+	Y+++	N*
Transsynaptic Recruitment	Y	Y+	Y+++
Substrates of Clinical Seizures	Y+	Y+	Y+++
2° Epileptogenesis	?	?	Y+++
Pharmacoresistance	Y+	Y+	Y++
Status Epilepticus	Y+	?	Y++
Behavioral Consequences	Y+	Y+	Y+++
Spontaneous Seizure Generation and Suppression	Y++	Y+++	N*

Y Yes, phenomena exist
N No, phenomena do not exist
++ Suitability of model for studying phenomena
*Might be present with overkindling

There is increasing evidence that a variety of genetic and acquired predisposing factors play an important role in the manifestations of human MTLE with HS. Several chapters in this book clearly demonstrate genetic influences on the progression of limbic kindling, but very few studies have systematically investigated these influences with respect to the chronic models. Just as there was a major paradigm shift with the realization that seizures produced in the normal brain were not the same as spontaneous seizures occurring in an abnormal epileptogenic brain, resulting in a shift in interest from acute seizure models to chronic epilepsy models, there may now need to be another paradigm shift: that an epileptogenic insult induced in a normal brain is not the same as an epileptogenic insult induced in a brain that is predisposed to epilepsy. Elucidation of the genetic and acquired disturbances that might predispose individuals to develop hippocampal sclerosis with specific precipitating injuries, or to develop epileptic seizures as a result of the progressive transsynaptic epileptogenic processes induced by these structural changes, will permit us to devise even better experimental animal models. Kindling could be the ideal model for investigating genetic and acquired factors that predispose the brain to become epileptic after cell loss and neuronal reorganization causes synchronous burst firing, because of the opportunities to bring this progressive transsynaptic epileptogenic process under laboratory control.

4. ACKNOWLEDGMENTS

Original research reported by the author was supported in part by Grants NS-02808, NS-15654, NS-33310, and GM-24839, from the National Institutes of Health, and Contract DE-AC03-76-SF00012 from the Department of Energy.

5. REFERENCES

1. J. Engel Jr., Etiology as a risk factor for medically refractory epilepsy: A case for early surgical intervention, *Neurology* **51**, 1243-1244 (1998).
2. J. Engel Jr., P. D. Williamson, and H.-G. Wieser, in: *Epilepsy: A Comprehensive Textbook*, edited by J. Engel Jr. and T. A. Pedley (Lippincott-Raven, New York, 1997), pp. 2417-2426.
3. G. W. Mathern, T. L. Babb, and D. L. Armstrong, in: *Epilepsy: A Comprehensive Textbook*, edited by J. Engel Jr. and T. A. Pedley (Lippincott-Raven, New York, 1997), pp. 135-155.
4. J. Engel Jr., Models of focal epilepsy. *Proceedings of the 27th International Congress of Clinical Neurophysiology* (in press).
5. J. Engel Jr., R. Bandler, N. C. Griffith, and S. Caldecott-Hazard, in: *Advances in Neurology, vol. 55*, edited by D. Smith, D. Treiman, and M. Trimble (Raven Press, New York, 1991), pp. 97-111.
6. P. D. Williamson and J. Engel Jr., in *Epilepsy: A Comprehensive Textbook*, edited by J. Engel Jr. and T.A. Pedley (Lippincott-Raven, New York, 1997), pp. 557-566.
7. M. W. Risinger, J. Engel Jr., P. C. Van Ness, T. R. Henry, and P. H. Crandall, Ictal localization of temporal lobe seizures with scalp/sphenoidal recordings, *Neurology* **39**, 1288-1293 (1989).
8. A. L. Velasco, C. L. Wilson, T. L. Babb, and J. Engel Jr. Functional and anatomic correlates of two frequently observed temporal lobe seizure-onset patterns, *Neural Plasticity* **7**, 49-63 (2000).
9. G. D. Cascino, C. R. Jack Jr., J. E. Parisi, F. W. Sharbrough, K. A. Hirschorn, F. B. Meyer, W. R. Marsh, and P. C. O'Brien, Magnetic resonance imaging-based volume studies in temporal lobe epilepsy: pathological considerations, *Ann. Neurol.* **30**, 31-36 (1991).
10. J. Engel Jr., W. J. Brown, D. E. Kuhl, M. E. Phelps, J. C. Mazziotta, and P. H. Crandall, Pathological findings underlying focal temporal lobe hypometabolism in partial epilepsy, *Ann. Neurol.* **12**, 518-528 (1982).
11. M. R. Newton, S. F. Berkovic, M. C. Austin, C. C. Rowe, W. J. McKay, and P. F. Bladin, SPECT in the localisation of extratemporal and temporal seizure foci, *J. Neurol. Neurosurg. Psychiatr.* **59**, 26-30 (1995).
12. J. S. Duncan, in *Epilepsy: A Comprehensive Textbook*, edited by J. Engel Jr. and T.A. Pedley (Lippincott-Raven, New York, 1997), pp. 1045-1051.
13. J. Engel Jr., Research on the human brain in an epilepsy surgery setting, *Epilepsy Research* **32**, 1-11 (1998).
14. H.-G. Wieser, Ç. Özkara, J. Engel Jr., A. W. Hauser, S. L. Moshé, G. Avanzini, C. Helmstaedter, T. R. Henry, and M. R. Sperling, Mesial temporal lobe epilepsy with hippocampal sclerosis: Report of the Commission on Neurosurgery, *Epilepsia* (in press).

15. M. A. Falconer, Genetic and related aetiological factors in temporal lobe epilepsy: A review, *Epilepsia* **12**, 13-31 (1971).
16. J. A. Kearney, N. W. Plummer, M. R. Smith, J. Kapur, T. R. Cummins, S. G. Waxman, A. L. Goldin, and M. H. Meisler, A gain-of-function mutation in the sodium channel gene Scn2a results in seizures and behavioral abnormalities, *Neuroscience* **102**, 307-317 (2001).
17. E. Kobayashi, L. M. Li, I. Lopes-Cendes, and F. Cendes, Magnetic resonance imaging evidence of hippocampal sclerosis in asymptomatic, first-degree relatives of patients with familial mesial temporal lobe epilepsy, *Arch. Neurol.* **59**, 1891-1894 (2002).
18. G. W. Mathern, T. L. Babb, B. G. Vickrey, M. Melendez, and J. K. Pretorius, The clinical-pathogenic mechanisms of hippocampal neuron loss and surgical outcomes in temporal lobe epilepsy, *Brain* **118**, 105-118 (1995).
19. C. L. Wilson, S. U. Khan, J. Engel Jr., M. Isokawa, T. L. Babb, and E. J. Behnke, Paired pulse suppression and facilitation in human epileptogenic hippocampal formation, *Epilepsy Research* **31**, 211-230 (1998).
20. A. Bragin, J. Engel Jr., C. L. Wilson, I. Fried, and G. W. Mathern, Hippocampal and entorhinal cortex high frequency oscillations (100-500 Hz) in kainic acid-treated rats with chronic seizures and human epileptic brain, *Epilepsia* **40**, 127-137 (1999).
21 T. Sutula and A. Pitkänen, editors, *Do Seizures Damage the Brain? Progress in Brain Research, Vol. 135* (Elsevier, Amsterdam, 2002).
22. F. Semah, M.-C. Picot, C. Adam, D. Broglin, A. Arzimanoglou, B. Bazin, D. Cavalcanti, and M. Baulac, Is the underlying cause of epilepsy a major prognostic factor for recurrence?, *Neurology* **51**, 1256-1262 (1998).
23. L. J. Stephen, P. Kwan, and M. J. Brodie, Does the cause of localisation-related epilepsy influence the response to antiepileptic drug treatment?, *Epilepsia* **42**, 357-362 (2001).
24. A. T. Berg, J. Langfitt, S. Shinnar, B. G. Vickrey, M. R. Sperling, T. Walczak, C. Bazil, S. V. Pacia, and S. S. Spencer, How long does it take for partial epilepsy to become intractable?, *Neurology* **60**, 186-190 (2003).
25. J. R. Hughes, Long-term clinical and EEG changes in patients with epilepsy, *Arch. Neurol.* **42**, 213-223 (1985).
26. R. Kalviainen and T. Salmenpera, in *Do Seizures Damage the Brain? Progress in Brain Research 135*, edited by T. Sutula and A. Pitkanen (Elsevier, Amsterdam, 2002), pp. 279-295.
27. J. Engel Jr., Experimental animal models of epilepsy: Classification and relevance to human epileptic phenomena, *Epilepsy Research (Suppl. 8)*, 9-20 (1992).
28. R. S. Fisher, W. van Emde Boas, W. Blume, C. Elger, J. Engel Jr., P. Genton, and P. Lee, Epileptic seizures and epilepsy: Definitions proposed by the International League against Epilepsy (ILAE) and the International Bureau for Epilepsy (IBE), *Epilepsia* (in press).
29. M. F. Levesque, N. Nakasato, H. V. Vinters, and T. L. Babb, Surgical treatment of limbic epilepsy associated with extratemporal lesions: The problem of dual pathology, *J. Neurosurg.* **75**, 364-370 (1991).
30. J. P. J. Pinel and L. I. Rovner, Experimental epileptogenesis: Kindling-induced epilepsy in rats, *Exp. Neurol.* **58**, 190-202 (1978).
31. W. Löscher and D. Honack, Profile of ucb L059, a novel anticonvulsant drug, in models of partial and generalized epilepsy in mice and rats, *Eur. J. Pharmacol.* **232**, 147-158 (1993).
32. A. Bragin, C. L. Wilson, J. Almajano, I. Mody, and J. Engel Jr., High frequency oscillations after status epilepticus: Epileptogenesis and seizure genesis, *Epilepsia* (in press).

Discussion

Wada: As I mentioned in my presentation, interictal discharge is often bilateral in the intractable temporal lobe epilepsy, yet there are patients whose epilepsy is intractable but who have no interictal spike discharge. In other words, what I am trying to say to you is that bilateral interictal spike discharge is not really a contributing factor to intractability as such. I had a patient who suffered from temporal lobe epilepsy, and 14 years ago she had a left temporal lobectomy. At the time of taking the corticogram, I couldn't find any interictal spikes. She was free of seizure until February of this year, she came back with a reoccurrence, with a slightly different semiology. Again there was no interictal spike discharge, and at the time of operation there was no spike discharge whatsoever. But she was intractable. It seems to me that there is no cause and effect in terms of spike proliferation. Second, you mentioned the aura. Quite often we regard an aura as where the seizures start. It is not in my view. Aura comes from the area that gives patients awareness of what is taking place. We have had a number of patients in whom we monitored the precise moment of aura occurrence. The point is that we often find that it is before the onset of EEG change but way into EEG discharge. So this is a peripheral issue. Epileptogenesis may be someplace else and spread to those areas that would produce awareness in the patient.

Engel: Your first point is a very important one, because there are multiple types of interictal spikes. Not all interictal spikes mean the same thing. In the 1970's, Bob Ackermann and I published the recycling experiment, where we restimulated the kindled rat every 5 minutes. Some rats would have a seizure the first time and then have prolonged period of seizure remission, where you couldn't get another seizure for all seven of the stimulations. Others would recycle, so that they would respond to each one, and we saw all variation of that. The interictal and postictal spikes that occurred after the first stimulation correlated with the degree of seizure suppression – the more spikes they had, the harder it was to get them to have another seizure. We postulated that some spikes actually represented a protective effect. That actually makes sense if you look at the neural correlate of the interictal spike and wave. The wave is actually inhibition, and the spike and wave is mostly wave. So when you see an interictal spike and wave it means there is actually inhibition and the brakes are on. But you are referring to Rasmussen's comment of spike chasing in the operating room, where they used to look at where the interictal spikes were when they did surgery, and then did the surgery based on the extent of the interictal spikes, but then they realized that you take out the whole brain if you do that because there are all these interictal spikes. Someone told Ted Rasmussen that you can't really do that, because you can't base the decision on the spike. Ted said that there are red spikes and green spikes, and you EEGers can't tell them apart. We think the fast ripples discriminate the red spikes from green spikes.

Wada: Great, I think the postictal spike tells what you are talking about. There is a progressive increase as the animal is kindled, and then as the animal acquires susceptibility the spiking just comes right down.

Engel: As far as the aura is concerned, we see patients who have beautiful ictal onset in one hippocampus. You take out that area and a lot of surrounding area when you do the anterior lobectomy, and you confirm the pathology, yet they continue to have auras. They are now free of disabling seizures, but continue to have auras. So where are the auras coming from?

Gale: You seem to imply that the injury in the chronic models was an important component. I wonder if you could comment on the study that was done by Zhang, Corcoran, and their colleagues showing that you can completely bock the injury and yet still go on to have the spontaneous seizures, whether its in the kainic acid or in the pilocarpine model.

Engel: How did you block the injury?

Corcoran: Xia Zhang primed the rats with an injection of kainate, but terminated the first seizure with pentobarbital.

Engel: So it was repeated injections.

Gale: It was a preexposure paradigm where the animals were predisposed so they would not get injured.

Engel: There are two different aspects of these excitotoxic insults. The quickest way to do it is use the excitotoxic agent to cause status and then the status causes injury. But I suppose you could use the excitotoxic agent to just produce the excitation, just like you would with PTZ or whatever, and do a chemical kindling. That is puzzling.
Gale: But these were not kindled; they just pre-exposed animals, like we do with electroshock. I am sure the same thing would happen with electroshock seizures. If you just pre-exposed animals before the status epilepticus event, they would be neuroprotected.. And if you then gave status epilepticus to a protected animal, so they have the status, they would not have the brain damage, yet they would go on to get the spontaneous seizures. So pre-exposure just subtracts out the brain damage and the sprouting and a lot of other things that happen in response to the damage as being causally related. There must be something else that is going on that is responsible for the spontaneous seizures.
Engel: It is visible brain damage. One of the things that we found, if you want to argue that fast ripples are the basis of epileptogencity in these models, is that the fast ripples occur before there is any mossy fiber sprouting. They occur sometimes within 2 or 3 days, and they only occur in animals that ultimately develop spontaneous seizures. So the changes we are seeing with the neuronal reorganization that is obvious with the Timm stain or other gross methods are probably later and help to solidify the phenomenon, but are not responsible for it. Maybe perforated synapses and other subtle changes that can happen overnight or within hours are more responsible.
Schwartzkroin: On your list of models I noticed that there was no genetic model of focal epilepsy.
Engel: There is no genetic model that I know of mesiotemporal lobe epilepsy, is there?
Schwartzkroin: In a way it depends on whether you require the kind of morphological abnormalities that you and Karen were just discussing. So for example, the Kcna1 potassium channel knockout mice that we have been studying appear to have a temporal lobe-like seizure phenomenon. But it is a generic channel deletion that affects the entire brain.
Engel: Yes, I didn't mention the fact that you might create knockin or knockout transgenic models. I was thinking in terms of naturally occurring genetic models, which I don't think model this condition.
Schwartzkroin: Let me go back to the interictal spike issue a bit and the fast ripples. In generating a model for temporal lobe epilepsy, do you think that it is necessary to have a situation in which interictal spiking occurs and/or that it occurs at a time before the spontaneous seizures occur?
Engel: I have no idea. That is, you can see fast ripples without spikes; so I don't think the spikes are necessary for fast ripples. It is just that when you have spikes in those areas, they have fast ripples on them. But we also see fast ripples without spikes. So I have no idea whether the spikes are necessary or not. At least in hippocampal epilepsy, I would predict that you have to have a small group of tightly connected abnormally synchronously bursting cells to begin with, or you don't get epilepsy. But that remains to be proven.
Schwartzkroin: One more quickie. In your slide showing the redefinition of epilepsy as Bob Post and others are generating it, it appears as though the repeated occurrence of spontaneous seizures is not necessary, that what you need is an enduring predisposition. But that makes kindling sound like it is much closer to that definition than it has been in the past.

Engel: Kindling definitely meets the definition of enduring disturbance definition, but what kindling does not meet is the requirement that at least one spontaneous seizure has to occur for a definition of epilepsy.

Moshé: The issue of when the initial insult may occur to produce e pilepsy may be answered in the next twenty years. In the second Palm Desert epilepsy surgery conference that Pete organized in 1991, we were given a mandate to study the development of mesiotemporal lobe epilepsy in children with febrile seizures. It took many years to organize the study because of we had the ethical issue of how we were going to take kids with febrile seizures that are supposedly benign and perform the MRI, which is potentially dangerous. In 1997, data came out from Darryl Lewis's group showing that few patients with prolonged febrile status epilepticus developed hippocampal changes visible by MRI. In 2002, Shlomo Shinnar organized a multiinstitutional study funded by NINDS. We aim at prospectively studying about 200 children with febrile status epilepticus for acute and chronic MRI changes, serologic data for herpes virus infections that may precipitate the status, EEG and neuropsychological data.

The goal is to correlate these findings with the development of mesial temporal lobe epilepsy. So for those people in the back of the room we anticipate that there are going to be some changes associated with feblile status epilepticues that may not be seizures but neuropsychological consequences. We will also look at predisposing factors,including genetics and we have no idea what we are going to look at, but thanks to NINDS, we can bank blood for future studies.

Engel: It is so hard to get data, because it is dangerous, with infants and small children having to be anesthetized to go in the MRI. You don't do that without a lot of concern.

Wada: I think that this is very important as well because I do know by now that there are at least 6 patients who are not intractable having mesiotemporal sclerosis. They just happened to come to me for a number of other reasons. So it is extremely important in those patients who had complicated febrile seizure, developing change in the limbic system.

Engel: Let me describe two other prospective studies. John Duncan has a study in London in which they are looking at all new onset seizures, and he tells me that the incidence of hippocampal sclerosis is very low, not the 25% that you expect to see. They have been following these people for a number years, and he uses this to argue that medial TLE is actually a rare disease, not a common disease, which Nico Moshé has also argued because he doesn't see it as often in the Bronx. But then I asked him, after you have followed these patients for 5 years, what percentage of your patients in your study have hippocampal atrophy, and he said 25%. The other is Allan Housser's study. He is doing MRIs on everybody in Iceland, and he is finding a lot of hippocampal atrophy, but in patients with things other than TLE, who have other types of epilepsy.

McIntyre: You mentioned that you can have ripples without having interictal spikes, but can you have interictal spikes without ripples on it?

Engel: Yes, most interictal spikes don't have ripples on them when they are projected or come from areas that are not capable of generating spontaneous seizures. So in the human or the intrarhippocampal kainate model, where all the seizures come from one side, only the spikes on that side have the fast ripples. The spikes on the other side don't.

McIntyre: So the focal spikes are rippling spikes.

Engel: Yes.

McIntyre: In one of your slides, you were asking the question why it is that humans sometimes have a big seizure but then auras and just little stuff. Could it have anything to do with the frequency of which they are having them? In your study with Ackermann many years ago, you would have regular kindled seizures, but when you would shorten the interval between, you would then start breaking that pattern all up and developing interictal spikes.
Engel: In humans there is no pattern like that. We see patients that never have generalized convulsions, and in the intrahippocampal kainate rat we see very frequent simple partial seizures, which we can recognize only because we have electrodes and we see the hypersynchronous discharges. Only rarely do they go on to become manifest as limbic or generalized seizures.
McIntyre: But do you think that has nothing to do with how frequently they might be occurring, because mass kindling, of course, really blocks seizure generation?
Engel: No. The reason we are now using pilocarpine and self-sustained status epilepticus (SSSE), because for the studies that we want to do, which is to look at the mechanism of ictal onset, is that these rats don't have seizures frequently enough to do that. So we have gone now to the pilocarpine and SSSE models, which have bilateral hippocampal sclerosis and very frequent seizures, and they almost all generalize.
Post: Can you drive cells at the ripple frequencies?
Engel: You can evoke the ripple frequency by stimulation.
Post: Can you stimulate that fast to kindle? In other words, could you in fact test out the idea that rippling is the key thing by producing ripples at the ripple frequency at the electrode focus?
Engel: That is an interesting thought. I have never thought about that and never tried it. But you can stimulate the perforant path, for instance, in an epileptic animal and you produce a response that has ripples on it, in the same areas of dentate where you get spontaneous ripples.

CLINICAL EVIDENCE OF EPILEPSY-RELATED PLASTICITY

Giuliano Avanzini and Silvana Franceschetti*

1. INTRODUCTION

The term "neural plasticity" was first used by Minea[1] to indicate the metamorphic phenomena of sensory neurons caused by the compression and transplantation of ganglia into various organs. More comprehensively, it describes the functional or structural reorganisation that occurs at molecular, cellular, or circuit level within the central nervous system as a consequence of physiological or pathological activities or lesional insults. When this happens during development, it may alter the developmental programme, but it can also cause rearrangements of an already mature neural structure.

Plastic adaptation mechanisms (i.e., activity-dependent changes in neural connectivity and synaptic efficacy) are known to occur in visual and other sensory systems, and may involve AMPA and NMDA receptor and local GABA-circuit regulation, Ca^{2+} and second messengers, neurotrophin, etc.

Some examples of compensatory plasticity are:

1. The cross-modal compensation demonstrated in subjects who be come blind before the age of 14 years;[2]
2. The enhancement of adult neurogenesis in the hippocampus[3,4] due to environmental stimulation and learning;
3. The compensatory reorganisation of intracortical circuitry or afferent sensory systems occurring in cortical dysplasias.

2. EPILEPSY-RELATED PLASTICITY

It is thought that plastic phenomena play an important role in the pathogenesis of some types of progressive epilepsy, including post-traumatic epilepsy, mesial temporal lobe epilepsy with hippocampal sclerosis (MTLE-HS), epileptic

Giuliano Avanzini and Silvana Franceschetti, Istituto Nazionale Neurologico C. Besta, Milano (Italy)

encephalopathies, and late-onset epilepsies associated with embryonic, perinatal, or early infancy diseases (e.g. brain dysgeneses).

In principle, progressive epilepsy may depend on the progression of the causative cerebral pathology or an "intrinsically" progressive potential of the underlying epileptogenic process. However, in the case of the epilepsies mentioned above, seizure-dependent plastic phenomena seem to play an important pathogenetic role and may give the underlying neurobiological mechanism the character of a self-sustained process.

Late seizures (i.e. post-traumatic epilepsy) occur in patients who experienced a traumatic brain injury with a standardized incidence rate of 3.1.[5] Recent experiments suggest an important pathogenetic role of chronic glial reactivity, which may be induced or aggravated by epileptiform activity itself.[6,7] The concept of seizure-dependent pathological plasticity is relevant to the definition of infantile epileptic encephalopathies: "conditions in which the epileptiform abnormalities themselves are believed to contribute to the progressive disturbance in cerebral function."[8]

The idea that established epilepsy may tend to progress towards an increasingly severe condition was first put forward by William R. Gowers in 1881.[9] However, Gowers's theory that "seizures beget seizures" is not tenable as it was formulated, because not all types of epilepsy are progressive, and not all types of seizures seem to be capable of starting a process leading to chronic epilepsy.

The concept of epilepsy-related plasticity applies to plastic changes following an epileptogenic insult and leading to persistent epileptic phenomenology (primary epileptogenesis), or plastic changes secondary to persistent epileptic activity and leading to the evolution of the epilepsy (secondary epileptogenesis). Furthermore, plastic adaptation phenomena occurring in maldeveloped epileptogenic areas should also be considered.

The concept of secondary epileptogenesis was first proposed by Frank Morrell[10] to account for the secondary development of mirror foci. In more general terms, it has been defined as a process whereby an actively discharging epileptogenic lesion (i.e., a primary focus) with massive synaptic connections to another normal ganglionic region gradually induces similar paroxysmal behaviour in the cellular elements of that otherwise normal network (i.e., a secondary focus).[11]

One typical example of such seizure-dependent plasticity is mesial temporal lobe epilepsy with hippocampal sclerosis (MTLE-HS), whose biological bases are thought to involve activity-dependent molecular and cellular changes leading to circuitry reorganisation within the CNS.

3. MESIAL TEMPORAL LOBE EPILEPSY WITH HIPPOCAMPAL SCLEROSIS (MTLE-HS)

MTLE-HS has been known since the 19th century. Ammon's horn abnormalities and hippocampal atrophy were respectively first described by Bouchet and Cazauvieilh in 1825[12] and by Meynert in 1867,[13] but its pathogenetic significance has been debated: according to Sommer in 1880,[14] Ammon's horn sclerosis should be viewed as a cause of seizures, whereas Pfleger in 1880[15] believed that hippocampal lesions should be viewed as a consequence of seizures.

The natural history of MTLE-HS includes an acute epileptic episode (typically a complex febrile convulsion) followed by a "latent period" of months or years that eventually leads to chronic epilepsy with temporal seizures.

Experimental pilocarpine, kainite, and kindling models have improved our understanding of the biological mechanisms by means of which MTLE-HS may develop from an early acute insult and further progress towards a more severe condition. The first two procedures involve the acute administration of pilocarpine or kainic acid, which induces severe epileptic symptoms in rodents followed (after a variable interval) by chronically recurring seizures; the kindling model is based on repeated subthreshold stimulations of susceptible limbic and neocortical structures that induce increasingly prolonged and severe seizures and epileptiform EEG discharges, and eventually lead to spontaneous seizures.[16] The kindling paradigm therefore brings the epileptogenic process under experimental control, and makes it possible to follow its development.

It is thought that "early" seizures occurring during the acute phase induce a cascade of events involving the activation of Ca2+-dependent proteases, protein kinase C, Ca^{2+}-calmodulin kinase systems, immediate early genes, and inflammatory reactions leading to neuronal injury and circuitry rearrangements. Important steps are Ca^{2+} entry and mitochondrial dysfunction, a process that may result in structural remodeling and lead to an epileptogenic aggregate (secondary epileptogenesis).

Evidence supporting this pathogenetic hypothesis has been provided by Sutula et al.,[17] who demonstrated that the kindling process is associated with the sprouting of a mossy fibre pathway that reorganises synaptic connections in the dentate gyrus, and then that a similar picture can be observed in the surgically resected hippocampus of patients with epilepsy.[18]

These findings have been confirmed by a number of other studies, including that of Babb et al.[19] from which Figure 1 has been taken. Mossy fibre sprouting had been previously observed after experimentally induced status epilepticus accompanied by extensive neural damage (e.g., in the kainic acid model[20]), but the kindling paradigm showed that repeated brief seizures can induce sprouting in the absence of initial extensive brain damage.[21]

In human MTLE, mossy fibre sprouting is consistently associated with hippocampal sclerosis and cell loss, and it is thought that the degeneration of mossy hippocampal hilus cells significantly contributes to circuitry rearrangement (schematically illustrated in Figure 1).

Experiments based on various animal models have shown that sprouted mossy fibres make synaptic contacts in ectopic locations, and thus provide an excitatory feedback circuit.[22-24] The excitatory effect of the aberrant recurrent fibres is further enhanced by the facilitation of NMDA receptor-mediated conductance, which has been demonstrated in dentate granule cells recorded in slices prepared from the surgically excised temporal lobe tissue of epileptic patients.[25] Recurrent axon collaterals also make synaptic contacts with inhibitory interneurons, thus leading to an enhanced inhibition[26] that, instead of preventing the generation of epileptic discharges, contributes to it by promoting synchrony.[27] The main dentate granule axon target is CA3 (which is therefore secondarily involved in the hyperexcitable state), but it is not known to what extent the sprouted collaterals of granule axons inside CA3 contribute to the enhanced excitability in postsynaptic neurons, or what the contribution of sprouting phenomena occurring elsewhere in the hippocampal-entorinal circuit, such as CA1, may be.[28]

In general, it can be said that, although the study of circuit reorganisation in MTLE has provided important insights into its biological bases, it has also left a number of unanswered questions.

Comparisons of human and animal studies have shown that brief seizure episodes can set in motion a cascade of events leading to sprouting and neosynaptogenesis, which may account for the tendency of MTLE to progress towards a condition of medical intractability.

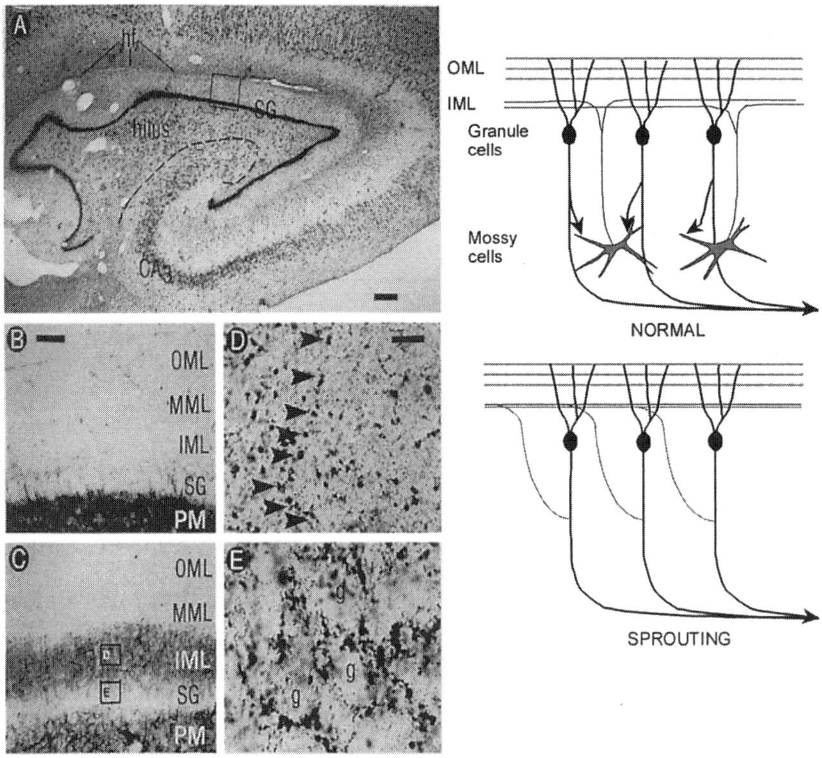

Figure 1. Evidence for circuitry rearrangements in human hippocampus surgically removed from patients with mesial temporal lobe epilepsy and hippocampal sclerosis. Left A: Coronal section of normal human hippocampus stained with Cresyl Violet. Dashed line segregates the CA4 pyramidal neurons from the hilus of the dentate gyrus (SG=stratum granulosum). B: adjacent section. Magnification of boxed area in A stained with the Timm method for heavy metals: the dense stain is strictly limited to the zinc containing granule cells of polymorph layer (PM), whereas supragranular layer (SG), inner, mesial, and outer molecular layers (IML, MML, OML) are completely devoid of staining. C: the corresponding area from a surgically removed hippocampus of a patient with temporal epilepsy and hippocampal sclerosis shows a second band of zinc-containing axons in the IML. D: Timm stain puncta in IML from boxed area in C. E: Timm stain puncta in SG from boxed area in C. (From Babb et al 1991[19]). Right: schematic representation of granule axon sprouting in hippocampal sclerosis. The newly formed axon collaterals occupy the inner molecular layer devoid of the mossy fibres due to the degeneration of the hilar mossy cells.

In both humans and experimental animals (the kainic and pilocarpine models), the time course is typically biphasic, with a more or less prolonged latent interval between the initial event and the chronic epileptic phase during which the activation of Ca^{2+}-dependent proteases, protein kinase C, Ca^{2+}/calmodulinkinase systems, and immediate early genes[29] promotes circuit remodelling. The most frequent initial event in the clinical history of MTLE is a prolonged febrile seizure during the first two years of life that can be assimilated to the status epilepticus induced by kainic acid and pilocarpine in experimental animals. Once established, the aberrant hippocampal circuitry creates a condition of hyperexcitability that leads to often difficult-to-treat chronic epilepsy.

However, this linear interpretation is not supported by a number of other observations. First of all, experimental interventions that prevent sprouting do not impair the acquisition of epileptic properties.[30] Secondly, although epileptiform discharges can be easily recorded from dentate granule cells in in vitro sclerotic hippocampal slices, and are putatively interpreted as being the result of newly formed glutamatergic synapses, in vivo recordings taken from the epileptogenic hippocampi of patients indicate that bursting neurons are rare, and that it is difficult to demonstrate their synchrony.[31,32] Thirdly, the role of the putatively seizure-dependent cell loss in determining the sprouting is still unclear, as is that of the excess zinc caused by the sprouting of zinc-containing mossy fibres upon glutamatergic and GABAergic synaptic transmission.[33]

Various aspects of the pathogenesis of MTLE-HS are unknown, such as the relevance of the seizure-stimulated modulations of adult neurogenesis in the dentate gyrus to human MTLE;[34] furthermore, the roles of immune factors, genetic predisposition, and developmental abnormalities suggested by recent data[35] need to be further investigated by means of prospective studies of patients presenting acute symptoms that may be a risk factor for the development of an epileptogenic process.

Recently, my colleague Laura Farina (collaborating with Robert Zimmerman's group in Philadelphia) has shown the value of early diffusion-weighted imaging (DWI) in prospective studies of the development of HS in children with prolonged new-onset seizures.[36] Within three days of the onset of prolonged psychomotor seizures, there is an increase in the signals shown on T2 and DWI images that indicates restricted diffusion throughout the affected hippocampus, and all of the patients showed hippocampal atrophy on follow-up 2–18 months later (Figure2).

4. EPILEPSY-RELATED PLASTICITY IN CORTICAL DYSPLASIA

Although epileptogenic dysgeneses are obviously already present at birth, the fact that seizures do not usually appear until later in childhood suggests that their expression requires some changes occurring during postnatal development. I will here discuss some evidence of primary (i.e., lesion-related) or secondary (i.e., epilepsy-related) epileptogenic changes in cortical circuits and compensatory plastic circuitry rearrangement in cortical dysplasias.

4.1 Circuitry rearrangement in human dysplastic tissue

The role of primary and secondary circuit abnormalities in determining epileptic phenomenology and its evolution towards a pharmacoresistance has been analysed in patients with Taylor's dysplasia by Spreafico et al.,[37,38] Garbelli et al.,[39] Najim et al,[40] and Crino et al.[41,42]

An increased expression of heteromeric associations of NR2 and NR1 NMDA receptor subunits is consistently observed, and leads to a receptor that produces larger focal currents upon activation. It is therefore highly likely that dysplastic neurons preferentially exhibiting this NMDA combination are differentially hyperexcitable. This hypothesis is supported by direct subdural electrocorticographic recordings of patients with cortical dysplasia in whom the cell density and distribution of NR2A/B correlated with the presence of in situ intrinsic epileptogenicity. The expression of membrane receptors is modulated by the functional status of nerve cells, and thus provides the system with a powerful plastic mechanism; however, whether this is relevant to the progression of epilepsy associated with cortical dysplasia has still to be determined.

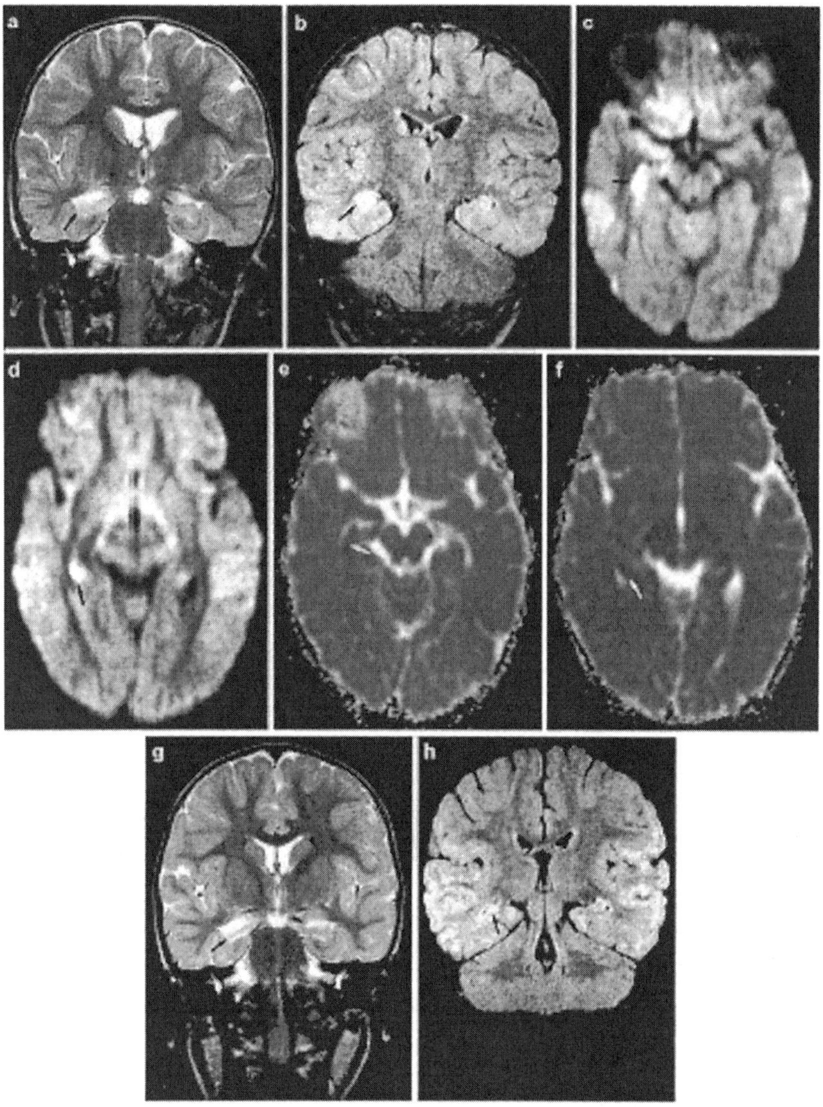

Figure 2. Longitudinal study of a patient. a, b Coronal T2-weighted spin-echo and fast-fluid attenuated inversion-recovery (FLAIR) images in the acute phase show swelling and increased signal in the right hippocampus(arrows) (a head; b body). c, d Diffusion-weighted images show increased signal (arrows) throughout the hippocampus. e, f Maps of apparent diffusion coeffcients show dark areas in the hippocampus (arrow). g, h Coronal T2-weighted spin-echo and FLAIR images 12 months after the first study show right hippocampal atrophy (arrows) with residual increased signal, loss of the hippocampal digitations, and slight enlargement of the parahippocampal sulcus (From Farina et al 2004[36]).

Immunoreactivity to the calcium-binding proteins characterising inhibitory interneurons is clearly reduced in the dysplastic cortex, whereas an increase in PV-immunoreactive terminals enveloping large round-shaped cells, balloon cells, and giant pyramidal neurons is consistently observed (Figure3).

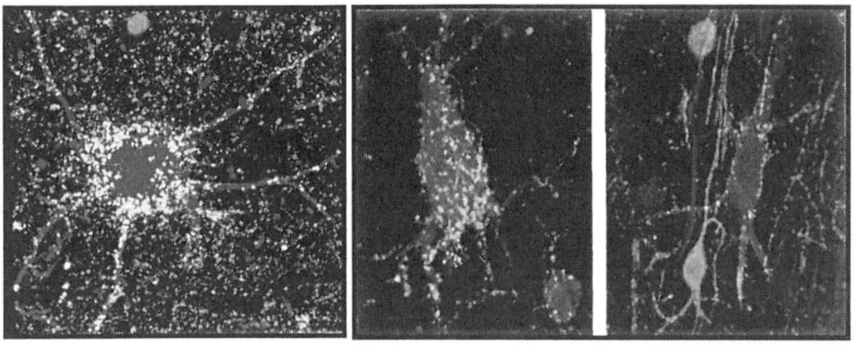

Figure 3. Confocal laser images of a balloon cell (A) and a giant pyramidal neuron (B) from a patient with Taylor's Dysplasia and of a normal pyramidal neuron (C) from a normal control. Sections processed for double labelingwith the neuronal marker SMI 311 (red signal,) and with GAD (A, green signal) or parvalbumin that is expressed in a subpopulation of GABAergic interneurons (B, C green signal). Overlapping of red and green signals yields a yellow color. Balloon cells (A) and dysmorphic pyramidal neurons (B) are completely surrounded by or adjacent to clusters of parvalbumin-positive or GAD-positive puncta. Note the difference with the normal pyramidal neuron depicted in C, where a GABAergic interneuron (green) is also visible. (Modified from Garbelli et al. 1999[39])

This exuberant GABAergic network has been interpreted as a selective response of residual PV-positive neurons aimed at compensating for the reduction in the number of GABAergic cells.[37] It can have an epileptogenic effect by pacing the activity of large populations of target excitatory neurons whose probability of synchronous post-inhibitory discharge is significantly enhanced.

The fact that GABAergic axons may gradually sprout during postnatal maturation could account for the delayed onset of dysplasia-associated epileptic manifestations.

It can be concluded that the plastic changes occurring secondarily in dysplastic tissue may interact with primary circuitry abnormalities and determine the clinical presentation of dysgeneses-associated human epilepsies.[35] It must be noted that one mechanism of action of a number of clinically effective antiepileptic drugs is to enhance GABA-mediated inhibition. In dysplastic tissue, the effect of GABAergic drugs can be substantially affected by the aberrant organisation of GABA circuits, and it can be assumed that paradoxical GABA-mediated epileptogenic activity is enhanced (rather than counteracted) by GABAergic drugs, thus partially accounting for the drug-resistant nature of human epilepsies associated with cerebral dysgeneses.

4.2 Reorganisation of sensory representation in cerebral dysgeneses

The developmental disorders leading to brain dysgeneses cause alterations in thalamo-cortical and intracortical connectivity that may affect the conduction and processing of sensory information. The functional consequences of different types of cerebral dysgeneses (polymicrogyria, periventricular nodular heterotopia, band heterotopia, diffuse pachygyria, focal cortical dysplasia, schizencephaly, and lyssencephaly) have been studied by means of multimodal evoked potentials and functional MRI (fMRI) by Scaioli et al.[43] and Villani et al.[44]

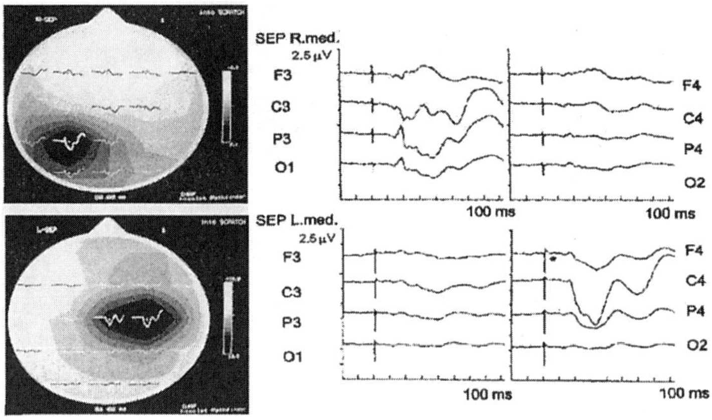

Figure 4. SEP recording. *LEFT:* Topographic distribution at the peak of the P25 component following right (top) and left side stimulation (bottom). The grey scale indicates the amplitude of positive polarity, and the white scale the amplitude of negative polarity. *RIGHT:* Traces recorded by parietal and frontal electrodes: after right side stimulation, parietal N20-P25 (P3) and frontal P22-N30 (F3 and F4) were normal; after left side stimulation, N20 was attenuated (compare leads P3 and P4), but P25 was enlarged, peaked on the central lead (C4), and shifted on the frontal lead (F4); frontal N30 was almost abolished (compare leads F3 and F4) (from Villani et al 2004[44]).

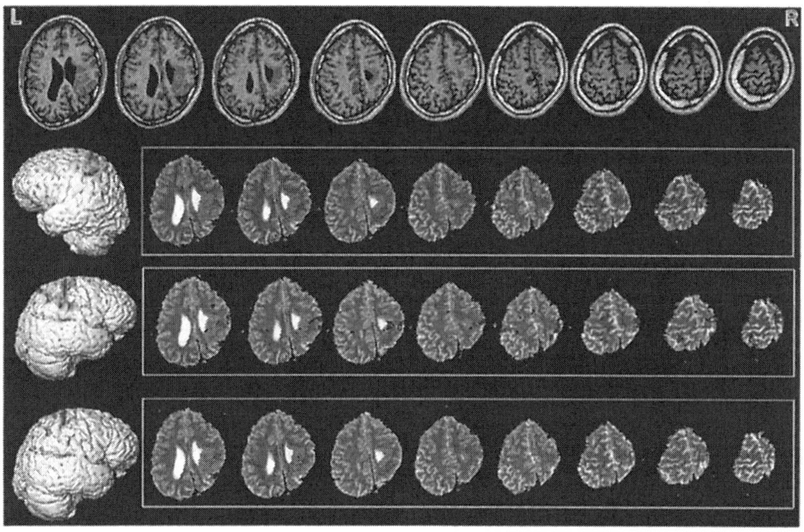

Figure 5. Structural and functional MR images using the neurological convention (left hemisphere on the left of each slice). First row: Transaxial T1-weighted images showing the upper part of the heterotopic nodule which, from the periventricular region, involves a large portion of the right hemisphere. First column: Cortical rendering with activation during the three tasks: right finger tapping (top), left finger tapping (middle), left hand surface stimulation (low). Note the abnormal gyration pattern of the posterior part of the right hemisphere. Second row: EPI average slices with activation clusters in the left pre- and post-central gyri during right finger tapping. Third row: EPI average slices with activation clusters during left finger tapping showing abnormally diffused activation mainly towards the frontal lobe; note the intense activation cluster located within the periventricular heterotopic tissue in the lowest two slices. Fourth row: EPI average slices during left hand surface stimulation: frontal displacement of the activation, with a small cluster within the periventricular heterotopic tissue in the lowest slice (from Villani et al 2004[44]).

With the exception of two patients with bilateral symmetrical periventricular nodular heterotopia, all of the subjects showed various kinds of visual (VEP) and/or somatosenory (SEP) evoked potential abnormalities, such as asymmetries and the severe deterioration of some components, that suggest a reorganisation of cortical topographic representation and distribution involving non-typical cortical areas.

A representative case of a patient with a giant subcortical nodular heterotopia and diffuse abnormal gyrations of the right temporo-parietal regions is shown in Figures 4 and 5. SEPs and fMRI revealed enlarged and displaced sensorimotor areas and the activation of the heterotopic nodule.[44]

The surgical outcomes of refractory epilepsy associated with cerebral dysgeneses mainly depend on the completeness of the excision of malformed tissue. Functional cortex mapping techniques may significantly improve presurgical planning by allowing extended resection and minimising post-operative deficits.

The topographic changes observed in our patients involved cortical sensory representation as a whole, or were limited to some of its sub-components in the form of asymmetrical distribution or ectopic cortical activation. This suggests that the pathological events taking place in the early stages of CNS development, and which cause different types of brain dysgenesis, lead to an extensive reorganisation of the cortical representation of sensory modalities. Whether these plastic changes are associated with epileptogenic changes in the excitability of the cortical regions upon which the aberrant afferents impinge could be an interesting subject of future investigation.

5. CONCLUSIONS

Plastic neural properties may play an important role in the biological processes leading to the establishment of an epileptogenic area as a consequence of an epileptogenic insult (i.e., primary epileptogenesis), but they may also explain the subsequent evolution of some epilepsies that is thought to depend on the effect of persistent epileptic activity (something that is currently known as secondary epileptogenesis). Furthermore, it is also necessary to consider the role of neural plasticity in impairing neural functions and compensating functional impairment. All of these phenomena can also overlap, thus giving rise to complex results.

Some evidence coming from clinical observations suggests the role of plastic properties in determining the progressive evolution of an epileptogenic process. The fact that the natural history of many human epilepsies is characterised by an acute insult followed by a more or less prolonged latent period and then by a persistent disorder involving recurrent tonic seizures suggests that, during the silent period, structural remodelling may lead to a mature epileptogenic aggregate. When the acute insult is associated with early seizures, their specific role in inducing late epilepsy through secondary seizure-dependent changes has to be considered, although the alternative possibility that the silent period simply reflects the outer fringe of a probabilistic spread of first seizure latency must also be taken into account.

Particular attention has been given to the plastic biological mechanisms underlying structural remodelling: i.e. the molecular and cellular changes leading to dentate granule axon sprouting. However, some of these mechanisms can lead to permanent epileptogenic hyperexcitability regardless of circuitry reorganisation.

Vreugdenhil et al[45] have recently shown that a substantial subset of pyramidal neurons isolated from the subiculum resected from drug-resistant epileptic patients have an enhanced persistent Na^+ current. As it is known that the persistent fraction of Na^+ currents is very effective in inducing burst firing in cortical neurons,[46] its enhancement may explain the increased susceptibility to seizures and reduced efficacy of antiepileptic drugs (AEDs) in counteracting Na^+-dependent epileptogenic activities. The lack of effect of AEDs may not merely depend on a quantitative disproportion, as there is recent evidence suggesting that the drugs acting on Na^+ currents can be less effective on channels that have undergone seizure-dependent plastic modulation.[47, 48] Data from our laboratory[49] suggest that pathophysiological events occurring in neuronal aggregates can induce plasticity as a result of second messenger activation (specifically PKC activation), and that this can change the effects of AEDs. The efficacy of an AED may thus vary depending on the specific neuromodulatory mechanism induced by epileptogenic activities in a neuronal network, as is shown by the fact that changes in PKC activity or the activation of other modulation pathways occur in a number of experimental models of epilepsy.[50-52] A better understanding of these interactions may not only help to explain the uneven antiepileptic potency of a given AED in different patients, but also the variability of its neurological side effects.

6. REFERENCES

1. Minea, Cercetarı experimentale asupra variatiunilor morfogice ab neuronului sensitiv, *Thesis bucaresti* (1907).
2 L.G. Cohen, R.A. Weeks, N. Sadato, P. Celnik, K. Ishii, and M. Hallett, Period of susceptibility for cross-modal plasticity in the blind, *Ann. Neurol.* **45**, 451-460 (1999).
3. H. van Praag, B.R. Christie, T. J. Sejnowski, and F.H. Gage, Running enhances neurogenesis, learning and long-term potentiation in mice, *Proc. Natl. Acad. Sci USA* **96**, 13427-13431 (1999).
4. E. Gould and P. Tanapat. Stress and hippocampal neurogenesis, *Biol. Psychiatry* **46**, 1472-1479 (1999).
5. J. F. Annegers, W. A. Hauser, S. P. Coan, and W.A Rocca, A population-based-study of seizures after traumatic brain injuries, *New Engl. J. Med.* **338** (1), 20-24 (1998).
6. O. Steward, E.R. Torre, R. Tomasulo, and E. Lothman, Neuronal activity up-regulates astroglial gene expression,. *Proc. Natl. Acad. Sci. USA* **88**, 6819-6823 (1991).
7. R. D'Ambrosio, J. P. Fairbanks, J. S. Fender, D. E. Born, D. L. Doyle, and J. W. Miller, Post-traumatic epilepsy following fluid percussion injury in the rat, *Brain* **127**, 304-314 (2004).
8. J. Engel, Jr.,. A proposed diagnostic scheme for people with epileptic seizures and with epilepsy. Report of the ILAE Task Force on Classification and Terminology, *Epilepsia* **42**, 796-803 (2001).
9. W.R. Gowers, Epilepsy and other chronic convulsive diseases (Churchill,London, 1881).
10. F.Morrell, Secondary epileptogenic lesions, *Epilepsia* **1,** 538-560 (1959-1960).
11. F.Morrell and L.deToledo-Morrell, From mirror focus to secondary epileptogenesis in man. An historical review, *Adv. Neurol.* **81**, 11-23 (1999).
12. C. Bouchet and Y. Cazauvielh, De l'épilepsie considérée dans ses rapports avec l'aliénatıon mentale, *Arch. Gen. Med. Par.* **9**, 510-542 (1825).
13. T. Meynert, Studien uber das pathologisch-anatomısche material der Wiener Irren-Anstalt, *Vierteljahrssch* **3**, 381-402 (1867).
14. W. Sommer, Erkrankung des ammonshorns als aetiologisches moment der epilepsie, *Arch. Psychiatr. Nervenkr.* **10**, 631-675 (1880).
15. L. Pfleger, Beobachtungen uber schrumpfung und sklerose des ammonshorns bei epilepsie, *Allgemeine Z. Psychiatr.* **36**, 359-365 (1880).
16. G. V. Goddard, Development of epileptic seizures through brain stimulation at low intensity, *Nature* **214,** 1020-1021 (1967).
17. T. Sutula, X. X. He, J. Cavazos, and G. Scott, Synaptic reorganization in the hippocampus induced by abnormal functional activity, *Science* **239**, 1147-1150 (1988).
18. T. Sutula ,G. Cascino , J. Cavazos , I. Parada, and L. Ramirez, Mossy fibers synaptic reorganization in the epileptic human temporal lobe, *Ann. Neurol.* **3**, 322-230 (1989).
19. T.L. Babb, W.R. Kupfer, J.K. Pretorius, P.H. Crandall, and M.F. Levesque, Synaptic reorganization by mossy fibers in human epileptic fascia dentate, *Neuroscience* **42**, 351-363 (1991).
20. D.L. Tauck and J.V. Nadler, Evıdence of functional mossy fiber sprouting in hippocampal formation of kainic acid-treated rats, *J. Neurosci.* **5**, 1016-1022 (1985).
21. A. Represa, G. L. G. LaSalle, and Y. Ben-Ari, Hippocampal plasticity in the kindling model of epilepsy in rats, *Neurosci. Lett* . **99,** 345-355 (1989).

22. M. Isokawa, M. Levesque, T. Babb, and J. Engel, Jr., Single mossy fiber axonal systems of human dentate granule cells: studies in hippocampal slices from patients with temporal lobe epilepsy, *J. Neurosci.* **13**, 1511-1522 (1993).
23. J. Franck, J. Pokorny , D. Kunkel, and P. Schwartzkroin, Physiologic and morphologic characteristics of granule cell circuitry in human epileptic hippocampus, *Epilepsia* **36**, 543-558 (1995).
24. M Okazaki, D. Evanson, and J. Nadler, Hippocampal mossy fiber sprouting and synapse formation after status epilepticus in rats. visualization after retrograde transport of biocytin, *J. Comp. Neurol.* **352**, 515-534 (1995)
25. M. Isokawa, M. Levesque, I. Fried, and J. Engel, Jr., Glutamate currents in morphology identified human dentate granule cells in temporal lobe epilepsy, *American Physiological Society*, 3355-3369 (1997).
26. Z. Nusser, N. Hàjos, P. Somogyi, and I. Mody, Increased number of synaptic GABAA receptors underlies potentiation at hippocampal inhibitory synapses, *Nature* **395**, 172-177 (1998).
27. J.Jr. Engel, P.D. Williamson, and H.G. Wieser, Mesial temporal lobe epilepsy, in: *Epilepsy: a Comprehensive Textbook,* edited by J. Engel, Jr. and T. A. Pedley (Lippincott-Raven Press Philadelphia,1997) pp. 2417-2426.
28 T. N. Lehmann, S. Gabriel , R. Kovacs, A. Eilers, A.Kivi, K. Schulze, W.R. Lan, H.J. Meencke, and U. Heinemann, Alterations of neuronal connectivity in area CA1 of hippocampal slices from temporal lobe epilepsy patients and from pilocarpine-treated epileptic rats, *Epilepsia* **41**(Suppl.6), 190-194 (2000)
29. Y. Ben-Ari and A. Represa, Brief seizure episodes induce long-term potentiation and mossy fibers sprouting in the hippocampus, *TINS* **13**, 312-318 (1990).
30. B.M. Longo and L.E. Mello, Blokade of pilocarpine- or kainate-induced mossy fiber sprouting by cycloheximide does not prevent subsequent epileptogenesis in rats, *Neurosci. Lett.* **226**, 163-166 (1997).
31. B.W. Colder, R.C. Frysinger, C.L. Wilson, R.M. Harper, and J. Engel, Jr., Decreased neuronal burst discharge near site of seizure onset in epileptic human temporal lobes, *Epilepsia* **37**, 113-121 (1996a).
32. B.W. Colder, C.L. Wilson, R.C. Frysinger , L.C. Chao, R.M Harper, and J. Engel, Jr., Neuronal synchrony in relation to burst discharge in epileptic human temporal lobes, *J. Neurophysiol.* **75**, 2496-2508 (1996b).
33. I. Mody, Ion channels in epilepsy, *Int. Rev.Neurobiol.* **42,** 199-225 (1998).
34. J. Parent, T. Yu, R. Leibowitz, D. Geschwind, R. Sloviter, and D Lowenstein, Dentate granule cell neurogenesis is increased by seizures and contributes to aberrant network reorganization in the adult rat hippocampus, *J. Neurosci.* **17**, 3727-3738 (1997).
35. HG. Wieser for the ILAE Commission on Neurosurgery of Epilepsy, Mesial temporal lobe epilepsy with hippocampal sclerosis, *Epilepsia* **45**, 695-714 (2004).
36. L. Farina, C. Bergqvist, R.A. Zimmerman, J. Haselgrove, J.V. Hunter, and L.T. Bilaniuk, Acute diffusion abnormalities in the hippocampus of children with new-onset seizures: the development of mesial temporal sclerosis, *Neuroradiology* **46**, 251-257 (2004).
37. R. Spreafico, G. Battaglia, P. Arcelli , F. Andermann , F. Dubeau , A. Palmini, A. Olivier, J. G. Villemure, D. Tampieri, G. Avanzini, and M. Avoli, Cortical dysplasia – an immunocytochemical study of three patients, *Neurology* **50**, 27-36 (1998).
38. R. Spreafico, L. Tassi, and N. Colombo, Inhibitory circuits in human dysplasia tissue, *Epilepsia* **41**, S168-S173 (2000).
39. R. Garbelli, C. Munari, S. De Biasi, L. Vitellaro-Zuccarello, C. Galli, M. Bramerio, R. Mai, G. Battaglia, and R. Spreafico, Taylor's cortical dysplasia: a confocal and ultrastructural immunohistochemical study, *Brain Pathology* **9**, 445-461(1999).
40. I.M. Najm, Z. Ying, T. Babb, A. Mohamed, J. Hadam, E. LaPresto, E. Wyllie, P. Kotagal, W. Bingaman, N. Foldvary, H. Morris, and H. O. Lüders, Epileptogenicity correlated with increased *N*-methyl-D-aspartate receptor subunit NR2A/B in human focal cortical dysplasia, *Epilepsia* **41**, 971-976 (2000).
41. P.B. Crino and K. Chou, Epilepsy and cortical dysplasias, *Curr. Treat. Options Neurol.* **2** (6), 543-552 (2000)
42. P.B. Crino, A.C. Duhaime, G. Baltuch, and R. White, Differential expression of glutamate and GABA-A receptor subunit mRNA in cortical dysplasia, *Neurology* **56,** 906-913 (2001).
43. V Scaioli, G. Battaglia, S. Curzi, L. D'Incerti, T. Granata, F. Panzica, R. Spreafico, and G. Avanzini, Evoked potentials in neuronal migration disorders (NMDs), in: *Recent Advances in Human Neurophysiology,* edited by I. Hashimoto and R. Kakigi (Elsevier Sciences BV, Amsterdam 1998), pp. 879-884.
44. F. Villani, P Vitali, V. Scaioli, G Rodriguez, M. Rosa, T. Granata, G. Avanzini, R. Spreafico, and L. Angelini, Subcortical nodular heterotopia. a functional MRI and somatosensory evoked potentials study, *Neurol. Sci.* **25**, 225-229 (2004).

45. M.Vreugdenhil, G.Hoogland, CW van Veelen, and W. J. Wadman, Persistent sodium current in subicular neurons isolated from patients with temporal lobe epilepsy, *Eur. J. Neurosci.* **19** (10), 2769-2778 (2004).
46. M. Mantegazza, S. Franceschetti, and G. Avanzini, Anemone toxin (ATX II)-induced increase in persistent sodium current effects on the firing properties of rat neocortical pyramidal neurones, *J. Physiol* **507**, 106-116 (1998).
47 M. Vreugdenhil and W. J. Wadman, Modulation of sodium currents in rat CA1 neurons by carbamazepine and valproate after kindling epileptogenesis, *Epilepsia* **40**, 1512-1522 (1999).
48. S. Remy, S. Gabriel, B. W. Urban, D. Dietrich, T. N. Lehmann, C. E. Elger, U. Heinemann, and H. Beck, A novel mechanism underlying drug resistance in chronic epilepsy, *Ann. Neurol.* **53**, 469-479 (2003).
49. G. Curia, P. Aracri, G. Sancini, M. Mantegazza, G. Avanzini, and S. Franceschetti, Protein-kinase C-dependent phosphorylation inhibits the effect of the antiepileptic drug topiramate on the persistent fraction of sodium currents, *Neuroscience* **127**, 63-68 (2004).
50. S. J. Chen, M. A. Desai, E. Klann , D. G. Winder, J. D. Sweatt, and P. J. Conn, Amygdala kindling alters protein kinase C activity in dentate gyrus, *J. Neurochem* **59**, 1761-1769 (1992).
51. K. Osonoe, S Ogata, Y. Iwata, and N Mori, Kindled amygdaloid seizures in rats cause immediate and transient increase in protein kinase activity followed by transient suppression of the activity, *Epilepsia* **35**, 850-854 (1994).
52. K. Akiyama, M. Ono, I. Kohira, A. Daigen, T. Ishihara, and S. Kuroda, Long-lasting increase in protein kinase C activity in the hippocampus of amygdala kindled rat, *Brain Res.* **679**, 212-230 (1995).

Discussion

Velisek: I agree with you absolutely that there is no evidence that the kindling paradigm would bring epilepsy in humans. But there are at least 2 reports that kindling occurs after electrical stimulation in humans. I recall Monore's report in 1977 and Sramka's report in the early 1980s on the stimulation of thalamic nuclei for otherwise intractable pain, which eventually led to the development of seizures.
Avanzini: You are right. There are several limited observations, also with electroconvulsive therapy. What I am saying is that it apparently is not a very common or consistent phenomenon, because otherwise it would be reported much more consistently in the literature.
Engel: Going back to the figure that showed the time course, that whole sequences is a very logical one. As Karen pointed out, it is not necessarily true that all those things are leading up to the hyperexcitability rather than a being a byproduct of it. If you accept the hypothesis that fast ripples really represent these synchronously bursting neurons that have been tightly wired together and are the basis of epileptogenesis, clearly they occur before the hard-wired synaptic changes. What we think is that these produce a kindling-like stimulus and that the hyper-synchronous discharges then kindle transsynaptically and that this is the process that leads to seizures, and not necessarily that seizures lead to more seizures. Rather, once this process is started, it continues. If this is necessary to kindle adjacent areas and produce behavioral seizures, then there is no reason to believe that once a behavioral seizure begins that this is the end of the process. It keeps going, as demonstrated by over-kindling and getting more severe seizures and secondary epileptogenesis. Logically there is no reason why, once a clinical seizure appears, the process should stop.
Avanzini: There are some observations that may speak to the contrary. For instance, in a population of patients not treated, like some studies in developing countries, you don't find a higher proportion of pharmacologically resistant epilepsy with respect to our population. This seems to tell us that this is a subpopulation with some very special risk factor, and not that the repetition of the seizure per se is inducing this.
Engel: I think that we are dealing with a broad spectrum of disease states. There are some forms of epilepsy that are benign and are self-limited and will go away, and

there are others that level out and don't really change very much, and there are others that will progress. The ones that we deal with at our epilepsy centers, the ones that get surgery, are the serious ones that go on and continue to progress. It is important to recognize those so that you can intervene early and stop that progression and prevent a lifetime of disability.

Avanzini: True. In any case, I agree with you that we cannot attribute meaning to any of these factors per se. They may all contribute, or they may be only a marker of what is happening, and we need that to understand better which are the real risk factors for generating this.

Engel: In fact as far as fast ripples are concerned, if they do reflect this kind of pathology, Mark Dichter actually went back and looked at his penicillin hippocampal studies that he did with Alden Spencer back in the 1970s and these animals, even though they were acute experiments, had seizures and fast ripples.

Schwartzkroin: It seems to me that we are getting into the lumpers and splitters issues again in the sense that you have been talking about dividing up this clinical population into a number of subpopulations that have different kinds of diseases. Some or all of these mechanisms may be causal in some of those populations, and some of them may be epiphenomena. I think that we are dealing with the same kinds of issues as we think of how to model whatever it is we want to model. When you develop a model, kindling for example, what is it that you are proposing to model? If we are proposing kindling as a model of temporal lobe epilepsy, then what is temporal lobe epilepsy? Obviously from what you have just said and even from what Pete said, is not a unitary, simple, identifiable disorder. It probably has lots of subpopulations. So it is important to identify what the target is for our model.

Avanzini: I think really agree with you, and I wanted to stress this point: Although hippocampal sclerosis is certainly an interesting marker, I think that it would be a disaster if it could be assumed to be the hallmark for epilepsy that should be operated on. This would be an oversimplification that would generate a lot of trouble in the future of surgery.

Engel: Let me tell you what the conclusions of our workshop were. The purpose of the workshop was to debate whether mesiotemporal lobe epilepsy with hippocampal sclerosis is a disease or a syndrome. The conclusion was that it isn't a disease *or* a syndrome. At best, it may be a whole group of syndromes. In answer to Phil's question and the debates that we have had for years, my position has always been that we don't model an entire human condition, that it is ridiculous to think that we could ever do that. What we need to do is break down the condition into a number of different phenomena and model the phenomena independently, like I tried to do with my table. I think there are good models for each of those processes, and I am very much a splitter. I can't understand lumpers, because I don't think that you get any answers by lumping it. I think that you have to break things down into their component parts and study them independently with the appropriate models for those independent phenomena.

Wada: Kindling does provide an excellent window. If you recognize those various aspects of epilepsy, you can use kindling as an appropriate model for each, a variety of conditions.

Engel: Yes, but not all of them, because it is not a model of human epilepsy.

Avanzini: If you think that hippocampal sclerosis is normal since the nineteenth century, the main question is it the cause or is it a consequence of epilepsy. After a huge investment for research in this field, the conclusion is not yet clear, whether hippocampal sclerosis is the cause or the effect. But this is the science.

PATHOLOGICAL SENSITIZATION OF THE DOPAMINE SYSTEM IN EXPERIMENTAL EPILEPTOGENESIS:

Implication for the mechanism of epileptic psychosis

Kiyoshi Morimoto and Mitsumoto Sato*

1. EPILEPSY AND PSYCHOSIS

In the long-term clinical course of epilepsy, there is an increased incidence of the development of schizophrenia-like paranoid symptoms, such as delusion and hallucination. The prevalence of this epileptic psychosis appears to be about 7–11%, which is much higher than in the general population.[22, 37] Since some patients with epileptic psychosis show an inverse correlation between the presence of epilepsy and the psychosis, this phenomenon was has been labeled "forced normalization" or "alternative psychosis" (see review of Krishnamoorthy and Trimble[18]). Slater et al.,[30] on the other hand, reported that the emergence of epileptic psychosis was related to the duration of epilepsy and to the brain damage, and was linked to temporal lobe epilepsy (TLE).

Thus far, the exact etiology of epileptic psychosis remains uncertain, but several risk factors have been implicated, including onset of epilepsy before the age 20 years, a history of epilepsy greater than 10 years, a history of complex partial seizures, and TLE that is focused on the left side.[37] In particular, neurodevelopmental abnormalities in the mesial temporal lobe (e.g., hamartomatous lesions or gangliogliomas) have been indicated.[25, 36] More recent studies suggest that therapeutic powerful antiepileptic drugs such as vigabatrin[10, 26], and a family history of psychosis[2] are also factors involved in the development of epileptic psychosis.

2. BEHAVIORAL ABNORMALITIES IN THE KINDLING MODEL

Another hypothesis is that enhanced activity of the mesolimbic dopamine (DA) system is responsible for the development of epileptic psychosis.[18] In experimental studies, kindling-induced chronic epileptogenesis produced long-lasting

* Kiyoshi Morimoto, Department of Neuropsychiatry, Faculty of Medicine, Kagawa University, Kagawa, 761-0793 Japan. Mitsumoto Sato, Department of Psychiatry, Graduate School of Tohoku Fukushi University, Sendai, 981-8522 Japan.

hypersensitivity to various DA agonists. Csernansky et al.,[7] for example, found that "superkindled" rats (a standard kindling paradigm with additional electrical stimulation) were supersensitive to the direct DA agonist apomorphine, while standard kindled rats were subsensitive, suggesting that the development of DA hypersensitivity in the epileptic brain depends on the number or magnitude of previously experienced seizure activity. Using the indirect DA agonists (enhancers of DA release), such as methamphetamine (MAP) and amphetamine, Sato[27] reported that MAP-induced stereotyped behavior was enhanced in amygdala kindled cats, and Ma and Leung[21] reported that MAP induced higher locomotor activity in partial hippocampal kindling rats. Kindling also caused a lasting increase in DA D2 receptor binding in the striatum and nucleus accumbens,[7, 8] an increase in the expression of DA D2 receptor mRNA in the striatum,[11] and a decrease in DA transporter binding in the striatum[13].

3. ENHANCEMENT OF DA ACTIVITY IN THE KAINATE MODEL

We have recently investigated alterations of central DA systems in the intra-amygdala kainate model of TLE.[3] This model is another chronic model for TLE, but unlike the kindling model, it is characterized by the recurrent appearance of spontaneous seizures and mesial temporal lobe neuropathology (e.g., pyramidal neuron loss and gliosis), which resembles hippocampal sclerosis as most commonly seen in human TLE.[34]

3.1. Enhancement of MAP-induced Locomotor Activity

In adult male Sprague-Dawley rats, kainate (2 μg/0.5 μl) was microinjected into the left amygdala via the chronically implanted guide cannula to induce status epilepticus. MAP (2 mg/kg, i.p.) was administered before and 1 month after the kainate treatment. MAP-induced locomotor activity was significantly enhanced in the kainate group compared with the baseline (pre-kainate) level ($P<0.001$, two-way ANOVA), while it was not significantly changed in the control group ($P=0.91$, Figure 1). When total counts of locomotor activity were measured during the first 4 h after MAP administration, they were significantly increased in the kainate group compared with the baseline and the control group. Similar enhancement of MAP-induced locomotor activity was also observed 2 weeks and 2 months after kainate–induced status epilepticus (Figure 2). Pretreatment with haloperidol (the antagonist of DA D2 receptors) significantly reduced the enhanced locomotor activity, and the enhanced locomotor activity recovered nearly to the baseline level.

In most of the kainate-treated animals, histological examination confirmed apparent neuron loss in the left hippocampus ipsilateral to the kainate injection. The increment in MAP-induced locomotor activity (total counts; "1 month after kainate" minus "baseline") was significantly correlated with hippocampal neuron density in the CA1 region of the kainate group ($r=0.67$, $P<0.01$, Figure 3). However, there was no such significant correlation in the CA3 region ($r=-0.07$, $P=0.83$).

3.2. Enhancement of MAP-induced DA Release

We measured striatal extracellular DA concentrations by *in vivo* microdialysis in freely moving rats 1 month after kainate-induced status epilepticus (Ando et al.,

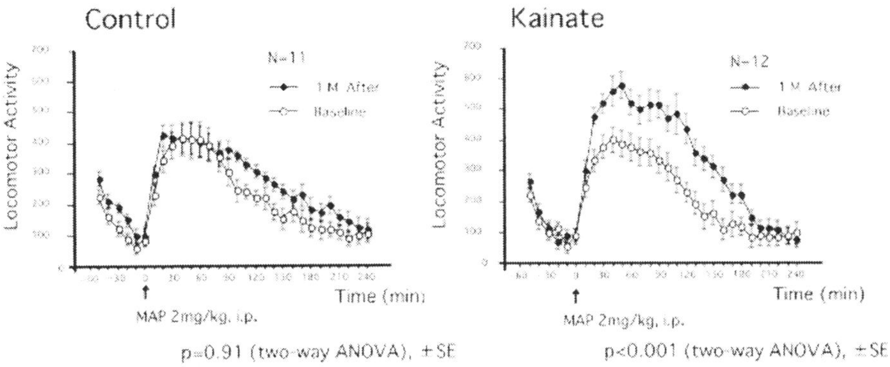

Figure 1. Enhancement of MAP-induced locomotor activity in the kainate model of TLE. MAP (2 mg/kg, i.p.) was administered before and 1 month after an intra-amygdala injection of vehicle (the control group; left) or kainate (the kainate group; right). Locomotor activity was measured by an infrared detector every 20 min for 60 min before, and for 240 min following MAP administration. Note the significant enhancement of MAP-induced locomotor activity in the kainate group, but not in the control group. Data from Ando et al.[3]

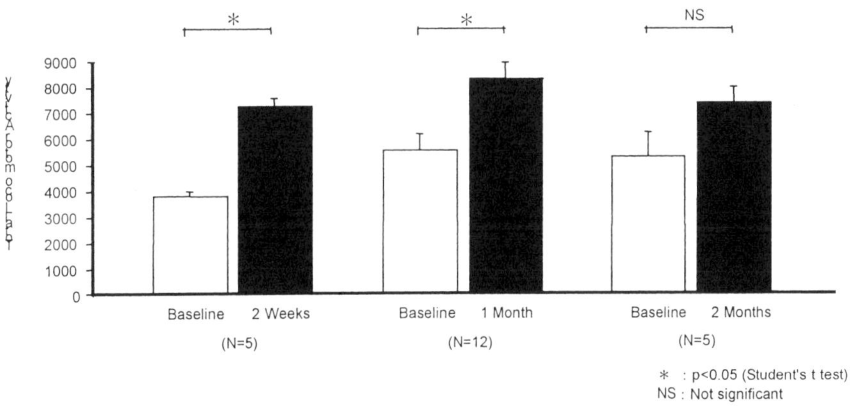

Figure 2. Time-course of enhancement of MAP-induced locomotor activity in the kainate model of TLE. MAP (2 mg/kg, i.p.) was administered before ("baseline") and 2 weeks, 1 month or 2 months after an intra-amygdala injection of kainate. Each value represents total counts of locomotor activity during the 4 h following MAP administration. Note that the total counts of locomotor activity were significantly increased at 2 weeks and 1 month after kainate injection, and similar enhancement was still observed at 2 months. Data from Ando et al. (unpublished)

2004). The basal value of the DA concentration (immediately before MAP administration) was significantly lower in the kainate group than in the controls (5.5 vs. 39.2 fmol/20-min sample, $P<0.05$, Figure 4). Following MAP administration, DA concentrations were significantly elevated in both groups. However, the maximal value of DA concentration following MAP administration tended to be higher in the kainate group than in the control group (395.2 vs. 828.6 fmol/20-min sample, Figure 4). When the time course of the MAP-induced increase in DA concentration (relative to the basal value) was compared, the kainate group tended to show a much higher elevation (P=0.09, two-way ANOVA, Figure 4). The maximal ratio of increase was 158-fold in the kainate group at 60 min following MAP administration, and 12-fold in the control group.

4. PATHOLOGICAL SENSITIZATION OF THE DA SYSTEM AND VTA KINDLING

In our study, it clearly appeared that the locomotor activity induced by systemic administration of MAP (an indirect DA agonist that facilitates DA release) was significantly enhanced during the chronic phase (1 month after status epilepticus) of the kainate model of TLE and this enhancement was antagonized by pretreatment with a relatively small dose of haloperidol (a non-selective DA D2 receptor antagonist). These results indicate that the TLE brain develops a hypersensitivity of the DA system, a state that would contribute to the pathophysiology of epileptic psychosis.

Since a number of studies have revealed that neonatal or adult hippocampal lesions result in DA hypersensitivity (see review of Lipska et al.[20]), the DA hypersensitivity seen in the kainate model seems to be attributable to hippocampal damage. In our study, however, there was a positive correlation between the number of surviving neurons in hippocampal area CA1 and the degree of enhanced MAP-induced locomotor activity (see Figure 3). This result means that more severe damage to hippocampal CA1 neurons by kainate-induced status epilepticus would result in a lesser degree of DA hypersensitivity. It is more likely that kainate-induced seizure activity is directly involved in causing pathological sensitization of the DA system.

4.1. Pathological Sensitization of the Mesolimbic DA System

Previous studies demonstrated that seizure activity transiently elevated extracellular DA levels at various brain sites, including the hippocampus, striatum, nucleus accumbens, and prefrontal cortex.[4, 9, 17, 31, 32] Furthermore, the seizure-induced elevation of DA release could be enhanced if seizures were induced repeatedly.[4, 9, 32] This phenomenon is comparable to the process of "behavioral sensitization", in which repeated administration of DA agonists augment abnormal behaviors with the concominant enhancement of DA release[14, 16] (also see review of Kalivas and Stewart[15]). In humans, repeated administration of these agents produces schizophrenia-like paranoid symptoms.[28, 29]

4.2. VTA Kindling

In the kainate model, the prolonged seizure activity associated with limbic status epilepticus might produce excessive DA release and have such sensitization-like effects on DA systems. In fact, our recent study has shown that kindling of the ventral tegmental area (VTA), a major source of mesolimbic DA

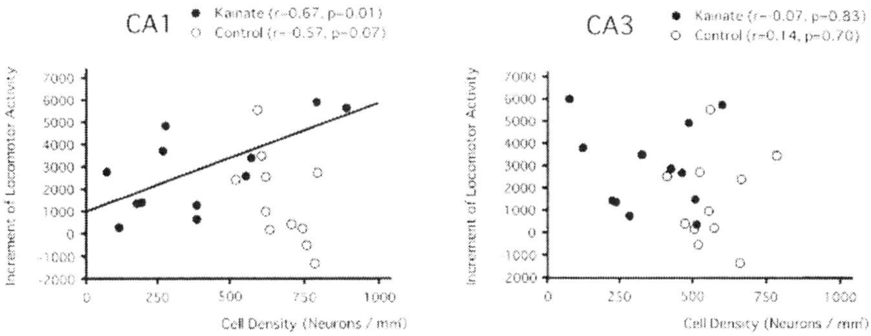

Figure 3. Correlation between the increase in MAP-induced locomotor activity and hippocampal cell density in the kainate model of TLE.
The number of pyramidal neurons/mm^2 was estimated in the hippocampal CA1 (left) and CA3 (right) regions 1 month after the intra-amygdala injection of kainate or vehicle. Note that the neuronal density was significantly decreased in the CA1 region (384.4 ± 75.6 vs. 653.3 ± 26.7 neurons/mm^2, $P<0.01$) and CA3 region (348.9 ± 48.9 vs. 591.1 ± 31.1 neurons/mm^2, $P<0.01$) of the kainate group, compared with those of the control group. There was a significant positive correlation between the increase in MAP-induced locomotor activity and hippocampal cell density in the CA1 region, but not in the CA3 region, only in the kainate group. Data from Ando et al.[3]

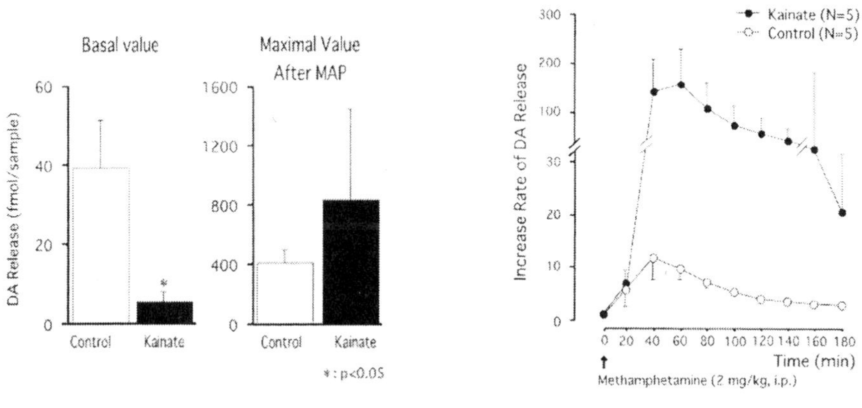

Figure 4. Striatal extracellular DA concentrations in the kainate model of TLE.
Left: Basal and maximal values of extracellular DA concentration following MAP administration. Extracellular DA concentrations were measured in the striatum by *in vivo* microdialysis at 1 month after the intra-amygdala injection of kainate or vehicle. The basal (left) and maximal values (right) represent DA concentrations measured immediately before and after MAP administration (2 mg/kg, i.p.), respectively. Note that the basal value was significantly lower, but the maximal value tended to be higher in the kainate group. Data from Ando et al.[3]
Right: Time-course of the MAP-induced elevation of striatal DA concentrations. The rate of increase in striatal DA concentrations following MAP administration (relative to the basal value) was compared between the kainate and control groups. Note the higher elevation of striatal DA concentrations in the kainate group ($P=0.09$, two-way ANOVA). Data from Ando et al. (unpublished).

pathways, could produce persistent enhancement of MAP-induced locomotor activity.[39] In this study, we applied two quantitative measurements of DA sensitivity before and 2 weeks after VTA kindling (20 times bilateral electrical stimulations (100 μA at 1 min intervals) delivered once per day for consecutive 14 days): behavioral responses induced by test VTA stimulation and methamphetamine (MAP)-induced locomotor activity. The total amount of MAP-induced locomotor activity was significantly increased after VTA kindling (Figure 5). However, the responses to electrical stimulation, which produced forward locomotion and exploration in a stimulus intensity-dependent manner, were unchanged.[38, 39] Consistent with these results, Glenthoj et al.[12] reported that when the VTA was stimulated electrically in rats, a significantly potentiated behavioral response was demonstrated after 70 stimulations, and Ben-Shahar and Ettenberg[5] reported that locomotor activities induced by amphetamine subcutaneous injection were augmented after self-stimulation of the VTA.

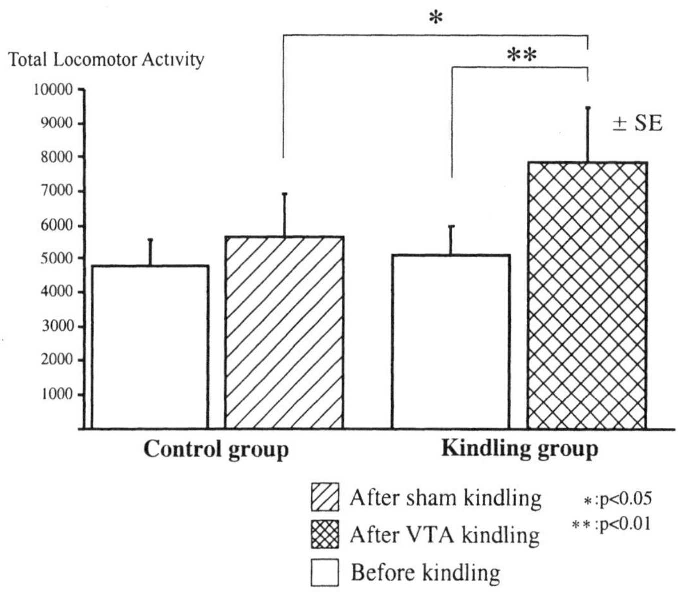

Figure 5. Enhancement of MAP-induced locomotor activity after VTA kindling.
MAP (2 mg/kg, i.p.) was administered before and 2 weeks after VTA kindling. Note that the total number of locomotor activity was significantly increased after VTA kindling.

Repeated administration of MAP, in turn, resulted in a reduction of the electrical threshold of the VTA for eliciting forward locomotion (Watanabe et al., 1998), suggesting the existence of bidirectional interactions between VTA kindling and behavioral sensitization. Kainate-induced status epilepticus might have indirectly kindled the mesolimbic DA system, which resulted in the DA hypersensitivity.

5. IMPLICATION FOR THE MECHANISM OF EPILEPTIC PSYCHOSIS

In our microdialysis study, MAP-induced striatal DA release was markedly enhanced in kainate-treated rats (approximately 14-fold relative to the control level), while the basal value of DA tended to be lower (approximately 1/7) in the chronic phase of the kainate model (see Figure 4). In contrast to this, there was no significant alteration of DA D1, D2, or D4 receptor binding in the striatum and nucleus accumbens following kainate administration.[35] Therefore, it is likely that the vulnerability of pre-synaptic DA to stress is responsible for eliciting psychotic symptoms in TLE.

Consistent with this hypothesis, the clinical finding in schizophrenic patients, obtained using DA receptor neuroimaging techniques, demonstrated that acute amphetamine administration aggravates psychotic symptoms concomitant with an enhanced DA release in the striatum.[1, 6, 19] An increase in the dopa decarboxylation rate was also found in the striatum of patients with epileptic psychosis and in those with schizophrenia, compared with non-psychotic epileptic patients or normal controls, suggesting the suppression of tonic release of striatal DA in these psychotic disorders.[23] Striatal post-synaptic DA D2 receptors, on the other hand, were down-regulated in patients with epileptic psychosis.[24]

6. CONCLUSIONS

In conclusion, DA hypersensitivity develops in the chronic phase of the kainite model of TLE via pathological sensitization or indirect kindling of the mesolimbic DA systems. Enhancement of pre-synaptic DA release may be the mechanism responsible for epileptic psychosis.

7. REFERENCES

1. A. Abi-Dargham, R. Gil, J. Krystal, R. M. Baldwin, J. P. Seibyl, M. Bowers, C. H. van Dyck, D. S. Charney, R. B. Innis, and M. Laruelle, Increased striatal dopamine transmission in schizophrenia: confirmation in a second cohort, *Am. J. Psychiatry* **155**, 761-767 (1998).
2. N. Adachi, M. Matsuura, Y. Okubo, Y. Oana, N. Takei, M. Kato, T. Hara, and T. Onuma, Predictive variables of interictal psychosis in epilepsy, *Neurology* **55**, 1310-1314 (2000).
3. N. Ando, K. Morimoto, T. Watanabe, T. Ninomiya, and H. Suwaki, Enhancement of central dopaminergic activity in the kainate model of temporal lobe epilepsy: Implication for the mechanism of epileptic psychosis, *Neuropsychopharmacology*, in press (2004).
4. A. Becker, G. Grecksch, W. Thiemann, and V. Hollt, Pentylenetetrazol-kindling modulates stimulated dopamine release in the nucleus accumbens of rats, *Pharmacol. Biochem. Behav.* **66**, 425-428 (2000).
5. O. Ben-Shahar and A. Ettenberg, Repeated stimulation of the ventral tegmental area sensitizes the hyperlocomotor response to amphetamine, *Pharmacol. Biochem. Behav.* **48**, 1005-`009 (2000).
6. A. Breier, T. P. Su, R. Saunders, R. E. Carson, B. S. Kolachana, A. de Bartolomeis, D. R. Weinberger, N. Weisenfeld, A. K. Malhotra, W. C.

Eckelman, and D. Pickar, Schizophrenia is associated with elevated amphetamine-induced synaptic dopamine concentrations: evidence from a novel positron emission tomography method, *Proc. Natl. Acad. Sci. U S A* **94**, 2569-2574 (1997).

7. J. G. Csernansky, J. Mellentin, L. Beauclair, and L. Lombrozo, Mesolimbic dopaminergic supersensitivity following electrical kindling of the amygdale, *Biol. Psychiatry* **23**, 285-94 (1988a).
8. J. G. Csernansky, S. Kerr, R. Pruthi, and E. S. Prosser, Mesolimbic dopamine receptor increases two weeks following hippocampal kindling, *Brain Res.* **449**, 357-360 (1988b).
9. L. Dazzi, M. Serra, M. L. Porceddu, A. Sanna, M. F. Chessa, and Biggio, Enhancement of basal and pentylenetetrazol (PTZ)-stimulated dopamine release in the brain of freely moving rats by PTZ-induced kindling, *Synapse* **26**, 351-358 (1997).
10. C. D. Ferrie, R. O. Robinson, and C. P. Panayiotopoulos, Psychotic and severe behavioural reactions with vigabatrin: a review, *Acta Neurol. Scand.* **93**, 1-8 (1996).
11. H. A. Gelbard and C. D. Applegate, Persistent increases in dopamine D2 receptor mRNA expression in basal ganglia following kindling, *Epilepsy Res.* **17**, 23-29 (1994).
12. B. Glenthoj, J. Mogensen, H. Laursen, S. Holm, and R. Hemmingsen, Electrical sensitization of the meso-limbic dopaminergic system in rats: a pathogenetic model for schizophrenia, *Brain Res.* **619**, 39-54 (1993).
13. I. Gordon, M. Mintz, E. Rosenne, and M. Rehavi, Long-term effects of amygdaloid kindling on striatal dopaminergic terminals, *Brain Res Bull.* **36**, 235-239 (1995).
14. T. Hamamura, K. Akiyama, K. Akimoto, K. Kashihara, K. Okumura, H. Ujike, and S. Otsuki, Co-administration of either a selective D1 or D2 dopamine antagonist with methamphetamine prevents methamphetamine-induced behavioral sensitization and neurochemical change, studied by in vivo intracerebral dialysis, *Brain Res.* **546**, 40-46 (1991).
15. P. W. Kalivas and J. Stewart, Dopamine transmission in the initiation and expression of drug- and stress-induced sensitization of motor activity, *Brain Res. Rev.* **16**, 223-244 (1991).
16. Y. Kazahaya, K. Akimoto, and S. Otsuki, Subchronic methamphetamine treatment enhances methamphetamine- or cocaine-induced dopamine efflux in vivo, *Biol. Psychiatry* **25**, 903-912 (1989).
17. G. M. Khan, I. Smolders, H. Lindekens, J. Manil, G. Ebinger, and Y. Michotte, Effects of diazepam on extracellular brain neurotransmitters in pilocarpine-induced seizures in rats, *Eur. J. Pharmacol.* **373**, 153-161 (1999).
18. E. S. Krishnamoorthy and M. R. Trimble, M. R. (1999) Forced normalization: clinical and therapeutic relevance, *Epilepsia* **40** Suppl 10, S57-64 (1999).
19. M. Laruelle, A. Abi-Dargham, C. H. van Dyck, R. Gil, C. D. D'Souza, J. Erdos, E. McCance, W. Rosenblatt, C. Fingado, S. S. Zoghbi, R. M. Baldwin, J. P. Seibyl, J. H. Krystal, D. S. Charney, and R. B. Innis, Single photon emission computerized tomography imaging of amphetamine- induced dopamine release in drug-free schizophrenic subjects, *Proc. Natl. Acad. Sci. U S A* **93**, 9235-9240 (1996).

20. B. K. Lipska and D. R. Weinberger, To model a psychiatric disorder in animals: schizophrenia as a reality test, *Neuropsychopharmacology* **23**, 223-239 (2000).
21. J. Ma and L. S. Leung, Schizophrenia-like behavioral changes after partial hippocampal kindling, *Brain Res.* **997**, 111-118 (2004).
22. P. J. McKenna, J. Kane, and K. Parrish, Psychotic syndromes in epilepsy, *Am. J. Psychiatry* **142**, 895-904 (1985).
23. J. Reith, C. Benkelfat, A. Sherwin, Y. Yasuhara, H. Kuwabara, F. Andermann, S. Bachneff, P. Cumming, M. Diksic, S. E. Dyve, et al., Elevated dopa decarboxylase activity in living brain of patients with psychosis, *Proc. Natl. Acad. Sci. U S A* **91**, 11651-11654 (1994).
24. H. A. Ring, M. R. Trimble, D. C. Costa, J. Moriarty, N. P. Verhoeff, and P. J. Ell, Striatal dopamine receptor binding in epileptic psychoses, *Biol. Psychiatry* **35**, 375-380 (1994).
25. G. W. Roberts, D. J. Done, C. Bruton, and T. J. Crow, A "mock up" of schizophrenia: temporal lobe epilepsy and schizophrenia-like psychosis, *Biol. Psychiatry* **28**, 127-143 (1990).
26. J. W. Sander, Y. M. Hart, M. R. Trimble, and S. D. Shorvon, Vigabatrin and psychosis, *J. Neurol. Neurosurg. Psychiatry* **54**, 435-439 (1991).
27. M. Sato, Long-lasting hypersensitivity to methamphetamine following amygdaloid kindling in cats: the relationship between limbic epilepsy and the psychotic state, *Biol. Psychiatry* **18**, 525-536 (1983).
28. M. Sato, C. C. Chen, K. Akiyama, and S. Otsuki, Acute exacerbation of paranoid psychotic state after long-term abstinence in patients with previous methamphetamine psychosis, *Biol. Psychiatry* **18**, 429-440 (1983).
29. M. Sato, Y. Numachi, and T. Hamamura, Relapse of paranoid psychotic state in methamphetamine model of schizophrenia, *Schizophrenia Bull.* **18**, 115-122 (1992).
30. E. Slater, A. W. Beard, and E. Glithero, The schizophrenialike psychoses of epilepsy, *Br. J. Psychiatry* **109**, 95-150 (1963).
31. I. Smolders, L. Bogaert, G. Ebinger, and Y. Michotte, Muscarinic modulation of striatal dopamine, glutamate, and GABA release, as measured with in vivo microdialysis, *J. Neurochem.* **68**, 1942-1948 (1997).
32. R. E. Strecker and M. E. Moneta, Electrical stimulation of the kindled hippocampus briefly increases extracellular dopamine in the nucleus accumbens, *Neurosci. Lett.* **176**, 173-177 (1994).
33. N. R. Swerdlow, N. Halim, F. M. Hanlon, A. Platten, and P. P. Auerbach, Lesion size and amphetamine hyperlocomotion after neonatal ventral hippocampal lesions: more is less, *Brain Res. Bull.* **55**, 71-77 (2001).
34. T. Tanaka, S. Tanaka, T. Fujita, K. Takano, H. Fukuda, K. Sako, and Y. Yonemasu, Experimental complex partial seizures induced by a microinjection of kainic acid into limbic structures, *Prog. Neurobiol.* **38**, 317-334 (1992).
35. F. I. Tarazi, A. Campbell, and R. J. Baldessarini, Effects of hippocampal kainic acid lesions on striatolimbic dopamine D1-, D2-, and D4-like receptors, *Neuroscience* **87**, 1-4 (1998).
36. D. C. Taylor, Factors influencing the occurrence of schizophrenia-like psychosis in patients with temporal lobe epilepsy, *Psychol. Med.* **5**, 249-254 (1975).
37. R. Torta and R. Keller, Behavioral, psychotic, and anxiety disorders in epilepsy: etiology, clinical features, and therapeutic implications, *Epilepsia* **40**, S2-20 (1999).

38. T. Watanabe, K. Morimoto, M. Nakamura, and H. Suwaki, Modification of behavioral responses induced by electrical stimulation of the ventral tegmental area in rats, *Behav. Brain Res.* **93**, 119-29 (1998).
39. T. Watanabe, K. Morimoto, M. Nakamura, T. Hirao, T. Ninomiya, N. Ando, and H. Suwaki, H. (2004) Kindling of the ventral tegmental area induces the supersensitivity of the central dopamine system, *Brain Res.* **1003**, 194-198 (2004).

Discussion

Burnham: Pharmacologically, whenever you repeatedly applied dopaminergic agonists, do you tend to get this sensitization?
Morimoto: Yes
Burnham: This is generally in line with your theory, but doesn't seem to go along with the idea of increased release.
Morimoto: Yes, I know. In previous studies we used repeated amphetamine or methamphetamine, and they produced behavioral sensitization. The effect is correlated with an abnormal increase in dopamine release, as shown in several pervious mircodialysis studies.
Phillips: Thank you very much for a lovely paper. Where did you put the kainic acid into the amygdala, was it central or basolateral?
Morimoto: I think the tip was in the basolateral amygdala, but the lesion very large, producing complete damage in the whole amygdala.
Phillips: I think the point that you are making about how the seizures might modulate the whole catecholamine and monoamine systems is very important. John Howland and I, and other colleagues in our lab, have been looking at the effects of just a single and in one instance multiple trains of stimulation of both the amygdala and the hippocampus, and we are seeing very profound changes in dopamine. What we do is place an electrode in the ventral subiculum, the outflow of the hippocampus, and a single train of 100 pulses causes an immediate increase in dopamine in both the nucleus accumbens and the medial prefrontal cortex. The remarkable thing is that this increase from just 10 seconds of stimulation is elevated for 30 to 40 minutes, and it is accompanied heightened locomotor activity. These dopamine systems are extremely sensitive to any sort of seizure-like event that might occur in other limbic structures. John showed that when he moved his stimulating electrode to the basolateral amygdala, he got exactly the same phenomenon. In one study that we haven't published, we repeated the stimulation in the hippocampus, daily, for three days and two things of great importance happened. One is that the basal level of dopamine in the nucleus accumbens was increased by 100% just by having three brief seizures in the hippocampus, and the progression of this could be blocked by blocking the ionotropic glutamate receptors. If these things can be happening with just one or two transient seizures, imagine what is happening in someone's brain who is having recurrent ictal events. I think this is a very important topic that you are addressing. Just one final comment: Mike Corcoran reminded me that at the First Kindling Conference I made the comment to Graham Goddard that when you looked at the then-current maps of structures that kindled, many of them were postsynaptic to the catecholamine system. I didn't know what that meant at the time, but in the intervening 30 years I can think a couple of things that might be happening. One is that this might be a way in which the psychopathological correlates of aberrant electrographic events manifest themselves in psychotic behavior by modulating the

indoleamine and catecholamine systems. The other has been mentioned several times, that perhaps the brain attempts to compensate for the seizure. One of the things that the dopamine system does is that at high concentrations it can inhibit the activity of target neurons. Perhaps this is an initial compensatory mechanism that the brain may have in order to try to limit the consequences of the seizure on the postsynaptic event. So you have these neurochemical modulators that can tone down the effects of seizure on the brain, and perhaps with repeated stimulation as these systems kindle, perhaps you are implying, that the modulatory role can be overcome as seizures become progressive over time.

Morimoto: Thanks for you comment. I think that this enhancement of the dopamine system in chronic epileptogenesis has some seizure-protecting effects.

Engel: Tony Phillips's comment at the First Kindling Conference stimulated me to start kindling, because I was working with Bob Ackermann and Bob Katzman, doing dopamine fluorescence, and I realized that I could start doing kindling and tag on to what they were doing and maybe find something interesting.

Phillips: If I did nothing else in my career, that was a great contribution.

Engel: Clinically, psychotic symptomotology in epilepsy can occur under four different conditions. It can occur during the ictal event itself. It can be a postitcal phenomenon, in which case there the is a lucid interval that can go on for a day or two before a prolonged period of psychosis. It can be interictal totally unrelated to the seizures. Or it can be preitcal, in the sense that it begins interictally but then the seizures will stop it. Are there any of those types of psychotic symptomotology that can be explained by the time course of dopaminergic activity in kindling?

Morimoto: It is very unclear. I think this phenomenon is related to the interitcal so-called alternative psychosis.

Post: People in Danny Weinberger's lab did ventral hippocampal lesions, and the results are different if you do them in neonates versus adult animals, which is kind of interesting for the notion of childhood onset that we are talking about here. That is, ventral hippocampal lesions done in the neonate result in adult hyperactivity, but if you do those same lesions as adults it doesn't happen. It is an interesting reciprocal to your increased excitability changes with kainate.

The question I have is, what do the VTA kindled animals look like and how hard is it to kindle them?

Morimoto: Actually, with kindling we never observed AD or a convulsive response. Usually when we did bilateral stimulation, it caused some behavioral abnormality such as forward locomotion and exploratory behavior like sniffing, walking around, and rearing. We did not observe an enhancement of such behaviors during kindling, but a previous study showed progressive enhancement of such behaviors. We have never found a seizure response with kindling.

Adamec: I am curious of the interictal behavior of these animals, peculiarly the ventral tegmental kindled animals. I am sure that you are aware that Stevenson and Livermore did a similar kind of study in the 1970s in cats. They used a kindling paradigm, and like you they couldn't find any evidence of epileptogenesis, but they reported profound behavioral changes. According Janice Stevenson, and these cats were very antisocial, showing a profound social withdrawal. I am curious to know whether you have tested that or seen anything like that in the rat, or whether you have considered doing something like a social interaction test in these animals, to see whether there is some change in spontaneous behavior other than the responsiveness to methamphetamine. Stevenson reported these behavioral changes, and they were reversible by neuroleptics.

Morimoto: I have never tested this sort of behavior, and I should do so. Thank you.

Sato: I would like to make two comments for future kindling studies from the clinical point of view of biological psychiatry. Kindling especially limbic kindling involves two persistent dysfunctions of the brain produced by repetition stimuli. The development of epileptogenesis is enough to induce recurrent spontaneous seizures and a change in response to dopamine agonists such as methamphetamine and cocaine. Kindling induced epileptogenesis is characterized by the development of secondary generalized seizures from partial seizures. However, it is interesting to note that the secondary generalization of partial seizure is much more stable in limbic kindling than cortical kindling in the cat. Moreover, positive transfer is found only among the limbic and mesolimbic structures, but is not found between different cortical areas of the cat. Thus, it seems possible that kindling may depend upon the kindled neuronal trace inside the limbic system. It will be interesting to examine the molecular genetics of the kindling trace, especially between the limbic system and structures related to seizure generalization, such as the midbrain reticular formation, thalamus, and claustrum. It may be critical to develop a new strategy to treat and prevent intractable partial epilepsy with secondary generalization including temporal lobe epilepsy. Secondly, lasting change in behavioral response to dopamine agonist is another point of interest in our clinical practice. We reported previously that the cats kindled from the amygdala showed remarkable abnormal behavior such as stereotopy and visual searching to a challenge dose of methamphetamine and cocaine. This finding suggests that amygdala kindling may involve lasting change also in the dopaminergic system, which may relate to behavioral disorders including interitcal psychosis in some patients with intractable epilepsy. Like behavioral sensitization induced by chronic dopamine agonists, repeated release of excessive dopamine in each brief seizure may produce dopaminergic hypersensitivity in animals kindled from the amygdala.

1. Phil Schwartzkroin
2. Gerry Chase Schwartzkroin
3. Michael Corcoran
4. Nico Moshé
5. Juhn Wada
6. Mary Wada
7. Pete Engel
8. Dan McIntyre
9. Ron Racine
10. Alexey Shandra
11. John Pinel
12. Merab Kokaia
13. Libor Velisek
14. Jana Veliskova
15. Bob Adamec
16. Claude Wasterlain
17. Mitsumoto Sato
18. Mac Burnham
19. Lisa Kalynchuk
20. Annamaria Vezzani
21. William Seidel
22. Tony Phillips
23. Michele Simanoto
24. Guiliano Avanzini
25. Gyorgy Buzsaki
26. Filiz Onat
27. Xia Zhang
28. Stan Leung
29. Heidrun Potschka
30. Bob Post
31. Kiyoshi Morimoto
32. Shigeru Chiba
33. Karen Gale
34. Rafael Gutierrez
35. Deb Saucier
36. Darren Hannesson
37. Aaron Sheerin
38. Steve Barnes
39. Wally Whiteheart
40. Pat Shinnick-Gallagher
41. Jeff Goodman
42. Alan Blau
43. John Howland
44. Ken Wolfe
45. Cam Teskey
46. Whytse Wadman
47. Dong Liang
48. Tom Sutula
49. Ed Bertram
50. Uwe Heinemann
51. Randy Stewart
52. Jim McNamara

INDEX

Absence seizures, 38, 40
  PTZ kindling, 50
  strain differences, 40
  effects of ethosuximide, 40
Acoustic startle
  after kindling, 85
ACPD, 200-201
Afterdischarge (AD) threshold
  relation to age, 2
  effects of betamethasone, 15
  thresholds, 2
  relation to strain differences, 22, 40
Age differences, 3, 153
  and AD threshold, 2
  and flurothyl seizures, 21
  and refractory periods, 6
  and spontaneous seizures, 4
  and strain differences, 19
AMPA
  infusions in MRF, 103-104
Anxiety-like behavior after kindling, 263-271, 285-294
  relation to behavioral baseline, 264-269
  relation to novelty of environment, 287-288
Apoptosis, 149
Astrocytes, 150
APT synthase
  changes after kindling, 192-195
Axon guidance factors, 234-236
Baboon, *see* Primates
Betamethasone, 15
BDNF
  deletion and kindling, 242
  involvement in kindling, 232-233
Bicuculline
  infusions in MRF, 100-102
Bradykinin, 209-217
  B_1 receptor density after kindling, 211
Brainstem
  involvement in kindling, 99-113
Burst discharge
  evoked by metabotropic glutamate agonists, 204
Calcium-calmodulin kinase II
  gene changes after kindling, 183
Cannabinoids, 333-342
Carbamazepine
  effects on neuronal firing, 66-67
  effects on sodium current, 64-66
  development of tolerance and cross-tolerance, 305-311
Claustrum (CLA)
  effects of CLA kindling on anxiety, 286
  effects of CLA lesions on kindling, 118
  persistence of CLA kindling, 118
  transfer between CLA and amygdala, 119
Cortical dysplasia, 389-393
Deep brain stimulation, *see* low-frequency stimulation
Degeneration, 94, 126-127, 149
Delayed-match-to-place (DMTP)
  effects of kindling, 276-277, 289-290
Dendritic spines
  changes with kindling, 127
Dentate gyrus
  increased seizure discharge after kindling, 92

Dentate nucleus of cerebellum
  role in PTZ kindling, 52
Deoxycorticosterone (DOC), 327-329
Differential display, 175-177
Dihydroxyprogesterone (DHP), 330-331
Diphenylhydantoin, *see* Phenytoin
Dopamine
  enhanced stimulant effects after kindling or kainate, 400
  sensitization during kindling, 399-405
Dorsomedial nucleus of thalamus, 138-143
Entorhinal cortex
  effects of kindling on spatial cognition, 278
  relations to amygdala altered by kindling, 91-97
  relations to hippocampus altered by kindling, 91-97
Environmental enrichment
  effects on anxiety after kindling, 290-291
Ephrins, 234-235
Epileptic status, 5, 7
  in primates, 363-364
Ethosuximide
  effects on kindling in GAERS, 40-42
Evoked potentials
  in cortex after kindling, 128
Extinction of conditioned effects on kindling, 296-297
Febrile convulsions, 27, 29
Flurothyl seizures
  relation to age, 21
  relation to strain differences, 21
Galanin, 219-227
  kindling in knockout mice, 220
  kindling in over-expressing mice, 221-222
Gamma-amino butyric acid (GABA)
  during development, 72
  in the mossy fibers, 71-72, 94
  transporter, 73
  release after kindling, 74, 84, 92
  $GABA_B$,
    autoreceptor function, 82
    heteroceptor function, 83
    in dentate gyrus, 83-84
    in field CA1, 84
    postsynaptic receptor function, 83-84
    seizures induced by antagonists, 86
GDNF
  involvement in kindling, 233
Gene expression, 181-188
  relation to strain differences, 175-177
Gene deletion
  BDNF and kindling, 242
  galanin and kindling, 220-221
  trkB and kindling, 242-244
Gene over-expression
  galanin and kindling, 221-224
  NPY and kindling, 251-254
Genetic Absence Epilepsy Rats from Strasbourg (GAERS), 38
Genetic predisposition, 372-373, *see also* Primates, Strain differences
Glia, *see* astrocytes
Glutamate
  metabotropic receptors and kindling, 199-208
  release after kindling, 211
Hemispheric differences, 264-267
Hippocampal lesions
  effects on PTZ kindling, 54
Hippocampal sclerosis, *see* MTLE
HU-210, effects on kindling and kindled seizures, 335-338
Immediate early genes
  expression after kindling, 149
Interictal spike discharge (IID),
  in primates, 361-363
Kainate receptors
  altered release of GABA after kindling, 92
Kindling-induced potentiation (KIP), 128-130
Kittens
  amygdaloid kindling, 1-8
Knockout, *see* Gene deletion
Lamotrigine
  development of tolerance and cross-tolerance, 307-310
Learning and memory,
  fear-based, 174
  spatial, 151, 273-283, 289-290
Long-term potentiation (LTP)
  induced by metabtorpic glutamate agonists, 204
  primed burst, 85
Low-frequency stimulation
  as an anticonvulsant, 343-353
  of NSUB, 119
Macaca fuscata, *see* Primates
Macaca mulatta, *see* Primates
Maternal behavior, 30
Medial dorsal nucleus of thalamus, 138-143
Mesencephalic reticular formation (MRF), *see also* reticular formation
Mesial temporal lobe epilepsy (MTLE)
  characteristics in patients, 369-375
  evidence for progression, 374-375
  relation to animal models, 375-384, 386-389
  relevance of kindling, 326-327, 376-379
Monkey, *see* Primates
Mossy fibers
  release of GABA, 71-72, 94
  release of galanin, 222-223
  sprouting, 230-236, 387-389
Movement representation
  in rat sensorimotor cortex after kindling, 129
Multidrug transporter P-glycoprotein
  relation to pharmacoresistance, 317-319
Negative transfer

from stimulation of the NSUB, 120-121
in primate kindling, 365-366
Neurogenesis, 149
Neuropeptide Y (NPY), 249-261
concentration changes after kindling, 249-250
receptor changes after kindling, 250-251
Neuropilin-2, 235-236
Neuroprotection
effects of kindled seizures, 153-154
Neurotrophins, 231-234, 241-247
NGF
involvement in kindling, 231-232
NMDA
alterations in synaptic current after kindling, 149
infusions into MRF, 104-105
responses altered after kindling, 93
NT-3
involvement in kindling, 233
Nucleus submedius of thalamus (NSUB)
effects of lesions on kindling, 120
susceptibility to kindling, 119
transfer to CLA and amygdala, 121
Object recognition
effects of febrile convulsions, 31
effects of hippocampal kindling, 275
Papio cynocephalus, *see* Primates
Papio hamadryas, *see* Primates
Papio papio, *see* Primates
Partial kindling, 81-86
Pavlovian (classical) conditioning, 295-303
Pentylenetetrazol (PTZ)
kindling, 50
Perirhinal cortex, 276-277, 286, 298-299
Permanence of kindling
role of Pavlovian conditioning, 299-300
Pharmacoresistance
in kindled rats, 315-323
Phenytoin
pharmacoresistance in kindled rats, 315-320
Play behavior, 31, 32
Population spike (PS)
enhancement after PTZ kindling, 54, 57
Prewar Gibson banjo, 1930
Primates, 361-369
Progesterone, 329-331
Protein
affected by kindling, 192
Proteomics
applied to kindling, 189-197
Psychosis
occurrence in epilepsy, 388
relation to dopamine sensitization, 403
Refractory periods
age differences, 6
Regulator of G-protein
genetic changes after kindling, 181-183
Reticular formation (MRF)
and kindling, 105-108
Secondary epileptogenesis
in primates, 364-365
Sex differences, 14
Sodium current
inactivation, 65
in kindled tissue, 64
in status, 64
Sodium-calcium exchange current, 201
Spike-wave discharge
effects of ethosuximide, 40, 42
with PTZ kindling, 50
Spontaneous seizures, 151-152
age differences, 4
in kittens, 4
in primates, 364-365
Sprouting, 149-150, 227-240
Steroid hormones, 15
Strain differences, 171-179
in behavior, 179
relation to age, 19
relation to GABA, 172-173
relation to KIP, 130
Subiculum
involvement in seizure discharge, 92
Thalamus
see Dorsomedial nucleus of thalamus
see Nucleus submedius of thalamus (NSUB)
Tolerance to anticonvulsants
contingent tolerance, 305-314
cross tolerance, 306-307
Tonotopic maps
of cat auditory cortex after kindling, 130-131
Transfer, 119, 121
role of Pavlovian conditioning, 300-301
TrkB
deletion and kindling, 242-244
Tuning curves
of cat auditory cortex after kindling, 130
Ultrasonic vocalization, 30
Ultrastructure
of kindled neocortex, 126-128
Ventral tegmental area, 400-402
Viral-vector transfection
of galanin, 224
of NPY, 252-254
Wet-dog shakes
after hippocampal kindling, 85
after neocortical kindling, 297-298